THE CSS *VIRGINIA*

Other History Press Books by John V. Quarstein

A History of Ironclads: The Power of Iron Over Wood (2007)

Old Point Comfort: Hospitality, Health and History on Virginia's Chesapeake Bay (2009)

The Monitor Boys: The Crew of the Union's First Ironclad (2011)

Big Bethel: The First Battle (2011)

THE CSS *VIRGINIA*

Sink Before Surrender

To Bill — Sink before Surrender

JOHN V. QUARSTEIN

Published by The History Press
Charleston, SC 29403
www.historypress.net

Cover image: *The Sinking of the* Cumberland, oil on canvas, ©James Gurney, 2010.

Hardcover edition 2012
Paperback edition 2013

Manufactured in the United States

ISBN 978.1.60949.580.0

Library of Congress Cataloging-in-Publication Data

Quarstein, John V.
The CSS Virginia : sink before surrender / John V. Quarstein.
p. cm.
Includes bibliographical references and index.
ISBN 978-1-60949-580-0
1. Hampton Roads, Battle of, Va., 1862. 2. Virginia (Ironclad) 3. Monitor (Ironclad) 4. United States. Navy--History--Civil War, 1861-1865. 5. Confederate States of America. Navy--History. 6. Armored vessels--United States--History--19th century. 7. Virginia--History--Civil War, 1861-1865--Naval operations. 8. United States--History--Civil War, 1861-1865--Naval operations. 9. Merrimack (Frigate) I. Title.
E473.2.Q36 2012
973.7'5--dc23
2011052413

Contents

Preface

Tales of the brave crew and officers of the CSS *Virginia* have captured my heart and mind since I was seven years old. My favorite exhibits at the Casemate Museum and The Mariners' Museum when I was a child were those depicting the tremendous battle between the *Monitor* and the *Virginia* (also known as the *Merrimack*). Each time I stood atop the parapet at Fort Monroe, I could envision the battle taking place before me. I even made models of the two ironclads and reenacted the battle on a pond at my family's farm. Of course, the *Virginia* won all of these battles.

By the time I was in fourth grade, I had become so enthralled with the Confederate ironclad that when I helped my teacher, Mrs. Mayo, create our class bulletin board about the Civil War, the CSS *Virginia* was the featured topic. One day shortly thereafter, Mrs. Mayo had to unexpectedly leave the room. She announced that a fellow student, Stephen, would be the class monitor. This struck a chord in my soul, and thinking myself a true southerner, I stood up and announced that I was the *Merrimack*. (And so, just another day in trouble for this excitable boy!)

I find myself almost fifty years later still dreaming about the CSS *Virginia*. One of the greatest aspects of living in the Hampton Roads region is the opportunity to tour the sites associated with the *Virginia* and her engagements with the Union fleet. Every day, I pass these places and imagine the great ironclads and the men who served on these experimental warships. Thinking about Franklin Buchanan's dynamic leadership that took the *Virginia* into combat on March 8, 1862, inspires me every time I gaze out onto Hampton

Monitor *and* Merrimack, oil on canvas by Warren Sheppard, circa 1940. *Courtesy of The Mariners' Museum, Newport News, Virginia.*

Roads. Often, I stop and reflect about the battles at Fort Monroe and on the Portsmouth waterfront. You can actually boat along the shore of Craney Island, where the *Virginia* was destroyed, or look out from Christopher Newport Park in Newport News and see where the *Virginia* dramatically sank the USS *Cumberland*. The *Cumberland* still lies on the bottom of the James River as a symbol of the power of iron over wood. However, one of my greatest thrills is standing atop the parapet of Fort Wool where President Abraham Lincoln stood, watching the *Virginia* chase the *Monitor* away from Sewell's Point on May 8, 1862. It is all very meaningful and inspiring.

I embarked on this book in an effort to transform my longstanding interest in the *Virginia* into a complete history of the ironclad. This volume, therefore, is intended to capture the entire Confederate ironclad experience in Hampton Roads, thereby serving as a thorough survey of the *Virginia*'s career from concept and construction to her battles and untimely destruction. The designers, commanders, builders, officers and crew all played critical roles in making the *Virginia* the mistress of Hampton Roads. The *Virginia*'s place in history is their legacy.

Acknowledgments

Sink Before Surrender" was a signal Flag Officer Franklin Buchanan advised his supporting gunboats, CSS *Raleigh* and *Beaufort*, that he would hoist if the *Virginia*'s initial attack against the Union fleet turned against the Confederates. My friend Joe Gutierrez and I have talked about this phrase for over thirty years. Back then, Joe worked at The Mariners' Museum, and we would often talk about the Battle of Hampton Roads. We were constantly amazed at Franklin Buchanan's daring and determined leadership. Although often impetuous, Buchanan's Nelsonian attributes epitomized his dynamic approach to naval warfare. Likewise, I believe that Joe and I, due to our many discussions about ironclads, Buchanan and the like, have endeavored to utilize some of these same heroic traits in our approach to historic preservation and interpretation. Throughout the years, Joe Gutierrez, now senior director for museum operations and education for the Jamestown-Yorktown Foundation, has truly enhanced my focus about museums and ironclads. I shall always be grateful for his advice, motivation and friendship.

This volume is a greatly expanded version of a book I wrote fifteen years ago for the H.E. Howard Virginia Regimental Series. Since then, I have learned so much more about the *Virginia* and her crew that I decided to rewrite and expand this volume.

The CSS Virginia*: Sink Before Surrender* could only have been achieved thanks to the assistance I received from several dedicated individuals. My sincere appreciation goes to my editor, Julie Murphy, of Circle C Communications. This is the fourth book Julie and I have worked on together for The History

Press. As always, Julie performed her duties in an outstanding fashion. As with my first book about the CSS *Virginia*, J. Michael Moore, a historian with Lee Hall Mansion, helped me organize this tome's images. The illustrations are from a variety of people and places, including The Mariners' Museum. Thanks go to Claudia Jew and Megan Steele for their assistance. Other images were obtained with help from the Hampton History Museum, Virginia War Museum, Museum of the Confederacy, Library of Virginia, Kirn Memorial Library, Library of Congress, U.S. Naval Historical Center and U.S. Army Institute of Military History, as well as from Charles V. Peery and my son, John Moran Quarstein. I must acknowledge James Gurney for his compelling work *The Sinking of the* Cumberland, which graces this book's front cover. It is indeed an honor to have such a magnificent painting to frame the moment when the *Virginia* changed naval warfare forever.

My primary research assistant for compiling the crew roster was G. Richard Hoffeditz, now curator of the Virginia War Museum. Dick traveled often to Richmond, Virginia, and Washington, D.C., searching for crew data. His assistance was invaluable.

I must also express my appreciation to Anna Holloway of The Mariners' Museum. Good fortune was mine when I was able to work with Anna on the Monitor Center project. Since its opening in 2007, I have had the privilege and pleasure of producing two other ironclad books for the Monitor Center: *The Monitor Boys: The Crew of the USS* Monitor and *A History of Ironclads: The Power of Iron Over Wood*. *The CSS* Virginia*: Sink Before Surrender* is a companion to these other volumes and completes my dream to provide those interested in the Civil War at sea the definitive studies about the ironclad revolution.

So many of my friends and colleagues have given me much-needed encouragement over the course of this endeavor. Most of all, I must thank Kimberlee Hesse for teaching me the fact that expressing the truth about the past is the key to embracing the future. Kimberlee's good thoughts, companionship and kindness truly prompted me to keep on target with this book. Thanks to all I have mentioned and others who gave me the vision to emulate Franklin Buchanan's daring strategy of targeting the USS *Cumberland*, proving the power of iron over wood.

Introduction

When the CSS *Virginia* slowly steamed down the Elizabeth River toward Hampton Roads on March 8, 1862, the tide of naval warfare turned from wooden sailing ships to armored, internally powered vessels. Little did the Confederate ironclad's crew realize that their makeshift warship would achieve the greatest Confederate naval victory of the war. The voyage was thought by most of the crew to be simply a shakedown cruise. Instead, the aggressive nature of the *Virginia*'s commander, Franklin Buchanan, made the voyage a test by fire that proved the power of iron over wood, thereby setting the stage for navies of the future.

The *Virginia*'s ability to beat the odds to become the first ironclad to enter Hampton Roads is a tremendous tribute to her designers, builders, officers and crew. Virtually everything about the *Virginia* was an improvisation and adaptation. The South had little choice if it wished to create a warship that might turn the tide of naval warfare in the Confederacy's favor. An ersatz ironclad with a crew of soldiers hastily trained for combat but commanded by some of the most outstanding naval officers in the South, the *Virginia* typified the Confederacy's efforts to wage a modern war with limited industrial resources.

One glaring problem virtually overlooked by the secessionist firebrands in early 1861 was maintaining the link between the agrarian South and European industrial nations. The support of Great Britain and France—the trade of cotton for cannons—was crucial to the independence of the Southern states. Accordingly, when the war erupted, Lincoln immediately recognized this weakness and sought to strangle the Confederacy via a blockade of Southern ports. Without a navy to defend its harbors and contest the Federal blockade, the young Confederacy appeared doomed to lose its all-important link to European manufactured goods.

Virginia's decision to leave the Union on April 17, 1861, gave the Confederate States the Tredegar Iron Works in Richmond and Gosport Navy Yard in Portsmouth. Tredegar was the largest facility in the South capable of rolling iron plate, and Gosport gave the Confederate shipbuilding program an immediate advantage. Gosport was one of the best-equipped yards in America, allowing the Confederacy to build a navy. Over one thousand cannons, machine shops, a granite dry dock and the scuttled USS *Merrimack* fell into Confederate hands. It was the charred hulk of the *Merrimack* that would give the Confederacy the immediate opportunity to use its limited resources to the utmost and challenge the U.S. Navy, albeit briefly, for naval superiority.

What the Confederacy could do with the *Merrimack* became the initial question. The answer was found in the astute mind of Confederate Secretary of the Navy Stephen Russell Mallory. Mallory, the prewar chairman of the United States Senate's Naval Affairs Committee, immediately realized that the Confederacy could never match the North's superior shipbuilding capabilities unless a new "class of vessels hitherto unknown to naval service" was introduced to tip the balance in favor of the South. Mallory's solution was to purchase and build a fleet of ironclad vessels. When the *Merrimack* was raised out of the murky depths of the Elizabeth River and placed into dry dock, it soon became apparent that the frigate offered the quickest way to produce an ironclad. Naval scientist Lieutenant John Mercer Brooke, Naval

CSS Virginia, engraving, circa 1880. *Courtesy of John Moran Quarstein.*

Constructor John Luke Porter and Chief Engineer William P. Williamson all concurred that the *Merrimack*'s hull and condemned engines could become the nucleus for an armor-clad warship. Work transforming the *Merrimack* began in July 1861 but lagged behind schedule.

Construction delays, iron plate production problems and disagreements between Brooke and Porter prompted Mallory to dispatch the brilliant Catesby ap Roger Jones to facilitate the ironclad project. Time was now of the essence as the Federals had begun their own ironclad construction program. If the *Merrimack* was to win the naval race, all of the Confederacy's meager resources needed to be directed to the ironclad project. Brooke and Jones struggled with iron production and the ship's armament, while H. Ashton Ramsay and his engineers labored to revitalize the old engines. Porter worked the laborers overtime throughout the fall and winter months to prepare the vessel for battle.

Somehow the *Merrimack* was launched, christened and commissioned on February 17, 1862. The *Virginia*, the ironclad's new name, was far from ready. Nevertheless, a crew was needed, and Jones, the ironclad's executive officer, and Captain Franklin Buchanan, then chief of the Office of Orders and Details, were able to assemble many of the Confederacy's most promising young officers to the ironclad. Perhaps one of the most difficult tasks—obtaining a crew—fell on the able shoulders of John Taylor Wood.

Wood searched throughout the Confederate units stationed in Norfolk and on the Peninsula for anyone with maritime or heavy artillery experience. Men like Richard Curtis, a prewar boatman, were detached from the 32nd Virginia Infantry for service on the *Virginia*. Captain Thomas Kevill of the United Artillery, along with thirty-nine of his men, volunteered to serve aboard the ironclad just days before she went into battle.

All that was missing was a commander, and Mallory selected perhaps the most aggressive officer in the Confederate navy to direct the ironclad's attack against the Union fleet. Franklin Buchanan knew that time for successful action in Hampton Roads was rapidly passing by. He resolved to immediately try his vessel in combat.

Only a few of the officers aboard the *Virginia* were aware of Buchanan's plans as the ironclad left Norfolk. Once at Craney Island, Buchanan announced his intentions to the crew and signaled to his consort gunboats, "Sink Before Surrender!" Buchanan's daring attack enabled the *Virginia* to achieve a stunning victory on March 8, 1862. The *Virginia* destroyed two Union warships and threatened the entire Union fleet. The *Virginia* had indeed won the race for naval supremacy.

Merrimack, Monitor *and* Minnesota, engraving, Alonzo Chappell, 1862. *Courtesy of John Moran Quarstein.*

Before Mallory could send the *Virginia* to shell Washington, D.C., or New York City, the Union ironclad *Monitor* arrived in Hampton Roads on the very evening of the greatest Confederate naval victory. The next day, the two ironclads fought to a draw during the first battle between iron ships of war. Both vessels claimed victory.

While the *Monitor* ended the *Virginia*'s brief dominance over the Union fleet in Hampton Roads, Federal naval leaders would still suffer from occasional bouts of "ram fever." Consequently, instead of supremacy, the *Virginia* achieved a naval balance in Hampton Roads. Even though the *Monitor* could protect the Union wooden ships from the Confederate ironclad, the *Virginia*'s strategic control of Hampton Roads thwarted McClellan's plan to use the James River to strike against the Confederate capital at Richmond. This strategic balance was the *Virginia*'s greatest victory. The very existence of the *Virginia* protected Norfolk as it aided the Confederate defense of the Peninsula. No Union warship dared to move against the Confederate Yorktown–Gloucester Point batteries while the *Virginia* remained as a perceived offensive threat. With the U.S. Navy paralyzed by the *Virginia*, and facing what he considered a comprehensive defensive system, McClellan believed that he had no other option but to besiege the Confederate Warwick–Yorktown line. McClellan's delays on the Lower Peninsula, prompted in part by the *Virginia*, helped save Richmond from Federal capture in 1862.

The *Virginia*'s presence, however, was not enough to delay the Federal advance up the Virginia Peninsula beyond early May. When the Confederate army retreated from the Warwick–Yorktown line, Norfolk was doomed to fall. The *Virginia*, hampered by her deep draft and other design problems, was destroyed by her own crew off Craney Island on the morning of May 11, 1862.

The *Virginia*'s brief career ended somewhat ingloriously, yet the Confederate ironclad achieved everlasting fame for her role as one of "the founders," as Franklin Buchanan wrote, "of iron-clad warfare at sea." Indeed, the *Virginia* and her antagonist, the *Monitor*, ushered in a new age of naval design when they fought in Hampton Roads. Although the brilliant Swedish-American engineer John Ericsson received most of the credit for the *Monitor* as the ship design of the future, it was the *Virginia*'s ramming of the *Cumberland* and the total destruction of the *Congress* that proved beyond a doubt the power of iron over wood.

The saga of the CSS *Virginia* is an amazing story. This makeshift ironclad destroyed two Union warships, fought another ironclad to a draw and secured strategic control over Hampton Roads for two months, which helped tip the balance in favor of the Confederacy during the Peninsula Campaign. The *Virginia* achieved more in her short life span than any other Confederate ironclad, which is an outstanding tribute to her officers' charisma, the skill of her designers and the dedication of her crew.

1
"A Magnificent Specimen of Naval Architecture"

Admirals throughout the world took serious notice on November 30, 1853, when the Russian Black Sea Fleet, under the command of Admiral Pavel Stepanovich Nakhimov, attacked a Turkish squadron off Sinope. The Russians, armed with guns firing explosive shells, utterly destroyed the Turkish wooden vessels. It was the first major naval engagement of the Crimean War and set in motion a revolution in ship design, culminating in the development of armor-clad warships. The engagement at Sinope sent a telling message to naval leaders: wooden ships could not withstand the destructive power of explosive shells. While the Europeans began their transition to an iron navy during the Crimean War with the construction of ironclad floating batteries, the U.S. Navy was slow to recognize the need to construct armored warships. Nevertheless, the United States, faced with the need to protect overseas economic interests and seaborne trade, sought to modernize its navy. The result would be the construction of six steam screw frigates, one of which was the USS *Merrimack*.

Thus, the story of the CSS *Virginia* actually begins on April 6, 1854, when the United States Congress authorized Secretary of the Navy James C. Dobbin to construct "six first-class steam frigates to be provided with screw propellers."[1] This new class of frigates was intended to be superior to any other warship in the world. Each was named in honor of an American river: *Merrimack*, *Wabash*, *Colorado*, *Roanoke*, *Minnesota* and *Niagara*. John Lenthall, chief of the United States Bureau of Naval Construction, was ordered to develop the overall concepts for the new class. Lenthall, called "the ablest naval

John Lenthall, Chief, U.S. Navy Bureau of Naval Construction, engraving, circa 1855. *Courtesy of John Moran Quarstein.*

architect in any country," designed the frigates to operate primarily under sail.[2] Steam engines were considered as auxiliary while entering port, during storms or when maneuvering in battle. Lenthall completed his plans on June 27, 1854, and forwarded them to Charlestown Navy Yard, near Boston, Massachusetts, where the first frigate, the *Merrimack*, would be built.

The *Merrimack*'s keel was laid on September 23, 1854. Construction was supervised by Commodore Francis Hoyt Gregory, commandant of the Charlestown Navy Yard, and directed by the yard's naval constructor, Edward H. Delano. Even though the *Merrimack* was considered a trial ship, nothing but the finest materials available were used in her construction. Her frame was constructed of live oak, originally procured for more traditional sailing warships, and she was built for speed under sail. Her hull was rather sheer but of traditional design. The only difference from a typical sailing warship of the era was her overhung stern. This design modification was necessary to accommodate the propeller, as there was a gap of more than six feet between the sternpost and rudderpost.

The *Merrimack* was launched on schedule on June 14, 1855. It was a grand affair with over twenty thousand spectators on hand to witness the event. The decks of the old ships of the line, USS *Ohio* and USS *Vermont*, were reserved for dignitaries, while others crowded along bridges and docks and aboard the estimated one hundred ships in the harbor to gain a view of the ceremony. Miss Mary E. Simmons, daughter of Master Carpenter Melvin Simmons, USN, sponsored and christened the *Merrimack*. In honor of the American Temperance Society (which had been founded in Boston in 1826), Miss Simmons christened the frigate with a bottle of water from the Merrimack River rather than the more traditional champagne. A thirty-one-gun salute, one for each state of the Union, was fired as the *Merrimack* slid down the ways into her "destined element" at 11:00 a.m. and "shot out

Launching of the Merrimack, engraving, 1857. *Courtesy of John Moran Quarstein.*

into the stream about half way to East Boston before stopping." The crowd reacted with "enthusiastic huzzas" as the frigate glided into the harbor. The *Boston Daily Evening Transcript* noted that the affair was "altogether the most beautiful and perfectly artistic" launch ever witnessed in Boston. The *Transcript* added a postscript that while "the *Merrimac* is in Dry Dock there will be a good opportunity for strangers and others to examine this new and important addition to our naval force."[3]

The *Merrimack* was then masted and rigged for sails. Her mainmast was 242 feet in length, divided into four sections. The foremast and mizzenmast were stepped in proportion. The frigate was designed to be fully rigged, with the area of ten principal sails being about thirty-two times the immersed midship section.[4] The *Merrimack* could unfurl 48,757 feet of canvas.

On July 21, 1855, the *Merrimack*'s machinery began to arrive at the navy yard for installation. The engines were designed by Robert Parrott and built at West Point Foundry in Cold Springs, New York. Two horizontal, back-acting engines formed the power plant, with

> *the cylinders being on opposite sides of the ship and located at diagonally opposite corners of a rectangle circumscribing the engines, the jet of the*

other cylinders, the two piston rods of each cylinder striding the crank shaft. The cylinders were 72 inches in diameter by 3 feet stroke of piston and were designed to make about 45 double strokes per minute. A three-ported slide valve placed horizontally on top of the cylinder and activated by a rock-shaft was used, expansion being obtained by the use of an independent cut-off valve of the gridiron type.[5]

The two engines were capable of delivering a total of 869 horsepower and reaching an estimated top speed under steam power alone, in smooth water, of 8.870 knots per hour. Actually, the engines never achieved that speed. Log entries noted that under sail and steam the *Merrimack* reached a top speed of 10.656 knots. The frigate managed just 6 knots excessively under steam.

Steam was provided by four 4-furnace (with one auxiliary), vertical glass tube, twenty-eight-ton boilers designed by Daniel B. Martin, engineer-in-chief of the U.S. Navy. Each boiler was fourteen feet wide, twelve feet deep and fifteen feet high, with an aggregate heating surface of 12,537 square feet. The boilers each had a series of brass tubes underneath for a fired

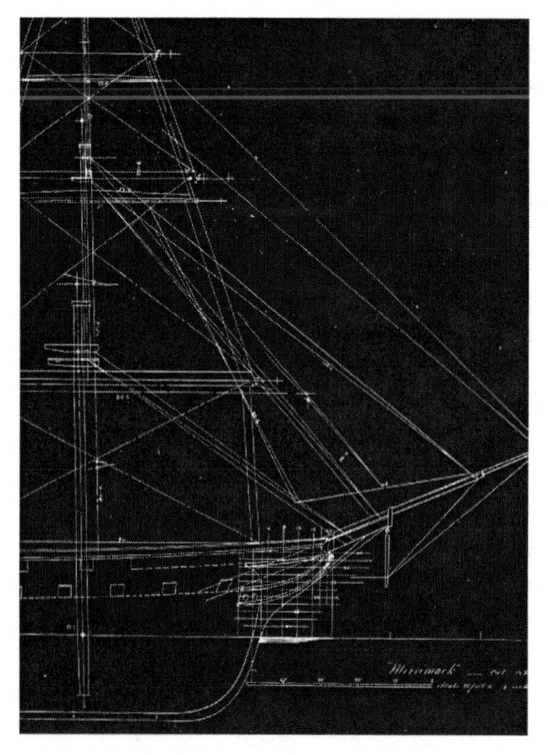

Merrimack sail plan, circa 1855. *Courtesy of the National Archives.*

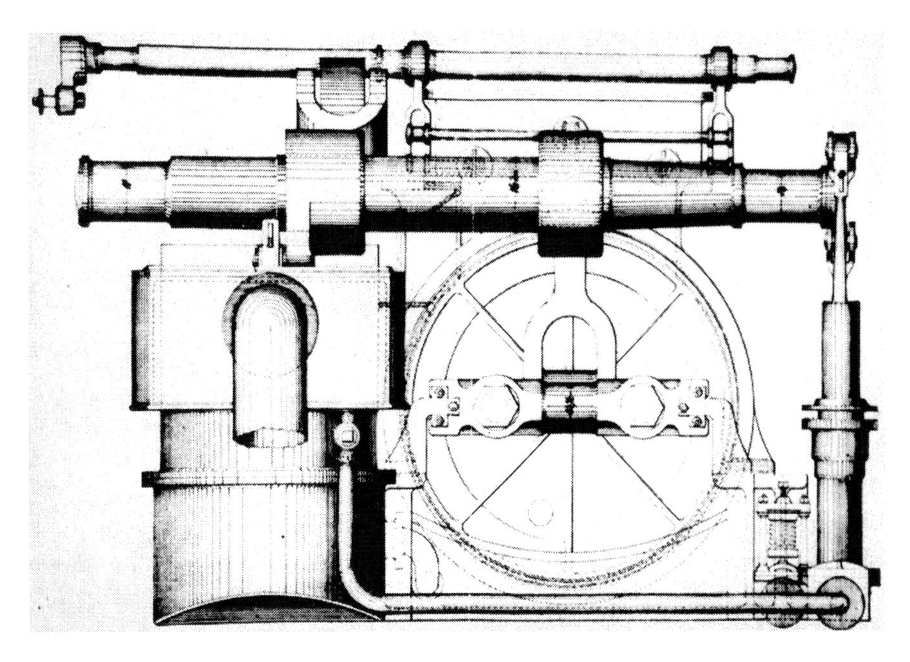

Merrimack's engine, engraving, circa 1857. *Courtesy of John Moran Quarstein.*

water heater; the superheated salt water was pumped through the tubes. Water was heated to 137 degrees Fahrenheit before entering the boilers. Two large steam-operated Worthington pumps were installed to keep the bilges dry or reversed to provide seawater for the boilers.[6]

All this energy only produced 869 horsepower at the propeller shaft. The engines required 103 horsepower to start, and when all this power was delivered to the segmented propeller shaft, 65 horsepower was lost from the friction of the propeller shaft turning on bearings mounted on unstable wooden supports, which dissipated even more power.

The engine system consumed a tremendous amount of space within the warship. The system had a combined weight of 129 tons and was 20.3 feet long, 22.6 feet wide and 12 feet high.

Perhaps the ship's most notable feature was her screw propeller. This propulsion method was invented by John Ericsson, although some give credit to Sir Francis Petit Smith. The screw propeller enabled the warship's engines to be installed beneath the waterline, thereby protecting the propulsion unit from enemy cannon fire. The *Merrimack*'s propeller was a two-blade bronze screw with spherical hub and blades designed by Robert Griffiths. Griffiths's concept endeavored to enhance the engine power while also operating under sail. His concept featured a narrow blade design and fitted the propeller in its socket with an automatic pitch gear to increase the pitch as the velocity increased. When operating under steam, the propeller was set at thirty-six degrees. The blades would be set at zero and locked vertically behind the sternpost to reduce drag for operation under sail. The propeller was huge: seventeen feet, four inches in height, and it weighed fifteen tons. At full speed, the propeller would turn at forty rpm. In addition to the sternpost locking device, the *Merrimack* was fitted with a bronze frame hoist system called a banjo. The *Merrimack*'s banjo could lift the propeller out of the water completely while operating under sail; it could also allow the propeller to be raised onto the deck for maintenance.[7]

Despite this advanced steam-powered, screw propeller system, the *Merrimack* was truly designed to rely on sail power for long sea voyages. The propeller's sternpost lock and banjo devices were major indicators of the preference for sail power. Also, the funnel was telescopic, which enabled the smokepipe to be lowered to reduce drag while operating under sail, as well as improving the frigate's appearance when in port. The U.S. Navy and its designers appear to have added steam power as an afterthought. The hull was designed for sail, and the purpose of the steam engine was to enable the frigate to have ease moving in and out of ports while maneuvering in combat.[8]

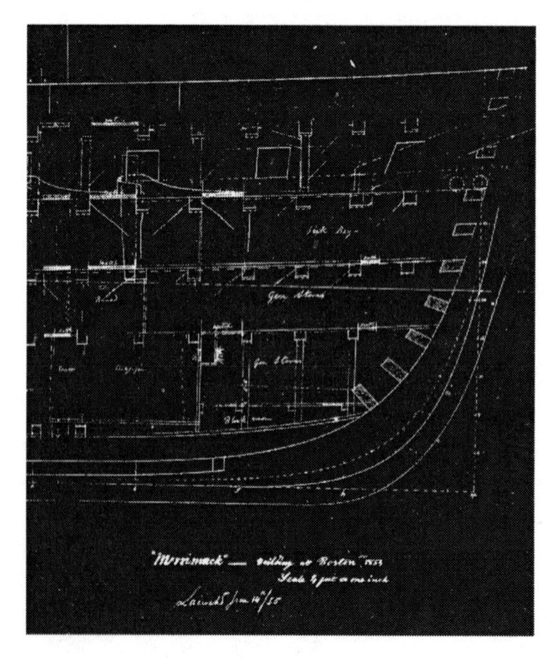

Merrimack hull plan, circa 1855.
Courtesy of the National Archives.

The frigate was completed in February 1856 at a cost of $685,842.19. She was 275 feet in length with a beam of 38 feet, 6 inches and a depth of 27 feet, 6 inches. The *Merrimack*'s tonnage was 3,800, with a 4,000-ton capacity. The frigate's draft was 24 feet, 3 inches. The *Merrimack*, based on the advice of the Delafield Report, was armed with the most advanced artillery available, making her one of the most powerful vessels afloat. Her armament was two 10-inch pivot guns, twenty-four IX-inch Dahlgrens and fourteen 11-inch guns. John Dahlgren, chief of the Bureau of Naval Ordnance, supervised the selection of the ship's armament. All of the *Merrimack*'s cannons were produced at the Alger Foundry in Boston.

The USS *Merrimack* was commissioned on February 20, 1856, and was immediately acclaimed as the pride of the U.S. Navy. She left Boston on February 25, 1856, for her sea trials under the command of Captain Garrett J. Pendergrast. A farewell hymn was written by Phineas Stowe, pastor of the First Baptist Mariner's Church in Boston. The hymn's first and final stanzas proclaimed:

Saviour, o'er the restless ocean
May the gospel banner wave,
And beneath its folds of beauty
Cheer the sailor—guide the brave.

Fare-thee-well, shall be our prayer,
We on earth may meet no more;
But we'll hope to dwell together,
On that calm and heavenly shore.

The *Merrimack* briefly returned to the Charlestown Navy Yard and then set sail for the Chesapeake Bay. She arrived at Annapolis on April 19, 1856, with great fanfare. President Franklin Pierce, accompanied by Secretary of the Navy James C. Dobbin, surveyed the frigate and proclaimed her "a magnificent specimen of naval architecture."[9] U.S. Naval Academy midshipmen paraded as the *Merrimack*'s powerful guns boomed in honor of the president and the many members of Congress who attended the gala affair.

Following the celebration in Annapolis, the *Merrimack* sailed to Gosport Navy Yard and then left for Havana. Her engines broke down, and the frigate made it to Cuba under sail. The *Merrimack*'s rudder then broke, and she was towed back to Boston. Once back in the Charlestown Navy

USS *Merrimack*, model, circa 1960. *Courtesy of The Mariners' Museum, Newport News, Virginia.*

Yard, her engines were repaired, and then the *Merrimack* sailed for Europe on September 9, 1856. Royal Navy authorities were very impressed by the *Merrimack* when she visited British ports. They recognized the steam screw frigate as the "finest vessel of war of her class that had ever been christened." The *Merrimack* returned to Charlestown Navy Yard on April 22, 1857, for an extensive overhaul. The frigate was recommissioned on September 1, 1857, and sailed to the Pacific Ocean on October 17 as the flagship of Commodore John Collins Long's Pacific Squadron.

As the first of her class, the USS *Merrimack* was an experimental vessel. Captain John Dahlgren sent his assistant, Lieutenant Catesby ap Roger Jones, on the frigate's first voyage to test her heavy cannons. The tests proved satisfactory; however, Jones noted that a "vessel with such great deadrise as the *Merrimack*'s could not offer a heavy battery with substantial stability… In a heavy sea…it will be difficult to handle her battery."[10] There were other, more serious technical issues aboard the *Merrimack*. The engines were a constant problem from the very beginning. The ship's log on March 2, 1856, noted the "ship rolling very deeply. Engines racing badly." The boilers were criticized for overheating, causing the engines to race dangerously.

The troublesome steam control valves malfunctioned regularly, which could lead to total engine shutdown. Problems with the engines prompted an even greater reliance on sail power. However, the sailing qualities of the frigate were also questioned. Many observers noted that the *Merrimack*'s bottom was too sharp for the ship's center of gravity and proportion of breadth. Catesby Jones commented that she "rolled very deeply—rolled badly."[11]

The *Merrimack*'s engines were poorly designed and configured. The large cylinders caused low pressure on the piston strokes. Furthermore, the vacuum was poor due to air leaks. The air pumps were inefficient, and if the pumps' sealings were not constantly repacked, the engines would simply stop. The engines vibrated because of how they were fitted into the hull, one designed as a sailing hull. They virtually shook the ship apart. There was no ventilation in the engine room, nor were the engines insulated. This caused a constant, humid, 120-degree temperature, making it almost impossible to operate the engines.

Not only were the engines unreliable, but also the propeller caused several problems. The narrow design resulted in a limited surface, causing excessive slippage through the water. The propeller's slow turning rate (forty rpm) provided inadequate power to make headway against strong currents. The banjo system constantly malfunctioned, perhaps due to the settling of either the rudder or the sternpost. Often the engineers could not start the engines or produce sufficient power "for hoisting the propellers."[12] The entire propulsion system was so undependable that during the *Merrimack*'s last cruise, two of the U.S. Navy's leading engineer officers, Alban C. Stimers and H. Ashton

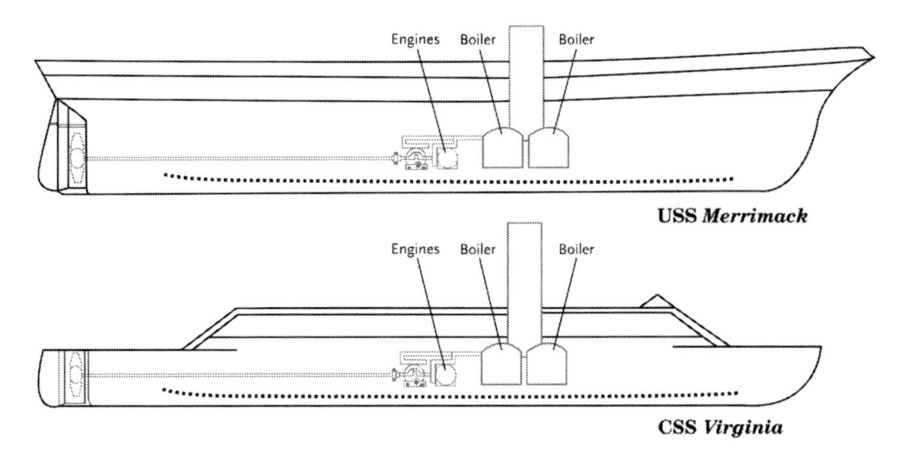

Merrimack/Virginia engines, illustration by Sara Kiddey, 2007. *Courtesy of The Mariners' Museum, Newport News, Virginia.*

Ramsay, were assigned to the frigate. When operating, the frigate's engines consumed 2,880 pounds of anthracite coal per hour, producing an actual top speed of six knots. It is no wonder that the *Merrimack*'s engines were described as "exclusively auxiliary to her sails, and to be used only in going in and out of port."[13]

When the USS *Merrimack* finished her Pacific Squadron service, she was ordered to Gosport Navy Yard to be placed in ordinary. The frigate was considered a total failure. Her engines were slow and unreliable and consumed too much coal. The *Merrimack* was not an effective gun platform because of her roll, caused by her hull design. Finally, the warship's draft was too great to enable her to enter most American ports. The *Merrimack* arrived on February 16, 1860, and was immediately anchored along the quay awaiting an overhaul of her engines. There, the frigate languished with her machinery dismantled.

2
Flashpoint—Gosport

The USS *Merrimack* appeared to be just another hard-luck ship when she sailed up the Elizabeth River to her final berth at Gosport Navy Yard. She joined several other vessels, like the frigate *United States* and the seventy-four-gun ship of the line *Delaware*, in ordinary. All of these ships were either awaiting various repairs or were simply outdated and rotting away along the quay. The yard itself, once a vibrant center of shipbuilding and experimentation, was also languishing from lack of work and poor morale as national events stumbled toward war.

Gosport Navy Yard, located across the Elizabeth River from the busy port of Norfolk, was one of the largest yards in the United States. Norfolk merchant Andrew Sprowle established the yard in 1767. Sprowle remained a loyalist, and the yard was confiscated by the Commonwealth of Virginia during the Revolutionary War and then burned by the British in 1779. The yard remained inactive until 1794, when the property was leased by the United States government. Captain Richard Dale served as the superintendent for this new government shipyard. When the U.S. Navy was formally established in 1798, it assumed operation of the yard and designated it as the Gosport Navy Yard. The unfortunate frigate USF *Chesapeake*, commissioned in 1799, was the first warship constructed at Gosport. The yard was purchased from the commonwealth by the U.S. Navy in 1801, and Captain Samuel Barron served as the first superintendent.

Gosport had many natural assets to recommend its expansion as a shipbuilding center. The deep-water Elizabeth River provided ready access to the excellent harbor of Hampton Roads, as well as access to the Atlantic Ocean via the Chesapeake Capes. The yard was located in a very defensible location with protection provided by several major fortifications. Furthermore, the yard's proximity to the growing industrial centers of Norfolk and Portsmouth, as well as the canal connection to the natural resources of eastern North Carolina, made Gosport a perfect location for the U.S. Navy to create one of the finest shipyards in the world. Gosport quickly fulfilled U.S. Navy expectations. The yard's international reputation for shipbuilding excellence was confirmed in 1820 when the enormous USS *Delaware* was completed.[1] Numerous venerable U.S. Navy vessels were entrusted to the yard for repair, including the USF *Constitution*, USS *Columbia* and USF *Constellation*. Gosport also played a leading role in the U.S. Navy's transition to steam. The USS *Roanoke* and USS *Colorado*, sister ships of the USS *Merrimack*, were completed by the yard in 1859. Gosport's growth into a major shipbuilding and repair center was due to the continual investment in the yard's facilities by the U.S. Navy. A granite dry dock, begun in 1825, was completed in 1837. The dry dock was described as having more than sufficient capacity for repairing "any size ship of the line."[2] The yard's ability to support fleet actions was also expanded in 1837 with the completion of the nearby Portsmouth Naval Hospital.[3]

The 1850s, however, were the heyday for yard improvements. The largest crane used for ship construction work in the United States was installed in 1856, and a steam pump was added to the dry dock during the same year. A large stone quay was finally finished in 1857, complete with two large cranes. A new foundry, ordnance building, boiler shop and gun and shot platforms were completed in 1859. Other improvements accomplished in 1859 included repairs and alterations to a majority of the existing buildings, as well as the installation of new equipment for the machine shop. All of these facilities were connected by an internal rail system. The Portsmouth Gas Works provided power for the machinery.[4]

The yard's growth prompted the development of nearby maritime industries. Mahaffey's Iron Works was founded in 1848, just outside Gosport's gates. Mahaffey's refitted the paddle-wheel sloop USS *Mississippi* with new boilers and machinery just after it opened for business. The ironworks completed the same work on the USS *Powhatan*. Another important facility established next to Gosport was the Page and Allen Shipyard in 1851. This yard launched three revenue cutters in 1856 and competed with government yards for the construction of steam screw frigates. The Gosport maritime

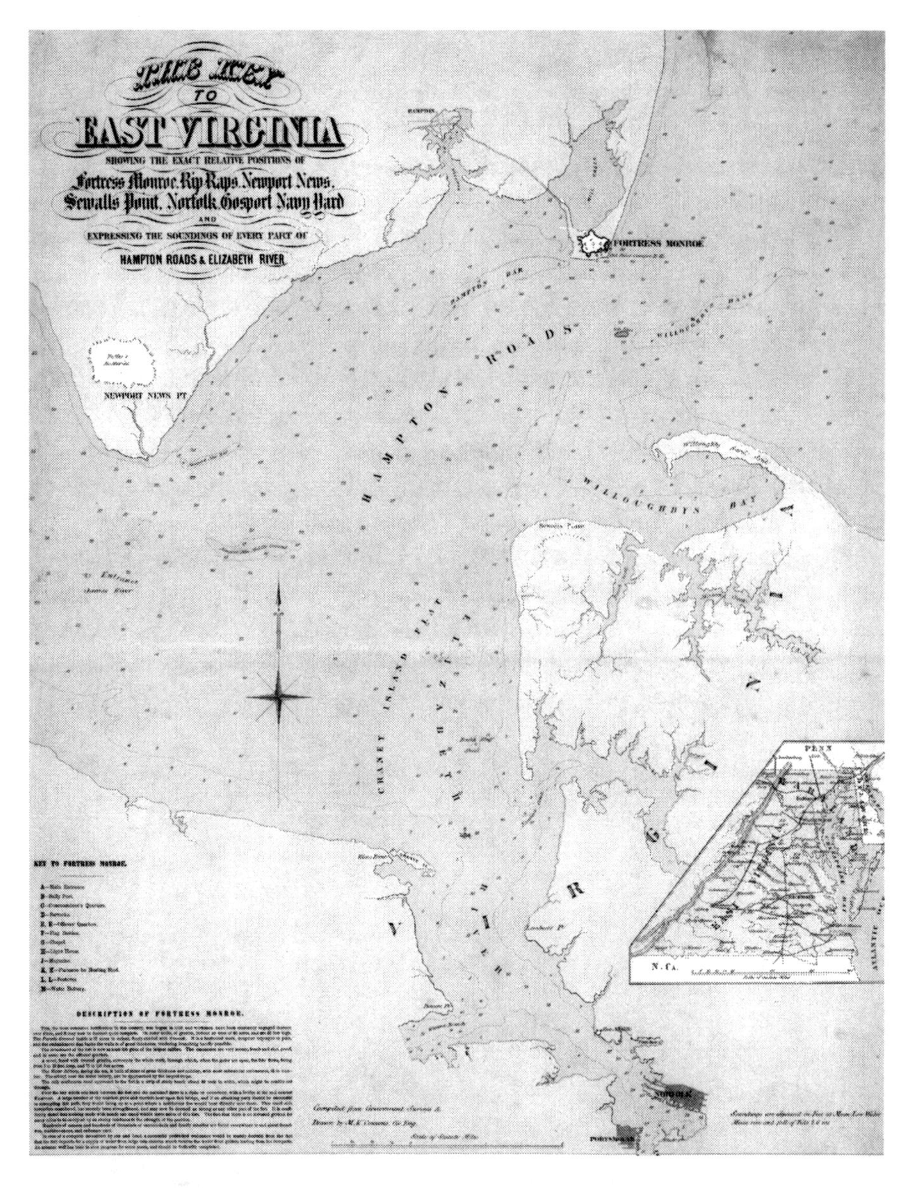

The Key to East Virginia, engraving, 1861. *Courtesy of the Library of Congress.*

community was bustling with activity by 1860. It had become the leading shipbuilding center in the South.

Gosport Navy Yard's excellent facilities prompted its use for training and testing activities. A school was established in the 1820s for training midshipmen, naval constructors and trade apprentices. Besides the granite dry dock, other innovations were employed at the yard, including an experimental diving bell apparatus. The diving bell was used in 1824 to repair the hull of the USS *Delaware* and construct the stone quay.[5] On April 17, 1852, Gosport's gun and shot testing platforms were the scene of special experiments testing the "capacity of iron vessels in resisting the force of shells and cannon balls." It was a gala affair with President Millard Fillmore, members of his cabinet and several congressmen on hand to review the experiments. The tests "proved that iron is not so invulnerable as many have heretofore supposed, and unsuited to such purposes."[6]

Gosport Navy Yard reached its peak during the late 1850s, employing over 1,400 workers. In 1855, the shipyard had over a dozen ships under repair or construction. The completion of the steam frigates USS *Roanoke* and USS *Colorado* in 1859 proved the yard's capabilities to construct modern warships for the U.S. Navy.[7] Despite all of these successes and improvements during the 1850s, the sectional crisis of 1860 cast a spell of gloom over Gosport Navy Yard. The workforce had dropped to fewer than seven hundred, and little work was accomplished. Only seventy-five men were repairing the USS *Germantown*, and construction of the new ship house had virtually stopped. Almost a dozen ships could be

Gosport Navy Yard, engraving, circa 1840. *Courtesy of John Moran Quarstein.*

Gosport Navy Yard, engraving, 1861. *Courtesy of John Moran Quarstein.*

found at the yard, either placed in ordinary or awaiting various types of repairs. These vessels included: the ship of the line USS *Pennsylvania*, 120 guns; the ship of the line USS *Columbus*, 74 guns; the ship of the line USS *Delaware*, 74 guns; the ship of the line USS *New York*, 74 guns; the frigate USF *United States*, 50 guns; the frigate USS *Columbia*, 50 guns; the frigate USS *Raritan*, 50 guns; the steam frigate USS *Merrimack*, 40 guns; the sloop of war USS *Plymouth*, 22 guns; the sloop of war USS *Germantown*, 22 guns; and the brig USS *Dolphin*, 4 guns. Many of these warships, however, were dismantled or in various states of repair. The USS *Pennsylvania* was stuck in the mud and served as a receiving ship, while the *Delaware* and *Columbus* were in the "Rotten Row" section of the lower wharf and deemed basically worthless.[8]

By early 1861, Gosport was in "disorder and confusion." In light of the impending crisis in Charleston, U.S. Secretary of the Navy Gideon Welles ordered the USS *Cumberland*, commanded by Flag Officer Garrett J. Pendergrast and already anchored near the yard awaiting repairs, to stay in Hampton Roads to protect Gosport and Fort Monroe. Welles believed that the *Cumberland*'s twenty-four guns could act as a deterrent to any secessionist move against Federal facilities in the Hampton Roads region.[9]

One reason for the unprepared state and utter confusion at Gosport Navy Yard was a lack of leadership. The problem primarily rested squarely on the shoulders of the yard's commandant, Flag Officer Charles Stewart McCauley. McCauley had served in the U.S. Navy since he was fifteen years old. He had fought on Lake Ontario during the War of 1812. In 1834, he was given command of his first ship and was promoted to captain in 1839. McCauley's long career included command of both the Pacific and South Atlantic Squadrons. He commanded the Washington Navy Yard during the Mexican War and later served as commander of the Home Squadron. When McCauley was appointed commandant of Gosport Navy Yard in 1860, it appeared to be a wise choice. The sixty-seven-year-old McCauley, however, was rumored to have taken to drink and was often ridiculed for being too old for active command. During the tense days of April 1861, many would question his decision-making abilities.

Tensions were rising throughout the South during the first week of April 1861. While the Upper South had yet to join the Confederacy, the Lincoln administration was alert to the threatening war clouds and the possibility of states like Virginia leaving the Union. Gideon Welles recognized that Gosport Navy Yard and the USS *Merrimack* were tempting targets for pro-secessionist Virginians. Accordingly, on April 10, 1861, Welles wrote to McCauley:

> *In view of the peculiar condition of the country, and of events that have already transpired, it becomes necessary that great vigilance should be exercised in guarding and protecting the public interests and property committed to your charge. It is therefore deemed important that the steamer* Merrimack *should be in condition to proceed to Philadelphia or to any other yard, should it be deemed necessary, or, in case of danger from unlawful attempts to take possession of her, that she may be placed beyond their reach.*

Welles added that McCauley was to do nothing to upset the Virginians, noting:

> *Indeed, it is desirable that all shipping and stores should be attended to, and should you think an additional force necessary, or that other precautions are required, you will immediately apprize the Department. In the meantime exercise your own judgment in discharging the responsibilities that resolves on you.*

Gideon Welles concluded that it is "desirable that there should be no steps taken to give needless alarm, but it may be best to order most of the shipping to sea or other stations."[10]

Gosport's commandant responded by telegram on April 11, stating that it would take a month to revitalize the *Merrimack*'s dismantled engines. Welles was shocked by McCauley's reply, calling the yard commandant "feeble and incompetent for the crisis," and sent the U.S. Navy's chief engineer, Benjamin Franklin Isherwood, to Gosport to prepare the *Merrimack* for sea. Isherwood estimated that it would take him a week to rework the *Merrimack*'s engines. Commander James Alden was ordered to accompany Isherwood and assume command of the frigate. Alden's orders instructed him to take the *Merrimack* to Philadelphia only if the evacuation of Gosport proved necessary. Alden and Isherwood arrived at Gosport Navy Yard on April 14, 1861. Isherwood immediately set to work around the clock restoring the frigate's machinery.[11]

While Benjamin Isherwood struggled with the *Merrimack*'s power plant, Welles continued to pressure McCauley from afar to preserve U.S. Navy property at Gosport. On April 16, Welles ordered McCauley that

Gosport Navy Yard, engraving, 1861. *Courtesy of John Moran Quarstein.*

"no time should be lost in getting her (the *Merrimack*'s) armament on board." Welles concluded his letter with the admonition that the "vessels and stores under your charge you will defend at any hazard, repelling by force, if necessary, any and all attempts to seize them, whether by mob violence, organized effort, or any assumed authority."[12] With war already at hand following the fall of Fort Sumter, Welles was very concerned about the U.S. Navy's impending need to enforce a blockade of the Southern coastline. Welles recognized that he required every available ship to interrupt the flow of commerce into Southern harbors. Since the *Merrimack* was one of only five first-class steam frigates in the U.S. Navy, the secretary of the navy demanded that the warship be taken "beyond the reach of seizure."[13]

Simultaneous with his missive to McCauley, Welles reiterated in a letter to the *Cumberland*'s commander, Captain Garrett J. Pendergrast, that the sloop must stay at Gosport because "events of recent occurrence and the threatening attitude of affairs in some parts of our country, call for the exercise of great vigilance and energy at Norfolk."[14] Pendergrast then moved the *Cumberland* up the Elizabeth River from an anchorage off the Portsmouth Naval Hospital to a new position just off Gosport Navy Yard. The sloop's firepower could now be utilized to either effectively defend the yard or to cover the release of ships so that they could be transferred to safer havens. Besides the *Merrimack*, the *Dolphin*, *Germantown* and *Plymouth* were the only warships considered in relatively good enough condition to warrant removal. Welles needed every available warship for blockade duty.

Pendergrast, however, left the yard's defense in McCauley's hands. Other than the *Cumberland*, the commandant had considerable resources available to organize some type of defense. Even though the 120-gun *Pennsylvania* was stuck in the mud and serving as a receiving ship, she was still armed with her full complement of cannons. Thus, the *Pennsylvania* could be used as a stationary battery, adding the weight of her broadside to that of the *Cumberland*. Despite the suspected loyalties of many of McCauley's officers (such as Portsmouth natives John Randolph Tucker and John Luke Porter), approximately six hundred marines and sailors were apparently ready to follow McCauley's commands. The yard contained an abundant supply of ammunition and weapons to arm these men. Gosport also featured a considerable defensive perimeter consisting of an eight-foot-high brick wall and the Elizabeth River.[15] Instead of organizing these resources or supporting the removal of vessels like the *Merrimack*, McCauley found excuses for inaction everywhere. McCauley believed that he was sitting on

U.S. Frigate Cumberland, engraving, Currier & Ives, 1842. *Courtesy of The Mariners' Museum, Newport News, Virginia.*

the powder keg that could force Virginia into war and remained passive as events headed toward an explosive conclusion.

President Abraham Lincoln's call for seventy-five thousand volunteers, following the fall of Fort Sumter, prompted Virginia to secede from the Union on April 17, 1861. Hampton Roads immediately became a major flashpoint. Virginians clamored to secure Federal property, which they believed to rightly belong to the commonwealth. Fort Monroe on Old Point Comfort, Fort Calhoun on the Rip Raps in the middle of Hampton Roads and the Gosport Navy Yard in Portsmouth were military assets that both the North and South wished to control. The two forts seemed out of Virginia's reach, but Gosport appeared ripe for conquest.

Meanwhile, Benjamin Isherwood completed the emergency repairs to the *Merrimack* on April 17 and reported to the yard commandant that the frigate would be ready to leave port the next day. On the morning of April 18, Isherwood had steam in the ship's boilers, and the jury-rigged engines seemed capable of taking the *Merrimack* at least as far as Fort Monroe. McCauley, however, advised Isherwood and the *Merrimack*'s commander, James Alden, that he had decided to retain the frigate in port. Alden and

Isherwood pleaded with McCauley, but to no avail. McCauley was the ranking officer, and the two men had little choice but to concede to the commandant's command. The *Merrimack*'s fires were banked, and the two officers immediately left for Washington to explain to Secretary Welles why the frigate had not left port.[16]

McCauley's indecision was caused by several factors. The commandant's actions were tempered by his initial interpretation of Welles's April 10 command that "there should be no steps taken to give needless alarm." Welles's order of April 16 instructing McCauley to remove all property from Gosport only added to the commandant's confusion. His decisions, or lack thereof, were further influenced by the many pro-Southern officers on his staff (thirteen of twenty would later serve the Confederacy). McCauley's leadership during the hour of crisis at Gosport was questioned by many loyal officers. They believed that the yard commandant was drinking heavily, which made him suspicious and an impediment to U.S. Navy goals.[17]

As the Federals struggled with McCauley's procrastination, local Southern firebrands were rapidly organizing their own effort to secure Gosport Navy Yard. Although many of the citizens of Norfolk and Portsmouth held strong pro-Unionist sentiments prior to the fall of Fort Sumter, Virginia's secession galvanized the communities into action. A Vigilant Committee was established while militia troops mustered in Norfolk and Portsmouth. The committee sank several ships in the channel off Sewell's Point to block Federal access to the navy yard and was rumored to be building "masked batteries" from which to shell the yard.[18]

On April 18, 1861, Governor John Letcher ordered Major General William Booth Taliaferro of the Virginia Militia to assume command of troops assembling in the Norfolk area. Letcher also appointed Catesby ap Roger Jones and Robert B. Pegram as captains in the newly formed Virginia State Navy and ordered these officers to Norfolk. Pegram was instructed by Letcher to "assume command of the naval station, with authority to organize naval defenses, enroll and enlist seamen and marines, and temporarily appoint warrant officers, and to do and perform whatever may be necessary to preserve and protect the property of the commonwealth and of the citizens of Virginia."[19]

Taliaferro, accompanied by his staff officers Captain Henry Heth and Major Nathaniel Tyler, arrived in Norfolk on the evening of April 18, 1861. The next morning, Taliaferro began organizing his meager resources of local militia units. Two artillery batteries, the Norfolk Light Artillery Blues and the Portsmouth Light Artillery, were available. However, these

Major General William Booth Taliaferro, Virginia Militia, seated, photograph, circa 1859. *Courtesy of the Museum of the Confederacy, Richmond, Virginia.*

units could only muster a combined firepower of eight cannons; the largest was a 12-pounder. Although his command numbered about 850 men, Taliaferro immediately opened aggressive negotiations with Commodore McCauley. Taliaferro, a Mexican War veteran and former member of the Virginia House of Delegates from Gloucester County, advised McCauley that he planned to assume possession of Gosport Navy Yard on behalf of the sovereign state of Virginia. Virginia's dignity demanded nothing less. When McCauley refused to concede, Taliaferro requested that more troops be sent to Portsmouth from South Carolina so Gosport could be taken by assault. Rumors prevailed throughout the yard about the impending arrival of more Southern soldiers. To further intimidate McCauley, William Mahone, a Virginia Military Institute graduate and president of the Norfolk & Petersburg Railroad, began running trains in and out of Portsmouth and Norfolk. When the trains came into town, they were filled with yelling citizens to create an illusion of massive troop arrivals.[20]

When the *Merrimack* failed to leave Gosport on April 18, Gideon Welles realized that more resolute action was required. Welles dispatched Flag Officer Hiram Paulding, a veteran of fifty years of naval service and the most senior officer in the U.S. Navy, on board the eight-gun steamer USS *Pawnee* with one hundred marines to take command at Gosport. The secretary also pleaded with Lieutenant General Winfield Scott, general-in-chief of the U.S. Army, to dispatch troops to defend Gosport and ordered Commander John A. Dahlgren to take a vessel loaded with explosives to the yard in case it had to be abandoned. Unfortunately, Welles did not advise McCauley about Paulding's relief force.

As Paulding prepared his force to leave Washington for Gosport, McCauley became increasingly unnerved by the events unfolding around him. The pro-Southern citizenry was in a bellicose mood and gathered outside the yard's

Fort Norfolk, engraving, 1861. *Courtesy of John Moran Quarstein.*

gates demanding access. The Virginia Militia continued its game of nerves, hoping to hoodwink McCauley into handing an intact facility over to the commonwealth. Even though the yard commandant rejected the demands, he feared that the newly constructed batteries and obstructions at the mouth of the Elizabeth River limited his options.

McCauley appears to have given up all hope of saving or defending the yard by April 20. Early that morning, he learned that militia troops had seized Fort Norfolk and with it a very useful magazine filled with more than 250,000 pounds of gunpowder.[21]

Yet there were some cooler heads amongst the seemingly trapped Federal forces. Commander James Alden surveyed the obstructions before leaving for Washington. The obstructions, primarily consisting of three light ships, were considered by Alden insufficient to halt Federal ship traffic. Lieutenant Thomas O. Selfridge of the USS *Cumberland* volunteered to take the brig *Dolphin* to Craney Island and stop the sinking of any more ships in the channel. McCauley rejected the plan. Selfridge also went to Norfolk under a flag of truce to learn about the Southern batteries. He met with Captain Robert Pegram and then Captain Henry Heth but could not glean any information about militia preparations to storm or shell the yard. Selfridge later climbed to the top of the *Cumberland*'s mainmast and observed that there were no batteries threatening Gosport.[22]

Saturday, April 20, 1861, dawned as yet another tense day along the Elizabeth River. McCauley saw danger everywhere and thought that he had few options at his disposal. The capture of the Fort Norfolk arsenal the previous evening only added to the pressure exerted by Taliaferro's militia and pro-Southern citizenry. The yard was in complete disorder. "The men were standing in grounds about the Yard," remembered John L. Porter, "and no one could work."[23] McCauley, believing his command surrounded with no relief in sight, made the fateful decision to abandon Gosport Navy Yard. At noon, he dismissed the workers and ordered the gates locked. Immediately thereafter, McCauley commanded loyal workmen, marines and sailors to scuttle ships and destroy property that could not be removed.

Local residents were in an uproar over the events that were unfolding before them. Rumors quickly spread throughout Portsmouth and Norfolk that "the authorities of the Yard were making preparations to destroy that establishment with fire."[24] William H. Peters and William Spooner, both former shipyard employees, hired a boat and sailed to the yard to investigate the rumors. What they saw shocked them beyond belief as workmen had already begun scuttling ships. Observing the once-proud sinking vessels and fearing that efforts to burn the yard might result in the fire spreading into town, the men returned to Portsmouth and hastily organized a town meeting. A Citizens Committee was established consisting of Peters and two yard employees, Samuel Watts and James Murdaugh, to entreat McCauley not to burn Gosport. The men were met at the yard's gates by Brigadier General George Blow of the Virginia Militia and two former officers of the USS *Cumberland*, Lieutenant John Maury and Paymaster John DeBree. Maury and DeBree, who had both just resigned their U.S. Navy commissions, advised the Citizens Committee that McCauley refused to meet with anyone on the subject and informed the committee that they were "not at liberty to talk on the subject."[25]

While the Citizens Committee was rebuffed in its effort to dissuade McCauley from burning the yard, others took a more active approach. One group of firebrands tried to capture the tiny side-wheel steamer *Yankee*, which had been commandeered by Pendergrast to move the *Cumberland* out of the Elizabeth River. The attempt failed, according to Private Daniel O'Connor of the *Cumberland*'s Marine Guard, because when "they seized her…they let her go very quick when we pointed our pivot gun at her." O'Connor remembered the rejected firebrands yelling "of massacring the whole of us or taking the navy yard that night and shipping but we intended to sell our lives as dear as possible."[26] Thomas O. Selfridge also remembered a somewhat more humorous incident. One particular steamer filled with rebels continued to hover near the yard shouting threats and abusive language at the loyal workmen who were destroying the yard. Selfridge and members of the *Cumberland*'s crew rigged an underwater line from the yard to a tug on the other side of the channel. When the rebel tug approached the yard for another round of abuse, the hawser line was pulled taut, raking the entire length of the tug, knocking off the vessel's smokestack and pushing several men overboard. After this episode, there were no other attempts to approach the yard by the Elizabeth River.[27]

As the Virginia Militia and local citizens watched in disbelief, McCauley's men went on with their duties of destruction. All the major warships in the harbor, except the *Cumberland* and the venerable *United States*, were scuttled by mid-afternoon. The *Columbia*, *Pennsylvania* and *Raritan*, anchored out in the river, were set on fire and burned almost down to their keels. The *Germantown*'s rigging was cut away, causing the shears to fall across the sloop. The sloop of war broke in two and sank. The scuttling was not as effective on most of the other ships. Due to the Elizabeth River's depth, many of the larger ships, such as the *Delaware* and *Columbus*, did not totally submerge. Crews simultaneously loaded the *Cumberland* with gunpowder, small arms and other supplies. Small arms that could not be saved were broken and thrown into the river. An effort was made to render useless the 1,085 heavy

Burning of Gosport Navy Yard, engraving, 1861. *Courtesy of John Moran Quarstein.*

cannons stored at the yard. Teams of men armed with sledgehammers attempted to break off the trunnions of cannons, but this proved to be a difficult, time-consuming and virtually impossible task.[28]

Meanwhile, Flag Officer Hiram Paulding's task force reached Fort Monroe on the afternoon of April 20. Paulding embarked 350 men of the 3rd Massachusetts Volunteer Infantry Regiment, commanded by Colonel David W. Waldrop, on board the *Pawnee* and steamed toward Norfolk. The task force arrived at Gosport about 8:00 p.m., but Paulding immediately realized that it was too late. Ships were already sinking, and the yard appeared to be in complete confusion.

Paulding assumed command from a demoralized McCauley but quickly recognized that there was little else that he could do other than finish the job that McCauley had initiated. Elements of the 3rd Massachusetts guarded the yard's outer walls so that the angry mob outside the gates would not attempt to disrupt the work of demolition crews. As an extra precaution, the *Cumberland* and *Pawnee* were anchored in the Elizabeth River to enable their cannons to rake the approaches to the yard. The local Southern volunteers could only observe from afar the flames' destructive work.

General Taliaferro made one last effort to save the yard. Even though his command had just been reinforced with four hundred Georgia soldiers who had arrived that evening from Petersburg, Taliaferro was in no position to stop the Federals from destroying the yard or leaving Gosport. Nevertheless, he attempted to stem the destruction with a paper ruse. Taliaferro sent a message to Paulding stating "that to save the effusion of blood, he would permit the *Cumberland* to leave port unmolested, if the destruction of property should be discontinued." Paulding immediately replied "that any act of violation on their part would devolve upon them the consequences." Nothing would stop Paulding's men from doing their duty as Private O'Connor reflected in a letter to his sister:

> *We sent a lieutenant with a flag of truce to the authorities to know on what conditions we could go out but they gave us no satisfaction. So we made up our minds intended to fight our way out or die in the attempt. Every face looked determined. We intended to burn the navy yard and shipping there for there was not water enough to sink the ships and blow down Gosport and Norfolk.*[29]

Once Paulding asserted his command, Union crews immediately went forward with their assigned demolition tasks. Paulding had stocked the *Pawnee* with a wide variety of combustibles, including forty barrels of gunpowder, eleven tanks of turpentine, twelve barrels of cotton waste and 181 flares, to destroy the yard. The mighty 120-gun *Pennsylvania*, now engulfed in a sheet of flames, dominated the eerie scene. No one had unloaded the ship's cannons, so the night was filled with occasional blasts from the four-decker's guns. All of the ships in the harbor were set ablaze with a determination to leave behind nothing of use. The *Merrimack* was torched by Lieutenant Henry A. Wise, who remembered:

> *I had hardly time to push off from her side, as I touched a match to the train of turpentine waste hanging from a port, when flames sprouted out in volume. Flames were belching from her lower decks, with her upper works, masts, and rigging burning at the same time.*[30]

As ships like the *Raritan* and *Columbia* burned in the harbor, buildings throughout the yard were set on fire. The two huge ship houses, A and B, were quickly engulfed by flames. Ship House A contained the partially completed 74-gun *New York*, which was consumed by fire as it sat on the stocks. Since everything would not burn, sailors and marines rushed through the yard laying powder trails to destroy the valuable machinery and facilities. When efforts to break off the trunnions of the 1,085 cannons in the yard with sledgehammers proved futile, the guns were spiked with wrought-iron nails. Two officers, Commander John Rodgers and Captain Horatio Gouverneur Wright, were assigned the critical task of mining the granite dry dock. Their work was purportedly foiled by a petty officer who did not wish the explosion to damage the nearby homes of his prewar friends.[31]

At 4:20 a.m., following the burning of the marine barracks, Paulding ordered his men to the ships. The tide was rising, and everything that could be done to destroy the yard's military value had been done. The men raced back to the ships, lighting the final fires behind them. McCauley, in tears, had to be escorted onto the *Cumberland* by his son. The commandant had wished to be left behind like Nero and be consumed in the fire. As the early morning sky turned bright with the flames, the *Pawnee* steamed out of the harbor, followed by the *Yankee* with the *Cumberland* in tow. The flames could be seen thirty miles away.

The small Union flotilla slowly made its way down the Elizabeth River past the feared but nonexistent Confederate batteries. The Federal ships crossed the obstructions sunk in the channel by Southern volunteers before dawn. "We were dragged over the obstruction," remembered Lieutenant Selfridge of the *Cumberland*, "and anchored off Fort Monroe."[32] The *Cumberland* immediately became the nucleus of the Federal squadron blocking Hampton Roads and the lower Chesapeake Bay.

The entire operation was over by 6:15 a.m., and only two men were lost by the U.S. Navy. Commander John Rodgers and Captain Horatio Wright failed to reach the retreating Federal ships and were captured by Taliaferro's troops. Governor John Letcher released the officers a few days later, allowing them to return to Washington. McCauley was charged with treason, cowardice and drunkenness for his part in the Gosport affair. He was censured by the U.S. Senate for acting irresponsibly yet continued to serve in the U.S. Navy until he retired as a commodore in 1867. The operation, despite the loss of Gosport and the *Merrimack*, was considered a success by Gideon Welles. The Confederates, however, were equally pleased with the events of April 20 and 21, 1861, having gained in one evening one of the finest shipyards in America.

3
Iron Against Wood

When the Virginia volunteers entered Gosport Navy Yard the morning after the blaze, evidence of destruction was everywhere. Citizens and militiamen rushed throughout the yard putting out fires. Two volunteers, David A. Williams of the Old Dominion Guards and Joseph F. Weaver of the Portsmouth Rifle Company, discovered powder trains leading into the granite dry dock. They immediately broke the connection. Lieutenant C.F.M. Spottswood of the Virginia State Navy, who had raised the first flag of Virginia over Gosport Navy Yard at dawn, hurried to the dry dock and discovered that twenty-six barrels of gunpowder had been placed in a culvert and along the head of the dock. Spottswood, who believed that the explosion would have destroyed the dry dock and every building in that section of the yard, flooded the dock, and the gunpowder floated harmlessly away.[1] Nevertheless, amongst all the rubble, scuttled ships and charred buildings, the young Confederacy was able to find the wherewithal to create a challenge to the U.S. Navy.

"The condition of the Navy Yard at Portsmouth as it appeared on the 21st and 22nd was melancholy to look upon," wrote Paymaster William H. Peters of the Virginia State Navy in his inventory of the yard following its abandonment by the Federals. "It was a mass of ruins. The extensive row of buildings on the north front of the yard which contained large quantities of manufactured articles and valuable materials was totally destroyed, together with their contents." The yard appeared, at first glance, devastated. All the ship houses were burned, and the ship of the line USS *New York* was nothing but smoldering timbers.[2]

Out in the Elizabeth River, all of the ships, except for the decrepit USF *United States*, were either scuttled or burned. The *Pennsylvania*, *Columbia* and *Dolphin* were all consumed by fire to the bottom of their hulls. The *Delaware* and *Columbus* were sunk at their moorings and considered worthless. Paymaster Peters, however, noted that several ships were not total losses. The *Germantown* was sunk but only burned down to her port-side bulwark, and the *Plymouth* was lying scuttled beneath the waves. He noted that the *Merrimack* was sunk and that the fire had burned the frigate down to her copper line, destroying the spar and gun decks and damaging the berthdeck.[3]

Obviously, the Federals left with such haste that their destructive work was far from complete. Much of the property was still intact. Over five warehouses filled with naval supplies, including $56,269 worth of clothing and $38,763 in food, survived the flames. While all the small arms had either been destroyed or taken away by the retreating Unionists, they had abandoned a tremendous array of ordnance. A total of 1,085 heavy cannons, many with carriages; a large number of shells and stands of grape; and over 250,000 pounds of powder were left behind. Many of the cannons were spiked but were only temporarily rendered useless, as they were quickly made serviceable by the Confederates. Peters noted in his report that the yard provided the South with an immense quantity "of munitions of war," and the Confederacy was "wholly indebted for our means of resistance to his [the Union's] loss and our acquisition of the Gosport navy-yard."[4]

Indeed, the Federals provided the Confederacy with military property worth $4,810,056.68. Numerous dwellings, the foundry, the machine shop and several workshops remained untouched by the blaze. A total of $1,448,223 worth of timber, canvas, iron sheets and bars, copper bolts and sheets, anchors and chains were left undamaged. More importantly, the retreating Federals failed to destroy the granite dry dock. Overnight and without a shot, the Confederacy had gained the infrastructure to construct the vessels to challenge the Federal blockade. The Richmond press gloated over the abundance of equipment and supplies, stating, "We have material enough to build a navy of iron-plated ships."[5]

Captain Robert Pegram of the Virginia State Navy assumed command of Gosport Navy Yard for the Commonwealth of Virginia on April 20, 1861. Pegram was replaced on April 22 by the fifty-year naval veteran Flag Officer French Forrest. Forrest was known as a "blusterer of the real old-tar school" and had gained considerable fame during his long service. He had joined the U.S. Navy on June 9, 1811. He fought with Oliver Hazard Perry at the Battle of Lake Erie and served aboard the USS *Hornet* when that vessel captured the

Gosport Navy Yard, photograph, 1862. *Courtesy of the National Archives.*

Gosport Navy Yard, engraving, circa 1861. *Courtesy of the Library of Virginia, Richmond, Virginia.*

HMS *Peacock*. Forrest's service during the Mexican War was very noteworthy; he commanded the *Cumberland* and *Raritan* during the U.S. Army's landing at Vera Cruz. Noted for his immaculate dress, stern countenance and flowing white hair, the Virginia native was one of the most senior officers to join the Confederate navy. Once in command at Gosport, he energetically set himself to the immense task of reorganizing the yard. Buildings were repaired, the Naval Hospital reopened and shops revitalized. Forrest even sent divers to investigate several of the hulls lying beneath the Elizabeth River. Thus, when the Commonwealth officially transferred the yard to the Confederacy on May 30, it was already humming with activity.[6]

Flag Officer French Forrest, CSN, photograph, 1861. *Courtesy of The Mariners' Museum, Newport News, Virginia.*

Forrest also recognized the virtual defenseless state of Norfolk and Portsmouth. Elements of the Federal fleet, including the USS *Cumberland*, were anchored in Hampton Roads, and the Union had retained control of Fort Monroe and the Rip Raps battery. Major General Benjamin Franklin Butler, commander of the Union Department of Virginia, had occupied and fortified Newport News Point on May 27, 1861. Camp Butler's four 8-inch Columbiads controlled the entrance to the James River, hindering the Confederate transportation link between Norfolk and Richmond. Forrest sought to counter this Union threat with the construction of earthworks along the south side of Hampton Roads. The yard commandant turned to Walter Gwynn to create an Elizabeth River defensive network. Gwynn, an 1822 U.S. Military Academy graduate, was an internationally esteemed railroad engineer and was considered the founder of the southeastern railroad network. When South Carolina left the Union, Gwynn joined the South Carolina Militia and constructed batteries in Charleston Harbor facing Fort Sumter. When that fort surrendered, Gwynn was named major general of the Virginia Militia and was sent to Norfolk. He participated in the capture of Gosport Navy Yard and then began constructing batteries at the mouth of the Elizabeth River. Eventually, Gwynn would use 196 of

Gosport Navy Yard Fortifications, engraving, 1861. *Courtesy of the Library of Virginia, Richmond, Virginia.*

the heavy cannons captured at Gosport to build batteries at Sewell's Point, Craney Island and Pig Point.

Gwynn realized that Union control of Fort Monroe had given the Federals an advantage in Hampton Roads. He urged that more cannons and men be made available to defend Gosport Navy Yard and other nearby points to prevent a Union blockade of Virginia's bays and rivers. As he supervised the work at Sewell's Point, General Gwynn watched the Federal steamers enter the Chesapeake Bay, bringing supplies and troops to Fort Monroe. He also noted how the Union naval squadron continued to increase its power. Gwynn and Flag Officer Garrett Pendergrast of the USS *Cumberland* corresponded about the capture of Virginian merchant ships. Pendergrast reminded Gwynn that it was war and that the Federals took advantage of their naval superiority to control the shipping lanes. Consequently, Gwynn redoubled his efforts defending the entrance to the Elizabeth River.

Union gunboats quickly tested the new Confederate fortifications. Gwynn was still in the process of arming the Sewell's Point Battery on May 18 when two Union gunboats, *Monticello* and *Thomas Freeborn*, shelled the Confederate defenses. When the Federals retired, Gwynn rushed Captain Peyton Colquitt's Columbus Light Guards, a newly arrived unit from Georgia, into action, mounting 32-pounders in the unfinished battery. The Federal gunboats returned the next day; however, they were greeted with counter fire and driven off. The USS *Yankee* continued to probe the Confederate river batteries over the next few weeks; however, the Federals quickly learned that the Confederate heavy ordnance mounted in earthen batteries provided Norfolk and Portsmouth ample protection against any naval attack. A stalemate now existed between Union and Confederate forces on either side of Hampton Roads.

Gosport's immediate importance was as a source of ordnance. The treasure-trove of cannons and gunpowder was quickly transferred to points throughout the seemingly defenseless Confederacy. Heavy guns were shipped to Charleston, Savannah, Memphis, New Orleans, Fredericksburg and Richmond for placement in fortifications then under construction. Since a majority of these guns were antiquated smoothbores, an effort to modernize them was undertaken by the yard's Ordnance Department. Department head Commander A.B. Fairfax coordinated Gosport's work rifling these outdated cannons. Several 32-pounders were rifled and strengthened by reinforcing their breeches with a strong iron band. This technique, which was developed by Lieutenant John Mercer Brooke to improve the range and accuracy of these smoothbores, was tested by Fairfax. One rifled 32-pounder was mounted on the tug *Harmony*, which steamed out into Hampton Roads. There, the little vessel traded shell fire with several Union vessels in a naval "repetition of the combat between David and Goliath."[7] While the experiments with banding proved a success, a better solution to obtain rifled ordnance was needed.

Meanwhile, Flag Officer Forrest contracted B. & J. Wrecking Company on May 18, 1861, to raise the *Merrimack* and tow her into the dry dock. Even though Forrest felt that the operation was a poor use of resources, he reported to General Robert E. Lee on May 30, "We have raised the *Merrimack* up and just pulling her in the drydock." The recovery operation cost $6,000. While questions were raised regarding what to do with this "burned and blackened hulk," Forrest also arranged for the salvage firm to raise the two scuttled sloops of war, *Germantown* and *Plymouth*.[8]

Gosport Navy Yard's capture was truly a godsend for the agrarian South. Since the Union blockade strangled Southern trade, Gosport provided the Confederacy with the capability to build ships to challenge the Union fleet. One of the first and most influential men to truly recognize Gosport's ability to achieve this goal was Confederate Secretary of the Navy Stephen Russell

Mallory. Perhaps one of Jefferson Davis's better cabinet appointments and one of only two men who served in the same cabinet post through the war, Mallory was born in 1812 in Trinidad. He moved to Key West, Florida, with his family when he was a young boy. Well educated at a Pietist school operated by Moravians, he became inspector of customs in Key West at nineteen years old. He became an attorney in 1839, fought as a volunteer in the Seminole War, served as a local judge and rose to political prominence. Mallory was elected senator from Florida in 1851 and became chairman of the Senate Committee on Naval Affairs in 1859. While serving as committee chairman, Mallory worked to modernize the U.S. Navy. He was successful obtaining appropriations to construct new screw propeller steam frigates and sloops of war, which became the wooden backbone of the Union navy during the Civil War. Mallory's efforts to champion the construction of an iron-cased battery, designed by Robert L. Stevens for the defense of New York Harbor, proved unsuccessful but brought him in contact with European efforts to construct armor-clads during the late 1850s. His tenure as chairman of the Committee on Naval Affairs prepared him to assume the tremendous task of creating a navy from nothing.

Mallory's appointment as secretary of the navy was not wholly popular and faced some serious opposition in the Confederate Congress. Many Southerners thought that he was too slow in supporting Florida's secession. Furthermore, his establishment of the Naval Retiring Board as chairman of the Committee on Naval Affairs made many senior officers, such as Matthew Fontaine Maury, sworn enemies of Mallory. Mallory's "stumpy, 'roly-poly' appearance" disguised his sharp mind and persuasive speaking skills. Nevertheless, as naval secretary, Mallory created a fleet to challenge the very navy he had helped to develop before the outbreak of war.[9]

Mallory realized that the South could never match the North's superior shipbuilding capabilities unless a novel weapon was introduced into the fray. The secretary advised Jefferson Davis that "we have a navy to build: and if in the construction of the several classes of ships we shall keep constantly in view the qualities of those ships which they may be called to encounter we shall have wisely provided for our naval success." He concluded his April 26 letter to Davis, stressing, "I propose to adopt a class of vessels hitherto unknown to naval service. The perfection of a warship would doubtless be a combination of the greatest known ocean speed with the greatest known floating battery and power of resistance."[10]

The Confederate secretary of the navy sought to employ a two-part program designed to attack the Northern merchant fleet on the high seas

Stephen Russell Mallory, secretary of the navy, CSA, photograph, circa 1860. *Courtesy of The Mariners' Museum, Newport News, Virginia.*

with commerce raiders while defending Southern harbors with the most advanced naval technology. Mallory proposed that a combination of iron-cased vessels armed with rifled cannons might be able to tip the naval balance of power in favor of the Confederacy. Consequently, Mallory wrote to Charles M. Conrad, chairman of the House Committee on Naval Affairs:

> *I regard the possession of an iron-armored ship as a matter of the first necessity. Such a vessel at this time could traverse the entire coast of the United States; prevent all blockades, and encounter, with a fair prospect of success, their entire Navy.*
>
> *If to cope with them upon the sea we follow their example and build wooden ships, we shall have to construct several at one time; for one or two ships would fall an easy prey to her comparatively numerous steam frigates. But inequality of numbers may be compensated by invulnerability; and thus not only does economy but naval success dictate the wisdom and expediency of fighting with iron against wood, without regard to first cost.*[11]

The concept of pitting "iron against wood" was not new to naval warfare. Other than the Vikings' habit of mounting their shields like armor on the sides of their long ships when moving from target to target, the first recorded iron ship of war can be attributed to an Italian vessel in 1538. The Koreans, however, should be credited with the first tactical use of an armored vessel. Korean Admiral Yi Sun-sin used "turtle ships" (kohbukson) to repel several Japanese invasions. A typical turtle ship averaged from 90 to 110 feet in length, with a 25- to 30-foot beam. The bulwarks were extended to support a roof and were covered with three-eighths-inch iron plate. The iron was affixed with knives and spearheads to repel boarders. The turtle ships were armed with several cannons and a barbed ram. These warships were fitted with one sail; however, propulsion was primarily achieved by oars. Although a defensive weapon, Admiral Yi Sun-sin was able to use these turtle-shaped vessels to destroy a Japanese invasion fleet in 1592. He repeated the feat on several other occasions until the Korean admiral achieved his greatest victory during the November 1598 Battle of Noryang. During this engagement, Yi Sun-sin attacked two Japanese squadrons using turtle ships and fire arrows. One Japanese squadron lost fifty ships, and the other was virtually destroyed. Admiral Yi Sun-sin was killed by a musket ball entering a gun port.[12]

European navies were slow to introduce iron into ship construction. Wooden sailing vessels had ruled the waves for centuries, and admirals seemed satisfied to fight sea battles in the traditional manner. The age of cannonballs bouncing off the sides of huge wooden ships of the line ended when Brigadier General Henri-Joseph Paixhans published his *Nouvelle Force Maritime et Artillerie* in 1822. Paixhans advocated a system of naval gunnery based on standardization of caliber and the use of shell guns. While he admitted that his concepts were not new, they did prove to be revolutionary. In 1824, Paixhans tested an 80-pounder shell gun against an old eighty-gun ship of the line, *Le Pacificateur*, at Brest. The battleship was virtually demolished by only sixteen shells. Besides demonstrating the tremendous destructive power of explosive shells, Paixhans argued that modern warships should be steam-powered, iron-plated and armed with like-caliber shell guns.[13]

In addition to Paixhans's work, the Sardinian army officer Giovanni Cavalli introduced the first effective rifled gun in 1845. Cavalli's gun featured a two-grooved rifled barrel with a ribbed, cylindrical, conical shell that could be hurled at a target with greater velocity, accuracy and penetrating power than that of smoothbore guns.[14]

All of these theories and new weapons were put to the test during the Crimean War. The stunning Russian naval victory at Sinope on November 30, 1853, proved the power of the new shell guns. Admiral Pavel Stepanovich Nakhimov's squadron attacked a Turkish fleet. Nakhimov's squadron of nine ships, many armed with 68-pounder shell guns, assaulted an anchored Turkish squadron of thirteen warships. After two hours, the Turkish fleet was destroyed. In an attempt to block Russian expansionism in the Black Sea, France and Great Britain declared war on Russia. On October 17, 1854, French and British ships engaged Russian forts defending Sevastopol. The Allies were repulsed. A stalemate continued until the French took a page from Paixhans's book and began the construction of floating iron batteries. Designed by Pierre Armand Guieysse, the *Lave* class featured formidable vessels mounting eighteen 68-pounders in a casemate covered by four-inch iron plating. The armor-clads had a length of 167 and a half feet, a draft of 8 feet and were propelled by a 225-horsepower engine, enabling a speed of four knots. Three of these batteries—*Devastation*, *Lave* and *Tonnante*—were towed into the Black Sea and used in the Allied naval assault against the Russian batteries at Kinburn on October 17, 1855. Anchoring just eight hundred yards from the forts, the French ironclads passed their trial by fire. After four hours of heavy cannonading, the armored floating batteries had suffered minimal damage, and only two men were killed when a Russian shell entered a gun port. In turn, the Russian forts were shelled into submission. Kinburn proved the value of armored vessels against fixed fortifications.[15]

The British also produced iron-plated floating batteries during the Crimean War. Although none of these vessels—*Glatton*, *Meteor*, *Thunder*, *Trusty* and *Aetna*—saw action during the war, the Royal Navy was quick to improve on the floating battery design. Four additional floating batteries—*Erebus*, *Terror*, *Thunderbolt* and *Aetna* (second version to replace the earlier ship accidently destroyed by fire)—were constructed.[16]

Even though the *Lave* class achieved great success at Kinburn, the floating batteries had two weaknesses: seaworthiness and speed. Accordingly, Stanislas Charles Henri Laurent Dupuy de Lome, chief constructor of the Imperial French Navy, used this Crimean War experience to develop a new ironclad design that combined speed, protection and firepower. The *Gloire*, launched on November 24, 1859, was a 253-foot wooden, steam-powered frigate covered with four-and-a-half-inch iron plate. Displacing almost six thousand tons, she mounted thirty-six 30-pounder shell guns and was capable of eleven knots. The Royal Navy, not to be outdone by the French, introduced an ambitious armor-clad production program. The HMS *Warrior* and HMS *Black Prince* were two of ten ironclads under construction in British shipyards by early 1861. The *Warrior* was an all-iron vessel protected by a four-and-a-half-inch armor belt. The ironclad mounted ten 110-pounder and four 70-pounder Armstrong breech-loading rifled guns, as well as twenty-six 68-pounder muzzleloading smoothbores. Capable of obtaining a combined (steam and sail) speed of 17.5 knots, the *Warrior* could escape what she could not destroy. As conceived, the *Warrior*'s superior speed, armor and armament gave the ironclad the ability to destroy any other ship in the world.[17]

All of the technological and tactical changes during the Crimean War were observed and recorded by a team of U.S. Army officers headed by Major Richard Delafield (Delafield's team included Major Alfred Mordecai and Captain George B. McClellan). The study produced by these observers, commonly called the Delafield Report, recommended modernizing the United States military establishment based on the Allied experiences during the Crimean War. Suggested improvements focused on advanced armaments and comprehensive fortifications required for coastal defense, as well as the need for ironclad vessels to protect American harbors and interests abroad. The Delafield Report, noting that Great Britain and France had used "their greatest exertions to devise the means of destroying the sea-coast casemated defenses of their enemy," advised that the U.S. Navy should construct steam-powered armored vessels armed with the most advanced ordnance to compete with modern European navies.[18]

The Delafield Report was published after the U.S. Navy had already initiated its modernization program with the *Merrimack* and her sister steam frigates. While the public questioned the navy's reluctance to construct ironclads, the refusal to enter into an arms race with the European powers was probably based on the U.S. Navy's prior experience with ironclads. In 1842, Congress had appropriated $250,000 for Robert Livingston Stevens to construct an armored steam vessel at Hoboken, New Jersey. The *Stevens Battery* was a revolutionary concept with an iron hull and no masts or sails. Designed to displace 1,500 tons, the ship was to mount seven guns capable of firing cylindrical-pointed, armor-piercing projectiles developed by Stevens. Begun in 1854, the ironclad was still incomplete in 1861.[19]

However, neither the U.S. Navy nor the U.S. Army neglected ordnance development. The military had developed several excellent types of heavy ordnance prior to 1861. Major George Bomford introduced his Columbiad seacoast gun, which combined the characteristics of cannon, mortar and howitzer. The Columbiad design was later improved by Captain Thomas J. Rodman. Rodman invented a production process of casting a gun hollow and then cooling the tube from the inside out with a constant stream of water. This strengthened the gun and enabled the production of huge Columbiads ranging in size from ten to twenty inches. A twenty-inch version was produced—nineteen feet in length and weighing 115,000 pounds. It was capable of hurling a 1,600-pound projectile over six miles.[20]

Since the Columbiad was not suited for shipboard use, the U.S. Navy developed its own powerful smoothbores. Commander John Adolphus Bernard Dahlgren, chief of naval ordnance, developed a solid cast cannon, cooled from the outside, that enabled an 11-inch gun to fire a shell weighing 170 pounds. The metal was thicker near the breech to strengthen the gun, giving the characteristic profile and the nickname "soda-bottle gun."[21] Despite the ability of these guns to safely fire (without bursting) heavy projectiles, the shot's velocity was rather slow. Neither the U.S. Army nor the U.S. Navy made a concerted effort to develop heavy rifled guns prior to the Civil War.

Mallory had been an active observer and advocate for many of these U.S. Navy advancements while serving as chairman of the Committee on Naval Affairs. This fortunate circumstance enabled him to immediately recognize the tools that the Confederacy required to challenge the U.S. Navy. Ironclads armed with rifled cannons were the obvious solution, and many Southern politicians and naval leaders supported Mallory's conclusion of pitting "iron against wood." Congressman William P. Chilton of Alabama advocated "that the Committee on Naval Affairs be instructed to inquire into the propriety of constructing by this Government of two iron-plated frigates and such iron-plated gunboats as may be necessary to protect the commerce and provide the safety of this Confederacy."[22] John Mercer Brooke wrote to Mallory advising that "an iron plated ship might be purchased in France loaded with arms and brought into port in spite of the wooden blockade."[23]

Brooke's letter stated the obvious. The best way for the industrially weak Confederacy to secure ironclads was to purchase armored vessels from either England or France. Accordingly, on May 10, 1861, the Confederate Congress appropriated $2 million for the acquisition "with the least possible delay, in France or England, [of] one or two steamers of the most modern improved description."[24] Mallory immediately dispatched Lieutenant James H. North overseas to purchase the *Gloire* or a similar vessel from France. If an ironclad was not available, North was instructed to make arrangements to construct an armored warship in England or France.[25]

North's selection as the Confederate overseas agent would prove to be one of Mallory's worse decisions in his quest for armored vessels. The naval agent's mission proved to be a complete disappointment, perhaps due to North's narrow-mindedness and lack of initiative.[26] Mallory conceded that North's efforts were fruitless when he wrote to the agent in September 1861 that the "Department regrets very much to learn that you were unable to purchase or contract for the construction of an ironclad war vessel, which could be invaluable to us at this time. But it is fully aware of the difficulties with which you had to contend. We will endeavor to construct such vessels in the Confederate states."[27]

Even before Mallory sent North to Europe, he had sought to establish an ironclad construction program within the Confederacy. Mallory knew that this effort faced numerous obstacles. The problem was apparent: how could an ironclad fleet be created without an industrial infrastructure? The secretary sought to resolve this problem by identifying the resources available to initiate the building of ironclads. On May 14, 1861, he instructed Captain Lawrence Rousseau to identify the availability of iron plate "of any given thickness from two and one-half to five inches."[28] Captain Duncan Nathaniel Ingraham was sent on a mission to ascertain if iron plates could be produced by foundries in Tennessee, Kentucky or Georgia. Ingraham and Rousseau both reported back to Mallory that, for a variety of reasons, iron plates could not be obtained for the navy. Mallory, however, did have a trump card. Tredegar Iron Works in Richmond did have the capability of rolling one- to two-inch iron plate and agreed to produce plate for the navy.[29]

Mallory faced the difficulties of building a new navy with tremendous determination despite the South's lack of resources. Once the source of iron plate had been found at Tredegar, he then sought to resolve the type of ironclad that could be manufactured within the Confederacy. On June 3, 1861, Mallory met with Lieutenant John Mercer Brooke to develop a workable ironclad design that could be utilized throughout the South considering the region's industrial limitations. Brooke conferred with Mallory on several occasions in early June, as he noted to his wife, Lizzie, "Mallory wants me to make some calculations in regard to floating batteries which I shall do today."[30]

Lieutenant John Mercer Brooke, CSN, engraving, circa 1862. *Courtesy of John Moran Quarstein.*

John Mercer Brooke was perhaps one of the most inventive and practical officers in the Confederate navy. Brooke, son of U.S. Army officer George M. Brooke, was born at Fort Brooke, Florida (named in honor of his father and located near present-day Tampa), on December 18, 1826. After his mother's death, Brooke was appointed midshipman in the U.S. Navy on April 30, 1841. After service at sea aboard ships including the *Delaware* and *Cyane*, he was sent on November 10, 1846, to attend the new Naval Academy. He graduated on July 21, 1847, and was promoted passed midshipman. Brooke was then assigned to the USF *United States* for a cruise in the Mediterranean and then, in 1849, detailed to the Washington Navy Yard. He joined the crew of the USS *Washington* to make observations to chart the eastern Atlantic. This work brought him into contact with the "Pathfinder of the Seas," Matthew Fontaine Maury. Maury was head of the Naval Observatory, and Brooke was an able pupil. His work with Maury was disrupted by service enforcing the end of the slave trade aboard the USS *Porpoise* in the African Squadron. When his father died, Brooke returned to the United States and was reassigned to the Naval Observatory. His first scientific success was the creation of a simple yet quite effective sounding device. The invention of the fathometer for typographic mapping of the ocean floor was exactly the tool Maury was looking for to test the deepest ocean bottoms. Brooke's fathometer earned him the Gold Medal of Science from the Academy of Berlin. He also invented a more efficient and effective boathook, which eliminated any twists in lines when detaching boats from davit lines. Brooke patented this device, and even though he was unable to patent the fathometer, the U.S. Navy eventually awarded a $5,000 prize for its invention.

Brooke's reputation as a scientist resulted in his 1854 appointment as astronomer and hydrographer aboard the USS *Vincennes* for a North Pacific survey. His fathometer tested depths up to 2,500 fathoms. He returned to the Naval Observatory in 1856 to produce his charts and then was assigned to plot the best shipping route between Hong Kong and San Francisco as commander of the schooner USS *Fenimore Cooper*. Brooke was able to conduct numerous soundings, discovered a guano-rich island he claimed for the United Sates and survived a severe gale. When the *Cooper* reached Okinawa, the schooner was blown ashore during a storm. Brooke discovered that his ship was in rotten condition, and he needed to secure another vessel. Eventually, Brooke and eight of his men returned to the United States on board the Japanese bark-rigged steamer *Kanin Masu*. It was a difficult voyage, and Brooke was forced to assume command of the vessel to ensure a safe return. Brooke was fated, however, not to enjoy his newly acquired fame. The secession crisis quickly prompted him to resign his commission and immediately set to work helping Secretary Mallory create a navy.

Brooke completed his preliminary drawings and reviewed them with Mallory. Mallory approved Brooke's concepts and assigned a draftsman from Gosport Navy Yard to assist him. This arrangement failed due to the draftsman, whom Brooke described as "lacking in confidence and energy, and…averse to performing unusual duty."[31] Mallory summoned Naval Constructor John Luke Porter and Chief Engineer William Price Williamson from Portsmouth to Richmond to discuss ironclad design concepts. On June 23, 1861, Brooke, Porter and Williamson met with Mallory. Porter had actually brought with him a model of an iron-cased, floating harbor defense battery he had created in 1846, while Brooke provided drawings he had made at Mallory's request. Both designs featured an inclined casemate to house cannons and engines placed below the waterline, but Brooke's concept submerged the bow and stern of the vessel to enhance buoyancy and speed. Since Mallory unrealistically wanted an oceangoing armored ship, Brooke's design was selected as the guideline for Confederate ironclads.[32]

Porter noted during the conference that his model "was intended for harbor defense only, and was of light draft."[33] While Porter's concept was

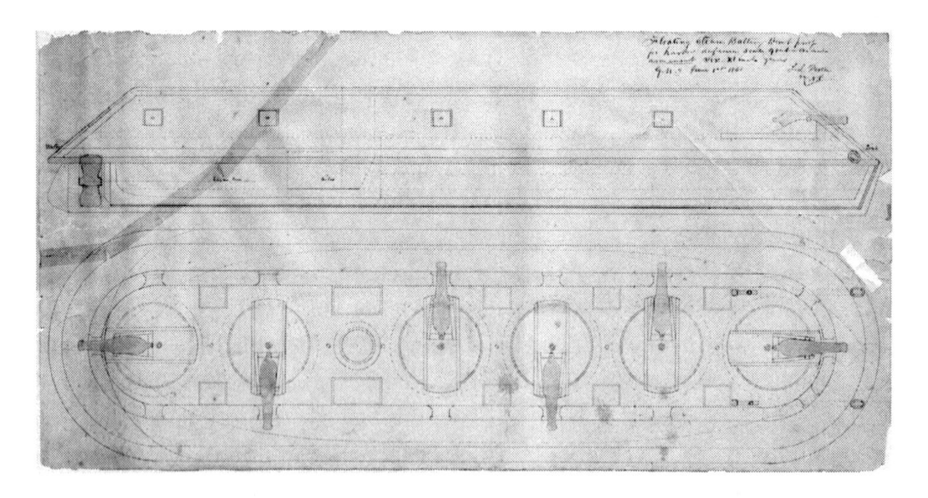

Floating steam battery bomb proof for harbor defense, plan, John Luke Porter, 1861. *Courtesy of The Mariners' Museum, Newport News, Virginia.*

similar to the French floating batteries that had attacked Kinburn, Brooke's ends were "shaped like those of any fast vessel, and in order to protect them from the enemy they were to be submerged 2 feet under the water, so that nothing could be seen afloat but the shield itself."[34] Brooke's "novel plan of submerging the ends of the ship and the eaves of the casemate," according to Mallory, was "the peculiar and distinctive feature…It was never before adopted."[35] Despite Brooke's fears "that Mr. P______ would, having an idea of his own make objection to my plan but he did not regarding it as an improvement," the committee accepted Brooke's concept of submerged ends "by unanimous consent."[36]

While Porter began drawing plans for a keel-up ironclad construction project, Brooke and Williamson sought to solve the power plant problems. They met with Joseph Reid Anderson of Tredegar Iron Works, but he was not encouraging. Williamson then noted that it "will take at least twelve months to build her engines unless we can utilize some of the machinery in the *Merrimac*." Williamson added that even though the old, previously condemned engines of the *Merrimack* could be reworked despite serious corrosion from the saltwater of the Elizabeth River, the machinery could not be transferred into another vessel. Porter countered, "I can adapt this model to the *Merrimac* and utilize the machinery in her." Brooke agreed, and despite his misgivings that the frigate's draft was too deep, he believed that the *Merrimack* provided the Confederacy with the best opportunity to quickly produce an ironclad.[37]

On June 24, 1861, the committee submitted a report to Mallory indicating that the *Merrimack* could be converted into an ironclad mounting ten guns. Committee members all believed that most of the machinery could be reused and estimated that the frigate could be transformed into an armored warship for $110,000.[38]

Mallory immediately approved the concept and, as an astute politician, presented his request for a project appropriation to the Confederate Congress. He reported that the reconstruction of the *Merrimack* as a frigate would cost $450,000, but the conversion to an armor-clad would cost only $172,523. The ironclad project was approved. Despite that shrewd success, Mallory did err with his shipbuilding program; he delegated responsibilities among several individuals. French Forrest, who did not really approve of the project, retained administrative control as yard commandant. Chief Engineer William Williamson was given the task of machinery revitalization, and Naval Constructor John L. Porter was charged with supervising the actual construction. John Mercer Brooke, a favorite of Mallory, managed the armor and armament for the ironclad, as well as acting as the inspecting agent for the entire project. Friction arose immediately between Brooke and Porter since much of the project overlapped. The acrimony began with the fact that both men claimed the vessel's design as their own and continued with Brooke's modifications throughout the project. Nevertheless, the Confederacy had its first ironclad under construction by July in its finest shipyard. It appeared that the Confederacy might indeed win the race for naval superiority despite its industrial weaknesses.

4
The *Virginia*

The charred Gosport Navy Yard now became the focus of the supreme effort to construct the Confederacy's first ironclad. At first glance it appeared that the transformation of the frigate into an armor-clad was an impossible task. "When she was raised," Acting Chief Engineer H. Ashton Ramsay commented, "she was nothing but a burned and blackened hulk."[1] The *Merrimack* may have been a dismal sight to behold, yet Stephen Mallory saw her as the key to the Confederate navy to contest the Federal blockade. Nothing, in Mallory's opinion, should delay the project. He wrote to Flag Officer French Forrest on July 11, 1861:

> *You will proceed with all practicable dispatch to make the changes in the* Merrimac, *and build, equip, and fit her in all respects according to her designs and plans of the Constructor and Engineer, Messrs. Porter and Williamson. As time is of the first importance in the matter, you will see that the work progresses without delay to completion.*[2]

The *Merrimack* was already in dry dock when Porter began removing all the burned portions of the ship. Naval Constructor John Luke Porter would actually complete all of the construction documents for the conversion and supervise the work. Porter was a Portsmouth native whose naval career had not been as successful as Brooke's.

John Luke Porter began to learn the shipbuilding trade in his father's shipyard. After his father died, the family shipyard was closed and Porter found work as a carpenter in various Portsmouth yards. Porter sought advancement as a constructor and in 1842 found employment in Pittsburgh, Pennsylvania, working on the iron-hulled, paddle-wheel steamer USRS *Allegheny*. It was an unusual vessel with submerged wheels beneath the hull. Porter worked with U.S. Navy engineer William Hunter on the design and construction; however, the *Allegheny* did not pass her initial sea trials and was placed in ordinary at the Washington Navy Yard.

Chief Naval Constructor John Luke Porter, CSN, photograph, circa 1870. *Courtesy of The Mariners' Museum, Newport News, Virginia.*

Porter's first attempt to pass the naval constructor exam was not successful, and he returned to Portsmouth as master shipwright at Gosport. He became active in politics and was elected president of the Portsmouth Town Council. Since Gosport was rapidly expanding in the 1850s, Porter worked on the *Constellation* overhaul that transformed the frigate into a sloop of war. After surviving yellow fever, he started work on the construction of the *Merrimack*-class steam screw frigates *Colorado* and *Roanoke*.

Porter passed his naval constructor exam in 1857 and was assigned to the Pensacola Navy Yard in Florida. His first major project was the construction of the screw sloop *Seminole*. The ship was launched in June 1860; immediately it was discovered that she had been built with inferior timbers. Porter was tried for neglect of duty. Though he was acquitted, Porter remained embittered. Reassigned to Gosport Navy Yard just before Virginia left the Union, he was the only naval constructor to join the Confederacy from the U.S. Navy. Despite his faults and problems with Brooke, Porter immersed himself in his work on the *Merrimack*.

Once the removal of all the upper works was accomplished, Porter then had the *Merrimack* full cut on a straight line from stem to stern. This was a very critical facet of the conversion project, as Porter described:

> *Having calculated the weight of the hull as I intended to fit her, with much care, and everything that was to go on her, as armor, guns, machinery, ammunition, stores, shot, coal, and shells, etc., I found that I could cut her hull down to nineteen feet fore and aft, and then have a surplus of fifty tons*

> *displacement, but when I drew my line I found I would cut one foot into her propeller, this I did not wish to do for several reasons, 1st, it would make more work and consume time, and 2nd, it would reduce her speed which I did not wish to do, consequently, I raised the line to twenty feet aft (holding the nineteen feet forward).*[3]

As soon as the *Merrimack* was cut down to about three feet above the waterline, Porter began the construction of the main deck. Designed to support the ironclad's ordnance and positioned two feet above the waterline, the area covered by the casemate was planked and supported by iron girders. Outside the casemate, the deck was constructed of solid pine beams bolted to the shield and upheld by iron girders. There was only one other deck—the berth deck—and her boilers and engines remained in their original positions. The power plant was located entirely below the waterline for protection against shot and consumed most of the twelve-and-a-half-foot-high space beneath the berth deck. The boilers, which had a heating surface of 12,527 square feet, were placed under the funnel in the middle of the hull. The engines were located behind the boilers with the propeller shaft running underneath to the stern. The engines were valued at $170,000.00.[4] Since Porter had removed the upper decks of the *Merrimack*, the transformed hull was actually shortened from 275 feet to 262 feet, 9 inches.

By August, work had begun on the casemate, which was the ironclad's most distinctive feature. The casemate design was the easiest approach to armored ship construction for the Confederates and was well within their limited industrial capabilities. Consequently, all other Confederate warships would feature this design. The casemate, also called the shield, began 29 feet from the bow and extended aft 170 feet. The fantail continued another 66 feet. The sides were sloped upward at a thirty-six-degree angle to deflect shot. Major John G. Barnard, chief engineer, Department of Washington, and a former superintendent of the U.S. Military Academy, published a treatise in early 1861 entitled *Notes on Sea Coast Defense* on sloped casemates based on English experiments. Barnard wrote that his design "is calculated to a much greater degree of invulnerability—that of inclining the iron-clad sides inwardly (to angle with a horizontal of 35 degrees or 40 degrees) by which arrangement shot will glance off with little injury to the side." Unfortunately, this acute slope allowed only 7 feet of headroom and a beam of 30 feet, which forced the cannons to be staggered along the opposing broadside to accommodate recoil and loading. The roof, also referred to

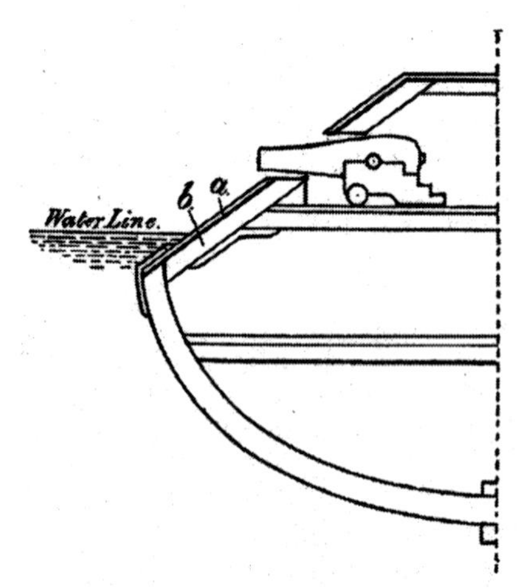

CROSS-SECTION OF "MERRIMAC," FROM A DRAWING BY JOHN L. PORTER, CONSTRUCTOR.

a— 4 inches of iron.
b—22 inches of wood.

Cross Section of Merrimac, engraving, circa 1880. *Courtesy of John Moran Quarstein.*

as the shield or spar deck, was grated with a two- by two-inch mesh to provide ventilation to the gun deck. The grating was made of two-inch iron bars, forged at Mahaffey's Iron Works, located just outside Gosport Navy Yard in Portsmouth, supported by ten- by twelve-inch yellow pine rafters with white oak veneer.[5]

One of the disagreements between Brooke and Porter concerned the hatchways leading to the shield deck. Porter had planned for two hatches, yet Brooke suggested that two additional hatchways be added. Brooke believed that the hatchways would permit the use of small arms by the crew to repel boarders during battle. A compromise was forged, and the ironclad had three hatchways installed with pivot shutter closures. The hatchways were positioned directly above those below for passing ammunition, coal and other supplies directly into the ship. Lanterns were placed along the gun deck to improve visibility.[6]

While the original plans called for one pilothouse to be installed at the forward end of the casemate, Porter had two cast-iron, conical houses made to enhance observation. They were hollow cast and twelve inches thick with four observation loopholes. Brooke thought the "sort of look-out-house" was poorly constructed and would "break into fragments if struck by a shot. He [Porter] has also closed the aperture so that now there will be great difficulty in repelling a boat attack." Only one pilothouse was mounted.[7]

The casemate was constructed of four inches of oak laid horizontally, eight inches of yellow pine laid vertically and twelve inches of white pine laid horizontally. Each course of wood was thoroughly caulked and bolted together. The solid timber was then eventually sheathed with eight-foot-long, two-inch-thick iron plate, each plate being eight inches in width. The first course was laid horizontally and the second (or top

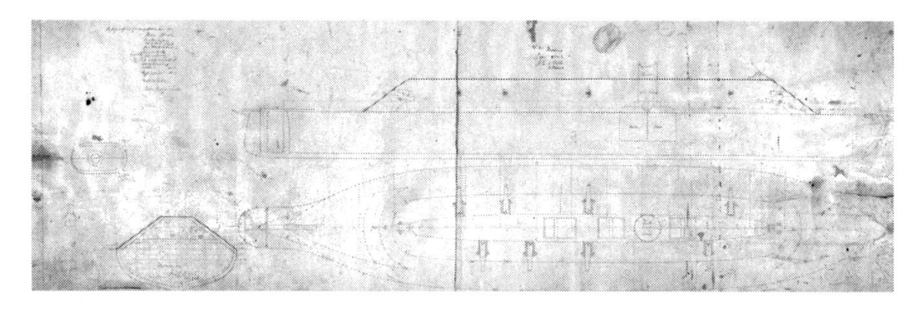

CSS *Merrimac*, plan, John Luke Porter, 1861. *Courtesy of The Mariners' Museum, Newport News, Virginia.*

layer) vertically. The plates were bolted together onto the shield. These long bolts, fabricated at Gosport, were almost two inches in diameter. The bolts crossed through the bulwark and were secured with stud nuts and washers that were clinched and countersunk so they would not be loosened by shot. The deck, designed to be almost awash with the sea, was covered with one inch of iron plate. On the forward deck, "a rough breakwater was built…to throw off water."[8] An additional course of one-inch iron plate extended three feet from the deck to a depth of three feet around the vessel. The joining of the casemate to the hull was an obvious weak point. Porter had devised the displacement that would submerge the knuckle two feet below the waterline. The casemate eaves were also extended two feet to provide additional protection from shot aimed at the ironclad's hull.

Acquisition of the iron plate for the *Merrimack* conversion project would prove to be a significant problem. On July 24, 1861, Tredegar Iron Works accepted the contract to produce the "iron work necessary for fitting up the *Merrimac*" at a cost of six and a half cents per pound. Brooke, who was assigned the dual task of coordinating the ironclad's armor and armament, stayed in Richmond to supervise the production of iron plate by Tredegar Iron Works. The original contract, based on Porter's specifications, detailed that the vessel should be sheathed by three layers of eight-foot-long, one-inch-thick iron plate. Tredegar had adequate facilities to roll and punch this size plate, and within a week of signing the contract, the firm was ready to ship plates to Gosport. Brooke had originally wanted thicker plates for the ironclad but had settled on this configuration because Tredegar was only capable of rolling one-inch plate. Unsure of the casemate's shot-proof qualities, Brooke resolved to test the protective qualities of three courses of one-inch plate at Jamestown Island.[9] Mallory concurred with Brooke's opinion and ordered him in mid-August "to test the iron plates now being prepared at the Tredegar Works to determine what their powers of resistance to shot and shell will be when placed on the *Merrimac* according to the plans adopted."[10]

Catesby ap Roger Jones, CSN, engraving, circa 1870. *Courtesy of The Mariners' Museum, Newport News, Virginia.*

Lieutenant Catesby ap Roger Jones commanded the fortifications on Jamestown Island. Jones was keenly interested in the ironclad project. He considered the *Merrimack* "the most important naval affair the country has to deal with and consequently am deeply interested in her success, and anxious that it may be complete."[11] Catesby Jones was a noted U.S. Navy ordnance expert under John Dahlgren and had served aboard the USS *Merrimack* to test the frigate's battery during her maiden voyage. David Dixon Porter wrote that Jones and Brooke were a major loss for the U.S. Navy.[12] Since Brooke wished to conduct the experiments in relative secrecy with Jones, Jamestown Island appeared to be the perfect proving ground.

Brooke and Jones prepared a target in early October 1861 to replicate the proposed shield for the *Merrimack*. Constructed with oak and pine, the target was almost two feet thick and measured twelve by twelve feet. The target was inclined at an angle of thirty-six degrees and covered with three layers of one-inch iron plate. An 8-inch Columbiad was positioned 327 yards from the target, and several rounds of solid shot were fired using a ten-pound powder charge. Brooke and Jones believed that this distance and shot would closely represent the shot enemy ships could fire at the Confederate ironclad. The test shield was shattered, with the ball entering five inches into the wood backing.[13]

The test was observed by several naval officers and officials: Commander Randolph Tucker; Lieutenants Robert D. Minor, David McCorkle, William P. Powell and James H. Rochelle; and Nelson Tift, the builder of the CSS *Mississippi* at New Orleans. Everyone agreed that one-inch plate was insufficient, and Jones also noted that if explosive shells had been used, the wood backing would have caught fire.[14]

Another target was created using two layers of two-inch iron plate. The 8-inch gun and one other 9-inch cannon were fired at the target. The outer plate shattered, but the interior plate was only cracked. The wood backing was not touched. The experiment proved Brooke's belief that the *Merrimack*'s casemate must be shielded with two-inch iron plate. Just to be sure, a third target was covered with two layers of railroad T-iron. This configuration was easily penetrated by both the 8-inch and 9-inch guns. Jones submitted the findings of these tests in an October 12, 1861 report to Mallory. He wrote that the sloped sides of the casemate would greatly enhance the shot-proof qualities of the shield, but the *Merrimack* must be clad with four inches of iron, preferably with two layers of two-inch plate.[15]

As Southern newspapers advertised that "the iron sheeting for the *Merrimac* has proved, under trials made recently at Jamestown Island with Columbiads, to be almost worthless," Brooke returned to Tredegar Iron Works to ascertain whether the mill could produce the thicker plates.[16] Brigadier General Joseph Reid Anderson, owner and president of Tredegar Iron Works, agreed to make the necessary improvements to fabricate two-inch plates. This miracle was achieved by widening the doors of the annealing furnaces and installing a new waterwheel. Additional workers were assigned to the rolling process, prompting Anderson to write, "We are now pressed almost beyond endurance for the heavy iron work to complete one of the war vessels now ready for operations." Anderson further noted that it "is a most fortunate thing that we could render the assistance to our little Navy."[17]

The thicker plates created yet another technical difficulty for Tredegar: how to mount the plates on the ironclad with bolts. The original contract had specified that the plates were to be "punched" for bolts, but Tredegar did not have the tools available to pierce cold iron two inches thick. The rolling mill foreman succeeded in punching bolt holes in a thicker plate, and Tredegar resorted to a drilling process that required additional time and money. Eventually, the cost per pound came to over seven and a half cents. The total cost was $123,715.[18]

Porter estimated that one thousand tons of iron plate would be needed for the project. Thus, a critical problem arose in the procurement of iron to make the plates. The South lacked an iron industry capable of meeting all its military needs. Tredegar had already used up its prewar supply of iron filling other orders and could not proceed with the project unless another source was identified. French Forrest scoured Gosport Yard for iron for the ironclad project. Over three hundred tons of scrap, including old tools and residual iron from the burned portions of the yard, were gathered and shipped to Richmond. Forrest eventually sent Tredegar over two thousand tons of obsolete guns. Iron was even secured from Cane Creek Furnace in Calhoun, Alabama. Yet this still was not enough. Mallory detached Brooke to "obtain from the President of the Danville Railroad Company by leave or otherwise, thirty tons of railroad iron for experimental purposes." Rails were taken from portions of the Baltimore & Ohio Railroad, which had been captured by Stonewall Jackson's troops in the Shenandoah Valley. Exposed, indefensible Southern railroads, such as the line connecting Winchester with Harpers Ferry, were torn up and the scavenged rails sent to Richmond. Brooke, along with the foreman of Tredegar's rolling mill, Uri Haskins, supervised and inspected the work.[19]

Once Tredegar began producing a sufficient supply of two-inch plate, another very serious problem surfaced: transporting the iron plate from Richmond to Portsmouth. The poor Southern railway system was already overburdened with military traffic, and the navy's demands were often overlooked or ignored. Porter estimated that only two of the one thousand tons of iron reached Gosport in October, even though Tredegar noted that "some 70 to 100 tons of the iron [were] now ready to ship." Flag Officer Forrest, frustrated by the delays arranging transport, assigned an agent, William Webb, to shepherd the finished plate to Gosport. Webb, operating in Richmond and Petersburg, faced tremendous difficulties. More often than not, he found that the cars assigned to the navy were misappropriated for the army or the tracks were simply overloaded with traffic.

One problem was getting the iron to Petersburg for shipment on the Petersburg & Norfolk Railroad. Often the plates had to be sent to Burkeville on the Richmond & Danville Railroad, thence to Petersburg via the Southside Railroad. Webb finally learned to use the longer Richmond–Weldon, North Carolina line when the Richmond–Petersburg line was unavailable. Once in Weldon, the plates were transshipped to Gosport via the Seaboard & Roanoke.

While the *Merrimack* was ready for sheathing in November, it was not until February 12, 1862, that the last shipment of iron plate arrived at Gosport Navy Yard. In all, Tredegar produced 723 tons of iron plate for the project.[20] All the delays associated with the production and delivery of iron plate frustrated everyone working on the conversion project. Ashton Ramsay lamented that in the time it took to receive the plates, "we could have rolled them in Norfolk and built four *Merrimacs*."[21]

Remodeling the Merrimack, engraving, J.O. Davidson, circa 1880. *Courtesy of John Moran Quarstein.*

Mallory wanted more than just an armor-plated ship; he desired an ironclad capable of firing shells and armor-piercing shot. The Confederate ironclads must feature, he reasoned, weapons technologically superior to anything that the Union could muster. "Rifled cannon are unknown to naval warfare," the secretary wrote, "but those guns [have] attained a range and accuracy beyond any other form of ordnance, both with shot and shell."[22] Mallory believed that armored ships armed with rifled cannon would give the South a "technical surprise," which could tip the naval balance in favor of the Confederacy. After all, he contended, modern naval battles would be "simply contests in which the question, not of victory, but of who shall go to the bottom first, is to be solved."[23]

The man to whom Mallory looked to develop rifle ordnance for the Confederate navy was John Mercer Brooke. Even before he had resolved plate production issues at Tredegar, Brooke had begun work on rifled guns. Brooke wrote in November 1861, "By order of the Secretary I designed two rifled cannon of 7-inch caliber for the *Merrimac*—one of them had been cast and is now nearly bored." The Brooke gun was one of the most powerful weapons produced by either side during the Civil War. Mallory thought the Brooke gun was superior "in strength, precision, and range" to any other cannon available in America.[24]

Brooke resolved the critical need to produce a safe, rifled cannon (not prone to bursting when fired) by supporting the powder-chamber area with a series of iron hoops. His gun looked somewhat like a Parrott gun; however, that is where the similarities between the guns ended. Instead of using the Parrott method of welding an iron band onto the tube in one piece, Brooke heated several heavy wrought-iron rings. When expanded, they were placed tightly around the breech. Once the iron cooled, it contracted over the breech, thereby forging a very strong and tight reinforcing band. Basically, Brooke's concept was similar to the British-made Blakeley and Armstrong guns. The built-up breech, created by the shrinkage hoops around the barrel, provided the metal with resilience to the internal force of exploding powder.[25]

Two models of the Brooke gun were designed and quickly placed into production at Tredegar Iron Works. The 7-inch model weighed 15,300 pounds and, during tests, could throw a 100-pound shell over four and a half miles using a 12-pound powder charge. The 6.4-inch version was developed as a broadside gun and weighed 10,675 pounds. These huge, powerful guns featured a distinctive rifling technique called the Brooke Rifling System. Each barrel was rifled with seven grooves made in the shape of elliptical bands. The grooves culminated in a cog from which the apex formed the next band. While working on the design of the Brooke rifle, Brooke also developed a brilliant system of converting old smoothbore cannons into rifles by forging bands over their breech.[26]

Once he had developed his rifled cannon, Brooke concentrated on the production of fuses, explosive shells and a wrought-iron elongated shot (bolt) for his rifled guns. The flat-headed bolt was prepared specifically for use against ironclads and could "literally punch a hole through armor plate."[27] Unfortunately for the Confederacy, when ordering projectiles for the *Merrimack*'s battery, Brooke and Catesby Jones decided to delay the production of the armor-piercing bolts. Both men believed that the Confederate ironclad would only face wooden ships when it attacked the Union fleet and instructed Tredegar to concentrate on fabricating explosive shells.[28] This decision would prove to be a major error.

Brooke's duties included preparing the armament for the *Merrimack*. Since Mallory wanted the ironclad to be armed with the finest heavy cannon, Brooke proposed that the *Merrimack* be armed with a broadside of six IX-inch Dahlgren smoothbores and two 6.4-inch rifles. The Dahlgrens were selected from the *Merrimack*'s original battery, and two of them were modified for hot shot. A special furnace was installed in the engine room to prepare shot for these guns during combat. The installation of the stern and bow 7-inch rifles caused another controversy between Brooke and Porter. Porter had overlooked this opportunity to enhance the ironclad's field of fire, and Brooke, as the project's inspecting officer and confidant of Mallory, insisted on this improvement. The bow and quarter ports were eventually pierced to enable the 7-inch rifles to serve as pivot guns. The casemate had a total of fourteen elliptical gun ports, four to a side and three to an end. The casemate's slope limited deck space, causing the broadside guns to be staggered. However, the casemate's interior layout still caused problems for the handling of cannon during combat and did not enable the creation of an integrated fire control system.[29]

The ram was another issue that only increased the animosity between Brooke and Porter. The advent of steam-powered armored ships in Europe revitalized this ancient naval weapon as a tool of future navies. Ramming as a decisive offensive tactic had been virtually abandoned with the rise of large sailing ships mounting artillery, yet steam power made this battlefield technique once again a viable weapon. Mallory was keenly aware that "even without guns the ship would be formidable as a ram." He likened ramming to a "bayonet charge of infantry." Considering the gunpowder shortage in the South, he recognized that Confederate ironclads could employ the ram as a technological weapon that could punch into the sides of Union wooden vessels. Nevertheless, the ram was an afterthought proposed by Mallory. Porter did not approve of the concept; however, Brooke insisted on the ram's installation. Reluctantly, Porter designed the 1,500-pound, cast-iron, wedge-shaped ram to be fabricated by Gosport's blacksmiths. The prow was designed to protrude three feet from the ironclad's bow. Bolted to the bow's stern head, it was further secured by iron braces. The ram nonetheless was poorly mounted. While hammering the bolts with heavy sledgehammers, a missed stroke thoroughly cracked one of the flanges holding the prow in place. Even though it was apparent that the ram was now improperly secured, nothing was done to correct the problem.[30]

The acrimony between Brooke and Porter continued throughout the project. "Lieutenant Brooke was constantly proposing alterations to her to the Secretary of the Navy," Porter wrote, "and as constantly and firmly opposed by myself." Brooke eventually proposed six major changes to the ironclad, and most of the naval officer's suggestions were valuable improvements.

One recommendation concerned the ironclad's steering gear. Porter had extended the vessel's fantail "in order to protect her rudder and propeller from being run into."[31] As the heavy wooden protective frame was constructed, he cut down the propeller and removed the banjo-lifting mechanism. The *Merrimack*'s original rudder was reused and fitted with a crosshead. The ironclad's rudder chains, however, were unprotected and "liable to be jammed by shot." Brooke ordered that the steering chains be moved into an iron channel to protect the mechanism from shell damage.[32]

Porter was overworked not only from his responsibilities with the *Merrimack* but also from providing designs for other Confederate shipbuilding projects. "I received but little encourage from anyone," Porter complained, "Hundreds—I may say thousands—asserted she would never float." Many observers believed that the ship was top-heavy. Captain Sidney Smith Lee asked, "Mr. Porter do you really think she will float?" Porter lamented that only Gosport's ordnance officer, Commander A.B. Fairfax, gave him any type of encouragement and hope for the ironclad's completion and success.[33]

Mallory's concerns about the project prompted him to assign the forty-year-old Virginia native Lieutenant Catesby ap Roger Jones as the *Merrimack*'s executive officer in November 1861. Jones's early assignment to the *Merrimack* expedited construction, in part, by mitigating the disagreements between Brooke and Porter. His duties included mounting the ironclad's ordnance, mustering a crew and preparing the vessel for sea. Jones had numerous problems to resolve. The conversion project was scheduled for launching in late November, but it was not until the end of January that the ironclad neared completion.[34]

One of the most difficult duties Jones faced when he assumed his position as the *Merrimack*'s executive officer was mustering an adequate crew to man the ironclad. Although sailors would prove hard to find, Jones had an excellent pool of former U.S. Navy officers from which to choose. Over a quarter of prewar naval officers resigned their U.S. Navy commissions to join the Confederacy. The ironclad project at Gosport was a choice assignment. Consequently, Jones was able to assemble an outstanding group of officers. John Taylor Wood, Hunter Davidson and Charles Carroll Simms were all former U.S. Navy officers of proven quality and experience who were detailed to the *Merrimack*. Hunter Davidson and John Taylor Wood were both prewar instructors at the U.S. Naval Academy. Davidson resigned his commission following the bombardment of Fort Sumter. While officers resigning prior to Lincoln's presidency were allowed to leave the service without threat, U.S. Secretary of the Navy Gideon Welles treated sympathizing resigning officers as deserters. Accordingly, Davidson was dismissed. He accepted a Confederate commission as a first lieutenant and was detailed to the North Carolina Squadron until transferred to the CSS *Patrick Henry*. Davidson was detailed to the *Merrimack* on December 8, 1861.

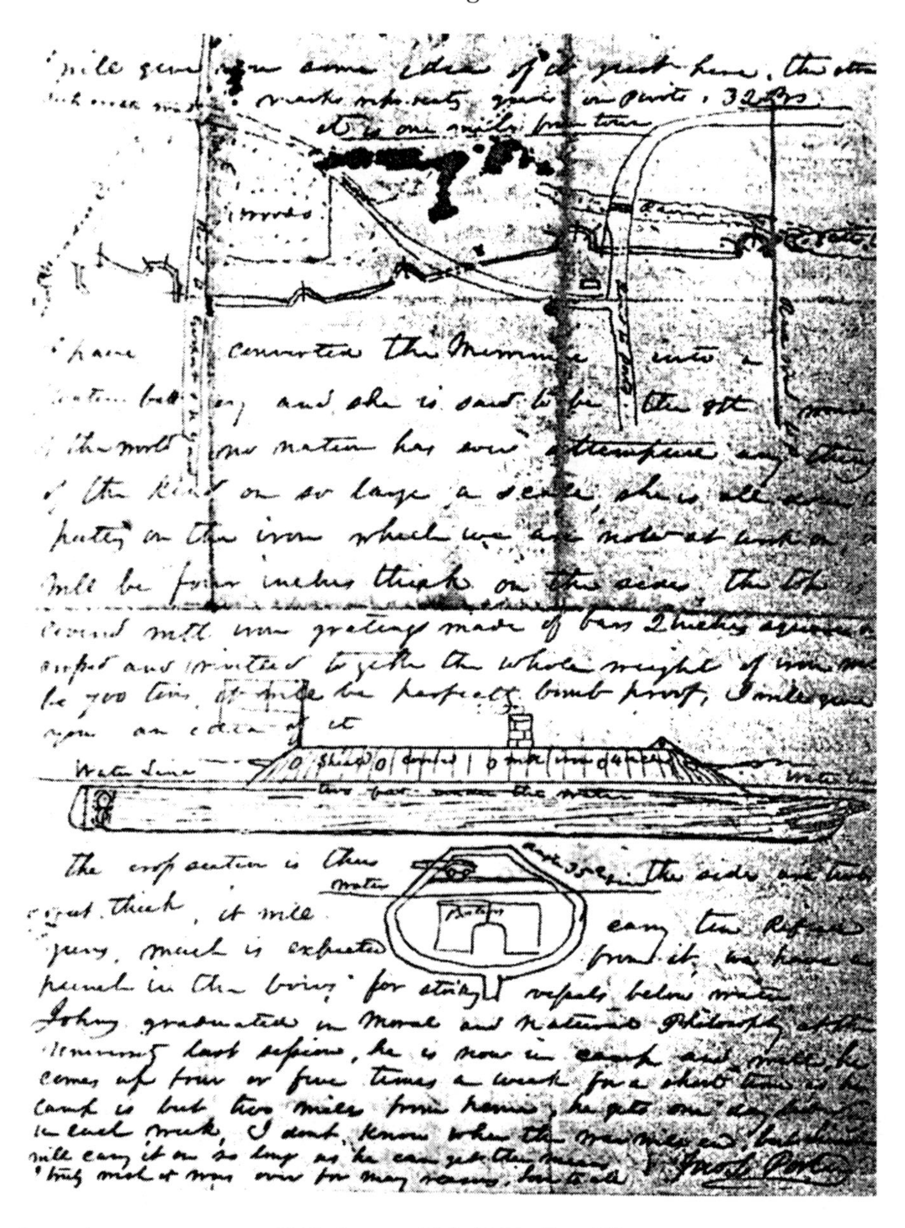

[illegible] will give you some idea of the yard here. the other [illegible] marks represents guns in ports, 32 lbs.
it is one mile from town

woods

[illegible] have converted the Merrimac into a floating battery and she is said to be the 8th wonder [illegible] the most [illegible] no nation has ever attempted anything of the kind on so large a scale. she is all [illegible] putting on the iron which we are now at work on, it will be four inches thick on the sides. the top [illegible] covered with iron grating made of bars 2 inches square lapped and riveted together the whole weight of iron will be 700 tons. it will be perfectly bomb proof. I will give you an idea of it

Water Line
two feet under the water

The construction is thus [illegible] the sides are [illegible] feet thick, it will carry ten Rifled [illegible] guns, much is expected from it, we have a [illegible] in the bow for sticking vessels below water. John graduated in Moral and Natural Philosophy at the [illegible] last session, he is now in camp and will, he comes up four or five times a week for a short time as his camp is but two miles from home, he gets one day [illegible] in each week, I don't know when the [illegible] will carry it on so long as he can get the means [illegible] truly [illegible] for many reasons, love to all. [illegible] Porter

Porter letter, pen on paper, 1862. *Courtesy of John Ridgley Porter III.*

Like so many of his fellow officers aboard the Confederate ironclad, John Taylor Wood had a distinguished prewar naval career. He was appointed midshipman on April 7, 1847, after training at the newly created Naval Academy at Annapolis, Maryland. He eventually was assigned to the USS

Lieutenant John Taylor Wood, CSN, engraving, circa 1880. *Courtesy of the Virginia War Museum, Newport News, Virginia.*

Ohio. The *Ohio* sailed to Mexico's west coast, and Wood served on the landing party that captured the port city of Mazatlan. After three years on the *Ohio*, Wood was granted a three-month leave of absence as his grandfather, General Zachary Taylor, had just been elected president. Wood enjoyed his time in Washington observing the corridors of power and was impressed by his grandfather's resolve to preserve the Union. In 1851, he was detailed to the brig USS *Porpoise* to disrupt the slave trade. While off the Gulf of Guinea, a Spanish slave ship was captured, and Wood was assigned as prize master. He then returned to the Naval Academy in Annapolis and graduated second in his class on October 1, 1852. After additional service aboard the *Cumberland* and *Wabash*, Wood used his family's influence to secure an appointment as a member of the U.S. Naval Academy faculty. He taught naval gunnery, seamanship and naval tactics. He purchased a small farm near the academy and began to raise a family. His life seemed idyllic.

But the secession crisis brought critical changes to his life. His parents, Colonel Robert C. Wood, MD, and Anne Mackall Taylor Wood, remained loyal to the Union. However, his uncle, Jefferson Davis; his cousin, Richard Taylor; and his brother, Robert C. Wood Jr., all opted for the Confederacy. John Taylor Wood lamented that "war, that terrible calamity, is upon us, and worst of all, among us." Even though he wished to stay neutral, orders for Wood to move to Rhode Island with the Naval Academy prompted him to resign on April 21. He intended to wait out the war on his Maryland farm. Instead, he was dismissed, effective April 2, 1861. He had to move several times to escape arrest. The Confederate victory at Manassas on July 21, 1861, led him to escape across the Potomac River with his family in a small boat. Wood was commissioned a first lieutenant on October 4, 1861, and was assigned to the batteries at Aquia Creek and Edamsport, Virginia, on the Potomac to enforce the blockade of Washington, D.C. He later reported that his battery sank several ships, including the 1,200-ton *Rappahannock*. On November 25, 1861, Wood was detailed to serve aboard the *Merrimack*.

Charles Simms had served under Franklin Buchanan at the Washington Navy Yard. He resigned his commission, only to be officially dismissed on the same day, April 22, 1861. He joined the Virginia State Navy and was appointed as a first lieutenant, CSN, on June 10, 1861. This talented officer served under Captain George Hollins in the daring capture of the Chesapeake Bay steamer *St. Nicholas*. Hollins had gathered numerous daring officers, including Dr. Algernon S. Garnett and Colonel Richard Thomas Zarvona of the Maryland Zouaves. All the men were disguised as civilians, and Zarvona actually dressed as a French lady, Madame La Forte, who was flirtatious and seemed scandalous. She traveled with three large trunks. Once the *St. Nicholas* was out in the Chesapeake Bay, on Hollins's signal, Zarvona went to his cabin and changed from his "Madame" gown into his Zouave uniform. He burst out onto the deck brandishing a cutlass and revolver. Simms and Garrett quickly gathered weapons from Zarvona's trunks, and on November 20, 1861, the *St. Nicholas* was captured. Hollins placed the ship under the command of Simms. On the same day, the *St. Nicholas* captured the brig *Monticello* and the schooners *Mary Pierce* and *Margaret*. Federal officials called it the worst kind of piracy.

Simms retained command of the *St. Nicholas* (renamed CSS *Richmond*) until detailed to the North Carolina Squadron. He commanded the CSS *Appomattox* during the Battle of Roanoke Island. Following the *Appomattox*'s forced destruction at the South Mills, North Carolina entrance to the Great Dismal Swamp Canal, Simms was assigned to the *Merrimack*.

Officers resigning from the U.S. Navy during President James Buchanan's administration were allowed to leave the U.S. Navy without stigma. Buchanan's secretary of the navy, Isaac Toucey, accepted all resignations without threatening dismissal or arrests for desertion. John Randolph Eggleston was a lieutenant assigned to the USS *Wyandotte* and resigned shortly after the first Deep South states left the Union on January 13, 1861. Eggleston was assigned to the New Orleans Station on the CSS *McRae* until detailed to the *Merrimack*. Acting Midshipman Robert Chester Foute of Tennessee resigned just after Abraham Lincoln's election on December 3, 1860. Foute holds the distinction of being the first officer to leave the U.S. Navy among his peers on the Confederate ironclad.

Several young officers destined to serve on the Confederate ironclad—such as Algernon S. Garnett, H. Ashton Ramsay and Walter Raleigh Butt—were at sea when the secession crisis turned to war. Assistant Surgeon Algernon

Garnett was aboard the USS *Wyandotte*. A Virginia native, he was allowed to resign on April 22, 1861, only to be later unceremoniously dismissed from the U.S. Navy on May 10, 1861. Garnett participated in the capture of the *St. Nicholas* and then served on the CSS *Patrick Henry*. He was the first medical officer assigned to the *Merrimack*.[35]

Henry Ashton Ramsay was a talented twenty-five-year-old engineer from Washington, D.C., in line for promotion to chief engineer, serving aboard the USS *Niagara* when the warship entered Boston Harbor from Hong Kong in April 1861. Since the frigate was immediately assigned to blockade Hampton Roads, the *Niagara*'s commander, Captain William McKean, called all the officers into his cabin on April 19 to sign the "iron-clad oath," professing loyalty to the Union. Ramsay and eleven other officers refused, as they "could not fight against their own blood." The next day, the pro-Southern officers were put ashore and were listed in Boston newspapers as traitors. Several men, including Ramsay, were "assailed by a mob in the streets of Boston" angered by the attack on Massachusetts troops in Baltimore. Ramsay was arrested by civil authorities and then held prisoner on a receiving ship in Boston Harbor for ten days. Dismissed from Federal service on April 6, 1861, Ashton Ramsay went south, joined the Confederate navy and was assigned as an engineer to Gosport Navy Yard. Ramsay worked on the *Merrimack* conversion project and was detailed to the ironclad on January 1, 1862, as an acting chief engineer.

A similar fate befell Portsmouth native Walter Raleigh Butt. Known as "Butsy" by his classmates at Annapolis, Butt was serving as a midshipman aboard the USS *Congress* in April 1861. He refused to take the oath and was imprisoned. Butt was held as a POW at Fort Lafayette, New York Harbor, and Fort Warren, Boston Harbor, until returned to the *Congress* at Newport News Point. He was paroled on December 21, 1861, exchanged in early January 1862 and immediately reported to the *Merrimack*.

Lieutenant Walter Raleigh Butt, CSN, photograph, circa 1864. *Courtesy of Charles Peery.*

Many Confederate naval officers sought after duty on this experimental warship. E.V. White, who served as acting third assistant engineer, "conceived the idea of becoming one of the number who should test the qualities of this mighty engine of destruction." White later wrote, "I made application for the position desired, received my appointment, January 16, 1862, and two days thereafter reported for duty."[36] E.A. Jack worked in the Gosport Navy Yard's Department of Steam Engineering prior to the war. When the conflict began, he enlisted in Company K, 9th Virginia Regiment. He eventually received orders to return to work at the navy yard, as the Confederacy was in need of skilled engineers. However, as he watched the transformation of the *Merrimack* into an ironclad, Jack yearned to serve on the warship and made an application to Chief Engineer William Williamson for a position. Jack received a warrant to serve as the ironclad's third assistant engineer and was assigned to the ship on November 29, 1861.[37]

Other officers would not be as fortunate. John Mercer Brooke had not only designed the ironclad and recommended Catesby Jones's assignment to the *Merrimack*, but he also actively lobbied to serve on the ironclad. Mallory noted that despite Brooke's desire to have "a crack at the Yankees," his assignment to the ironclad would interfere with his ordnance work for the navy. Yet he left the decision to Brooke. Brooke would eventually realize that he could not do both and stayed in Richmond.[38]

Shipping a crew was a totally different matter and would not be accomplished until the ironclad was ready for battle. Jones needed at least 320 men to man the vessel in combat, and the pool of available, experienced sailors was limited. The prewar South did not have a large merchant fleet. Many of the men with sea experience had already enlisted in the army with patriotic zeal at the war's onset. Since the army was reluctant to reduce its ranks for the fledgling Confederate navy, the crew was "obtained with great difficulty."[39]

An initial effort was made to recruit seamen with U.S. Navy and merchant experience. Recruiting stations, called naval rendezvous, were established in Norfolk and Richmond to enlist sailors in the more traditional manner. Several sailors were obtained for the ironclad through these stations. Seamen were offered between twelve (landsmen) and eighteen (seamen) dollars per month plus four cents a day for grog.[40] Prewar seaman Peter Williams, a native of England and formerly a private in the 8th Alabama, enlisted in the navy for the war at the Richmond Naval Rendezvous on February 16, 1862. Another Williams, John, would also serve on the

Merrimack as a quarter-gunner after transferring from the 15th Louisiana into the navy at the Richmond Naval Rendezvous. Several other men from Louisiana, including boatswain's mate Andrew G. Peterson, enlisted in the navy at the New Orleans Naval Rendezvous when the war began and were transferred to Portsmouth for service on the *Merrimack*.

"There was a sprinkling of old man-of-war's men," wrote eight-year U.S. Navy veteran Lieutenant John R. Eggleston about the ironclad's crew, "whose value at the time could not be overestimated."[41] Several U.S. Navy enlisted men deserted their ships or resigned their warrants to serve the South. Master Tailor Albert C. Griswold deserted the USS *Cumberland* just before the sloop left Gosport on the morning of April 21, 1861. One seaman, Benjamin R. Sheriff, had enlisted in the U.S. Navy on April 15, 1861, and deserted from the USRS *Allegheny* on April 30, 1861. Sheriff later enlisted in the Confederate navy to serve as a quarter gunner on the *Merrimack*. Master's Mate Charles B. Oliver was dismissed after tendering his resignation from the U.S. Navy in April 1861 and then enlisted in the Confederate navy two months later. Oliver was appointed gunner and was assigned to the CSRS *Confederate States* (formerly the USF *United States*) until transferred to the *Merrimack*. Daniel Knowles was another U.S. Navy veteran who volunteered to serve on the Confederate ironclad. While a classboy on the USS *Albany* during the Mexican War, Knowles was twice wounded in action and received a pension for his honorable duty. Another veteran sailor was Charles Hasker. Hasker served in the Royal Navy as a youth and then immigrated to the United States. He was appointed a boatswain, U.S. Navy, and was later dismissed in June 1861.

Hasker was one of several sailors who had previously been assigned to the North Carolina Squadron. The squadron was a combination of converted canal tugs and other coaster steamers armed with at least one forward cannon. Aptly called the "Mosquito Fleet," the squadron was commanded by Flag Officer William F. Lynch. On October 1, 1861 the fleet scored its only major success at the Battle of Loggerhead Inlet. The gunboats CSS *Curlew*, CSS *Raleigh* and CSS *Junaluski* captured the USS *Fanny*, along with a large quantity of quartermaster and commissary stores. The *Fanny* was a screw steamer armed with one 32-pounder smoothbore and one 8-pounder rifle. Hasker was assigned as boatswain on the recommissioned CSS *Fanny* and served aboard the gunboat during the Battle of Roanoke Island, February 7 and 8, 1862, and Elizabeth City, North Carolina, February 10, 1862. The *Fanny* was damaged during the Elizabeth City engagement, ran aground and was destroyed. Hasker and other gunboat crew members were able to escape to Norfolk aboard the CSS *Beaufort* and CSS *Raleigh*. These gunboats were able to enter the South Mills lock of the Great Dismal Swamp Canal.

The CSS *Appomattox*, another gunboat, could not fit into the lock. Consequently, her commander, Lieutenant Charles Simms, scuttled her in a manner that blocked any Federal entry into the canal. Besides Hasker, many other survivors of the North Carolina Squadron were available for reassignment to the *Merrimack*, including Emsley H. Ives, CSS *Ellis*; James C. Cronin, CSS *Seabird*; James Mercer, CSS *Ellis*; and John A. Rosler, CSS *Forrest*.

John Cunningham was another experienced seaman to ship aboard the *Merrimack*. Cunningham had served on the ill-fated Confederate privateer CSS *Petrel* and was captured during the engagement with the USS *St. Lawrence* on July 10, 1861. Imprisoned at Fort Delaware, Cunningham escaped to enroll as an ordinary seaman on the *Merrimack*.

Despite this nucleus of trained sailors, the Confederate ironclad still lacked enough seamen to effectively operate the warship as the vessel neared completion. The best remedy was to recruit seamen or artillerymen from nearby Confederate army units. The War Department, at the Confederate navy's urging, issued special orders approving such transfers; however, recruitment was slow. Captain Franklin Buchanan, chief of the Office of Orders and Details, wrote on February 10, 1862, that the "*Merrimack* has not yet received her crew, notwithstanding all my efforts to procure them from the Army."[42]

Catesby Jones had already detached Lieutenant John Taylor Wood in January 1862 to search for recruits from nearby army commands. Wood, a former U.S. Navy lieutenant and an instructor at the U.S. Naval Academy, was a wise selection for this duty. Perhaps one of Wood's greatest assets was his family connections. He was a grandson of President Zachary Taylor, a nephew by marriage of Confederate president Jefferson Davis and cousin to Brigadier General Richard Taylor. Wood met with Major General John Bankhead Magruder, commander of the Army of the Peninsula, to obtain volunteers. Magruder, who also needed more soldiers to man his Peninsula defensive lines, acquiesced to Wood's request and provided two hundred volunteers. Wood selected eighty men with "some experience as seamen or gunners."[43] However, when some of the men arrived at Gosport for training, Wood commented to Franklin Buchanan that they "are certainly a very different class of men from those I selected." Wood concluded, "I find that but two of the men selected by myself were sent; the others are men I did not see, nor even visit their encampment."[44] Robert Minor wrote:

> *Some of the "so-called" volunteers had bad characters from their commanding officers, who could not manage them, and were brought on board in double irons. Jones immediately had their irons struck off, and informed them that he would have no forced volunteers on board, and that if they wished to remain they could do so and start fair with the other men, and make a character for themselves. This course proved eminently judicious, as some of them were the best men on board.*[45]

Among these forced volunteers were Corporal James Cullington, 1st North Carolina, who had recently (November 15, 1861) been reduced to the rank of private; Private John Leonard, 1st North Carolina, who on December 22, 1861, had been sentenced to thirty days' hard labor and the loss of ten dollars' pay; and Private Charles Dumphrey, 14th Virginia, who had been placed under arrest on charges of insubordination while on guard duty and sentenced to twenty days' hard labor on December 20, 1861, by order of court-martial. George May, a prewar sailor from New Orleans, Louisiana, had been court-martialed on October 15, 1861. May was convicted for failure to obey lawful order and showing contempt for a superior officer. He was sentenced to confinement and five days' hard labor. Other castoffs included soldiers who were constantly absent from duty due to illness. Private Andrew Lang, 4th Georgia, was sick with fever from September 19 to 29, 1861; he survived cholera after being hospitalized from October 5 to 16 and 18 to 21, 1861. He reported unable to report for duty on January 18–19, 1862, due to rheumatism. Lamb also suffered from sciatica on two occasions, January 22–28 and February 2–6, 1862. Amazingly, the soldier-turned-sailor never reported sick while aboard the *Merrimack*.

Private Richard Curtis, Wythe Rifles, photograph, 1861. *Courtesy of the Hampton History Museum, Hampton, Virginia.*

Despite complaints that many of the volunteers were disorderly, ill trained or sickly, Wood later wrote that the new sailors "proved themselves to be as gallant and trusty a body of men as anyone would wish to command."[46] After searching for suitable men during recruiting trips to the Peninsula, Petersburg and Richmond, Wood culled volunteers from units stationed throughout eastern Virginia, including the 1st North Carolina, 13th North Carolina, 14th North Carolina, as well as the 1st South Carolina, 3rd Georgia, 4th Georgia, 3rd Alabama, 2nd Louisiana, 15th Louisiana, Cobb's Legion, 6th Virginia, 12th Virginia and 32nd Virginia. Private Richard Curtis, a prewar boatman and veteran of the Battle of Big Bethel, "transferred to the *Merrimack* by order of General Magruder" from Company A (the Wythe Rifles), 32nd Virginia, to serve as an ordinary seaman. New Jersey native and prewar sailor John J. Sturges was discharged from the 9th Virginia to enlist as an ordinary seaman. Many volunteers, like farmer Alford Stroup of the 14th North Carolina and paperhanger Allen Gilmore of the 12th Virginia, served on the ironclad as landsmen. Ephraim K. McLaughlin had previously enlisted in the 1st Battery, Maryland Artillery, and provided a substitute, S.R. Berry, before shipping on the *Merrimack* as an acting master's mate. Thomas Traylor was discharged from the 12th Virginia as he was underage, yet this young apprentice shoemaker then enlisted as a landsman.

William G. Burke, whose prewar occupation was listed as a comedian, originally enlisted at Harpers Ferry, Virginia, as a private in Company F, 2nd Virginia Infantry. While promoted to corporal, Burke was discharged from the Confederate army on September 7, 1861, and appointed as a master's mate, CSN. Burke was assigned to Gosport Navy Yard and then Richmond Station. He was discharged on November 25, 1861, only to enlist as a seaman on the CSS *Virginia*. John Davis left Baltimore during the secession crisis and enlisted in Company E, 32nd Virginia Infantry. Davis transferred to ship aboard the CSS *Jamestown* on December 21, 1861, and eventually became a landsman aboard the Confederate ironclad. Jefferson Tinsley enlisted at Covington, Georgia as a private in Company H, 3rd Georgia Infantry. On January 13, 1862, Tinsley "transferred to steamer *Merrimac*" and served as a first-class fireman.

As the ship's crew was slowly being filled, an area of great concern was the *Merrimack*'s old power plant. The *Merrimack* was selected to become the nucleus of the Confederacy's first ironclad, primarily because she already had engines. Despite the fire that had consumed her upper works, Chief Engineer William P. Williamson declared that the "boiler and heavy and costly parts of the engine [were] but little injured."[47] Although the engines could be salvaged, a "more ill-contrived or unreliable pair of engines" could not have been found anywhere else.[48] H. Ashton Ramsay, who served as an assistant engineer on the USS *Merrimack* under Alban Stimers at Panama and was assigned as acting chief engineer of the ironclad on January 1, 1862, commented:

> *From my past experience with the engines of this vessel, I am of the opinion that they can not be relied upon. During a cruise of two years whilst I was attached to this ship in the United States Service they were continually breaking down, at times when least expected. When she returned, the Chief Engineer reported that all experiments to improve their working and reliability had failed, and as the defects were radical, embracing the entire engines, recommended that they should be removed from the vessel; such as the intention of the U.S. Government before she fell into our hands.*[49]

The engines were indeed "radically defective," a circumstance made only worse by the fire that consumed the frigate and the vessel's several weeks' emersion in the salty Elizabeth River. According to U.S. Navy Chief Engineer Benjamin Isherwood, the problems with the motive system were centered on the overall design. The propeller shaft was poorly supported and lacked stability. The cylinders produced low pressure because of air leaks, and the engines "were too light in construction for direct acting screw engines." The crankshafts were weak, and the system often overheated, requiring, according to Isherwood, "a constant dousing of water to the brasses…because of heat."[50]

Ashton Ramsay believed that

> *her engines were radically defective in their condensing apparatus; the average vacuum maintained being about one-half what it should be after the engines had worked a short time. The links, mechanism for operating the valves and reversing the engines, were also defective. The arrangement of steam chests for main and expansion valves was not well designed, involving as it did a large loss of steam…the boilers…economical in fuel, but very slow combustion and sluggish in draught. The engines were of a type known as horizontal back-acting, with two cylinders, each 72" diameter and three feet stroke of piston, and as 18 pounds was the average pressure of steam the whole power the engine was capable of developing was only 1000 horse, about one-fourth of what would be considered the duty of this size engine now, with the high pressure of steam carried.*[51]

In spite of these problems, Williamson, Ramsay and other engineers and mechanics were able to return and keep the system in working order.

Williamson, however, was the key to the revitalization of the ironclad's power plant. Born in Norfolk on July 26, 1810, he studied mechanical engineering in New York before working at Gosport Navy Yard. In 1842, he became supervisor of Gosport's machine works and then was appointed a chief engineer in the U.S. Navy. Williamson worked closely with John Porter on the construction of the *Colorado*. When he refused to take the oath of allegiance, Williamson was imprisoned and then dismissed. Williamson was named the senior engineer in the Confederate navy, and once he completed his work on the *Merrimack*'s old engines, he was assigned as head of the Department of Steam Engineering. "To Engineer Williamson," John L. Porter wrote, "who had the exclusive control of the machinery, great credit is due for having improved the propeller and engines as to improve the speed of the ship three knots per hour."[52] The extensive repairs may have upgraded the engines' power and speed, yet they were still temperamental and undependable. Ramsay learned that just to start took great care. First, he slowly warmed them with steam. He then would allow the steam to pass from the valve chest into the condenser to expel the air. Once the engines were warmed, crew members had to bar the crankshaft around a few times by force. Ramsay would then open the condenser injection valve and move the steam throttle forward to turn the propeller. The engines, however, were still liable to break down at any moment. Ramsay noted that if one engine stopped, a vacuum caused by this stoppage forced the other engine to cease functioning.[53] They were, in Catesby Jones's opinion, "radically defective."[54]

The ironclad was almost two months behind schedule in January 1862, and the Confederates knew that they were running out of time. Many Southerners believed that there were just too many repairs to finish the conversion before a Union strike against Norfolk might be orchestrated. "Public opinion generally about here," Porter lamented, "said she would never come out of the dock."[55] Catesby Jones was extremely frustrated by all of the "many vexatious delays attending the fitting and equipment of the ship" and commented to Brooke in a letter on January 24, 1862, that "somebody ought to be hung."[56]

Porter was under great pressure from Mallory and Brooke to complete the ironclad. The project was wearing heavily on Porter's state of mind as he wrote a friend, "You have no idea what I have suffered in mind since I commenced her...Some of her inboard arrangements are of the most intricate character, and have caused me many sleepless nights in making them."[57] On January 11, 1862, Gosport's blacksmiths and finishers signed a public testament volunteering "to do any work that will expedite the completion

Captain Reuben Triplett Thom, CSMC, photograph, circa 1870. *Courtesy of the U.S. Naval Historical Center, Washington, D.C.*

of the *Merrimac*, free of charge, and continue on until eight o'clock every night."[58] Boiler makers and machinists were impressed from nearby workshops and civilian yards to join the estimated 1,500 men rushing to advance the ironclad toward completion.

This final push was just what was required. Workmen finished the iron plating on January 27, 1862. The much-needed lubricating oil to work the old engines and other machinery arrived from Richmond on February 7. On February 11, a midshipman with a team of fifty crew members began stowing food and other supplies aboard the warship. On February 13, 1862, the ironclad, despite all the misgivings that the warship was top-heavy and would immediately turn bottom-up, floated in the dry dock. Jones, who had showed "a want of faith in her ability to float," noted immediately that there were serious problems with the still incomplete warship.[59]

Flag Officer French Forrest, nevertheless, ordered the ironclad to be launched and christened on February 17, 1862. The event was an unimpressive affair. Workmen were still feverishly completing the conversion, and the mood was one of desperation. "There were no invitations to governors and other distinguished men, no sponsor nor maid of honor, no bottle of wine, no brass band, no blowing of steam whistles, no great crowds to witness this memorable event," remembered William Cline. Cline noted that the only officer of the newly named CSS *Virginia* on hand to observe the momentous event was Captain Reuben Thom, commander of the ironclad's marine detachment.[60] That afternoon, Forrest ordered executive officer Catesby Jones "to receive on board the *Virginia* immediately after dinner today, all the officers and men attached to the vessel with their baggage, hammocks, etc., and have the ship put into order throughout. She will remain where she is to coal and receive her powder." The flag officer added, "You will report to me when you and your officers are onboard and use every available effort to get the ship in order, as this day she is put into commission."[61]

5
The Race for Hampton Roads

The good ship was almost ready for the fray," noted Assistant Engineer E.A. Jack as he observed the CSS *Virginia* along the wharf, taking on stores and equipment.[1] Actually, the ironclad was far from being complete, and a major flaw had been discovered. Porter had miscalculated the vessel's displacement, which caused the ship to ride too high in the water. Jones advised Brooke on February 25, 1862, that the "water is just now above the eaves." The worried executive officer added hopefully, "We have yet to take our powder, and most of the shells, and 150 tons of coal which is thought will weigh it down a foot or more."[2]

Porter had forgotten to subtract the weight of the old frigate's masts, upper decks, sails and rigging when calculating the ironclad's displacement. Consequently, when the *Virginia* slid into the Elizabeth River, the armored shield barely reached below the water surface. Porter believed that he had done the best he could in devising the displacement of this experimental warship. Instead of accepting fault, he blamed Brooke's alterations to the propeller and armor. Regardless of who was at fault, Catesby Jones was furious about the ironclad's vulnerability. Jones wrote to Brooke:

> *The ship will be too light, or I should say, she is not sufficiently protected below the water. Our draft will be a foot less than was first intended, yet I was this morning ordered not to put anymore ballast in fear of the bottom. The eaves of the roof will not be more than six inches immersed,*

> *which in smooth water would not be enough; a slight ripple would leave it bare except the one-inch iron that extends some feet below. We are least protected where we most need it. The constructor should have put on six inches where we now have one.*[3]

Porter wrote that by cutting down the burnt parts of the ship it "gave me two hundred tons more than I required for the displacement." This circumstance troubled Brooke, as he had suggested to "put six inches of iron on bow and stern" to enhance the ironclad's shot-proof abilities, but Porter had asserted that the ironclad could not carry the weight.[4] Porter, consequently, was severely criticized for this displacement flaw, which many officers agreed to be an error in the constructor's calculations. It would prove to be one of the ironclad's most serious weaknesses. News of the problem quickly reached the public. "This is a bad piece of work," proclaimed the *Mobile Register*, "but the *Day Book* seeks to comfort us by assurance that, if unfit for sea, she will make an invaluable floating battery for the protection of Norfolk, better good for something than nothing."[5]

The Confederates were desperate to finish the ironclad. The Union army was on the move, and it was obvious to the Confederates that Southside Hampton Roads, especially Gosport Navy Yard, was a target. A huge fleet of warships and transports assembled in Hampton Roads on January 8, 1862. Local Confederate commanders, Major General Benjamin Huger in Norfolk and Major General John Bankhead Magruder on the Peninsula, scrambled to prepare their defenses against the expected Union onslaught. The anticipated assault never materialized, as Flag Officer Louis Malesherbes Goldsborough, commander of the North Atlantic Blockading Squadron, left Hampton Roads on January 11 and 12, 1862, with Brigadier General Ambrose Burnside's amphibious corps to attack Roanoke Island.

Major General Benjamin Huger, CSA, photograph, 1862. *Courtesy of the Museum of the Confederacy, Richmond, Virginia.*

Roanoke Island was captured on February 8, 1862. The Confederates were shocked by this defeat, as Brigadier General Henry A. Wise wrote:

> *Such is the importance and value, in a military point of view, of Roanoke Island, that it ought to have been defended by all the means in the power of the Government. It was the key to all the rear defenses of Norfolk. It unlocked two sounds...eight rivers...four canals...and two railroads...It guarded more than four-fifths of all Norfolk's supplies of corn, pork, and forage, and it cut the command of General Huger off from all of its most efficient transportation. It endangers the subsistence of his whole army, threatens the navy-yard at Gosport, and to cut off Norfolk from Richmond and both from railroad communication with the South. It lodges the enemy in a safe harbor from the storms of Hatteras, gives them a rendezvous, and a large, rich range of supplies, and the command of the seaboard from Oregon Inlet to Cape Henry.*[6]

Wise, a former governor of Virginia, was furious that Huger had not taken stronger steps to defend Roanoke Island. The Confederate navy lost faith in Huger's ability to effectively defend Norfolk from this threat from the South. French Forrest issued small arms to his workmen and redoubled work on the ironclad. The *Virginia* was recognized as the key to the defense of eastern Virginia.

The Roanoke Island operation was all part of a grand scheme developed by Union general-in-chief George Brinton McClellan to end the Civil War in the spring of 1862. McClellan, a self-assured thirty-five-year-old major general and West Point graduate, sought to exploit the mobility provided by the Union's naval superiority and move the Army of the Potomac from Washington, D.C., to outside the Confederate capital at Richmond. Since the Confederate field forces were distributed throughout Virginia, McClellan conceived a plan to move his army by way of Urbanna and the York River to the gates of Richmond before the Confederacy could concentrate its armies against him. McClellan noted:

> *I had determined not to follow the line of operations leading by land from Washington to Richmond, but to conduct a sufficient force by water to Urbanna, and thence by a rapid march to West Point, hoping thus to cut off the garrison of Yorktown and all the Confederates in the Peninsula; then using the James River as a line of supply, to move the entire Army of the Potomac across that river to the rear of Richmond.*[7]

McClellan believed this indirect approach would produce fewer casualties than a direct march from Washington overland to Richmond and effectively end the war with his rapid capture of the Confederate capital. Instead of a direct confrontation, McClellan's strategy employed siege artillery and rapid troop maneuvers to achieve victory.

One problem that delayed McClellan's proposed operation was the CSS *Virginia*. The Federals had long been aware of the Confederate effort to convert the *Merrimack* into an ironclad. Despite the initial intelligence provided by Flag Officer Silas Horton Stringham, the first commander of the North Atlantic Blockading Squadron, advising Gideon Welles that the burned *Merrimack* "was taken into dry dock, examined, and pronounced worthless…[h]er machinery was all destroyed,"[8] the Federals in Hampton Roads soon began reporting the *Merrimack* as a very real threat.

On October 6, 1861, Major General John Ellis Wool, stationed at Fort Monroe as commander of the Union Department of Virginia, wrote to Lieutenant General Winfield Scott:

> *I also mentioned to you that Newport News was threatened by the Confederates in order to aid in getting to sea two steamers in James River, and also the* Merrimac *at Norfolk. This ship is constructed to resist cannon shot. I have this morning seen the flag-officer in the Roads, who more than confirms all that I have said on the subject, and that it is settled that an attempt will be very soon made to get these vessels to sea. In order to facilitate this movement an attempt will be made to get possession of Newport News. We have guns, but not artillerists sufficient to man them, at Newport News. I hope you will at once send back the four companies of artillery recently sent to Washington. If you do not send these, or some other companies of artillery to supply their places, I trust you will not hold me responsible for any disaster that may befall us at Newport News. The danger, I assure you, is imminent. This subject I presented to the President in Cabinet council, when I assured them of the intentions of the rebels, and that it was their design to attack Newport News, and, as it was reported, very soon. The flag-officer is satisfied that such will be the case. He says there is no mistake as to their intentions. He further expresses his apprehension that they will succeed in capturing Newport News and that the steamers may get to sea. He also says the* Merrimac *is so constructed that no cannon shot can make an impression upon her.*[9]

Flag Officer Goldsborough had replaced Flag Officer Stringham as commander of the North Atlantic Blockading Squadron in September 1861. Goldsborough was born in Washington, D.C., and entered the U.S. Navy as a midshipman at the age of seven. A veteran of the Seminole and Mexican Wars and a former superintendent of the U.S. Naval Academy in Annapolis, Goldsborough was a huge (reports indicate that he weighed well over three hundred pounds) and intimidating man with a powerful temper. One naval officer noted that Goldsborough possessed "manners so rough, so that he would almost frighten a subordinate out of his wits."[10] When a group of midshipmen set fire to his backyard privy as a practical joke, the flag officer reportedly raged, "I'll hang them! Yes, I'll hang them! So help me God, I will." Since assuming command of the squadron, Goldsborough continuously bombarded Gideon Welles with reports about the Confederate ironclad. On October 17, 1861, Goldsborough wrote to Welles, noting, "I have received further minute reliable information with regard to the preparation of the *Merrimack* for an attack on Newport News and these roads, and I am quite satisfied that unless her stability be compromised by her heavy top works of wood and iron and her weight of batteries, she will in all probability, prove to be exceedingly formidable."[11]

To counter any possible threat from the Confederate ironclad, Goldsborough turned the Hampton Roads blockading force into a strong complement of warships. At the fleet's heart were the two sister steam screw frigates USS *Minnesota* and USS *Roanoke*. Each carried an impressive broadside. The *Minnesota*'s battery consisted of one X-inch and twenty-eight IX-inch Dahlgren smoothbores, fourteen 8-inch guns, two 24-pounders and two 12-pounders. She was Goldsborough's flagship. The *Roanoke*'s broadside of forty-six guns was equally impressive. The station also included three sailing warships: the fifty-gun USS *Congress*, the fifty-two-gun USS *St. Lawrence* and the twenty-four-gun sloop of war USS *Cumberland*. The *Cumberland* featured perhaps the most powerful armament in Goldsborough's Hampton Roads station. Formerly a forty-four-gun sailing frigate and flagship of the Mediterranean and African Squadrons, in 1856, the *Cumberland* had been razed down to a sloop and armed with twenty-two IX-inch Dahlgren smoothbores, one X-inch Dahlgren smoothbore and a powerful 70-pounder rifle.

The squadron was supported by the steamer *Cambridge*, the store ship *Brandywine*, three coal ships, a hospital ship, five tugboats, a side-wheel steamer and a sailing bark. Goldsborough, before leaving for North Carolina with Burnside's expedition to capture Roanoke Island, had planned to confront

USS Minnesota, engraving, circa 1860. *Courtesy of The Mariners' Museum, Newport News, Virginia.*

the Confederate ironclad when it entered Hampton Roads by bringing all his vessels together against the enemy warship. The flag officer, who had once urged the Union to build a fleet of thirty ironclads to "have the enemy thrown upon his knees," knew that only by pounding the Confederate armorclad with all the heavy guns of his fleet could he hope for any margin of success.[12]

"Nothing I think," wrote Goldsborough, "but very close work can be of service in accomplishing the destruction of the *Merrimack* and even of that a great deal may be necessary."[13] There were concerns, of course, that the Union ships, without the support of an ironclad, might be unable to confront the ironplated Confederate warship. Wool's fears for the defense of Newport News Point prompted Goldsborough to station the USS *Cumberland* and USS *Congress* to guard the entrance to the James River and Camp Butler. Commander William Smith, captain of the USS *Congress*, noted that his frigate's exposed position made her extremely vulnerable to an attack by the Confederate ironclad. Smith wrote that even though the *Congress* "had been a model in her day," since all of her cannon were older smoothbores, "we should only be a good target for them, as none of our guns could throw a shot to them." There was also a critical shortage of veteran seamen aboard all the Union warships in Hampton Roads. Smith recognized this serious deficiency when he wrote on January 21, 1862, that the "present position of this ship (threatened daily with attacks from Norfolk by the *Merrimack* and the other steamers, and from the James River by the *Patrick Henry*, and *Thomas Jefferson*, and other steamers, submarine batteries, torpedoes, fire rafts) makes it very important that we shall have our entire crew."[14] "I regret to say it is impossible to assist him," Gideon Welles advised Captain John Marston, "as men do not enlist as fast as required." Welles added, "If this frigate is not capable of defending herself at Newport News, the Department suggests whether one of the steam frigates should not relieve her."[15] General Wool endeavored to relieve Smith's concerns by assigning Captain William J. McIntire's Company D, 99th New York Infantry, to the frigate. McIntire's company of eighty-eight officers and soldiers-turned-sailors were assigned as gunners aboard the *Congress*.

The USS *Roanoke* not only suffered from an acute shortage of veteran seamen, but she also required major repairs. The *Roanoke*'s engines were useless due to a broken crankshaft. "When I think of this ship's crippled condition—no engine and 180 of her crew deficient," wrote Marston, "it makes me sick at heart."[16] Seaman Joseph McDonald lamented, "We sailors couldn't understand why the government should leave such a powerful ship in a condition like that."[17] The very threat of the impending excursion forced the Federals to maintain a strong defensive posture. Since Goldsborough was in North Carolina with Burnside, John Marston, captain of the *Roanoke*, was acting commander of the Hampton Roads station. Marston complained to Welles that the vessels in Hampton Roads could not be overhauled "as long as the *Merrimack* is held as a rod over us."[18]

The USS *Minnesota* was the most powerful Union warship in Hampton Roads. She was the sister ship to both the *Roanoke* and *Merrimack*. The *Minnesota* was built at the Washington Navy Yard and commissioned in 1857. The frigate's commander solved his recruitment requirements by enlisting contrabands, African Americans who had escaped bondage and established communities near Fort Monroe and Camp Butler. This continued the U.S. Navy's long-standing tradition of recruiting African Americans for service. The *Minnesota*'s aft pivot gun was manned by an all-African American crew. G.J.H. Van Brunt, captain of the *Minnesota*, would later write that the "Negroes fought energetically and bravely—none more so. They evidently felt that they were thus working at the deliverance of their race."[19]

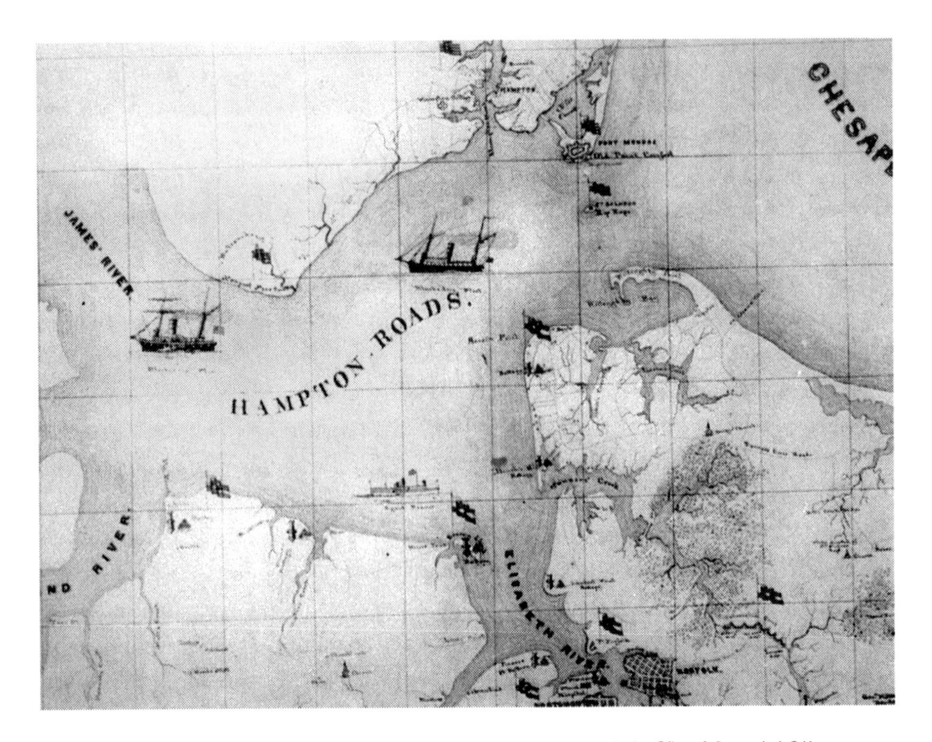

Map of Hampton Roads, pen-and-ink sketch, 1861. *Courtesy of the Kirn Memorial Library, Norfolk, Virginia.*

Hampton Roads was a natural amphitheater for the impending battle. Even though the Confederates controlled the Southside with major fortifications at Sewell's Point, Craney Island and Pig Point, the Federals maintained the northern shore. Camp Butler at Newport News Point effectively sealed off the entrance to the James River. Fort Monroe and its companion work, Fort Calhoun on the Rip Raps, provided a powerful gatekeeper at Hampton Roads' outlet to the Chesapeake Bay. Accordingly, Marston placed his capital ships to help seal off Hampton Roads in conjunction with these fortifications. If the Confederate ironclad sought to reach the Chesapeake Bay, the *Minnesota*, *Roanoke* and *St. Lawrence* were anchored just off Old Point Comfort. The *Congress* and *Cumberland* were moored off Newport News Point to block any move of the ironclad to Richmond. Two armed tugs, the *Dragon* and *Zouave*, were positioned to tow these sailing ships at Newport News Point into action if needed.

The Federal fleet was patiently waiting for the Confederate ironclad, as Captain Marston noted, "I am anxiously expecting her and believe

I am ready."[20] "Rumors of her expected appearance came so often," recalled Thomas O. Selfridge of the *Cumberland*, "that at last it became a standing joke with the ship's company."[21] Marston wrote to Gideon Welles on February 21, 1862, that "by a dispatch from General Wool, I learn that the *Merrimack* will positively attack Newport News within five days, acting in conjunction with the *Jamestown* and *Yorktown* from the James River, and the attack will be at night."[22] The crewmen were ready for "a chance for active operations."[23] They had been on alert throughout the winter and drilled "until every man knew not only the duties of his own station at quarters, but those of every station as well," remembered Master Moses S. Stuyvesant of the *Cumberland*.[24] Captain Gershom Jacques Van Brunt, commander of the USS *Minnesota*, summed up the frustration waiting for the Confederate ironclad when he wrote to Goldsborough, "We have nothing new here; all is quiet. The *Merrimack* is still invisible to us, but report says she is ready to come out. I sincerely wish she would; I am quite tired of hearing of her." Van Brunt added that "the sooner she gives us the opportunity to test her strength the better."[25] Nevertheless, the Confederate ironclad still did not appear.

Secretary of the Navy Gideon Welles, engraving, circa 1861. *Courtesy of the Virginia War Museum, Newport News, Virginia.*

Union intelligence was very accurate and appeared to have inside knowledge of Confederate plans. The Federals were constantly receiving information about the impending attack. The Southern press, which continually boasted about the power of its new ironclad, was only one of the problems. The *Mobile Register* proclaimed:

> *It would seem that the hull of the* Merrimac *is being converted into an iron-cased battery. If so, she would be a floating fortress that will be able to defeat the whole Navy of the United States and bombard its cities. Her great size, strength, and powerful engines and speed, combined with the invulnerability secured by the iron casting, will make the dispersal or*

> *the destruction of the blockading fleet an easy task for her. Her immense tonnage will enable her to carry an armor proof against any projectile and she could entertain herself by throwing bombs into Fortress Monroe, even without risk. We hope soon to hear that she is ready to commence her avenging career on the seas.*[26]

Confederate naval leaders were concerned how the public interpreted such articles and whether they considered them written to boost public confidence in the Confederate navy and its conversion project at Gosport. "The *Mobile Register* contains information in relation to the *Merrimac* of much value to the enemy," John Mercer Brooke wrote in his diary, "Editors are doing infinite harm in that way. I shall begin to think that even the South can not tolerate a free press."[27] Brooke was also concerned that Southerners, their expectations heightened by a boastful press, would expect too much of the ironclad. He was somewhat relieved when the *Norfolk Day Book* announced the failure of the iron plate tests on Jamestown Island but still feared that the Federals would begin building their own armor-clads to counter news of the Confederate ironclad.

Southern newspapers continued their coverage of the *Merrimack*'s conversion. J. Nicholson, who was allowed to go aboard and survey the ironclad in dry dock, provided a very thorough review detailing the vessel's progress and capabilities in the *Lynchburg Virginian*:

> *To get into it would be impossible, and to make a hole in it would be impossible, and to make a hole in it with anything, or machinery just the same. It will be covered with oak plank, two courses, each four inches thick, and then encased in railroad iron. Do you suppose a cannonball can have the courage to go through all of that. She will carry ten tremendous guns.*[28]

French Forrest was so enraged by this latest in a series of articles about the conversion project that he ordered Gosport's gates closed to all visitors. Workers were ordered to sign in and out of the yard. Failure to do so would result in the loss of a day's wages.[29]

The Confederates became very concerned about security. Information leaked almost daily to the Federals across Hampton Roads. Escaped slaves, exchanged prisoners of war, Northern sympathizers and other informants kept the Federals filled with up-to-date knowledge about the Confederate ironclad project. The City of Portsmouth established a patrol boat system to stop slaves from reaching Union lines with information about the ironclad. The patrol boats soon uncovered another espionage technique. Corked bottles containing secret messages were sent by sympathizers or spies via the Elizabeth River's outgoing tide toward Fort Monroe. Forrest could only redouble his security around the ironclad. On December 7, 1861, Captain Reuben T. Thom's Company C, Confederate States Marines, arrived at Gosport Navy Yard and was immediately deployed as guards throughout the shipyard. Thom, a Mexican War veteran with the 1st Alabama Volunteers and 13th U.S. Infantry, had joined the Confederate States Marine Corps (CSMC) at Montgomery, Alabama, just nine days after the corps was authorized by the Confederate Congress. Thom was the first officer appointed to the CSMC. He was initially assigned to recruiting duty and organized one of the first CSMC companies. Captain Thom served as commander of Company C, totaling forty-four men, at Pensacola, Florida, until organizing another detachment of marines to join in the July 6, 1861 capture of Ship Island in the Gulf of Mexico. Thom's command helped to drive off an attack on Ship Island three days later by the USS *Massachusetts*. Lieutenant Alexander F. Warley, CSN, praised Thom, noting that where "work was to be done there was the captain to be found and his men working as I never saw raw recruits work before."[30] Thom returned to Pensacola and was eventually assigned to Gosport with Company C. Reuben Thom detached marines to various gunboats, such as the CSS *Jamestown*, as guards aboard the receiving ship *Confederate States* and retained fifty-four men to act as sentries for the ironclad.[31]

Department of Norfolk commander Benjamin Huger saw spies and malcontents everywhere. "My object is to get rid of a disaffected and troublesome population," Huger wrote to Richmond, "most of whom are idle and would be liable to turn against us if we were in any danger of defeat."[32] Huger suggested that martial law might even be necessary to control the local population.

Despite the censorship, guards and threats of martial law, the leaks continued. Simultaneous with the ironclad's first float test, an article appeared in the *New York Times* commenting that the "work on the *Merrimac* was stopped on Saturday last, and she is now at the Navy Yard, drawing so much water that she could not get out even if she was ready for sea."[33] Since Norfolk was an exchange point for prisoners of war, as well as civilians traveling north and south, the flow of information could not be stopped. Colonel LeGrand B. Cannon, General Wool's chief of staff, remembered how one message from a Unionist workman in Gosport Navy Yard was delivered sewn inside the lining of an accomplice's jacket. The message stated that the Confederate ironclad would soon attack the *Congress* and *Cumberland* at Newport News Point. Cannon rushed the vital knowledge to Washington, D.C. Lincoln reportedly introduced the information at a cabinet meeting, which prompted Assistant Secretary of the Navy G. Vasa Fox to quip, "Mr. President, you need not give yourself any trouble whatever about that vessel."[34]

Fox was confident because the USS *Monitor* was nearing completion. The Union command was ready for the ironclad to see service. "Hurry her for sea," Gustavus Fox wrote to the ironclad's designer, John Ericsson, "as the *Merrimack* is nearly ready at Norfolk and we wish to send her there." Gideon Welles had hoped to get the *Monitor* to Hampton Roads while the Confederate ironclad was still under construction. Welles believed that the *Monitor* could easily steam up the Elizabeth River past the Confederate batteries and destroy the *Merrimack* as it sat in dry dock.[35]

George McClellan was also mindful of the *Merrimack* and her impact on his forthcoming campaign. The commanding general was under heavy pressure from President Abraham Lincoln to strike against the Confederate capital. His delays and proposals to move against Richmond via Urbanna and the James River concerned Lincoln, who preferred a more direct approach that would simultaneously protect Washington, D.C. McClellan asserted in early February that "so much am I in favor of the southern line

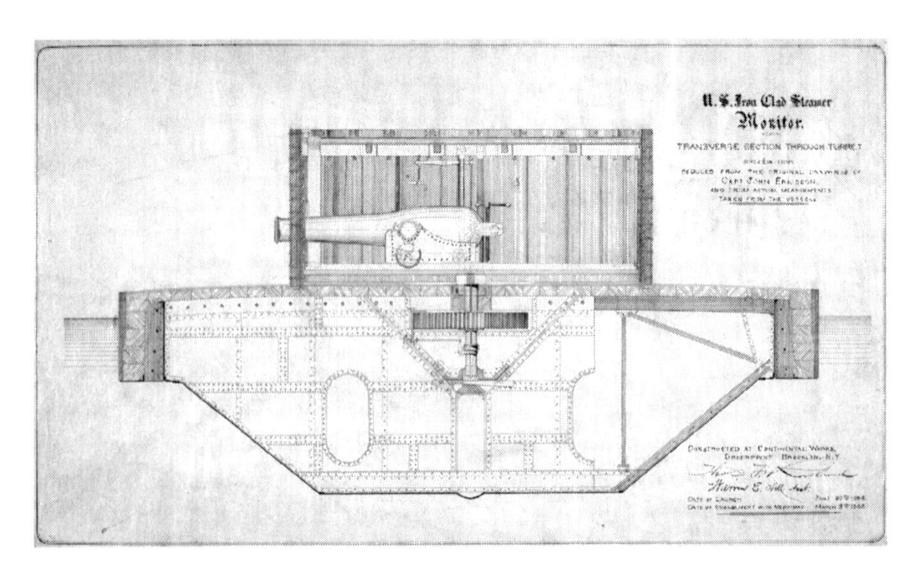

U.S. ironclad steamer *Monitor*, plan, John Ericsson, 1862. *Courtesy of The Mariners' Museum, Newport News, Virginia.*

of operations, that I would prefer the move from Fortress Monroe as a base as a certain though less brilliant movement than that from Urbanna to an attack upon Manassas.[36] McClellan viewed Fort Monroe and Camp Butler as good staging areas, but only if the U.S. Navy would protect his army. While he had organized the Roanoke Island operation to isolate Norfolk, he still recognized the Confederate ironclad as a major threat. McClellan's experience preparing the Delafield Report made him well aware of the capabilities of armored warships, and he sought to gain assurances from the U.S. Navy that his army could safely use the James River. Instead of fearing the *Merrimack*, Brigadier General John Gross Barnard, chief engineer of the Army of the Potomac, believed in taking an aggressive approach. Barnard later lamented:

> *In the winter of 1861–1862 Norfolk could and should have been taken. The Navy demanded it, the country demanded it, and the means were ample. By its capture the career of the* Merrimack, *which proved so disastrous to our subsequent operations, would have been prevented. The preparation of this vessel was known, and the Navy Department was not without forebodings of the mischief it would do.*[37]

McClellan refused to consider any direct attack against Norfolk, as such an assault would probably result in heavy casualties. The navy also appeared content to wait for the arrival of its own ironclads before organizing any operation to capture the Confederate shipbuilding center.

The Confederates were well aware of the Federal efforts to build several ironclads in East Coast shipyards. Northern publications were as indiscreet as Southern newspapers. The February 4, 1862 *New York Times* announced the *Monitor*'s launching with headlines, while an issue of *Scientific American* contained construction details of Ericsson's ironclad. John Mercer Brooke's summertime fears had come true. Once the Federals learned about the *Merrimack* conversion project, they turned the superior Northern industrial power to constructing ironclads. Brooke could only hope that the Confederate head start would be sufficient to attain naval superiority and break the blockade.[38]

The Confederates were now straining every available resource to finish the *Virginia*. Yard commandant French Forrest was charged with the responsibility of securing sufficient oil, coal and gunpowder for the ironclad. Forrest had to beg, borrow or trade these necessary supplies from every conceivable source. Additional oil was obtained from a Union barge that had run aground during a storm. Even though the old, cantankerous engines consumed coal at an excessive rate, only 150 tons were loaded. This helped lower the ironclad even farther in the water and compensate for Porter's miscalculations. Gunpowder proved to be the most difficult resource to procure. While eighteen thousand pounds were required to adequately prepare the ship for battle, only one thousand pounds had been received from Richmond by late February. Forrest requisitioned powder from the receiving ship *Confederate States* but obtained most of the gunpowder from supplies the flag officer had transferred to Fort Norfolk and other batteries under Benjamin Huger's command.

Forrest also began preparing for the anticipated sortie of the *Virginia* by clearing a path in the double row of obstructions at the mouth of the Elizabeth River. The flag officer moved the damaged sloop *Germantown* to the river's entrance to act as a gatekeeper. The *Germantown*, burned by the Federals when they abandoned Gosport, was cut down and transformed into a powerless floating battery. The sloop was armed with seven cannons and fitted with a sand-filled bulkhead seven feet thick. The CSRS *Confederate States*, mounting nineteen cannons, was also moved down the Elizabeth River to act as a floating battery.

Work continued feverishly on the *Virginia* during the last weeks of February. Two boats were mounted on each side of the casemate, and numerous other minor details kept crews working until eight o'clock each evening. Brooke desired to protect the guns and crew from enemy shot by fitting each gun port with iron shutters. Compound shutters were produced at Gosport, but Brooke found them unsatisfactory and ordered solid, cast-iron shutters from Tredegar. Tredegar did not complete the shutters until mid-February, so the bow and stern gun ports were the only parts produced in this manner. The shutters, each fabricated of three-inch iron plate, were designed to be opened and closed by chains. They worked off a pivot point bolted into each lower side of the gun ports. This configuration was necessary to ensure that if the chains were broken by shot, the shutters could still be opened.

The *Virginia* still lacked a captain. Mallory knew the man he wanted for the command: Franklin Buchanan. Yet he hesitated to name him the ship's captain because of the time-honored seniority system, which placed two men, French Forrest and Victor Randolph, senior to Buchanan. Forrest actively sought the position, and Catesby Jones was "actually oppressed with undue expectations"[39] of being named the ironclad's commander. Both men would be disappointed. Mallory sidestepped the issue by naming Buchanan commander of the James River defenses. His flagship, of course, would be

CSS Virginia, pencil sketch, 1862. *Courtesy of The Mariners' Museum, Newport News, Virginia.*

the *Virginia*. Buchanan was an excellent choice for command of the ironclad, and his selection, according to Ashton Ramsay, "was hailed with great satisfaction by all of us."[40] Ramsay thought that Buchanan was "one of the grandest men who ever drew a breath of salt air."[41] "A typical product of the old-time quarter deck," John Eggleston wrote of Buchanan, "as indomitably courageous as Nelson, and as arbitrary. I don't think the junior officer or sailor ever lived with nerve sufficient to disobey an order given by the old man in person." Eggleston remembered that during the Japanese expedition, "while going up the Canton River in charge of a Chinese pilot the vessel struck the ground, Buchanan, who was standing by the pilot, turned on him so fiercely that the Chinaman jumped overboard."[42]

A Maryland native and grandson of a signer of the Declaration of Independence, Buchanan was born on September 17, 1800. He was appointed midshipman in the U.S. Navy in January 1815. Franklin Buchanan was the first superintendent of the U.S. Naval Academy at Annapolis, and during the Mexican War's siege of Vera Cruz, he commanded the sloop of war *Germantown*. He was commander of Matthew C. Perry's flagship, the USS *Susquehanna*, when Perry opened Japan to American trade. Promoted to the rank of captain in 1855, Buchanan (fondly known as "Old Buck") commanded the Washington Navy Yard when the Civil War erupted. When it appeared that Maryland would leave the Union following the April 19, 1861 Baltimore Riot, Buchanan resigned his commission. Since Maryland did not secede, Buchanan strove for reinstatement, only to be denied by Gideon Welles. Buchanan was then named a captain in the Confederate

Captain Franklin Buchanan, USN, engraving, circa 1855. *Courtesy of The Mariners' Museum, Newport News, Virginia.*

navy and assigned to the Office of Orders and Detail until detached on February 24, 1862, to prepare the Confederate ironclad for combat.

Mallory expected great things from the *Virginia* with Buchanan in command. The secretary's orders detailing the flag officer to command the James River defenses included his hopes for the ironclad's use:

> *You will hoist your flag on the* Virginia, *or any other vessel of your squadron, which will, for the present, embrace the* Virginia, Patrick Henry, Jamestown, Teaser, Raleigh, *and* Beaufort.
>
> The Virginia *is a novelty in naval construction, is untried and her powers unknown, and the Department will give specific orders as to her attack upon the enemy. Her powers as a ram are regarded as very formidable, and it is hoped that you may be able to test them.*
>
> *Like the bayonet charge of infantry, this mode of attack, while the most distinctive, will command itself to you in the present scarcity of ammunition. It is one also that may be rendered destructive at night against the enemy at anchor.*
>
> *Even without guns the ship would be formidable as a ram.*
>
> *Could you pass Old Point and make a dashing cruise on the Potomac as far as Washington, its effect upon the public mind would be important to the cause.*

The condition of the country, and the painful reverses we have just suffered, demand our utmost exertions, and convinced as I am that the opportunity and the means for striking a decided blow for our Navy are now for the first time presented, I congratulate you upon it, and know that your judgment and gallantry will meet all just expectations.

Mallory realized that the Confederacy was reeling from defeats in Tennessee and along the North Carolina sounds. It was up to the Confederate navy to turn the tide, and the *Virginia* was the weapon that could achieve a stunning victory. Thus, he closed his orders to Buchanan with the admonition, "Action—prompt and successful action—now would be of serious importance to our cause."[43]

Buchanan had visited Gosport once on December 20, 1861, before his assignment and observed the *Merrimack* in dry dock. Several officers believed that he already knew of his future assignment at that time, which may account for his active role detailing some of the finest naval officers in the South to serve aboard the ironclad.[44] Buchanan, however, did not report to Gosport until early March. Catesby Jones, who was still dismayed over not receiving the command himself, lamented that Buchanan had not arrived. He noted, "The want of interest and energy in completing the *Merrimac* is disheartening." "Her Captain should be here and so I wrote him a month ago," Jones lamented to John Mercer Brooke.[45]

Buchanan, whose replacement as head of the Office of Orders and Detail had not yet been named, was busy in Richmond developing a plan of attack against the Federals. The Union defenses in Hampton Roads, five capital ships and three major defensive works—a total of over six hundred cannons—appeared formidable, yet Buchanan was undaunted by these odds. "Old Buck" had been assigned to the *Virginia* because of his aggressive nature and was prepared to immediately introduce his technological surprise into Hampton Roads. Buchanan realized that even though his vessel might be unable to break through the blockade and then move up the Chesapeake Bay toward Washington or Baltimore, it could wreak havoc amongst the Federal wooden warships in Hampton Roads. His ability to achieve control of Hampton Roads would end Union dominance of the harbor. More importantly, Buchanan understood that Hampton Roads was a key base for Union amphibious operations, and to effectively block any move up the James River toward Richmond, he must strike against the Union-held Newport News Point. Buchanan also knew that he must act swiftly or his moment of opportunity in Hampton Roads would be lost.

Consequently, Mallory's orders for prompt action were not lost on Buchanan. Buchanan sought to coordinate a joint attack against the Federal land and naval forces at Newport News Point with Major General John Bankhead Magruder's Army of the Peninsula. Buchanan envisioned such an attack dislodging the Union hold on Hampton Roads. Meanwhile, Magruder had been bombarding Richmond for assistance against an impending Union assault up the Peninsula toward Richmond. The Army of the Peninsula commander recognized that his James River flank was vulnerable and could be turned by the Federal fleet. Thus, he called on the Confederate navy for assistance. Sometime in late February, Magruder met with Buchanan to discuss a strategy to strike at Newport News Point with a simultaneous land and sea operation. It was a well-conceived plan. The *Virginia*, supported by her escorts the CSS *Beaufort* and CSS *Raleigh*, would steam from the mouth of the Elizabeth River to the James River, while Commander John Randolph Tucker's James River Squadron, consisting of the CSS *Patrick Henry*, the CSS *Jamestown* and the CSS *Teaser*, would move down the James River from Mulberry Island. The Army of the Peninsula would then assault Camp Butler. Buchanan's tactical plan was intended to have far-reaching strategic implications. The flag officer hoped that this three-pronged attack would wrestle Newport News Point from Union control, thereby denying McClellan's use of this excellent staging area.

"It is my intention to be off Newport News early on Friday morning next unless some accident occurs to the *Virginia* to prevent it, this I do not anticipate," Buchanan advised Magruder on March 2, 1862. "You may therefore look out for me at the time named," he added. "My plan is to destroy the Frigates first, if possible, and then turn my attention to the battery on shore. I sincerely hope that acting together we may be successful in destroying many of the enemy."[46] Even though Magruder had initially agreed to this plan, he eventually backed off, advising Richmond:

> *I am...satisfied that no one ship can produce such an impression upon the troops at Newport News as to cause them to evacuate the fort. The demoralization to our troops under similar circumstances has been produced by a concentration of fire from many ships at many points. No important advantages can be obtained by the* Merrimac *further than to demonstrate her power, which, as she is liable to be injured by a chance shot at this critical time, had better be reserved to defeat the enemy's serious efforts against* Norfolk *and* James River.[47]

Magruder was emphatic that his army could not support Buchanan's proposed assault on Newport News Point, writing that "the roads are almost impassable for artillery...I do not think the movement advisable."[48] Prince John, as Magruder was fondly called, did not appear to grasp the major advantage the *Virginia* would give the Confederacy in Hampton Roads. Instead of using this technological advantage as an offensive weapon, Magruder recommended that the *Virginia* should be used as a floating battery and "be stationed a little above Newport News, to prevent the gunboats coming up the swash channel leading into the Warwick River and turning the right flank of his line."[49] Fearing an assault by the increasing Union presence on the tip of the Peninsula, Magruder noted that he was preparing to pull his little army back to his second line (Warwick River Line) of defense and concluded his correspondence concerning the *Virginia*:

> *It is too late to co-operate with my army in any manner below with the* Merrimac, *even if the roads will admit it, which they will not, for the enemy is very re-inforced both at Newport News and Fortress Monroe with infantry and six batteries of light artillery.*
>
> *It would have been glorious if you could have run into these as they were being landed from a Baltimore boat and a commercial transport.*

"Any dependence," Magruder concluded, "upon me, so far as Newport News is concerned, is at an end."[50]

Franklin Buchanan was undaunted by Magruder's refusal to participate in his attack against Newport News Point and ignored his suggestion that the *Virginia* be used as a floating battery in the James River to protect the Confederate Army of the Peninsula's right flank. Buchanan was determined to strike at the enemy's fleet as quickly as possible.

The *Virginia* was still not ready for combat when Buchanan finally arrived at Gosport. "She is no means ready for service," Buchanan lamented. "She requires eighteen thousand two hundred pounds of powder, howitzers are not fitted and mounted on the upper deck to repel boats and boarders and none of the port shutters are fitted on the ship. Much of the powder has now arrived, and other matters shall not detain us."[51] Buchanan reported to Mallory on March 4 that the howitzers "may be ready by Thursday. The shutters for the two bow and quarter ports, I will have temporarily placed there to keep out shot and shells. The last of our powder and shells will be received on board on Wednesday. I feel confident," Buchanan added, "that the acts of the *Virginia* will give proof of the desire of her officers and crew to meet the views of the Department as far as practical."[52]

French Forrest obtained the last requisition of gunpowder from Fort Norfolk on March 6, 1862, as well as additional men to man the ironclad. Despite the determined efforts of Catesby Jones and John Taylor Wood, the *Virginia* still had not mustered an adequate crew. Each 7-inch Brooke rifle required twenty-six men to serve the gun, as well as to pivot the rifle to enable it to fire from one of its three gun ports. The ironclad was neither a comfortable nor a healthy ship. "There has not been a dry spot aboard of her, leaks everywhere," Jones recalled, and she is "as uncomfortable as possible."[53] "The quarters for the crew were damp, ill-ventilated and unhealthy," remembered William Norris. "One-third of the men were always on the sick-list and upon being transferred to the hospital, they would convalesce immediately."[54] Due to unhealthy conditions, the officers sought frequent replacements. Fortunately, several soldiers from Major General Benjamin Huger's command volunteered at the last moment for service on the *Virginia*.

The most significant group of men to volunteer in the days before the *Virginia*'s sortie against the Federal fleet was the United Artillery (Company E, 41st Virginia Infantry Regiment). Buchanan's urgent call for volunteers in early March 1862 prompted Captain Thomas Kevill and the entire complement of the United Artillery, thirty-nine men, "to go on board the ironclad steamer *Virginia*"[55] from Fort Norfolk on March 6.

The United Artillery was organized prior to the bombardment of Fort Sumter by Thomas Kevill. A native of Ireland and a Norfolk clothing merchant, Kevill created this artillery company from the ranks of the Volunteer Fire Company he had established in 1860. Kevill was named captain and led his men to occupy Fort Norfolk on April 19, 1861. The United Artillery remained at the fort until it transferred to the *Virginia*. This action was just what recruiting needed. Not only were all the men trained artillerists, but many also had a seafaring or business background. Edward Lakin, the company's first lieutenant, originally from New York, was a tavern keeper prior to the war. William Bramwell Colonna, a farmer from Northampton County, had originally sailed to Charleston, South Carolina, to enlist in the 1st South Carolina Infantry in February 1861. When his enlistment expired, he joined the United Artillery. Like Colonna, Andrew Joseph Dalton was imbued with Southern patriotism. Dalton left his job as a master printer with the *Norfolk Herald* and went to Charleston, South Carolina, to enlist in Company C, 1st South Carolina, on February 10, 1861.

Captain Thomas Kevill, CSA, oil on canvas, circa 1930. *Courtesy of The Mariners' Museum, Newport News, Virginia.*

He manned an artillery piece that fired on Fort Sumter when the war erupted in Charleston Harbor. When his enlistment expired, Dalton returned to Norfolk and enlisted in the United Artillery on March 4, 1862. John Krendal Belote, an oysterman from Gloucester County, Virginia, joined Company E the day the United Artillery occupied Fort Norfolk. James E. Barry Jr., once a member of the prewar militia unit, the Norfolk Light Artillery Blues, was a wealthy crockery merchant. He also mustered into the company to participate in Fort Norfolk's capture. Private Isaac Walling, a professional diver, assisted the Baker Wrecking Company in raising the *Merrimack* from the bottom of the Elizabeth River. Another private, William Duncan, had previously deserted to sail on a privateer. When the vessel never left port, he returned to his company. The United Artillery had already served the Confederacy in an outstanding fashion when the company captured Fort Norfolk with its over 250,000 pounds of gunpowder. Confederate control of the fort threatened Gosport and helped sway Flag Officer McCauley's decision to abandon the navy yard. Now it was time for the artillerists to once again rise to the occasion to aid the Confederacy's military actions. The majority of the United Artillery was assigned as gunners, and Captain Kevill assumed command of Gun #9.

As everything was readied aboard the *Virginia*, Buchanan considered exactly how his ironclad would strike at the enemy without the Confederate army's support. He advised Commander John R. Tucker of the James River Squadron, "It is my intention…to appear before the enemy off Newport News at daylight on Friday morning next." Buchanan planned to move his ironclad down the Elizabeth River during the evening of March 6 for a surprise attack. "You will," Buchanan instructed, "be prepared to join me. My object is to destroy the frigates *Congress* and *Cumberland*, if possible, and then turn my attention to the destruction of the battery on the shore and the gunboats."[56]

Buchanan was anxious to strike, as he had learned from a March 3 edition of the *New York Times* that the *Monitor* had passed her sea trials. He knew that the Union ironclad would soon leave New York, and his warship must attack before the enemy's vessel arrived in Hampton Roads. Unfortunately, neither the weather nor the pilots cooperated. A heavy gale struck eastern Virginia on March 6. Buchanan knew that his ironclad was so unseaworthy that it could only operate in the calm waters of the harbor. The gale forced a delay that further complicated Buchanan's plan, as the pilots refused to guide the huge ironclad down the narrow Elizabeth River at night. The *Virginia* "drew 22 feet of water, was in every respect ill-proportioned and top-heavy; and what with her immense length and wretched engines," wrote Confederate secret agent Major William Norris. "She was a little more manageable than a timber-raft."[57] "The pilots," according to Catesby Jones, "of whom there were five, having been previously consulted…all preparations were made, including lights at obstructions…claimed that they could not pilot the ship during the night."[58]

Lieutenant Robert Dabney Minor, CSN, photograph, circa 1861. *Courtesy of the Museum of the Confederacy, Richmond, Virginia.*

Rebuffed by nature and human frailties, Buchanan sent his flag lieutenant, Robert Dabney Minor, to the mouth of the Elizabeth River on March 7 to observe the Union fleet dispositions. Minor was an outstanding, brave and resolute officer who, after fourteen years of U.S. Navy service, had joined the Virginia State Navy in April 1861. He had already served the Confederate navy well, participating with Captain George N. Hollins and Colonel Richard Thomas Zarvona in the daring June 1861 capture of the USS *St. Nicholas*, brig *Monticello* and schooner *Mary Pierce*. Minor then worked under his brother George at the Bureau of Ordnance and Hydrography. During this time, he assisted Matthew Fontaine Maury with electric torpedo experiments. In October 1861, Minor tested a clumsy set of mines arranged together in a snare device against the Union squadron in Hampton Roads. The attempt failed, and Minor was assigned shortly thereafter to the CSS *Virginia*.[59] Minor returned from his reconnaissance mission with an enthusiastic report of the opportunity for a Confederate victory. "I have great hopes in our success," he wrote to his wife. "I reconnoitered the enemy off Newport News and Old Point and was glad to be able to report that they were not in such force as I had been led to suppose."[60]

The enforced delay on March 7 enabled Buchanan to load the final shipment of gunpowder, as well as receive some additional correspondence. Another letter from Magruder arrived once again stating the general's rationale for not supporting Buchanan's operation. Magruder noted to General Samuel Cooper:

> *The* Merrimac *will make no impression on Newport News, in my opinion, and if she succeeds in sinking the ships lying there it would do us little or no good, but if had attacked the* Baltic *and other transports filled with troops*

> *in those waters her success would have been certain and of incalculable advantage to us. Please ask the Secretary of War to impress these views on the Navy Department...the policy of merely being present near Fortress Monroe and Newport News when the latter is bombarded is exceedingly doubtful, as it would incur a risk of disaster without any corresponding advantage, and especially as withstanding the recent departures, while my own is diminished by more than ½ disposable for field service. In any event I could render no assistance to the* Merrimac *merely by my presence.*[61]

Inasmuch as Magruder failed to realize the ironclad's strategic power and combat capabilities, Stephen Mallory failed to recognize the *Virginia*'s limitations. The secretary of the navy sent Buchanan a confidential letter prompting the flag officer to seek greater glory for the Confederate navy with an attack on New York. "I submit for your consideration the attack of New York by the *Virginia*," Mallory wrote. "Once in the bay, she could shell and burn the city and the shipping. Such an event would eclipse all the glories of the combats of the sea," and, the secretary concluded, "peace would inevitably follow. Bankers would withdraw their capital from the city. The Brooklyn Navy Yard and its magazines and all the lower part of the city would be destroyed, and such an event, by a single ship, would do more to achieve our immediate independence than would the results of many campaigns."[62] Mallory, at the final hour, now envisioned a new role for the pride of the Confederate navy as a blockade-busting Leviathan destined to win the war in a single stroke.

Buchanan apparently discounted these letters, at least for the time being, and remained focused on his attack against Newport News Point. He knew that the *Virginia* was a powerful technological tool only able to achieve dominance over Hampton Roads and not an oceangoing cruiser capable of striking against Northern cities. Harbor defense, albeit with an aggressive mode of action, was Buchanan's method of achieving a resounding victory to renew Southern hopes for independence.

As each day in March passed, the Union feared the impending attack of the Confederate ironclad. "She will be almost certain to commit great depredations to our armed and unarmed vessels in Hampton Roads," wrote steam ram expert Lieutenant Colonel Charles Ellet.[63] Goldsborough longed for the *Monitor* to arrive in Hampton Roads as a counter to the Confederate ironclad. "I ask therefore if it would not be well to send the *Ericsson*," he wrote Welles, "to contend with that vessel on her own terms. I hope the Dept. will be able to send the *Ericsson* soon to Hampton Roads to grapple with the *Merrimac* and lay her out as cold as a wedge."[64] Union intelligence was excellent, despite the efforts of Huger and Forrest to plug information leaks.

Welles had originally ordered the *Monitor* to proceed to Hampton Roads. However, on March 6, as the Union ironclad left New York, he altered his plans to combat the Confederate ironclad and directed the *Monitor* to steam directly to Washington, D.C. On March 7, 1862, he ordered Marston to send the *St. Lawrence*, *Cumberland* and *Congress* to the Potomac River. The Union secretary of war's new orders would leave only steamships in Hampton Roads. Either Welles had succumbed to McClellan's request to silence the Confederate batteries blocking the Potomac River or he had learned of Mallory's orders that the *Virginia* strike at Washington. Regardless of his motivation, the Union command was deeply concerned with what the Confederate ironclad might do whenever it ventured forth from the Elizabeth River.

6
Like a Huge Half-Submerged Crocodile

March 8, 1862, began as "calm and peaceful as a May day," remembered Hardin B. Littlepage.[1] The gale had passed during the night, leaving behind a cloudless, mild day "common in southern Virginia during early Spring." Hampton Roads was smooth as glass, and the light wind barely produced a ripple. The conditions were perfect for the unseaworthy Confederate ironclad, and Buchanan was determined not to miss this opportunity to strike at the Federal fleet.

Buchanan intended to be out in Hampton Roads by 1:30 p.m. to obtain the maximum benefits of high tide. His huge ironclad required twenty-two feet of water to operate, and Buchanan knew that he needed every advantage available to operate within the shallow confines of Hampton Roads. A thick coating of tallow, or "ship's grease," was sloshed over the iron-sided casemate in the thought, according to Catesby Jones, "that it would increase the tendency of the projectiles to glance."[2] The ironclad was cleared for action, and all the workmen, some of whom were still making final adjustments to the armor and mechanical parts, were ordered off the vessel. At 11:00 a.m., Buchanan hoisted his flag officer's red pennant over the ironclad and commanded the *Virginia* to cast off from the dock.

The *Virginia* now began her ten-mile "shakedown cruise" down the Elizabeth River. Few of the ironclad's crew realized Buchanan's true intentions. Catesby Jones was the only officer on board the ironclad with whom Buchanan had shared his plans. "We thought we were going on an ordinary trial trip," remembered Lieutenant John R. Eggleston.[3] However, a few of the officers had some idea of Buchanan's aggressive intent. "Most of them had taken, as they supposed, a last farewell of wives, children, friends, and had set in order their worldly affairs," recounted William Norris. "All the lieutenants (Jones excepted) shortly before, and for the first time, had in their respective churches—Protestant and Catholic—publicly partaken of the Holy Sacrament of the Lord's Supper."[4]

Buchanan had shared his offensive strategy with Stephen Mallory, John Bankhead Magruder, Catesby Jones, French Forrest, John Randolph Tucker and the commanders of his escort gunboats, Lieutenant Joseph W. Alexander of the CSS *Raleigh* and Lieutenant William Harwar Parker of the CSS *Beaufort*. On March 7, Buchanan met with Alexander and Parker. The flag officer had detailed his plans, noting that if the battle turned against them, he would hoist a new signal, "Sink Before Surrender."[5]

As the *Beaufort* steamed alongside the ironclad on March 8, Parker recalled that a "great stillness came over the land." The flotilla was joined by "everything that would float, from the army tugboat to the oysterman's skiff…on its way down to the same point loaded to the water's edge with spectators." Forrest was aboard the tug *Harmony*, anxious to watch the heroic events unfold before him. William Parker was enthralled by the scene, writing:

> *As we steamed down the harbor we were saluted by the waving of caps and handkerchiefs, but no voice broke the silence of the scene; all hearts were too full for utterance, an attempt at cheering would have ended in tears, for all realized the fact that there was to be tried the great experiment of the ram and ironclad in naval warfare. There were many who thought as soon as the* Merrimack *rammed a vessel, she would sink with all hands encased in an iron-plated coffin.*[6]

"In an instant the whole city was in an uproar," wrote Private James Keenan of the 2nd Georgia Infantry, "women, children, men on horseback and on foot were running down towards the river from every conceivable direction shouting 'the *Merrimac* is going down'…I saw the huge monster swing loose from her moorings and making her way down the river…A good portion of her crew was on top and received the enthusiastic cheers from the excited populace without a single response."[7] Both sides of the Elizabeth were "thronged with people" when the news passed that the "*Virginia* is coming up the river."[8] "Most of them," ship surgeon Dinwiddie B. Phillips commented, "perhaps, attracted by our novel appearance,

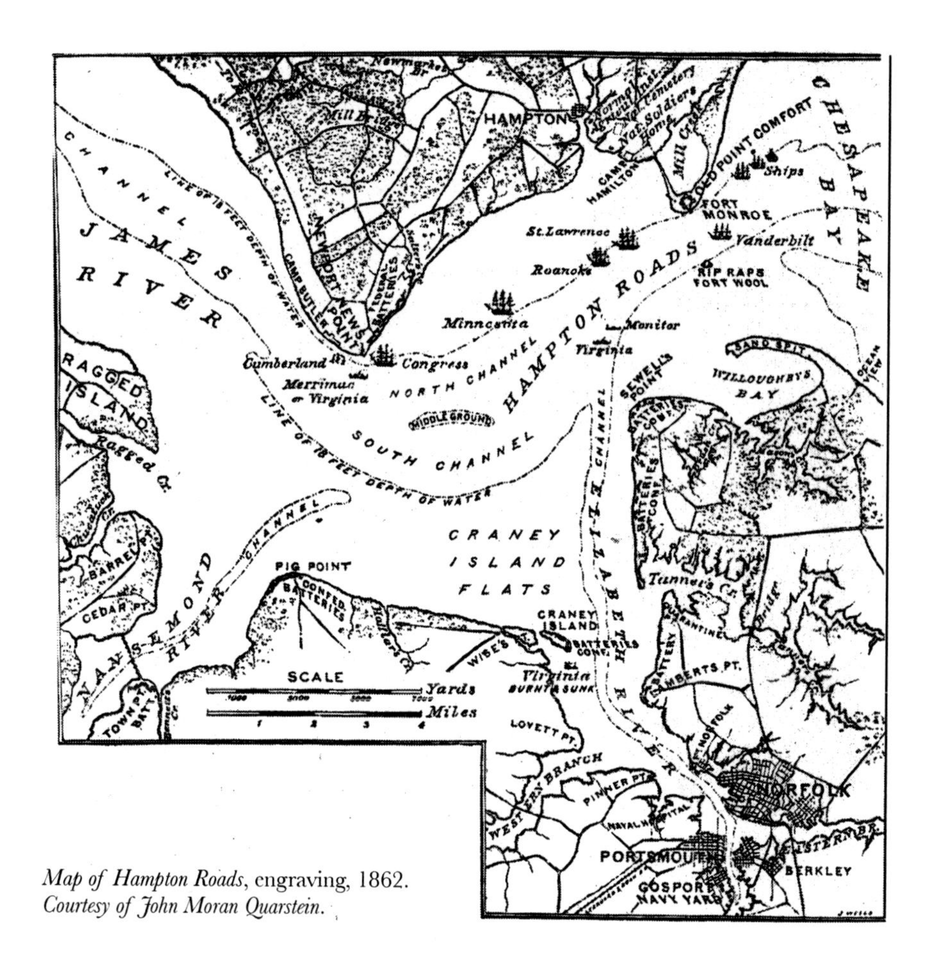

Map of Hampton Roads, engraving, 1862. *Courtesy of John Moran Quarstein.*

and desirous of witnessing our movements through the water." "Few, if any," Phillips added, "entertained an exalted idea of our efficiency, and many predicted a total failure."[9] Hardin Littlepage recalled, "One man, I remember called out to us, 'Go on with your old metallic coffin! She will never amount to anything else!' "[10]

The *Norfolk Day Book* perhaps reflected the overwhelming hope that the Confederacy would secure a great victory with its untried weapon:

> *It was a gallant sight to see the ironclad leviathan gliding noiselessly through the water, flying the red pennon of her Commander at the fore flagstaff and the gay Confederate ensign aft. Not the least impressive thought which she suggested was that her gallant crew, under a Commander and Officers worthy to direct their destiny and defend the flag she bore, went thus boldly*

> *with smiles and huzzas to solve a new problem in maritime warfare—to make the "trial trip of the Virginia the trial of battle."*[11]

The *Virginia* appeared fearsome to many onlookers, yet there were still problems on board the ironclad. "From the start we saw that she was slow, not over five knots," Lieutenant John Taylor Wood later commented. "She steered so badly that, with her great length it took thirty to forty minutes to turn...She was as unmanageable as a water-logged vessel."[12] The huge ironclad's keel was running so close to the river bottom that the ship's rudder could not steer her. The *Virginia* took a towline from the *Beaufort* to help her round a bend in the river near Craney Island.

"The engines did very well," noted Chief Engineer Ashton Ramsay, "better than I anticipated...the pilot told me she was going well on to 9 knots."[13] While most of the ironclad's officers agreed that "the engines were our weak point," Buchanan had reviewed the engines' condition and ability to withstand his proposed tactics earlier in the morning. Just before leaving the dock, he asked Ramsay, "What would happen to your engines and boilers if there should be a collision?" Ramsay replied that all the machinery was securely braced and the ten-mile trip downriver would be sufficient to test the engines' reliability. Buchanan was satisfied with Ramsay's opinion and declared, "I am going to ram the *Cumberland*. I am told she has the new rifled guns, the only ones in their whole fleet we have cause to fear. The moment we are out in the Roads, I'm going to make right for her and ram her."[14]

As the ironclad neared Craney Island, Ramsay reported to Buchanan on the top deck that the engines were operating beyond his expectations. "Very good," Buchanan replied. The flag officer "had been anxious concerning the engines as they had a bad reputation," Ramsay recalled, "but he now felt better satisfied, and would rely on me to keep him advised of the exact condition of the machinery at all times during the engagement, dwelling on the importance of the utmost vigilance in regard to it, and I remember his telling me, in case of my feeling a collision, not to wait for the signal to reverse, but to do so and with all the power of the machinery, as in the excitement of the moment he might forget to give the signal, or be incapacitated from doing so."[15]

Buchanan ordered his men to their midday meal as the ironclad steamed toward Hampton Roads. A caterer provided the officers with a picnic of "cold tongue and biscuit." Ramsay was called up to the ward room to share in the repast, and as he moved through the warship to his lunch,

> *passing along the gun-deck for this purpose, I was particularly struck with the countenances of the guns' crews as they stood motionless at their posts with ramrod or sponge in hand. The appearance of a ship cleared away for action was nothing new to me…but these men now looked pale and determined, standing straight and stiff, showing their nerves were wrought to a high degree of tension. From the time of leaving the yard…I had not reflected much on the possible results of the deadly conflict we were about to engage in. The appearance of these men; with the surrounding warlike preparations, brought it all to my mind. Here we were, with an untried, experimental ship…about to make an attack on a fleet of the very best material in the U.S. Navy…10 guns against 200; 300 men against 3000…yet this was our hazardous enterprise.*[16]

When Ramsay finally reached the wardroom, he noticed Assistant Surgeon Algernon S. Garnett laying out his surgical instruments in preparation for the coming battle. "The sight," Ramsay reflected, "took away my appetite."[17]

Buchanan, however, was ready for battle, and he now stepped onto the gun deck from his station in the forward hatchway behind the pilothouse to address the crew. It was common knowledge that his brother, Thomas McKean Buchanan, was aboard one of the Union vessels about to be attacked. Due to a desire to dispel any questions about his loyalty, Buchanan sought to confirm his commitment to destroy the Federal fleet with his men. The flag officer, with a Nelsonian flourish, reportedly informed the crew, "Sailors in a few minutes you will have the long awaited opportunity to show your devotion to your country and our cause. Remember that you are about to strike for your country and your homes, your wives, and your children. The Confederacy expects every man to do his duty, beat to quarters!"[18]

Buchanan reminded everyone that the "whole world is watching you today" and commanded them, "Go to your guns."[19] "He also told us that the Confederates had complained that they were not taken near enough to the enemy," Hardin Littlepage remembered Buchanan's comments, "and assured us that there should be no such complaint this time, for he intended to head directly for the *Cumberland*."[20]

By 1:30 p.m., the *Virginia* had entered Hampton Roads at high tide. The crew could see the entire Federal fleet arrayed in a line that stretched from Newport News Point to Fort Monroe. Two French ships, the *Gassendi* and *Catinat*, were also in the harbor awaiting the events as neutral observers. The steam corvette *Catinat* had arrived in Hampton Roads on January 8, 1862, using the station to maintain official contact with the Confederates in

CSS Virginia *Passing Craney Island*, engraving, circa 1880. *Courtesy of John Moran Quarstein.*

Norfolk and Richmond. She was later joined by the *Gassendi*, captained by Commander Ange Simon Gautier. The French often visited Gosport to view the Confederate ironclad yet were discreet with the information they gleaned from Southern naval officers, despite the pointed Union inquiries. After a visit to Norfolk two days prior, the Federals noticed that the French had moved their ships out of the main channel to a position near the entrance to Hampton Creek on the morning of March 8. Union officers wondered if this was a telltale sign of impending action. Their curiosity was also peaked by the new position of Commander John Randolph Tucker's squadron at the entrance to the Warwick River near the tip of Mulberry Island. The Confederate gunboats were just six miles upriver from Newport News Point and clearly visible to the crew of the USS *Cumberland*.

Despite these signs of impending action and all the news gathered by informants, the Confederates appeared to achieve a tactical surprise. "Nothing indicated," John Taylor Wood wrote, "that we were expected."[21] The Confederates could see clothes hanging from the rigging of the Federal frigates, "but for a long time no official notice was taken of us," Dinwiddie Phillips remembered. Buchanan steamed his ironclad into the channel in the direction of Fort Monroe and then turned the *Virginia* toward his targets at Newport News Point. The lack of any apparent reaction to the *Virginia*'s appearance gave time for some reflection among several Confederate officers who had previously served on the *Congress* and *Cumberland*. Lieutenant John Eggleston fondly recalled his time spent aboard the *Congress*, "my floating home for nearly three years." He reflected, "Little did I think then that I should ever lift a hand for her destruction."[22]

"The 8th of March, 1862, came and a finer morning I never saw in that southern latitude," wrote Seaman William Reblen of the USS *Cumberland*. "The sun came up smiling in all its splendor. It was 'up hammocks' that morning and then 'holystone' and wash decks as usual. After breakfast (it being Saturday) it was 'up all bags,' and every old tar went through his bag, mending and getting ready for Sunday's muster 'round the capstan."[23] The hours crept lazily along, and "sea and shore in this region saw nothing to vary the monotony of the scene," wrote a *Boston Journal* correspondent. He continued:

> *Now and then a soldier might be heard complaining that his detachment of the loyal army was having no part in the glorious victories which everywhere were crowning American valor with such brilliant success, or a sailor might be noted, on ship board telling how much he hoped the* Merrimac *would show herself, and how certainly she would be sunk by our war vessels.*[24]

Camp Butler—Newport News Point, engraving, 1862. *Courtesy of John Moran Quarstein.*

The peaceful Saturday was broken when the Confederate vessels were observed steaming out of the Elizabeth River about 12:45 p.m.

"I would not try to describe what happened at that moment in Hampton Roads," Commander Ange Simon Gautier of the *Gassendi* recalled when the *Virginia* finally entered the harbor in full view of the Union warships. "Everyone seemed to work for himself: no signal, no order, was seen at any note," wrote Gautier.[25] Everyone seemed to take notice of the *Virginia* all at once. One crew member of the *Congress* supposedly announced to the deck officer, "I wish you would take a glass and have a look over there, Sir. I believe that that thing is a-comin' down at last."[26] Another Union sailor remembered that day and his first sight of the *Virginia*: "Pretty soon that great black thing, different from any vessel ever seen before, poked her nose around Sewell's Point."[27] The Union tug, *Zouave*, steamed from Newport News toward Pig Point to investigate. "It did not take us long to find out," noted Acting Master Henry Reaney of the *Zouave*, "that it was the long-awaited Confederate ironclad. We had not gone over two miles," Reaney continued, "when we saw what to all appearances looked like the roof of a very big barn belching forth smoke as from a chimney on fire."[28] The *Zouave* fired several shells from its 30-pounder Parrott at the huge ironclad and returned to the Newport News anchorage.

"Suddenly, huge volumes of smoke began to pour from the funnels of the frigates *Minnesota* and *Roanoke* at Old Point," Ashton Ramsay recalled. "They had seen us and were getting up steam. Bright colored signal flags were run up and down the masts of all of the ships of the Federal Fleet," Ramsay continued. "The *Congress* shook out her topsails, down came the clotheslines on the *Cumberland* and boats were lowered and dropped astern."[29] "Men jumped lively,"[30] a crew member of the tug *Dragon* recalled as the Federal fleet prepared for action. Brigadier General Joseph King Fenno Mansfield, commander of Camp Butler, telegraphed Major General John E. Wool at Fort Monroe alerting him that "the *Merrimack* is being towed down by two steamers past Cranby [*sic*] Island towards Sewell's Point." Shortly thereafter, he reported again, "The *Merrimack* is close at hand."[31]

As the crews and officers of the *Congress* and *Cumberland* readied themselves for combat, other vessels in the Federal fleet struggled to reach Newport News Point. The *Roanoke*, her engines disabled, was taken under tow, and the *Minnesota* got underway. A tug was ordered out to the *St. Lawrence*, anchored just outside Hampton Roads in the Chesapeake Bay, to tow the fifty-gun sailing frigate into action. Buchanan knew that these warships would soon be brought into action. He had already "urged us to hurry with the work before us," Littlepage recalled Buchanan's talk with his officers en route to Hampton Roads. "That is, the destruction of the *Cumberland* and *Congress*, as the heaviest of the enemy's ships were following in our wake."[32]

It took the *Virginia* over an hour to steam across Hampton Roads. Once within range, the Union ships and shore batteries began shelling the ironclad. The Federal shot and shell harmlessly bounced off the *Virginia*. The shot "had no effect on her," as Thomas O. Selfridge of the *Cumberland* recounted, "but glanced off like pebble stones."[33] At 2:20 p.m., the Confederates opened fire at the Union ships. The *Beaufort*, flying her unusual pennant from the Battle of Roanoke Island, fired the first Confederate shot from her 32-pounder bow rifle at the *Congress*. Buchanan, however, waited until the range was less than 1,500 yards. Then he ordered Lieutenant Charles C. Simms to fire the 7-inch Brooke bow rifle at the *Cumberland*. The *Virginia*'s first shot hit the *Cumberland* at the starboard rail, showering splinters across the deck and tumbling several marines. "The groans of these men," remembered Thomas Selfridge, "the first to fall, as they were carried below, was something new" to many of the crew.[34] The second shell fired by Simms hit just below the sloop's forward pivot gun. It was the *Cumberland*'s heaviest gun. The gun was disabled and the entire gun crew decimated. Dead and wounded were everywhere. "No one flinched," Selfridge recalled, "but went on loading and firing, taking the place of some comrade, killed or wounded, as they had been told to do."[35] One Northern correspondent wrote that he saw "from the ship's scuppers running streams of crimson gore."[36]

The *Virginia* had now come abreast of the *Congress*. "The *Merrimac* was steaming slowly towards us," wrote one of the frigate's crew members, "and every eye on the vessel was on her. Not a word was spoken, and the silence that prevailed was awful. The time seemed hours before she reached us."[37] Lieutenant John Eggleston, commander of the ironclad's two hot shot guns, remembered:

> *The view from my station was restricted to the gun port, some three by four feet. For a time only the wide waters of the bay and the distant shores were visible, till suddenly the port became the frame of the picture of a great ship. It was the* Congress *only about a hundred yards distant. But for an instant was she visible, for suddenly there leaped from her sides the flash of thirty-five guns, and as many shot and shell were hurled against our armor only to be thrown from it high into the air. As by a miracle, no projectile entered into the wide-open ports.*[38]

Virginia *Attacking the* Cumberland, oil on canvas, by Alexander Charles Stuart, circa 1880. *Courtesy of The Mariners' Museum, Newport News, Virginia.*

The sound of the *Congress*'s broadside "was a terrible noise," remembered Richard Curtis of the forward Brooke gun crew, "and most of us gave a start when our Commander, Charles C. Simms, said, 'be quiet, men, I have received as heavy a fire in open air.'" Simms's comments steadied the men, and the "broadside had no effect on the iron sides of our ship."[39]

As the Confederate ironclad passed the hapless frigate, she unleashed her starboard broadside of four guns at the *Congress*. The effect was devastating. One shell went through a gun port, dismounting the gun and "sweeping the men about it back into a heap, bruised and bleeding."[40] Hot shot from Lieutenant Eggleston's gun rumbled through the frigate starting two fires, one of which threatened to ignite the *Congress*'s powder magazine. The *Virginia* was armed with hot shot guns just for this purpose. They were a deadly weapon against wooden ships and were placed on board the Confederate ironclad to give it yet another technological advantage. "I commanded the two hot-shot guns directly under the main hatch, and just over the furnace," Lieutenant Eggleston later noted. "The hot shot was hoisted from below in an iron bucket, placed by means of togs in the muzzle of the gun, slightly elevated and allowed to roll against the well-soaked wad that rested against the powder. Another soaked wad kept the shot in place."[41] The combination of hot shot and explosive shells was too much for the *Congress*. She was critically damaged by the *Virginia*'s salvo; however, Buchanan did not pause to finish off the stricken prey, as nothing would delay the flag officer's intended rendezvous with the USS *Cumberland*. "The pilot at the wheel has drawn a bead upon the *Cumberland*," wrote William Norris, "and holds her true as the needle for the doomed ship."[42]

The *Virginia* continued toward the sloop "like some devilish and superhuman monster, or the horrid creature of a nightmare." The *Cumberland* kept up her fire against the oncoming ironclad, but her shot "struck and glanced off, having no more effect than peas from a pop-gun." The *Virginia* was pounded with shot as she approached the sloop, creating a terrific din within the casemate. The *Cumberland*'s initial broadside shot away her flagstaff and the *Virginia*'s Stars and Bars fell onto the grating overhead the gun deck. When John R. Eggleston realized that the flag was no longer flying, he dashed up atop the ironclad and, amidst the Union sloop's shell fire, affixed the flag to the funnel. Even though the *Cumberland*'s cannonade was strong, it could not stop the Confederate ironclad's destructive mission. "The shot struck our sloping sides," Ramsay remembered, and "were deflected upward to burst harmlessly in the air, or rolled down and fell hissing into the water, dashing the spray up into our ports."[43] Buchanan was standing in the hatchway behind the pilothouse. Untouched by the hail of shot and shell, he purportedly called out to the *Cumberland* to surrender. The sloop's executive officer, Lieutenant George Upham Morris, replied, "Never!" Just before the *Virginia* reached the *Cumberland*, Richard Curtis peered out a gun port and "saw a sight that has been ever since indelibly stamped on my mind: all on the starboard side of the *Cumberland* was lined with officers and men with rifles and boarding pikes, all ready to repel us, thinking we intended to board her; I saw an officer, hat off and his sword raised, cheering on his men; this was the sight of a moment."[44]

The *Cumberland* appeared doomed as the ironclad rushed toward her at six knots. "At her prow I could see the iron ram projecting," the *Cumberland*'s pilot A.B. Smith remembered, "straight forward, somewhat above the water's edge, and apparently a mass of iron." Smith sadly reflected that "it was impossible for our vessel to get out of her way."[45]

"Like a huge half-submerged crocodile," the *Virginia* broke through the anti-torpedo obstructions surrounding the *Cumberland* and rammed the sloop on the starboard side of her bow.[46] Lieutenant Jones recalled that "the noise was heard above the din of battle."[47] The 1,500-pound, cast-iron ram punched a hole into the *Cumberland*'s berth deck. According to Lieutenant John Taylor Wood, it was "wide enough to drive in a horse and cart."[48]

The *Cumberland* was mortally wounded, the ramming made only worse by a simultaneous shot from the *Virginia*'s bow rifle, which killed ten men. "We've sank the *Cumberland*," Lieutenant Robert Minor shouted, running down the gun deck and waving his cap. Minor would later write, "The crash into the *Cumberland* was terrific in its results. Our cleaver fairly opened her side."[49] Tons of water was now gushing into the *Cumberland*, and she began to sink very rapidly to starboard, thus trapping the *Virginia*'s ram within her.

"Soon, however, I heard the reports of our own guns, and then there came a tremor throughout the whole ship," wrote Assistant Engineer E.A. Jack, "and I was never thrown from the coal bucket upon which I was sitting. This is when we drove into the *Cumberland* with our own ram. Then, the cracking and breaking of her timbers told full well how fatal to her that collision was. Then, there was a settling motion of our vessel that aroused suspicion that our ship had been injured too, and was sinking."[50] Just before ramming the *Cumberland*, Buchanan ordered the ironclad's engines to be reversed, yet there was an "awful pause," remembered Ashton Ramsay. The *Virginia* was caught by the weight of the sinking *Cumberland* as her engines labored to free her. Ramsay wrote, "The vessel was shaken in every fiber." Ramsay then, as the *Virginia*'s bow began to depress into the water, heard an explosion in the engine room. He thought the boilers had burst, struggling

Sinking of the USS Cumberland, engraving, Currier & Ives, 1862. *Courtesy of John Moran Quarstein.*

to back the *Virginia* away from the *Cumberland*. Instead, it was only an enemy shell exploding in the funnel. The crisis was quickly over, as the current turned the ironclad alongside the *Cumberland*. The ram, which was faultily mounted, broke off when the weight of the stricken sloop rested upon it, and the *Virginia* was freed. "Like a wasp we could sting but once," wrote Ramsay, "leaving the sting in the wound."[51] Hardin Littlepage remembered that as the *Cumberland* rested on the *Virginia*'s ram, "we felt great uneasiness lest she might carry us down with her, as she was filling rapidly. As soon, however, as there became a slight angle of inclination, the *Cumberland* glided off, and settled rapidly to her top sail yards."[52]

Lieutenant Thomas O. Selfridge later lamented that he failed to seize the initiative to drop the sloop's starboard anchor down onto the *Virginia*'s deck as the ironclad stood alongside the *Cumberland*. Selfridge believed the anchor could have acted as a grappling hook and pulled the ironclad under the James River as the *Cumberland* sank. This moment of opportunity, limited by the mounting casualties, quickly slipped away from the Union officer. The *Virginia*'s engines finally reversed. The ram broke off, and the ironclad was freed from the sinking sloop.[53]

Buchanan now positioned the ironclad parallel to the *Cumberland*, and for the next half hour they exchanged cannon fire. Both ships were engulfed in smoke as the *Cumberland* sent "solid broadsides in quick succession…into the *Merrimac* at a distance of not more than one hundred yards." Selfridge was "fighting mad when I saw the shells from my guns were producing no effect upon the iron sides of the *Merrimac*."[54] Unbeknownst to the *Cumberland*'s crew, the sloop inflicted serious damage to the *Virginia* during this phase of the battle. "She did us more damage than all of the rest of the fleet and batteries," wrote Ashton Ramsay, "put together."[55] When the ironclad rammed the sloop, the *Virginia* received, at the same time, a tremendous broadside from the *Cumberland*," remembered Hardin Littlepage.[56] Richard Curtis, a member of the bow rifle crew, recalled, "As the gun recoiled back, the Captain of the gun cried 'sponge,' and then Brave Dunbar, one of our gun crew and a good friend…jumped over the breechin, threw his head partly out of the port and was instantly killed…he fell at my feet—the first man killed."[57] A shell had "struck the sill of the bow-port and exploded," Catesby Jones noted. "The fragments killed two and wounded a number."[58] The ironclad's smokestack was riddled by this broadside. The damaged funnel caused the gun deck to be filled with smoke, "making it so dense we could hardly breathe." The stack's condition lessened the draft of the ironclad's boilers and caused the *Virginia* to lower her speed. One shot cut the ironclad's anchor chain, which whipped inboard, wounding several men. The *Cumberland*'s three broadsides swept away the *Virginia*'s starboard cutter, guard howitzers, stanchions and iron railings.[59]

As the *Virginia* swung away from the *Cumberland*, Ashton Ramsay recalled:

> *We were now exposed to perhaps the heaviest fire ever concentrated on one ship and at the closest quarters, for besides having to receive the broadsides of the* Cumberland *and* Congress, *Newport News batteries only a few cables' lengths off were pouring a deadly fire into us, and the sharp-shooters picking off every visible man. Our flag was shot down several times and was finally secured to the vents in the smokestack by Lieutenant Eggleston, who gallantly climbed up and secured it amidst the hail of shot.*[60]

The *Cumberland*'s gunners aimed at the *Virginia*'s gun ports, hoping to send a solid shot inside the casemate. "Our after nine-inch gun was loaded and ready for firing, when its muzzle was struck by a shell," Jones reported, "which broke it off and fired the gun. Another gun also had its muzzle shot off; it was broken so short that at each subsequent discharge its port was set on fire." "Lieutenant Hunter Davidson, in direct command of the disabled guns," John Eggleston exclaimed, "continued to fight with what was left of them while the battle lasted."[61] Louis Waldeck was killed and several others were wounded when Davidson's gun lost its muzzle from a shot from the *Cumberland*. Midshipman Henry Marmaduke "received several painful wounds," Buchanan noted of this incident, but "manfully fought his gun until the close."[62] The *Cumberland*'s broadsides, however, did not penetrate the *Virginia*'s armor. "The damage to the armor was slight," Jones noted, "had it been concentrated at the water-line we would have been seriously hurt, if not sunk."[63]

Despite the effective broadsides from the *Cumberland*, the ironclad kept up its punishing cannonade against the sloop. "On our gun deck the men were fighting like demons," Ashton Ramsay recalled. "There was no thought or time for the wounded and dying as they tugged away at their guns, training, and sighting their pieces while the orders rang out: 'Sponge, load, fire.'" Ramsay observed the officers shouting to the gun crews: "Keep away from the side ports, don't lean against the shield, look out for sharpshooters!" rang the warnings. "Some of our men who failed to heed them and leaned back against the shield were stunned and carried below, bleeding at the ears."[64]

The engine room was also a scene of determination while the battle raged above. "We were constantly busy with the operation of the condensing

Gun Deck of Cumberland, engraving, J.O. Davidson, circa 1880. *Courtesy of John Moran Quarstein.*

engines," wrote Assistant Engineer E.A. Jack. "The four boilers had to be fired and the linkgear of the engines was hard to operate. To reverse, after ramming, two men had to force the reversing gear to activate. For one of the furnaces, a frame of wrought iron was made upon which shot for the smoothbore could be placed, heated and tooled. Trivets were supplied to handle and convey them to the guns. We also constantly worked the pumps to relieve the ship of bilge." Jack noted from his station "down in the fire room, twenty-five feet under the water mark from battle. We were passed information on the battle through the ash chute. I knew the fight had begun by the dull reports of the enemy artillery, and an occasional sharp crack and tremor of the ship told me we had been struck."[65]

The *Virginia*'s sloped sides, coated with grease to help deflect shot, began to crackle and pop from the heat and flames caused by exploding shells. Midshipman Hardin B. Littlepage wrote that the ironclad seemed to be "frying from one end to the other." Littlepage later recounted one exciting exchange between crew members Jack Cronin and John Hunt: "Jack, don't this smell like hell?" "It certainly does, and I think that we'll all be there in a few minutes."[66]

It was, indeed, hell on the *Cumberland.* Master Moses S. Stuyvesant remembered it as "a scene of carnage and destruction never to be recalled without horror."[67] "The shot and shell from the *Merrimack* crashed through the wooden sides of the *Cumberland* as if they had been made of paper," remembered Acting Master's Mate Charles O'Neil, "carrying huge splinters with them and dealing death and destruction on every hand." O'Neil was spattered with "the blood and brains" of Master's Mate John M. Harrington when a shell whizzed past. He remembered how the "*Cumberland*'s once clean and beautiful deck was slippery with blood, blackened with powder and looked like a slaughterhouse."[68]

Lieutenant George Upham Morris, the *Cumberland*'s executive officer and acting commander, strove to save the Union vessel. Command of the *Cumberland* fell upon Morris on March 8, 1862, because her captain, William Radford, was assigned to court-martial duty aboard the USS *Roanoke.* He rushed to his ship on horseback when the *Virginia* entered Hampton Roads but arrived too late. Meanwhile, Morris attempted to turn the *Cumberland* on her anchor cable to either bring more guns into action against the ironclad or cut the cable in an effort to save the ship by running her around. It was too late; water had already reached the *Cumberland*'s berth deck, and the ship was clearly doomed. About 3:35 p.m., Morris gave the order to abandon ship, exhorting the remaining crew members to "give them a Broadside boys, as She goes."[69] "She went down bravely, with her colors flying,"[70] Catesby Jones remembered. E.V. White wrote that the *Cumberland* fought in a manner "exhibiting a heroism worthy of all praise and which entitled her to the renown that has since that day been attached to her name."[71] The *Cumberland*'s masts protruded above the waves, the flag marking the spot where 121 Union sailors had gallantly perished.

"The smoke from the *Cumberland*'s and the *Virginia*'s guns settled over the *Virginia*," Hardin Littlepage later wrote, "completely hiding her from view. We heard them cheering on the *Congress* some of the other frigates into action...the *Congress* was under the impression that the *Cumberland* had sunk us. So, of course, the *Congress* set up a lusty cheering."[72] As the ironclad was shrouded by smoke during the action, many of the thousands of onlookers were confused about the battle's outcome. Brigadier General Raleigh Edward Colston watched the engagement from Ragged Island, across the James River from Newport News Point. "We could see every flash of the guns and the clouds of white smoke,"[73] he remembered. Other Confederates, such as Private Keenan, observing the battle were concerned that "the *Merrimac* never halted nor fired a gun in reply to the *Cumberland* which was firing away with desperation. You may be able to partly imagine the great anxiety which prevailed along the shore now lined with thousands of anxious spectators. Everyone said, 'Why don't the *Merrimac* fire, the *Cumberland* will sink her.'"[74] The Southerners were soon overjoyed when it became obvious that the *Virginia* had destroyed her opponent. "I could hardly believe my eyes when I saw the masts of the *Cumberland* begin to sway wildly," Raleigh Colston commented.[75]

Having destroyed the *Cumberland*, Buchanan now turned his ironclad toward the USS *Congress*. The *Virginia*, because of its draft and poor steering, was forced to go up the James River to complete its turn. This maneuver took over thirty minutes and was only accomplished with the assistance of the *Raleigh* and *Beaufort*. As Franklin Buchanan reported:

> *We were sometime in getting our proper position in consequence of the shoalness of the water and the great difficulty of managing the ship when in or near mud. To succeed in my object I was obliged to run the ship a short distance above the batteries on James River in order to wind her. During all the time her keel was in the mud; of course she moved but slowly. Thus we were subjected twice to the heavy guns of all the batteries in passing up and down the river, but it could not be avoided. We silenced several of the batteries and did much injury on the shore. A large transport steamer alongside of the wharf was blown up, one schooner sunk, and another captured and sent to Norfolk. The loss of life on shore we have no way of ascertaining.*[76]

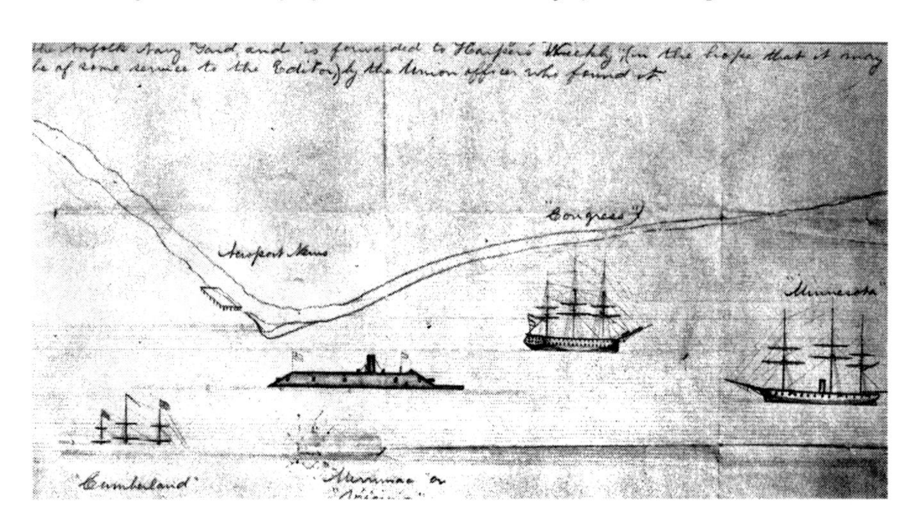

Virginia, Congress *and* Minnesota, pencil sketch, 1862. *Courtesy of the Chrysler Museum of Art, Norfolk, Virginia.*

Accurate fire from the ironclad disabled several Union cannons and destroyed two Union transports, the *Reindeer* and the *Whilden*, as well as an unidentified schooner, moored along a wharf. The *Reindeer*, a water tank transport, was captured. A shell was sent through General Mansfield's headquarters, showering Camp Butler's commander with splinters. The *Congress*'s crew cheered as the *Virginia* steamed upriver, thinking that the ironclad was leaving Newport News Point. "They were soon sadly undeceived," Buchanan later wrote, "for a few minutes after we opened upon her again."[77]

Lieutenant John Taylor Wood, commander of the stern pivot gun, had not fired his 7-inch Brooke rifle during the contest with the *Cumberland*. The *Congress* entered his sights as the ironclad moved upriver, and Wood sent three shells into the frigate's stern with deadly effect. Wood's shells found their mark, "breaking the muzzle on one and dismounting the other of our stern guns," wrote William McIntire.[78] The other shell crashed through the frigate's stern and wardroom until exploding on the berth deck amidst a line of ship's boys passing powder forward from the aft-magazine.

Meanwhile, the *Congress*, which was already seriously damaged, endeavored to escape the ironclad. The tug *Zouave* helped to run the *Congress* onto the shoal under the supposed protection of the Federal shore batteries.

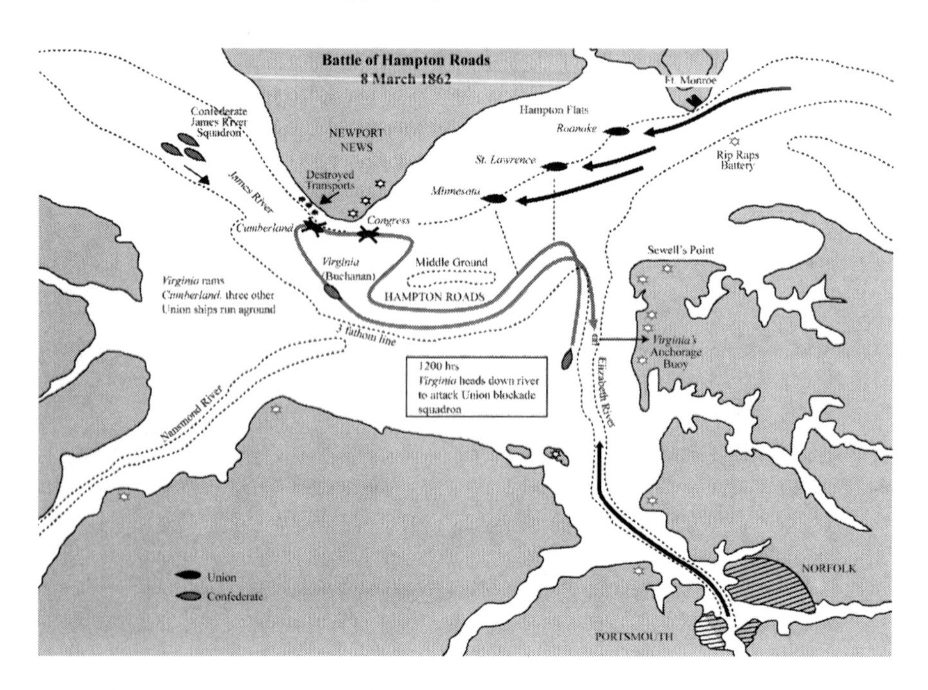

Battle of Hampton Roads, March 8, 1862, illustration, by Sara Kiddey. *Courtesy of The Mariners' Museum, Newport News, Virginia.*

Lieutenant Joseph Smith also thought the shallow water might make his ship safe from the *Virginia*'s deadly ram. The *Congress* was unfortunately unable to position herself parallel to shore to help protect the frigate from the approaching *Virginia*. Instead, the *Congress* ran aground with her bow facing Newport News Point, leaving only her stern guns available to return fire.

The three other Union capital ships in Hampton Roads, USS *Minnesota*, USS *Roanoke* and USS *St. Lawrence*, had all run aground rushing to the aid of the Federal forces at Newport News Point. The *Roanoke* was freed from the shoals, but seeing that wooden ships were powerless to stop the rampaging Confederate ironclad, Marston retreated toward Old Point Comfort and the protection of Fort Monroe. When the *St. Lawrence* was freed from the shoal by the steamship *Cambridge*, her commander, Captain Hugh Y. Purviance, tried to move his old fifty-two-gun frigate toward Newport News Point. The *St. Lawrence* grounded yet again near the *Minnesota*. Van Brunt's warship remained lodged on the Middle Ground Shoal one and a half miles from the *Congress*. All efforts to free the *Minnesota* failed. As the *Virginia* began its final approach against the stranded *Congress*, the ironclad was joined by Commander John Randolph Tucker's James River Squadron. Tucker commanded three vessels: CSS *Patrick Henry*, CSS *Teaser* and CSS *Jamestown*. When the *Virginia* attacked the *Cumberland*, these vessels started steaming down the James River from the Swash Channel near the mouth of the Warwick River.

Tucker took his small squadron past the Union batteries. The *Teaser* and *Jamestown* escaped without damage, whereas the *Patrick Henry* received a disabling shot through her steam drum. The *Patrick Henry*, formerly the bay steamer *Yorktown*, was armed with ten guns and featured a one-inch iron shield to protect her engines. She was the most powerful Confederate ship in Virginia waters until the emergence of the CSS *Virginia*. While the *Patrick Henry* received emergency repairs, the *Teaser* and *Jamestown* joined the *Virginia* and her escorts in the final attack on the *Congress*. The *Raleigh* and *Beaufort* were already shelling the *Congress* while the *Virginia* completed its turn. The *Raleigh* was put out of action when her only gun was disabled, and the *Beaufort*'s fire had little impact on the frigate.

Just before 4:00 p.m., the *Virginia* steamed within two hundred yards of the *Congress* and began shelling the helpless frigate. The *Congress*'s stern was quickly demolished, and the main deck was "literally reeking with slaughter." Blood from the frigate's deck, wrote Henry Reaney of the *Zouave*, poured "onto our deck like water on wash-deck morning." The *Zouave* was hit by several shells from the ironclad, which destroyed the tug's figurehead and pilothouse. Reaney, once he realized that little could be done to save the *Congress*, cut the line to the doomed frigate and steamed toward the *Minnesota*. A shell hit the *Zouave*'s rudderpost, virtually disabling the vessel. Somehow, the tug was able to escape the scene.[79]

Lieutenant Joseph B. Smith, acting commander of the *Congress*, struggled to keep his ship in action. Several large fires were raging throughout the ship, and over a quarter of the crew were killed or wounded. At 4:20 p.m., Smith was struck by a shell fragment, which tore off his head and a portion of his shoulder. The ship's command now rested on the shoulders of Lieutenant Austin Pendergrast. Pendergrast, in consultation with the ship's former captain, Commander William Smith, who was still on board, agreed that the *Congress* was helpless and that surrender was the only reasonable alternative.

The Confederates were jubilant when they saw the white flag rise above the *Congress*, and Buchanan immediately commanded his gun crews to cease fire. He then ordered the *Raleigh* and *Beaufort* to go alongside the frigate to "take the officers and wounded men prisoners, to permit the others to escape to shore, and to burn the ship."[80] Lieutenant William H. Parker, the *Beaufort*'s commander, steamed his gunboat alongside the stricken vessel to accept its surrender. Parker was visibly shocked when he learned about the death of his Annapolis classmate and close friend Joseph Smith. As he looked upon the scene of destruction about the frigate's deck, he was overheard by a prisoner to say, "My God, this is terrible. I wish this war was over."[81] Parker accepted the frigate's surrender from Austin Pendergrast and William Smith. Pendergrast offered Parker a sailor's cutlass he had just picked up off the deck of the *Congress* as a token of submission, but Parker demanded that both officers surrender their personal sidearms. Parker told Pendergrast his "orders and asked him to get his officers and wounded men on board as quickly as possible as I wanted to burn the ship. He said there were 60 wounded men on board the frigate and begged me not to burn the vessel. I told him my orders were peremptory."[82] As the two Union officers went to retrieve their weapons and organize the removal of the wounded, Parker ordered the *Jamestown* to the other side of the *Congress* to speed the evacuation. "I had scarcely given him the order," Parker later wrote, "when a tremendous fire was opened on us from the shore." Several Confederates were killed, and Parker recalled that at "the first discharge every man on the deck of the *Beaufort*...was either killed or wounded. Four bullets passed through my clothing; one of which carried off my cap cover and eye glass, and another slightly wounded me in the left knee."[83]

Brigadier General Joseph K.F. Mansfield, a forty-year army veteran and West Point graduate, witnessed the entire fight and reported to General Wool at Fort Monroe late that afternoon:

> *We want powder by the barrel. We want blankets sent up tonight for the crews of the* Cumberland *and* Congress. *The* Merrimack *has had it all her own way this side of Signal Point and will probably burn the* Congress *now aground with white flag flying and our sailors are swimming ashore.*[84]

Though the *Congress* may have surrendered, Mansfield certainly had not, and he was determined to strike back at the Confederates. Mansfield, who was then questioned about the propriety of firing on the Confederates, reportedly snapped, "I know the d–d ship has surrendered, but we haven't." He ordered detachments from the 20th Indiana and 1st New York Mounted Rifles with three rifled cannons down to the beach to open fire on the Confederate gunboats. One of the cannons was manned by survivors of the *Cumberland* and commanded by Master Moses Stuyvesant.[85] As his men continued to remove the wounded from the *Congress*, Parker shouted, "Make haste, those scoundrels on the shore are firing at me now!" Parker "blew the steam-whistle, and my men came tumbling on board." The *Beaufort*, suffering more than ten casualties, backed away from the *Congress* with thirty prisoners on board and steamed toward Craney Island. "The sides and masts of the *Beaufort* looked like the top of a pepper-box from the bullets," Parker wrote, "which went in one side and out the other."[86]

Buchanan was enraged by the *Beaufort*'s unauthorized retreat. He did not think highly of Lieutenant Parker, who had attended the Naval Academy at Annapolis while Buchanan was superintendent, stating that Parker lacked "judgment and discretion" and was "unfit to command."[87] Parker had not informed Buchanan of the situation on board the *Congress*, and "Old Buck" was particularly displeased that Parker had not set fire to the frigate. Robert Dabney Minor, the flag lieutenant on the *Virginia*, noticed that Buchanan was in a rage and that "the old gentleman was very anxious to destroy" the *Congress*. Lieutenant Minor volunteered to go in the *Virginia*'s only remaining cutter to complete the frigate's destruction. He set out in the boat, covered by the one-gun tug *Teaser*, with eight men at the oars. "I did not think the Yankees on the shore would fire at me on my errand to the *Congress*," Minor later wrote, "but when in about two hundred and fifty yards of her they opened fire on me from the shore…

Sinking of the Congress, engraving, J.O. Davidson, circa 1880. *Courtesy of John Moran Quarstein.*

and the way the balls danced around my little boat and crew was lively beyond all measure." Minor was knocked down into the boat, shot in the chest. Two of his men were wounded. "I was only down a second or two, and steering my crippled boat for the *Teaser*." The *Teaser*, commanded by Lieutenant William A. Webb, picked up the survivors and returned them to the *Virginia*. Minor recalled, "It had already been reported that they were firing upon me, and the flag officer, seeing it, deliberately backed our dear old craft close astern of the *Congress* and poured gun after gun, hot shot, and incendiary shells into her."[88]

Franklin Buchanan's battle blood was now boiling. Lieutenant John R. Eggleston heard Buchanan shout, "Destroy that _______ ship! She's firing on our white flag! [89] Buchanan, who was noted for his excitable nature, then recklessly climbed up on top of the *Virginia* to gain a better view of the action. He was so enraged by the events that he called for a rifle and began firing at the troops on the shore. Buchanan did not go unnoticed, as one member of the 1st New York Mounted Rifles remembered that "whenever a man came up on top of the *Merrimack* we shot at him all together."[90] Such an obvious target, Buchanan was shot in the thigh. The bullet grazed his femoral artery. He was carried below and ordered Catesby Jones to "plug hot shot into her and don't leave her until she's afire."[91]

"Brave, cool, determined old Jones fought the action out in his quiet way," noted Minor, "giving them thunder all the time."[92] Jones, "who knew the ship from her keel upwards and who had been made responsible for the efficiency of her battery," assumed command of the *Virginia* and continued firing on the *Congress* until she was blazing from stem to stern.[93] "Dearly did they pay for their unparalleled treachery," wrote Eggleston. "We raked her fore and aft with hot shot and shell."[94]

With the *Congress* blazing, there was still more work for the Confederate ironclad to accomplish in Hampton Roads. It was after 5:00 p.m., and the sun began to set. The tide had receded so that the *Virginia* could not venture out of the channel without risk of grounding. Nevertheless, Jones headed the *Virginia* toward the stranded *Minnesota*. The *Virginia*, accompanied by the *Jamestown* and *Patrick Henry*, moved to within a mile of the Union steam frigate and began shelling the vessel. Van Brunt frantically tried to free his warship from the shoal as "a solid shot tore through the hammock netting within a few yards of me," remembered Engineer Thomas Rae of the *Minnesota*, "killing two men and strewing the brains, even the head of one, all over the deck."[95] The *Minnesota* "seemed a huge monster at bay," Raleigh Colston commented. "The entire horizon was lighted by the continual flashes of the artillery," Colston continued. "Clouds of white smoke rose in spiral columns to the skies, illuminated by the evening sunlight, while land and water seemed to tremble under the thunders of the cannonade."[96]

Jones had hoped to send hot shot into the helpless *Minnesota*, but the broadsides from the *Cumberland* had disabled the only port side hot shot gun. The ironclad would have to turn around to bring her starboard hot shot gun to bear on the *Minnesota*, but time and tide were against such a maneuver. The *Virgini*a was leaking at the bow, and the pilots were all clamoring to return to the deeper water of the Elizabeth River. After an hour of long-range bombardment, often using ricochet fire to reach the *Minnesota* and *St. Lawrence*, the *Virginia* steamed back to her moorings near Sewell's Point. It had grown "so dark that we could not see to point the guns with accuracy," reflected Catesby Jones.[97] "As we moved across the Roads...we had a lively time before we anchored for the night," wrote William Drake. The *St. Lawrence* and *Minnesota* continued "such a rain of shot on the *Merrimac* as would disable her. Vain hope," Drake recalled,

> *as the darkness came on the shells began to explode about us, lighting up the expanse of water; both solid shot and shell frequently striking the armored side, to glance high up in the air. Whole shot passed through the big smokestack, and it was perforated with many fragments. Passing through the storm in safety, the* Merrimac *anchored at Sewell's Point after darkness had long set in. We had been standing at our guns since 11 'o'clock, and you will agree that we deserved supper.*[98]

By 8:00 p.m., the battle in Hampton Roads was over, but Jones was determined to renew the attack the next morning.

7

Aftermath

It was a great victory," recalled Robert Minor. "The IRON and the HEAVY GUNS did the work."[1] In one afternoon, the *Virginia* devastated the Union fleet and achieved tactical control of Hampton Roads. The CSS *Virginia* had truly stunned the Federal fleet and had left in her wake a path of destruction proving the power of iron over wood. The March 8, 1862 battle would remain the worst U.S. Navy defeat until Pearl Harbor.

Henry Wadsworth Longfellow wrote a dirge for the USS *Cumberland* shortly after learning of the vessel's destruction:

"Strike your flag!" the rebel cries,
In his arrogant old plantation strain.
"Never!" our gallant Morris replies;
It is better to sink than to yield!"
And the whole air peeled
with the cheers of our men.

Then, like a kraken huge and black,
She crushed our ribs in her iron grasp!
Down went the Cumberland *all a wreck.*
With a sudden shudder of death.
And the cannon's breath
For her dying gasp.

Next morn, as the sun rose over the bay,
Still floated our flag at main mast head.
Lord, how beautiful was Thy day!
Every waft of the air
Was a whisper of prayer,
Or a dirge for the dead.

Ho! bravehearts that went down in the seas;
Ye are at peace in the troubled stream;
Ho! brave land! with hearts like these,
Thy flag, that is rent in twain,
Shall be one again,
And without a seam![2]

The *Cumberland*'s flag sent a chilling message across Hampton Roads. Even though the Union wooden ships were apparently powerless to stop the Confederate ironclad, the wooden frigates remaining in the harbor were determined to continue the fight.

The Union losses on March 8, however, were staggering: two transports destroyed, one schooner captured, two capital warships sunk, one steam frigate damaged, one sailing frigate slightly damaged, one tug badly damaged and almost three hundred casualties. Following the destruction of the *Congress* and *Cumberland*, only the USS *Roanoke* reached the relative safety of Fort Monroe. The *Minnesota* and *St. Lawrence* were still stranded on the shoal.

The sailing frigate *St. Lawrence* had suffered some minor damage, particularly an unexploded shell lodged in her mainmast. However, the grounded *Minnesota* was hard hit by almost a dozen shells. The mainmast was damaged and her wooden sides penetrated by shot. Most of the *Minnesota*'s officers and crew did not care "to look forward to the morrow, as there was but one termination possible as far as we knew then," Thomas Rae reflected.[3] Captain Gershom Jacques Van Brunt was struggling to free his ship from the shoal while the other Union ships fled Hampton Roads. Commander Gautier of the *Gassendi* recalled the scene of fear:

All seemed very desperate the evening of the 8th, and a general panic seemed to possess everyone…Some ships changed their anchorages and all held themselves in readiness to stand out to sea at the first movement of the enemy: everything was in an uproar at Fort Monroe; ferries, gunboats, and

Virginia *Sinking the* Cumberland, *March 8, 1862*, oil on canvas, by Benjamin A. Richardson, 1862. *Courtesy of the Chrysler Museum of Art, Norfolk, Virginia.*

> *tugboats came and went in all directions; drums and bugles were beating and ringing with unwonted ardor; Fort Monroe and the Rip-Raps battery exchanged signals through the night without intermission.*[4]

General Wool reported the scene of destruction to Washington, lamenting the loss of the *Congress* and *Cumberland*. He feared that the grounded warships *Minnesota* and *St. Lawrence* "probably both will be taken" on the next day.[5] Colonel LeGrand B. Cannon, Wool's chief of staff, believed that the Union prospects were "gloomy" and questioned the U.S. Army's ability to maintain its fortifications on the Peninsula. He noted that the land-based, outdated armament at Fort Monroe was "as useless as musket-balls against the ironclad." Cannon complained that the "success of the *Merrimack* gave her control of the Roads, and if she could get sufficient elevation to her guns, she had the ability to shell and destroy the vast stores in and about the fort without the least power on our part to resist her."[6] McClellan immediately responded to Wool's telegram:

> *If the rebels obtain full command of the water it would seem impossible for you to hold Newport News. You are therefore authorized to evacuate that*

> *place, drawing the garrison in upon Fort Monroe which I need not say to so brave an officer is to be held at all hazards, as I will risk everything to sustain you should you be attacked by superior forces...By authorizing you to withdraw from Newport News I do not mean to give you the order to do so, but to relieve you from the grave sense of responsibility which every good officer feels in such a case—I would only evacuate Newport News when it became clear that the rebels would certainly obtain complete control of the water and render it untenable. Do not run the risk of placing its garrison under the necessity of surrendering. You will also please inform me fully of your views and wishes—the practicality and necessity of reinforcing you and c. The performances of the* Merrimac *places a new aspect upon everything, and may very probably change my whole plan of campaign, just on the eve of execution.*[7]

McClellan even telegraphed Major General John A. Dix in Baltimore regarding the threat of attack by the Confederate ironclad: "See that Fort Carroll is placed in a condition for defense as rapidly as possible in case the *Merrimac* should run by Fort Monroe."[8]

The Union command was obviously greatly stunned by the defeat. President Abraham Lincoln viewed the March 8 events as the greatest Union calamity since Bull Run. Lincoln told Captain John A. Dahlgren that it was "frightful news." Dahlgren recounted that Lincoln sadly expressed to him early the next morning that the "*Merrimac* had come out yesterday, smashed the *Cumberland*, and compelled the *Congress* to surrender...The President did not know whether we might not have a visit here...I could give but little comfort; such a thing might be prevented but not met."[9]

Secretary of War Edwin W. Stanton became "almost frantic." According to Gideon Welles, he was the "most frightened man on that gloomy day," stating that "the *Merrimac*...would destroy every vessel in the service, could lay every city on the coast under contribution, could take Fortress Monroe; McClellan's mistaken purpose to advance must be abandoned." Stanton feared, as Welles noted, that the Confederate ironclad would soon "come up the Potomac and disperse Congress, destroy the Capitol and public buildings; or she might go to New York and Boston and destroy those cities."[10]

An emergency cabinet meeting was called by Lincoln on the morning of March 9, 1862, to discuss the threat posed by the Confederate ironclad. Welles observed that everyone was extremely concerned, and the general opinion, as expressed by the president's secretary, John G. Hay, was that the enemy's ironclad "would capture our fleet, take Fort Monroe [and] be in Washington before night." Welles reported that Edwin Stanton looked out the window toward the Potomac, stating, "Not unlikely we shall have shell or cannon-ball from her guns in the White House before we leave this room."[11] Despite Welles's report that a Union ironclad was en route to Hampton Roads, the cabinet despaired for the future. Stanton ordered General McClellan to sink stone-laden boats in the Potomac to block the advance of the Confederate ironclad.[12]

E.V. White recorded one Northern newspaper correspondent's comments that clearly captured the initial Union response to the *Virginia*'s victory:

> *The swift work done by the* Merrimac *on this occasion spread consternation throughout the Northern States. The blockade of the Atlantic Coast maintained at that time could not last long before this mighty and invulnerable engine of destruction. New York, Boston, and Washington would soon be threatened. The most alarming crisis of the Civil War was at hand. As the sun went down that night over Hampton Roads, every Union heart in the fleet and in the fortress throbbed with despair. There was no gleam of hope. The* Merrimac *was impervious to balls, and could go where she pleased. In the morning it would be easy work for her to destroy our whole fleet. She could then shell Newport News and Fortress Monroe at her leisure, setting everything combustible in flames, and driving every man from the guns. As the news of the terrible disaster was flashed over the country by the telegraph wires all faces wore an expression of consternation…The panic cannot be described. There was absolutely nothing to prevent the* Merrimac *from ascending the Potomac and laying the capital in ashes…The* Merrimac *could laugh at forts.*[13]

The panic that seized Northern cities with a sense of impending doom was based on the short sortie of the *Virginia* against outdated wooden sailing ships. Little did the Northern leaders know that Franklin Buchanan considered the *Virginia* so unseaworthy that it could not leave Hampton Roads. William Norris concluded that the Confederate ironclad "was not weatherly enough to move in Hampton Roads, at all times, with safety, and she never should have been found more than three hours' sail from a machine shop."[14] "Should she encounter a gale," Buchanan believed, "or a very heavy swell, I think it more than probable she would founder."[15] The *Virginia*'s great draft, poor maneuverability and slow speed precluded any dash to break the blockade and limited the ironclad to harbor defense duties.

The Federals did not understand the *Virginia*'s weaknesses. Instead, as E.V. White recorded, the Union command saw the *Virginia* as

> *the experiment of an hour* [which] *had wrought an entire change in naval architecture and in defensive fortifications throughout the world. Wooden frigates had almost ceased to be of any value. The blow which sunk the* Cumberland *demolished also the fleets of England and France. All navies went down with that frigate into the abyss together. It is not too much to say that such a night of anxiety, of terror, of bewilderment, as followed the triumphant return of the* Merrimac *to her anchorage behind Craney Island, this world has seldom witnessed before.*[16]

When the *Virginia* reached her mooring off Sewell's Point shortly after 8:00 p.m., civilians and soldiers lined the shore to celebrate and learn more about how the ironclad achieved her great victory. Lieutenant Douglas French Forrest immediately caught a train to Richmond and late that evening presented a report detailing the Confederate's astounding victory to President Jefferson Davis. Catesby Jones instead focused on preparing the *Virginia* for the next day's action. Jones inspected the ironclad for damage and discovered a small leak in the bow but did not notice that the ram was missing because of the dark. He merely thought that the ram was twisted from its collision with the *Cumberland*. Despite the missing anchors, boats, flagstaffs, railings and howitzers, most of which were lost during the fight with the *Cumberland*, Jones believed that the *Virginia* was ready once again to venture out against the Federal fleet of wooden ships.[17] The Confederate ironclad had stood the test of battle rather well during her maiden voyage. "Our loss in killed and wounded was twenty-one," wrote John Taylor Wood. "The armor was hardly damaged, though at one time our ship was the focus on which were directed at least one hundred heavy guns, afloat and ashore."[18]

While the Confederates rejoiced over their victory, Flag Officer Buchanan lay depressed and wounded in his cabin, lamenting that "my brother Paymaster Buchanan was on board the *Congress*."[19] Unbeknownst to Franklin Buchanan, Lieutenant Commander Thomas McKean Buchanan had somehow survived the battle. The seriously wounded Flag Officer Buchanan and Lieutenant Minor were finally convinced to leave the ironclad. Chief Surgeon Dinwiddie Phillips supervised this task, and as he returned to the *Virginia*, he surveyed the ship's damage. He counted ninety-eight indentations in the iron plate from enemy shot and noted that the smokestack was so riddled that it "would have permitted a flock of crows to fly through without inconvenience."[20]

Commander Thomas McKean Buchanan, USN, photograph, circa 1862. *Courtesy of the U.S. Army Military History Institute, Carlisle, Pennsylvania.*

Considering the damages she had wrought on the Union fleet, the *Virginia*'s losses were slight. Buchanan and Minor were the most seriously wounded; however, the *Virginia* had suffered a total of twenty-one casualties during the battle. Two men were killed. Charles Dunbar was killed by shell fragments just before the *Virginia* rammed the *Cumberland*. The shell struck the sill of the bow port when the 7-inch Brooke gun was being reloaded. Dunbar died at the feet of his gun mate Richard Curtis. As the engagement with the *Cumberland* continued, a shot during a massive broadside from the sloop struck and knocked off the muzzle of Gun #2 (6.4-inch Brooke rifle). Louis Waldeck was probably killed at this moment. This same shot also wounded Midshipman Henry H. Marmaduke in the arm. While several men suffered minor injuries during the engagement, only five men besides Buchanan, Minor and Marmaduke were listed in Surgeon Dinwiddie Phillips's report after the battle. John H. Leonard, Emsley H. Ives and William G. Burke were among the wounded, but their injuries were not described. Two United Artillery men, John C. Capps and Andrew J. Dalton, were wounded by "musket balls coming thru gun port." These men were serving at Gun #9, commanded by Thomas Kevill. Another United Artillery artillerist, Daniel Knowles, was reportedly wounded in the head during the battle. Knowles was one of approximately ten men suffering minor injuries during the battle but not included in the surgeon's report. Knowles probably either received his injury during the loss of Gun #9's muzzle or, more likely, from standing too close to the casemate's shield when it was struck by shot. He could have been one of the several men who, according to Ashton Ramsay, "leaned against the shield, were stunned and carried below, bleeding at the ears."[21]

Buchanan's wound resulted in significant changes to the ironclad's command structure in preparation for the next day's sortie. Jones remained

The IX-inch Dahlgren damaged during the engagement between the CSS *Virginia* and the USS *Cumberland*, 1862. *Courtesy of The Mariners' Museum, Newport News, Virginia.*

acting commander of the warship. It was an excellent decision. "To that brave and intelligent officer," Buchanan wrote about Catesby Jones, "I am greatly indebted for the success achieved." Lieutenant Charles C. Simms was elevated to the vacant post of executive officer. Simms "fully sustained his well-earned reputation" on March 8 while commanding the forward Brooke rifle. The *Virginia*, according to Catesby Jones, was fortunate to have "the services of so experienced, energetic, and zealous an officer." Hunter Davidson assumed command of the bow pivot 7-inch Brooke gun from Simms. This assignment added greatly to Davidson's responsibilities; however, he had fought his guns on March 8 with "great precision" and warranted this added duty. Buchanan noted that throughout the battle, Davidson's "buoyant and cheerful bearing and voice were contagious and inspiring." Davidson was assisted with his three guns by Midshipman Henry Marmaduke, Boatswain Charles Hasker and Gunner Charles B. Oliver. These men were all U.S. Navy veterans and had "discharged well all duties required of them" during the action with the *Congress* and *Cumberland*. Buchanan gave special attention to Oliver's service, stating that the "gunner was indefatigable in his efforts; his experience and exertions have contributed very materially to the efficiency of the battery."[22]

The flag officer was very generous in his praise of the conduct of the *Virginia*'s officers and men. He applauded Wood's "zeal," Minor's "general cool and gallant bearing" and Eggleston's "judgment [which] exerted a happy influence in his division." Laurels were spread throughout the ironclad. Buchanan praised "Captain Thom, whose tranquil mien gave evidence that the hottest fire was no novelty to him," and remarked how Lieutenant Butt "was gay and smiling" throughout the action. Special commendation was given to Ashton Ramsay and his assistants, as the "engines and machinery, upon which so much depended, performed much better than was expected." Since he was one of the wounded, Buchanan heaped appreciation on the ship's surgeons, Dinwiddie Phillips and Algernon Garnett, who

> *were prompt and attentive in the discharge of their duties; their kind and considerate care of the wounded, and the skill and ability displayed in the treatment, won for them the esteem and gratitude of all who came under their charge, and justly entitled them to the confidence of officers and crew.*[23]

Tremendous tributes were also bestowed on Buchanan for the great victory on March 8, 1862. The accolades were well deserved. Buchanan's leadership ensured that not only had the Confederates achieved a tactical surprise but that they had also struck a mortal blow against the Federal fleet. He had told his men that they "must not be content with only doing your duty, but you must do more than your duty!" The flag officer then pointed to the Union fleet and exclaimed, "Those ships must be taken…Go to your guns!" This

Gun Deck of CSS Virginia, engraving, circa 1880. *Courtesy of John Moran Quarstein.*

exhortation had given confidence to the crew to steam their untried and outgunned experimental warship against the entire Union fleet in Hampton Roads. The impossible was achieved, and when a great victory was in the grasp of Buchanan, he was grievously wounded. The flag officer's injury sent a shockwave through the ironclad. Recognizing the need to exhort the men again to continue the fight, Buchanan sent the wounded Robert Minor to deliver a message to the men as they battled the *Congress*. "Tell Mr. Jones to fight the ship to the last. Tell the men that I am not mortally wounded, and hope to be with them soon." Buchanan's message was met with cheers, which "resounded far above the cannon's roar, and every man was again quickly at his post, dealing death and destruction with their heavy guns."[24]

The *Virginia*'s victory on March 8 prompted Flag Officer Josiah Tattnall to write to Buchanan on March 12:

> *I congratulate you, my dear friend, with all my heart and soul, on the glory you have gained for the Confederacy and yourself. The whole affair is unexampled, and will carry your name to every corner of the Christian world, and be on the tongue of every man that deals in salt water. That which I admire most in the whole affair is the bold confidence with which you undertook an untried thing...You don't know how much you have aided in removing the gloom which recent military events had cast over us.*[25]

Franklin Buchanan (left) and Josiah Tattnall, photograph, circa 1870. *Courtesy of the Virginia War Museum, Newport News, Virginia.*

The *Virginia*'s officers and crew, however, did not have time to contemplate their victory. Once at her Sewell's Point moorings, the ironclad was inspected and readied for the next day's events. The crew did not receive their supper until 10:00 p.m. The men were exhausted, "begrimed with powder, dirt and blood."[26] Many immediately sought rest, while others, still tense from the battle, watched the final scene of the battle illuminate Hampton Roads.

The glowing *Congress* provided a spectacular conclusion to the events of March 8. Ashton Ramsay remembered the scene:

> *All the evening we stood on deck watching the brilliant display of the burning ship. Every part of her was on fire at the same time, the red-tongued flames running up shrouds, masts, and stays, and extending out to the yard arms. She should in bold relief against the black background, lighting up the Roads and reflecting her lurid lights on the bosom of the now placid and hushed waters. Every now and then the flames would reach one of the loaded cannon and a shell would hiss at random through the darkness. About midnight came the grand finale. The magazines exploded, shooting up a huge column of firebrands hundreds of feet in the air, and then the burning hulk burst asunder and melted into the waters, while the calm night spread her sable mantle over Hampton Roads.*[27]

"The magazines blew up with a terrific noise," recollected Private William H. Osborne while observing the burning *Congress* from Newport News Point. Osborne remembered:

> *This event had been anticipated by the garrison, and the shores and adjacent camps were crowded with awe-struck gazers. The whole upper works of the frigate had, hours before, been reduced to ashes by the devouring flames; the masts and spars, blackened and charred, had fallen into and across the burning hull; these were sent high into the air with other debris, and as blast succeeded blast, were suddenly arrested in their descent and again sent heavenward. The spectacle thus presented was awfully grand; a column of fire and sulphurous smoke, fifty feet in diameter at its base and not less than two hundred feet high, dividing in its centre into thousands of smaller jets, and falling in myriads of bunches and grains of fire, like the sprays of a gigantic fountain, lighted up the camp and bay for miles.*
>
> *The yards and rigging of the* Minnesota *and* St. Lawrence *were filled with men armed with fire-buckets lest the falling sparks should ignite the tarred ropes of these vessels, and unite them in one general conflagration.*

Destruction of the USS Congress, engraving, 1862. *Courtesy of John Moran Quarstein.*

> *The sides of the hapless Congress were thrown open by the last explosion, and the next morning, all that could be seen of the once proud ship were a few blackened ribs, a short distance above the surface of the water.*[28]

When members of Cobb's Legion were crossing the James River near Jamestown, over thirty miles upriver from Hampton Roads, they noticed a bright flash downriver, which was followed shortly thereafter by the sound of a tremendous explosion. Confederate soldiers in Suffolk also reported seeing and hearing the last gasp of the *Congress*. Raleigh Colston, an eyewitness from his vantage point on Ragged Island, noted that the frigate's death was "one of the grandest episodes of this splendid yet somber drama."[29] As he watched the *Congress* glow from his position at Camp Jackson between Pig Point and Craney Island, Private Ossiam D. Gorman of the 4th Georgia Regiment wrote a poem about the day's events:

> *The victory's ours, and let the world*
> *Record Buchanan's name with pride;*
> *The crew is brave, the banner bright,*
> *That ruled the day when Hutter died.*[30]

While Colston marveled as the *Congress*'s spars and ropes "glittered against the dark sky in dazzling lines of fire," one of the *Virginia*'s "pilots chanced, about 11:00 p.m.," Catesby Jones later wrote, "to be looking in the direction of the *Congress* when there passed a strange-looking craft, brought out in bold relief by the brilliant light of the burning ship, which he at once proclaimed to be the *Ericsson*."[31] The *Monitor* had finally arrived, but it was a day late! The *Virginia* had won the race for Hampton Roads; however, the hard-fought Confederate victory of March 8 would be quickly put to the test.

8
Enter the *Monitor*

When daylight came on Sunday, March 9, 1862, the *Minnesota*, despite all efforts to float her, was still aground. There appeared nothing that the warship could do but wait for the *Virginia*'s arrival. The frigate's commander, Captain Gershom Jacques Van Brunt, was prepared to destroy his ship rather than allow it to be captured by the Confederate ironclad.

Meanwhile, Catesby Jones and his men had breakfast just after dawn. "We began the day with two jiggers of whiskey," an elated William Cline later wrote, "and a hearty breakfast."[1] The *Virginia* got underway from its Sewell's Point mooring about 6:00 a.m. on March 9, accompanied by the *Patrick Henry*, *Jamestown* and *Teaser*. Due to heavy fog, the small fleet delayed entering Hampton Roads until nearly 8:00 a.m. The day turned "as bright and beautiful as the day proceeding it," remembered Richard Curtis. "The broad waters of Hampton Roads were as smooth as glass, not a ripple on its surface, an ideal day to go to Church, but alas it was soon to be broken by the roar of cannon and angry men seeking each other's lives."[2]

Jones saw that the *Minnesota* was still stranded on the shoal as the *Virginia* closed within range. At 8:30 a.m., Hunter Davidson's forward Brooke rifle sent the first shot of the day at a range of one thousand yards through the frigate's rigging. Another shot quickly followed, "exploding on the inside of the ship, causing considerable destruction and setting the ship on fire."[3] The ironclad's crew expected to make short work of the *Minnesota*. Ashton Ramsay recounted, "We approached her slowly, feeling our way cautiously along the edge of the channel, when suddenly, to our astonishment, a black object that looked like the historic description, 'a barrel-head afloat with a cheesebox on top of it,' moved slowly out from under the *Minnesota* and boldly confronted us."[4]

Daylight confirmed the pilot's late-night report of the *Monitor*'s arrival. As the *Monitor* steamed away from the *Minnesota* toward the *Virginia*, Jones, who had been following the Federal ironclad's construction in Northern newspapers, calmly noted to his officers that there "was an iron battery near" the *Minnesota*.[5] Most of the Confederates, however, were thoroughly amazed by the sight of the Union ironclad. Hardin Littlepage wrote of the crew's first view of the *Monitor*, "We were taken wholly by surprise." "We could see nothing but the resemblance of a large cheese box," John Eggleston recounted. Eggleston, peering through his gun port, thought the ship he viewed was "the strangest looking craft we had ever seen before." Lieutenant Hunter Davidson thought at first that "the *Minnesota*'s crew are leaving on a raft." "Such a craft as the eyes of a Seaman never looked upon before—an immense shingle floating in the water, with a gigantic cheesebox rising from its center," Lieutenant James H. Rochelle of the *Patrick Henry* noted. "No sails, no wheels, no smokestack, no guns. What could it be?" Davidson, called by Seaman Richard Curtis "one of the smartest and bravest officers on the *Merrimac*," remarked to the crewmen around him, "By George, it is the Ericsson Battery, look out for her hot work." Jones and Davidson were among "a few visionary characters," Rochelle remarked, who "intimated that it might be the *Monitor*." When the Union ironclad moved to interrupt the *Virginia*'s course toward the *Minnesota*, John Taylor Wood reflected, "She could not possibly have made her appearance at a more inopportune time."[6]

Lieutenant Hunter Davidson, CSN, photograph, circa 1862. *Courtesy of the Hampton History Museum, Hampton, Virginia.*

The USS *Monitor*'s appearance in Hampton Roads the night before was a virtual miracle. While the *Monitor* was not designed merely to counter the threat of the Confederate ironclad, it was fortunate that the Union ironclad arrived in time to disrupt the *Virginia*'s destructive work.

She was a completely new concept of naval design created by Swedish-American inventor John Ericsson. While the *Virginia* was a brilliant adaptation of materials at hand, the *Monitor* was an engineering marvel, containing several patents created by Ericsson. The ironclad was 173 feet in length, weighed 776 tons and had a beam of 41.5 feet. Her draft was 11 feet with a freeboard of less than 1 foot. The *Monitor* was virtually awash with the sea. All of the ship's machinery, magazine and quarters were positioned below the waterline. The turret and pilothouse were the only features protruding from the deck. The *Monitor*'s most impressive feature was her steam-powered, rotating, circular turret mounting two XI-inch Dahlgren smoothbores. The turret was constructed of eight layers of one-inch-thick curved, rolled plates made in Baltimore, Maryland. The gun ports were equipped with iron shutters. The turret had an interior diameter of 20 feet and a height of 9 feet. A pilothouse was the only other main feature protruding from the deck. It was a rectangular box of iron, standing 3 feet above the deck and made of nine nine- by twelve-inch iron bars bolted at the corners. A half-inch observation slit was included below the upper tier of iron bars. The anchor well was located forward of the pilothouse.[7]

Lieutenant John Lorimer Worden was selected as the *Monitor*'s commander. Worden had served in the U.S. Navy since 1834 and had been a prisoner of the Confederates after conducting a secret mission to Fort Pickens in Pensacola Bay, Florida. Recently exchanged, Worden accepted the command of the experimental warship and commented, "After a hasty examination of her [I was] induced to believe that she may prove a success. At all events, I am quite willing to be an agent in testing her capabilities."[8]

Lieutenant John L. Worden, USN, engraving, 1862. *Courtesy of The Mariners' Museum, Newport News, Virginia.*

The *Monitor* was commissioned on February 25, 1862. Lieutenant Worden assembled a handpicked crew, which would eventually number sixteen officers and forty-nine men. Despite John Ericsson's claims that the "impregnable and aggressive nature of this structure will admonish the leaders

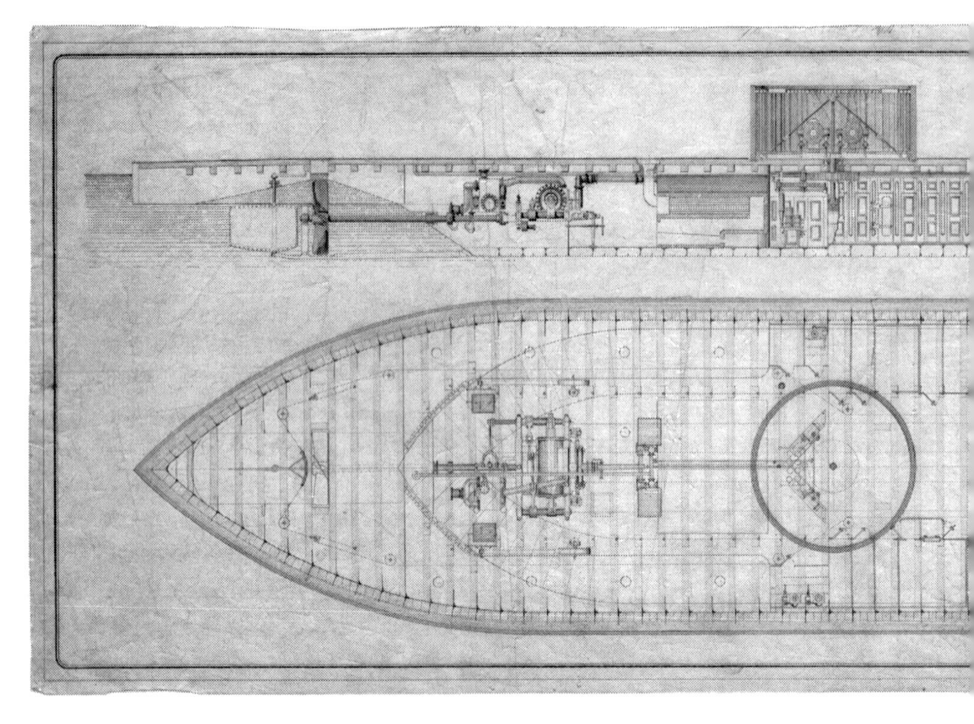

U.S. ironclad steamer *Monitor*, plan, John Ericsson, 1862. *Courtesy of The Mariners' Museum, Newport News, Virginia.*

of the Southern Rebellion that the batteries on the banks of their rivers will no longer present barriers to the entrance of the Union forces," many others were unsure of the ironclad's abilities.[9] Quartermaster Peter Truscott noted that "she was a little bit the strangest craft I had ever seen." Seaman David R. Ellis made perhaps the most telling remark about the *Monitor* as she readied to leave New York, commenting, "She had not been pronounced seaworthy, and no one could safely judge of her fighting qualities."[10]

Nevertheless, the Federals were anxious to bring the new ironclad into action. When the *Monitor* was launched, Assistant Navy Secretary Gustavus Vasa Fox wrote to Ericsson, "I congratulate you and trust she will be a success. Hurry her for sea, as the *Merrimack* is nearly ready at Norfolk and we wish to send her there."[11]

On the afternoon of March 6, 1862, the USS *Monitor* left New York under tow by the steam tug *Seth Low*, accompanied by the steamers *Sachem* and *Camtuck*. The *Monitor* encountered severe storms off the New Jersey coast and almost sank en route to Hampton Roads. Somehow, the little ironclad

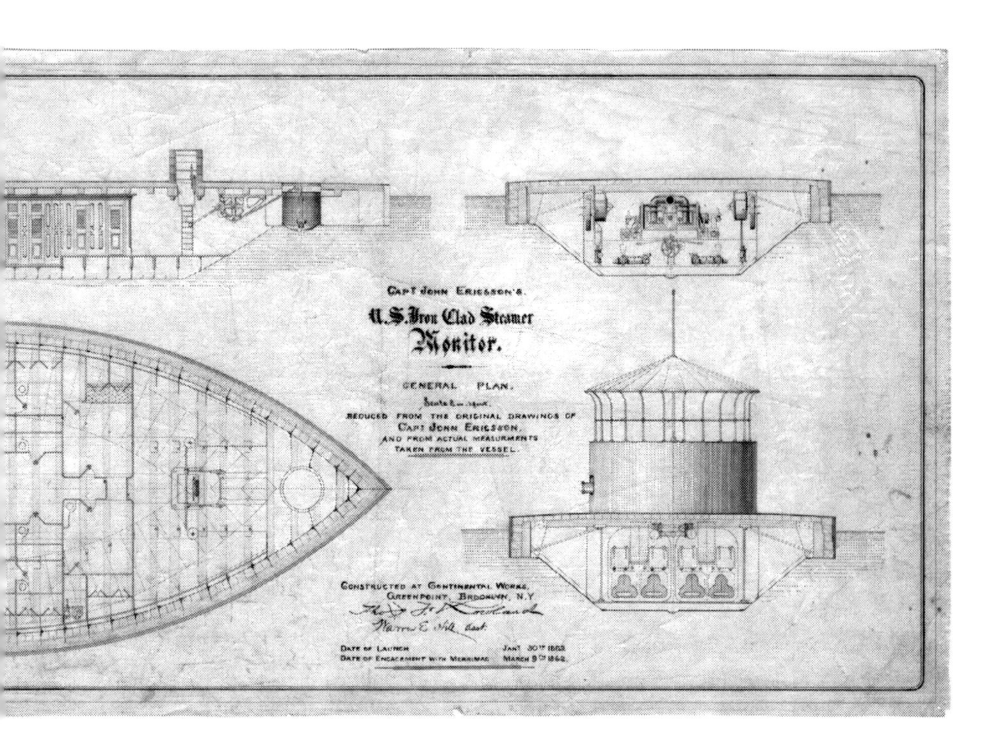

survived the angry sea. Lieutenant Samuel Dana Greene, the *Monitor*'s executive officer, later wrote about the stormy trip from New York: "I think I lived 10 good years."[12]

The *Monitor* entered the Chesapeake Bay late in the afternoon of March 8, 1862, and steamed toward Fort Monroe. As the ironclad neared Hampton Roads, the crew heard sounds of distant cannon fire. Acting Assistant Paymaster William F. Keeler recalled:

> *The shelling seemed to let up. All manner of ships, sail and steam, were running out of Hampton Roads, leaving like a covey of frightened quails and their lights danced over the water in all directions. A huge glow glimmered red and yellow in the Roads, fire leaping high in the air around what appeared to be burning masts and spars.*[13]

The *Monitor*'s crew was shocked by the destruction and chaos left behind by the *Virginia*. "Our hearts were so very full," Dana Greene wrote, "and we vowed vengeance on the *Merrimac*."[14]

Worden immediately reported to Captain John Marston of the USS *Roanoke*, acting commander of the Union naval forces in Hampton Roads.

USS Monitor *and USS* Minnesota, engraving, J.O. Davidson, circa 1880. *Courtesy of John Moran Quarstein.*

Marston had received a telegram from Gideon Welles instructing him to immediately send the *Monitor* up the Potomac to defend Washington against the Confederate ironclad. He recognized that the best defense for the nation's capital was in Hampton Roads. Thus, Marston ordered Worden to station near the *Minnesota* and protect her from the *Virginia*.

Van Brunt was not impressed by the *Monitor* and doubted that the ironclad could aid the *Minnesota* in any manner whatsoever. "The idea," William Keeler recalled, "of assistance or protection being offered to the huge thing by the little pygmy at her side seemed absolutely ridiculous."[15] Worden, however, knew that his ironclad was the only thing that could save the *Minnesota*. When the *Virginia* entered Hampton Roads on a course toward the *Minnesota*, many members of the *Monitor*'s crew were standing atop the ironclad's turret. "There was no mistaking her slanting, rakish outlines,"[16] remembered Peter Truscott. Worden ordered his men to their battle stations, stating, "Gentlemen, that is the *Merrimac*, you had better go below." As Worden passed through the turret to his station in the pilothouse, he noticed one of the XI-inch Dahlgrens being loaded. He encouraged the men by saying, "Send them that with our compliments, my lads." The *Monitor* then steamed away from the *Minnesota* to intercept the *Virginia*'s attack against the frigate. Lieutenant Worden advised Dana Greene, commander of the turret during the action, "not to fire till I give the word, to be cool and deliberate, to take sure aim and not waste a shot."[17] The *Monitor* opened fire on the Confederate ironclad at 8:45 a.m., and for the next four hours, the two ironclads pounded each other with shot and shell. "When we came out for the second day's fight, thinking we would clean up the *Minnesota*, *Roanoke*, and *St. Lawrence*," remembered Acting Assistant Engineer Elsberry V. White, "we were not entirely surprised to find the *Monitor*, for, while we did not know exactly what to expect, we knew some kind of craft had come in during the night. However we had no doubt that we could handle her easily." White, from his vantage point in the pilothouse where he relayed messages to the engineer room during the battle, recalled:

> *As we came in range of the* Minnesota, *a puff of smoke and the whistle of a shot over our heads let us know that she had no intention of striking her colors without a fight. Lieutenant Jones laid his course directly for her, thinking to bring his rifled guns to bear upon her quickly finishing the first part of the fight so that he could then engage the* St. Lawrence *and* Roanoke *in turn. But we were to find that before the day ended we would not be able to carry out our plans.*
>
> *At this time we noticed a volume of smoke coming up from the opposite side of the* Minnesota *and there emerged the queerest looking craft afloat. Through our glasses we could see she was ironclad, sharp at both ends and appeared to be almost awash. Mounted amidships was a turret with ports and, as we looked, the turret began to revolve until her forward gun bore directly on us and, run out, it resembled a cheese box on a raft.*
>
> *We didn't have long to wait before she fired. Her first shot fell a little short and sent up a geyser of water that fell on our top and rolled off. We then fired our forward rifle and scored a direct hit on her turret, but with no apparent effect. Her next shot was better and caught us amidships with a resounding wham, but while the old boat shuddered, there seemed to be no appreciable damage. By this time we were getting pretty close, and both crafts were firing as fast as the guns could be served. The men were stripped to the waist and were working like mad. Powder smoke filled the entire ship so that we could see but a short distance and its acrid fumes made breathing difficult.*[18]

The *Virginia* entered the battle totally unprepared to engage another ironclad. She had only explosive shells, hot shot and canister to use against wooden ships. The decision not to produce armor-penetrating bolts for the Brooke rifles now haunted Jones. Chief Engineer Ashton Ramsay wrote, "If we had known we were to meet her, we would have at least been supplied with solid shot for our rifled cannons."[19] E.A. Jack echoed this problem when he wrote, "Our only hope to penetrate the *Monitor*'s shield was in the rifled cannon, but as the only projectiles for those were percussion shells, there was barely a chance that we might penetrate our adversary's defense by a lucky shot."[20]

Jones decided to concentrate on the *Minnesota* and ordered the pilots to take the huge ironclad to within a half-mile of the frigate. "Knowing we had nothing but shells to fight her with," Hardin Littlepage lamented about the *Monitor*'s appearance, "we paid little attention to her as we did not care to develop the situation."[21] Inasmuch as Jones hoped to focus on the destruction of the *Minnesota*, the pilots refused to take the ironclad closer than within a mile of the *Minnesota*. Since the *Monitor* "came nearer we concluded to test our shell upon her, and the action was continued, sometimes at very close quarters," Hardin Littlepage later recorded.[22] Jones, recognizing that his shot and shell would have little impact on the ironclad, evolved a new strategy to try to ram or board the *Monitor*. Otherwise, he could only hope that a chance shot might disable the Union ironclad or that another opportunity might arise to strike against the *Minnesota*.

During the entire engagement, the *Monitor*, according to E.V. White, "was hit by 23 projectiles." White noted:

> *The balls from the* Merrimac, *especially those fired almost muzzle to muzzle, produced some results. Three cylindro-conical balls fired from the rifled guns made an indentation nearly four inches deep on the armor plating. Two of them made an equally deep indentation on the inside of the turret, and a man leaning against the inside walls at the place receiving the blow was thrown forward and wounded…The other shots which reached the* Monitor, *and were for the most part round, did not appear to me to have produced a very great effect, those especially which struck the sides perpendicularly; two, however, struck the side at the edge of the deck, lifting and tearing it, causing the iron plating to give way and breaking three of them. The others only produced insignificant effects.*[23]

The battle was primarily fought at a range of less than one hundred yards. Often the ships almost touched each other as each ironclad sought to gain an advantage. Worden hoped that by firing his heavy shot, 168-pound spherical

USS Monitor–*CSS* Virginia, oil on canvas, by Alexander Charles Stuart, circa 1890. *Courtesy of The Mariners' Museum, Newport News, Virginia.*

projectiles using 15 pounds of powder from his XI-inch Dahlgrens, it would loosen or break the *Virginia*'s iron plates.

The *Monitor*'s gun crews were handicapped by the U.S. Navy powder charge requirements. Tests would later prove that Dahlgrens could handle double charges (thirty pounds), and this load would have probably been sufficient to propel a shot through the *Virginia*'s casemate. "One shot struck directly over the outboard delivery," wrote E.A. Jack. "That was our weak spot. The shot broke the backers to the shield and sent a splinter into our engine room with about enough force to carry it halfway across the ship."[24] The shot "so bent in our iron plating that the massive oak timbers were cracked," noted E.V. White.[25] Fortunately for the *Virginia*, only a few iron plates were damaged during the engagement.

The Confederates were actually very surprised that the *Monitor*'s guns did not inflict greater damage on the *Virginia*'s shield or strike against the Confederate warship's waterline. "Not a single shot struck us at the water-line, where the ship was utterly unprotected," John Taylor Wood commented, "and where one would have been fatal. Or had the fire been concentrated on any one spot, the shield would have been pierced; or had larger charges been used, the result would have been the same. Most of her shot struck us obliquely." Wood added, "breaking the iron of both courses, but not injuring the wood backing. When struck at right angles, the backing would be broken, but not penetrated."[26] "Another shot or two from the *Monitor*, following up two or more that she placed between my two guns on the starboard side," reflected Hunter Davidson, "would have brought down the shield about our ears." The Confederates were totally unaware of the serious problems encountered by the *Monitor*'s crew managing the turret yet were relieved that the Union ironclad did not take advantage of their ironclad's unseaworthiness and other weaknesses. As William Norris noted:

> *Now, the enormous weight of her shield and battery, kept the* Virginia, *all the time, just hovering between floating and sinking. She was sluggish, sodden and entirely irresponsive to the breathing of the sea. In a very slight roughening of the water, a sailor could tell in a moment, by the feel of her, under his feet that it was a touch and go matter whether she staid up, or went down; a very few tons of water through the home made by two, or even one, well aimed shot from the splendid eleven-inch gun of the* Monitor, *and the* Virginia *would have gone to the bottom in five minutes. With such a gun, and at such a short range, it would be no great feat for the intelligent side-boy to plant his shot every time in the space covered by an ordinary straw-hat. The* Virginia *was so large a mark that almost every shot struck here somewhere; but they were scattered over the whole shield and on both sides, and were therefore harmless. To point her gun in our direction, and fire on the instant, without aim or motive, appeared to be the object. The turret revolving rapidly, the gun disappears only to repeat in five or six minutes the same hurried and necessarily aimless, unmeaning fire; not a shot appeared to have any "motif."*

"She fired," Norris concluded, "all told, during the fight forty-one shots (taking her time about one fire in six minutes), and any three of them properly aimed would have sunk us, and yet the nearest shot to the water-line was over four feet."[27]

The *Monitor*'s small size and quickness frustrated the Confederates, who tried to fire at the *Monitor*'s gun ports but found that the turret revolved too fast. Lieutenant Eggleston complained that "we never got sight of her guns except when they were about to fire into us."[28] Eggleston was later chided during the battle by Catesby Jones for not firing his gun at the Union ironclad. He replied to the *Virginia*'s acting commander, "It is quite a waste of ammunition to fire at her. Our powder is precious, sir, and I find I can do the *Monitor* as much damage by snapping my finger at her every five minutes."[29]

The *Monitor*'s turret truly amazed the Confederates, yet gun crews were dismayed as they "saw our shells burst into fragments against her turret."[30] In turn, the Federal shot continued to bounce off the sloped, iron sides of the *Virginia*. The scene aboard the *Virginia* was likened to a page from Dante's *Inferno*, as Chief Engineer Ashton Ramsay later wrote:

> *On our gun deck, all was bustle, smoke, grimy figures and stern commands, while down in the engine and boiler rooms the 16 furnaces were belching out fire and smoke and the firemen standing in front of them, like so many gladiators, tugged away with devil's claw and slice-bar, inducing by their exertions more and more and more intense combustion and heat. The noise of the crackling, roaring fires, escaping steam, and the loud and labored pulsations of the engines, together with the roar of battle above and the thud and vibrations of the huge masses of iron which were hurled against us, produced a scene and sound to be compared only with the poet's picture of the lower regions.*[31]

The two ironclads continued to circle each other, both ships striving to find a weak point in their opponent's armor. John Eggleston remembered that the "*Monitor* circled around and around us, receiving our fire as she went, and delivering her own."[32] The *Monitor* had, however, effectively blocked the *Virginia*'s path to the motionless *Minnesota*. Van Brunt noted how the Union ironclad, "much to my astonishment, laid herself right alongside of the *Merrimac*, and the contrast was of a pygmy to a giant."[33] After over two hours of combat, Worden ordered the *Monitor* out of action to replenish ammunition in the turret. Cannonballs were hoisted up from a storage bin below deck through a scuttle, which required the turret to be stationary. It was a slow and laborious process, during which time Worden actually went out on deck to inspect the *Monitor* for any damage.

Catesby Jones immediately took advantage of the lull and moved the *Virginia* toward the *Minnesota*. The Confederate ironclad, leaking at its bow due to the loss of its ram from the day before, now ran aground on the Middle Ground Shoal and was unable to defend itself. The *Virginia* was in serious danger. The *Monitor* approached, and according to Ashton Ramsay, the Union ironclad

> *began to sound every chink in our armor—everyone but that which was actually vulnerable, had she known it. The coal consumption of the two day's fight had lightened our prow until our unprotected submerged deck*

Battle between the Monitor *and* Merrimac, engraving, circa 1862. *Courtesy of John Moran Quarstein.*

> *was almost awash. The armor on our sides below the waterline had been extended but about three feet, owing to our hasty departure before the work was finished. Lightened as we were, these extended portions rendered us no longer an ironclad, and the* Monitor *might have pierced us between wind and water had she depressed her gun.*
>
> *Fearing that she might discover our vulnerable "heel of Achilles," we had to take all chances. We lashed down the safety valves, heaped quick-burning combustibles into the already raging fires, and brought the boilers to a pressure that would have been unsafe under ordinary circumstances.*
>
> *The propeller churned the mud and water furiously, but the ship did not stir. We piled on cotton waste, splits of wood, anything that would burn faster than coal. It seemed impossible that the boilers could stand the pressure we were crowding upon them.*[34]

"Everyone was watching and waiting," reflected Dinwiddie Phillips while on the gun deck, "with an impatience which may well be imagined to be relieved of the horrible night-mare of inactivity, from which we all suffered." Phillips remembered that then the crew "felt the old vessel give one Samsonian effort."[35] "Just as we were beginning to despair," Ramsay recalled, "there was a perceptible movement, and the *Merrimack* slowly dragged herself off the shoal by main strength."[36]

Lieutenant Catesby ap Roger Jones, CSN, photograph, circa 1870. *Courtesy of the U.S. Naval Historical Center, Washington, D.C.*

Somehow, Ramsay achieved a miracle and coaxed enough power out of the *Virginia*'s unreliable engines to drag the heavy ironclad off the shoal. Catesby Jones, frustrated by the ship's ineffectual fire against the Federal ironclad, decided to ram the *Monitor*. "It was a last resort," E.V. White wrote. "Seeing that our shots were ineffective, I was directed to convey to the engine room for orders for every man to be at his post."[37] It took over thirty minutes for Jones to maneuver his ironclad into ramming position. "The ship," John Taylor Wood later wrote, "was as unwieldy as Noah's Ark."[38] Finally, the *Virginia* began its half-mile run and steamed straight at the *Monitor*.

Worden braced his ironclad for the *Virginia*'s ramming attack when he saw the Confederate ironclad begin its lumbering approach. "Look out now," he advised William Keeler, "they're going to run us down, give them both guns."[39] The Federals actually feared that the *Virginia*'s ram might penetrate the *Monitor*'s thinly armored hull. "This was the critical moment," Keeler thought as he raced through the ironclad toward the turret, "one that I feared from the beginning of the fight—if she could so easily pierce the heavy oak beams of the *Cumberland*, she surely could go through the ½-inch iron plates of our lower hull."[40]

The more nimble Union vessel was able to veer away right before the Confederate ironclad struck. Thus, the *Monitor* was hit only with a glancing blow. The huge ironclad rocked the *Monitor* with a "heavy jar," William Keeler remembered, "nearly throwing us from our feet."[41] The Confederates, according to Dinwiddie Phillips, thought they had made the Union ironclad "reel beneath our terrible blow."[42] E.V. White believed that the *Virginia* "came near running her under the water; not that we struck her exactly at right angles, but with our starboard bow drove against her with a determination of sending her to the bottom."[43]

Jones's order to reverse the engines before striking the *Monitor* lessened the collision's impact. Since the *Virginia*'s ram, unbeknownst to Jones, was imbedded in the *Cumberland*, the ramming caused minimal damage to the *Monitor*. "Whether because our prow was gone or because we eased up too soon," noted John Eggleston, "we did not do her any apparent injury."[44] Jones may not have tried this tactic had he known that the *Virginia*'s ram was missing.

The only visual evidence of the ramming was that several wooden splinters from the *Virginia*'s hull were stuck on a bolthead on the *Monitor*'s deck and a minor indentation in the ironclad's iron plate. The *Virginia* suffered some damage from this tactical maneuver. The collision caused another leak in the Confederate vessel's hull. A temporary repair was made, but Ramsay needed to keep the pumps running to stem the flow of water. Ramsay noted, "I had the means to keep her free, even if she had a ten-inch shot in her hull, as there were two, large Worthington pumps and the bilge injections of the engines which could be set to work."[45] The greatest damage, however, occurred when the *Virginia* hit the Union ironclad and Dana Greene fired both Dahlgrens almost simultaneously at the Confederate ironclad. The shots struck the *Virginia*'s casemate just above the stern pivot gun port, forcing the shield in two to three inches. "All the crews of the after guns were knocked over by the concussion, and bled from the nose or ears. Another shot at the same place," John Taylor Wood noted, "would have penetrated."[46]

Jones also considered the feasibility of boarding the *Monitor*. A group of volunteers was organized to leap onto the *Monitor*'s deck, cover the pilothouse with a coat to blind the ship and toss specially prepared grenades into the turret and down the funnels. While alongside the *Monitor*, moments after the ramming attempt, "boarders were called away," remembered John Taylor Wood.[47] The *Monitor* quickly slipped past the *Virginia*, and the Confederates were unable to launch such a bold and desperate attack. Worden had anticipated such an attempt and had ordered Greene to double shot the Dahlgrens with canister to repel the boarders.

Many crewmen of the *Virginia* independently conceived the same tactic during the battle. Hardin Littlepage remembered

> *that a man at my gun by the name of Hunt jumped in the port and I ordered him to get back as it was useless to expose himself in that way. He said he wanted to get aboard of the bloody little iron tub, and that he would put his pea-jacket around the pilot-house so she could not tell which way she was going. It would not have been a bad idea, as we realized afterwards, provided he could have gotten back aboard of the* Virginia.[48]

Another plan was hatched by Hunter Davidson. According to Richard Curtis, "The quarter gunner had placed all along the side our ship loaded

Springfield rifles," and "Lieutenant Davidson, the commander of our division, said to me 'take one of those guns and shoot the first man that you see on board of that ship.'" Curtis and another gunner, Benjamin Sheriff, took up rifles and

> *took our positions at the bow port...both on our knees, but not in prayer. By this time we had struck the* Monitor *and I was looking right into the port of the* Monitor *for that man, Sheriff kept saying to me look out Curtis, look out Curtis, which I was doing with all my might; while looking out for that man I saw one of her guns coming slowly out of her ports and looking me squarely in the face, Sheriff and myself thought it was time to move, which we did quickly. Saw no man, fired no gun; the* Monitor *fired her gun and the ball came very near coming into our port.*[49]

The *Monitor*'s evasive action during the *Virginia*'s ramming attack enabled Jones to once again maneuver toward the *Minnesota*. Several shots were sent against the stranded frigate, starting a fire on the *Minnesota*. One shell struck the tug *Dragon*. The *Dragon*'s boiler burst, and the tug, which had been alongside the *Minnesota* to tow that vessel to safety, sunk.

Worden was once again able to steer his ship between the Confederate ironclad and the Union frigate. He now decided to ram the *Virginia*, seeking to strike the larger ironclad's vulnerable propeller and rudder. "Our rudder and propeller were wholly unprotected," William Norris noted, "and a slight blow from her stern would have disabled both and ended the fight."[50] Furthermore, the *Virginia* was riding high in the water, and Worden could see the propeller churning. The *Monitor*'s captain realized that an effective strike could disable the propeller or rudder, thereby leaving the Confederate ironclad adrift and subject to capture. The *Monitor* steamed toward the *Virginia*'s fantail but missed her target at the very last moment due to a malfunctioning steering system.

As the Union ship passed the stern of the *Virginia*, Lieutenant John Taylor Wood fired his 7-inch Brooke gun at the *Monitor*'s pilothouse. It was a direct hit. Wood's shell blew off one of the wrought-iron bars that formed the pilothouse just as Worden was peering out the observation slit. The explosion created "a flash of light and a cloud of smoke," which blinded Worden. Worden fell back from the damaged slit and exclaimed, "My eyes, I am blind."[51] Despite his blindness, Worden could sense the bright light and cool air now coming into the pilothouse and believed that the command center was destroyed. The *Monitor*'s commander, with an amazing presence

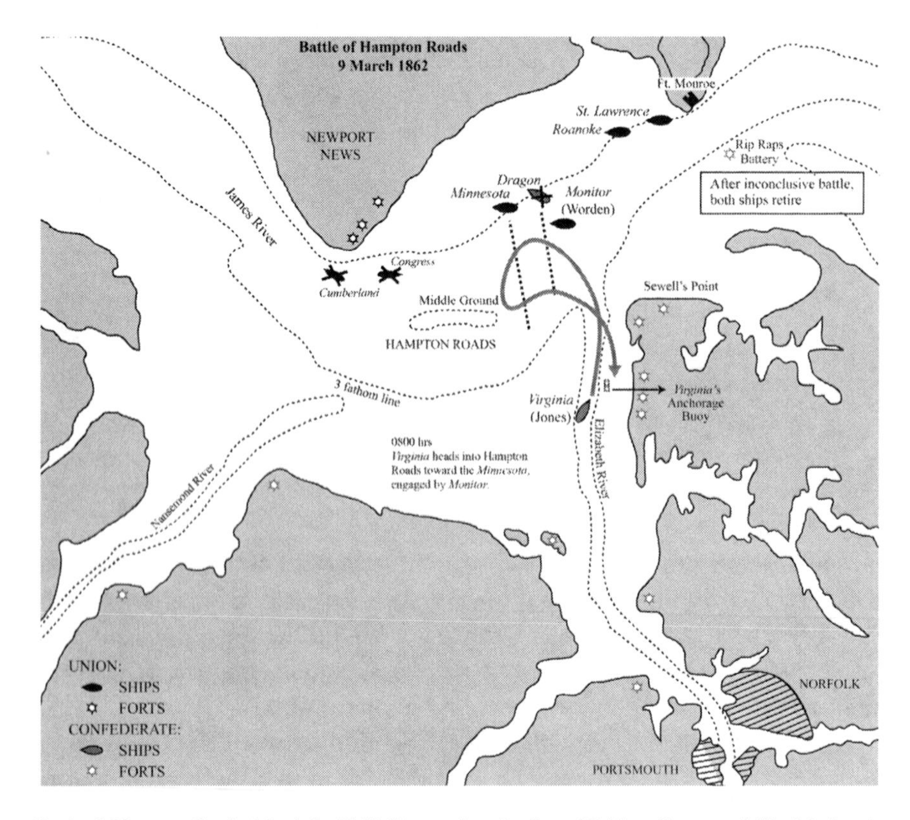

Battle of Hampton Roads, March 9, 1862, illustration, by Sara Kiddey. *Courtesy of The Mariners' Museum, Newport News, Virginia.*

of mind, ordered the helmsman to turn the ironclad to starboard, and the ship veered off onto a shoal.

Several officers helped Worden out of the pilothouse and sent for the ship's executive officer, Samuel Dana Greene. Greene finally made his way from the turret and found Worden, "a ghostly sight with his eyes closed and the blood apparently rushing from every pore in the upper part of his face."[52] Greene, along with ship surgeon Daniel C. Longue, escorted Worden to his cabin. Worden then told Greene that he was seriously wounded and directed the young lieutenant to take command of the ironclad. Dana Greene asked Worden what he should do. Worden replied, "Gentleman, I leave it with you, do what you think best. I cannot see, but do not mind me. Save the *Minnesota* if you can."[53]

The *Minnesota*'s captain immediately noticed that something was wrong with the *Monitor*. Van Brunt feared for the worst and began making preparations to scuttle his ship. Catesby Jones could also see that the *Monitor* was damaged. He believed that the "*Monitor* has given up the fight and ran into shoal water."[54] Many Confederates saw that the Union ironclad had "retreated out of our range into shoal water, where she was safe from our pursuit." John Eggleston proudly added that the *Monitor* "was not only whipped, but she stayed whipped."[55] The *Virginia*'s commander decided to resume the attack against the *Minnesota* "still hard aground, and out of range of the direct fire of our smooth-bore guns." "I was firing mine at ricochet—that is," noted Eggleston, "with the gun level—so that the shot would skip along the surface like pebbles the boys 'skell' along a pond. But, Davidson, with his rifle guns, just forward of me was actually 'plumping' the target with direct fire…At the time, we could not see that we were inflicting any serious damage on her."[56]

Despite the Confederates' desire to strike at the *Minnesota*, the pilots warned Catesby Jones that the tide was falling fast and the huge ironclad should move back into the Elizabeth River. Jones lamented to Ashton Ramsay that the "pilots will not place us nearer to the *Minnesota*, and we cannot run the risk of getting aground again."[57] Catesby Jones then walked along the gun deck conferring with his officers. He quickly reviewed with each of them the circumstances of the *Monitor*'s apparent retreat and the inability of the ship's deep draught to move any closer than a mile to the *Minnesota*. Jones summarized the situation by stating, "This ship is leaking from the loss of her prow; the men are exhausted by being so long at their guns; the tide is ebbing, so that we shall have to remain here all night unless we leave at once. I propose to return to Norfolk for repairs. What is your opinion?" A majority argued for the return to Norfolk. Eggleston remembered:

> *I answered: "If things are as you say, I agree with you." So did the other lieutenants, with the exception of Lieutenant John Taylor Wood. He stepped over from his gun to mine for a moment, and said, "I proposed to Jones to run down to Fortress Monroe and clean up the Yankee ships there or run them out to sea."*

Eggleston noted that such a move "was too hazardous to be considered by a cool-headed commander like Jones."[58] Jones, commenting that "had there been any sign of the *Monitor*'s willingness to renew the contest we would have remained to fight her," turned the *Virginia* away from the *Minnesota* and steamed toward Sewell's Point.[59]

The *Virginia* fired its last shell at the *Monitor* from John Taylor Wood's stern Brooke rifle at 12:10 p.m. Just as the *Virginia* steamed away, Dana Greene reentered the *Monitor*'s pilothouse and, upon seeing that the damage was minimal, ordered the Union ironclad back into action. It had taken almost thirty minutes for Greene to assume command, and the lull prompted the *Virginia* to break off action. Greene mistook the *Virginia*'s course toward Sewell's Point as a sign of defeat and proclaimed, "We had evidently finished the *Merrimac*." The young lieutenant, however, did not order the *Monitor* to follow up its believed victory by attacking the *Virginia* as it left Hampton Roads. He later explained:

> *I knew if another shot should strike our pilot house in the same place, our steering apparatus would be disabled and we should be at the mercy of the Batteries on Sewell's Point. The* Merrimac *was retreating towards the latter place. We had strict orders to act on the defensive and protect the* Minnesota. *We had evidently finished the* Merrimac *as far as the* Minnesota *was concerned, our pilot house was damaged and we had strict orders not to follow the* Merrimac *up; therefore, after the* Merrimac *had retreated, I went to the* Minnesota *and remained by her until she was afloat. General Wool and Secretary Fox both have complimented me very highly for acting as I did and said it was the strict military plan to follow. This is the reason we did not sink the* Merrimac, *and everyone on her says we acted exactly right.*[60]

Alban Stimers immediately telegraphed John Ericsson with the message, "You have saved this place to the nation by furnishing us with the means to whip an ironclad frigate that was, until our arrival, having it all her way with our most powerful vessels."[61]

When Worden was told that the *Monitor* was victorious and had saved the *Minnesota*, he merely said from his bed, "Then I can die happy." The injured officer was immediately transferred by his close friend Lieutenant Henry A. Wise and sent on an overnight boat to Baltimore for the best available medical attention. Worden was considered the true hero of the battle by the Northern press and was lionized throughout the Union. Eventually, he was taken to Washington to recover, and President Abraham Lincoln actually visited him in his sickbed at Lieutenant Wise's home. The wounded Worden said to Lincoln, "You do me great honor, Mr. President, and I am only sorry I can't see you." Lincoln replied, "You have done me more honor, sir, than I can ever do to you."[62]

Once the battle was over, small boats flocked to the *Monitor* to congratulate the crew on their splendid victory. Assistant Secretary of the Navy Gustavus Vasa Fox, who observed the entire engagement, went on board the Union ironclad and told the *Monitor*'s officers, "Well, gentlemen, you don't look as though you just went through one of the greatest naval conflicts on record." "No, sir," Greene replied, "we haven't done much fighting, merely drilling the men at the guns a little."[63]

"Cheer after cheer went up from the frigates and small craft for the glorious little *Monitor* and happy, indeed," Greene wrote in a letter to his parents, "we did all feel. I was the captain of the vessel that had saved Newport News, Hampton Roads, Fortress Monroe (as General Wool himself said) and perhaps your Northern ports."[64]

Likewise, the Confederates were jubilant over their victory. "As the *Merrimack* passed up the Elizabeth River," Ashton Ramsay remembered, "trailing the large ensign of the *Congress* under our own stars and bars, she was the recipient of a perfect ovation—cheering, waving of handkerchiefs, flags, hats, and caps; people hallooing themselves hoarse, hundreds of small boats lining our course up the river, and approaching as near as they dared the battle-scarred monster that had passed through such a fiery ordeal."[65] Lieutenant William Parker of the CSS *Beaufort* was amazed how the "whole city was alive with joy and excitement." "No conqueror of ancient Rome," Robert Foute believed, "ever enjoyed a prouder triumph than that which greeted us." "Will anyone ever forget," William Norris eloquently observed, "that Sunday afternoon ovation to our glory-covered sailors—a whole people wild with delight?"[66]

Midshipman Chester Foute, CSN, photograph, circa 1864. *Courtesy of Charles Peery.*

Immediately upon arriving at Gosport, John Taylor Wood reported to Franklin Buchanan lying in his hospital bed. Buchanan dictated a dispatch, and Wood sped off to Richmond via the Norfolk & Petersburg Railroad. Late that evening, he presented the blood-soaked flag of the *Congress* to President Jefferson Davis along with news of the battle. Wood later wrote:

> *The news of our victory was received everywhere in the South with the most enthusiastic rejoicing. Coming, as it did, after a number of disasters in the south and west, it was particularly grateful. Then again, under the circumstances, so little was expected from the navy that this success was entirely unlooked for. So, from one extreme to the other, the most extravagant anticipations were formed of what the ship could do. For instance: the blockade could be raised, Washington leveled to the ground, and New York lay under contribution, and so on.*

Wood advised Davis and Mallory that the two-day battle was the "most momentous naval conflict ever witnessed" and that the engagement had "revolutionized the navies of the world." He concluded his report with a realistic appraisal: "As to the future, I said that in the *Monitor* we may have met our equal, and that the results of another engagement would be very doubtful."[67]

The Confederate government was extremely ecstatic about the victory. Buchanan was immediately appointed an admiral, while Confederate Secretary of State Judah Benjamin sent propaganda messages to European nations, stating:

> *The success of our iron-clad steamer* Virginia *(late the* Merrimac*) in destroying three first-class frigates in her first battle, evinces our ability to break for ourselves the much-vaunted blockade, and ere the lapse of ninety days we hope to drive from our waters the whole blockading fleet. In less than that time we shall have several powerful iron-clad steamers of light draft that will be able to sweep from the coast all their so-called gunboats.*[68]

Wood returned to Gosport the next day, March 10, and attended a church service held on the *Virginia*'s deck "to return thanks to Almighty God for this great victory and deliverance." The "eloquent divine" Reverend J.H.D. Wingfield of Portsmouth's Trinity Episcopal Church sermonized that the "sunshine of a favoring Providence beams upon every countenance," and "the fierce weapons of our insolent invaders are broken."[69] The service, Wood reflected, was "impressive."

Despite all the fanfare, several of the *Virginia*'s officers and battle observers were not as confident about the Confederate victory. "We could not doubt that the *Merrimac* would, either by shot or by ramming," wrote John S. Wise, "make short work of the cheese-box; but as time wore on, we began to realize that the newcomer was a tough customer." Wise was surprised that the *Virginia* steamed back to Sewell's Point without destroying at least part of the Union fleet, lamenting that "dear old Buchanan would have never done it."[70]

Ashton Ramsay echoed these words when he wrote:

> *Had not Buchanan been incapacitated, the next day starting out with the first dawn of light, flushed with the victory of the day before, he would doubtless have forced the* Minnesota *to capitulate before the* Monitor *came on the ground, and then either run the* Monitor *down or forced her outside, into deep water, where she would not have had the advantage afforded by her light draft of water.*[71]

While Ramsay believed that "Jones was a clear-headed, cool and determined man, and his reasoning doubtless good," the young engineer was very dismayed by the *Virginia*'s withdrawal from Hampton Roads following the *Monitor*'s apparent retreat. "I remember feeling as if a wet blanket had been thrown over me," Ramsay later lamented, "for after the success we had already achieved, I felt as if we could accomplish almost anything, and it seemed to me as if we were abandoning the fruits of our victory, to leave the Roads without forcing the *Minnesota* to haul down her colors.[72] Catesby Jones was also disappointed and viewed the two-day battle as only a partial success. He stated to William Norris that the "destruction of those wooden vessels was a matter of course especially so, being at anchor, but in not capturing the ironclad, I feel as if we had done nothing." "Give me that vessel," Jones added, "and I will sink this one in twenty minutes."[73]

More than twenty-five thousand soldiers, sailors and civilians had witnessed the battle, and many, such as Raleigh Colston, believed that "the grandeur of the spectacle was so fascinating."[74] While John Bankhead Magruder commented, "The naval attack produced no effect upon the fort except to increase its garrison,"[75] other observers agreed that the battle was "one of the greatest Naval engagements." The *Norfolk Day Book* added that this "successful and terrible work will create a revolution in naval warfare, and henceforth iron will be the king of the seas."[76] The battle received immediate international attention. The *London Times* reported, "Whereas we had available for immediate purposes one hundred and forty-nine first-class war-ships, we have now two, these being the *Warrior* and her sister *Ironside*." Great Britain hurriedly committed itself to building a fleet of ironclad vessels. Lord Paget, secretary of the admiralty, noted in Parliament, "No more wooden ships would be built." Not only were wooden ships doomed

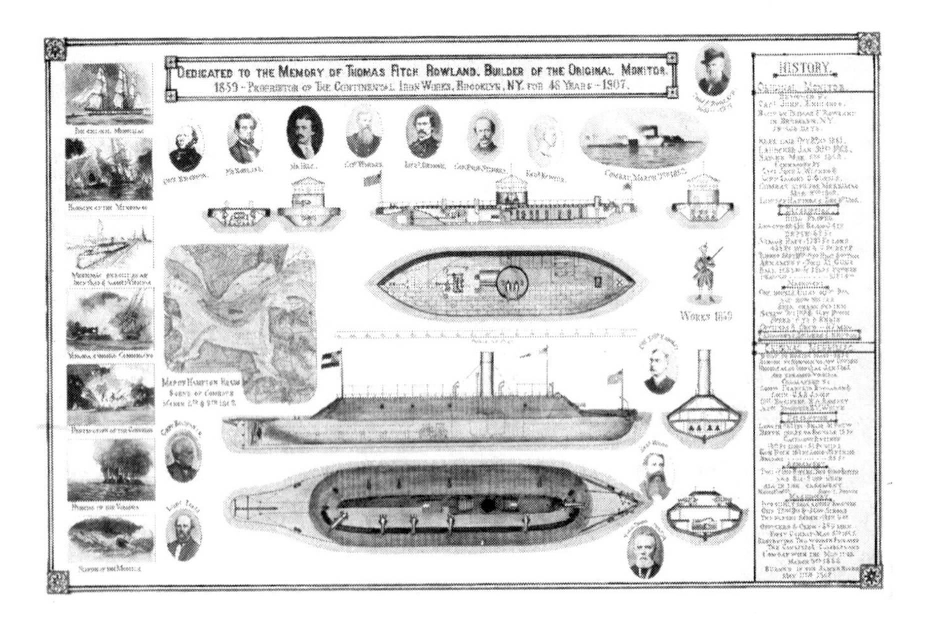

Battle of the Ironclads Panoramic, lithograph, Thomas Rowland, 1907. *Courtesy of The Mariners' Museum, Newport News, Virginia.*

by this horrific two-day naval battle in Hampton Roads, but the strength of coastal defense batteries was now questioned. "The recent conflict at Newport News shows conclusively that water batteries," reported the chief of the Confederate army's Engineer Bureau, Alfred Rives, "especially those near deep water, cannot injure materially properly constructed iron-clad vessels, nor contend with them."[77]

Neither ship had been seriously damaged during the four-hour battle. The Union ironclad fired forty-one shots at the *Virginia*; approximately twenty struck the Confederate ironclad. Other than Worden, who would eventually recover sight in his right eye but would be permanently disfigured, there were not any serious casualties on either vessel. Accordingly, in terms of combat losses, the *Virginia* can be considered the victor. The battle had little impact on the Confederate ironclad. Only a few men suffered concussions, several iron plates were cracked and she developed a new leak in her bow. In turn, the *Monitor*'s commander was severely wounded and unable to captain his vessel. The damage to the pilothouse was significant. Another shot striking the *Monitor*'s command center might have disabled the vessel.

Even though both sides claimed victory, the battle, according to Ashton Ramsay, "was a drawn one."[78] The *Monitor*, despite her damaged pilothouse and wounded commander, had indeed won a tactical victory. The Union ironclad had stopped the *Virginia* from destroying the *Minnesota*, relieving Federal fears of a broken blockade and an attack on Northern cities. The battle, however, had ever more immediate ramifications than serving as a major turning point in naval warfare. The undefeated *Virginia* controlled Hampton Roads. The Confederate ironclad blocked the entrance to the James River, thereby defending the water approach to Norfolk and Richmond. The mere existence of the *Virginia* would have a powerful influence on Major General George Brinton McClellan's strategic initiative to capture the Confederate capital in Richmond by way of the Virginia Peninsula. Brigadier General John Gross Barnard, chief engineer of the Army of the Potomac, lamented that the "*Merrimac*...proved so disastrous to our subsequent operations."[79]

9
"Mistress of the Roads"

The *Virginia*'s strategic victory closed the James River to the Federals just as General George McClellan decided to march toward Richmond by way of the Virginia Peninsula. McClellan's original plan entailed the use of both the James and York Rivers. Gunboats could guard his flanks, while steamers could transport supplies as McClellan's troops moved up the Peninsula. The entire concept was now in jeopardy because of the Confederate ironclad. McClellan wired Assistant Secretary of the Navy Gustavus Fox at Fort Monroe: "Can I rely on the *Monitor* to keep the *Merrimack* in check, so that I can make Fort Monroe a base of operations?" "The *Monitor* is more than a match for the *Merrimack*, but she might be disabled in the next encounter," Fox replied. "The *Monitor* may, and I think will, destroy the *Merrimack* in the next fight; but this is hope, not certainty."[1] Others were not quite as sure of the *Monitor*'s ability to defeat the Confederate ironclad. "If the *Merrimac* were successful," John Dahlgren wrote, "no one could anticipate the consequences to our side."[2] Montgomery C. Meigs, the U.S. Army's quartermaster general, agreed with Dahlgren and wrote the Union naval officer that "I have seen nothing yet to satisfy me that in the next engagement the *Monitor* will not be sunk."[3]

Major General John Ellis Wool, commander of the Union Department of Virginia, concurred that the *Monitor* might not survive another encounter with the *Virginia*. He had already received orders on March 9 from McClellan to hold Fort Monroe "at all hazard" against the Confederate ironclad. Wool immediately began to block the *Virginia*'s access into the Chesapeake Bay by mounting a 15-inch Rodman gun, nicknamed the "Lincoln gun," next to the 12-inch "Union gun" on the beach near the Old Point Comfort Lighthouse. The Union general was sure that the *Monitor*, positioned in the channel near the Rip Raps and supported by these two powerful guns, would effectively close the mouth of Hampton Roads to any sortie by the *Virginia*.

Regardless of Wool's preparations, the *Virginia*, according to Brigadier General John G. Barnard, "paralyzes the movement of this army by whatever route is adopted."[4] McClellan held a council of war on March 12, 1862, with his corps commanders, and they agreed to proceed with the campaign only using the York River. The U.S. Navy was expected to neutralize the *Virginia* and help destroy the Confederate batteries on the York River. Yet the *Virginia* mesmerized Flag Officer Louis Goldsborough and influenced the U.S. Navy's ability to support McClellan's campaign. "The James River was declared by the naval authorities closed to the operations of their vessels by the combined influence of the enemy's batteries on its banks and the Confederate steamers *Merrimac*, *Yorktown*, *Jamestown*, and *Teaser*," McClellan later wrote. He added:

> *Flag Officer Goldsborough...regarded it (and no doubt justly) as his highest and most imperative duty to watch and neutralize the* Merrimac, *and as he designed using his most powerful vessels in a contest with her, he did not feel able to attack the water batteries at Yorktown and Gloucester. All this was contrary to what had been previously stated to me and materially affected my plans. At no time during the operations against Yorktown was the Navy prepared to lend us any material assistance in its reduction until after our land batteries partially silenced the works.*[5]

Even though Secretary Welles believed the *Monitor* to be the superior vessel, he thought that the Union ironclad "might easily be put out of action in her next engagement and that it was unwise to place too great dependence on her." Goldsborough told Gustavus Fox that the *Monitor* "is scarcely enough for the *Merrimac*" and prayed, "Would to God we had another ironclad vessel on hand."[6] Goldsborough feared that the Confederate ironclad would soon again strike the Union fleet. Until the other Union ironclads under construction arrived, Goldsborough was content to remain on the defensive.

The *Monitor* was recognized as the only real defense against the Confederate ironclad. William Keeler noted, "We shall remain here as guardians of Fortress Monroe and the small amount of shipping which will remain in harbour, have been ordered off in apprehension of the reappearance of the *Merrimac*. We shall remain here to meet her. We are very willing and anxious for another interview."[7] The *Monitor*'s crew might have been ready to reengage the *Virginia*; the Federal command was not. Gideon Welles sent a telegram on March 10, stating, "It is directed by the President that the *Monitor* be not too much exposed and that in no event shall any attempt be made to proceed with her unattended to Norfolk."[8]

Meanwhile, the *Monitor* went through a change of command. Dana Greene, who was considered too young and inexperienced to succeed the wounded Worden as commander of the *Monitor*, was relieved by Lieutenant Thomas O. Selfridge, formerly of the USS *Cumberland*, the day after the battle. On March 12, Selfridge was replaced by Lieutenant William N. Jeffers. The thirty-eight-year-old Jeffers, an 1846 Annapolis graduate and ordnance expert, was not popular with the crew.

The *Monitor* also received some immediate alterations to prepare her for the next encounter with the *Virginia*. The pilothouse was considered one of the *Monitor*'s greatest weaknesses, and Chief Engineer Alban Stimers placed a shell of solid oak covered with three inches of wrought iron, laid in three layers, around the structure. The sides were reconfigured from perpendicular to a slope of thirty degrees to deflect shot. Worden advised from his recovery bed that the Union ironclad was susceptible to enemy boarding. Accordingly, small arms, muskets, grenades and cutlasses were stored in the turret to help repel any boarders.

Stephen Mallory's faith in the Confederate ironclad was vindicated by the *Virginia*'s actions on March 8 and 9, 1862. The Confederate secretary of the navy believed that the *Virginia* had won "the most remarkable victory which naval annals record" and dreamed that the ironclad would soon steam to New York, where "she could shell and burn the city and shipping. Such an event would eclipse all glories of the combat of the sea…and would strike a blow from which the enemy could never recover."[9] The *Virginia*'s officers, however, were not as ecstatic about their ironclad's performance on March 9. John Taylor Wood advised Mallory that in "the *Monitor* we had met our equal."[10] Buchanan wrote to Mallory on March 19 in response to the secretary's letter. "The *Virginia* is yet an experiment, and by no means invulnerable as has already been proven in her conflict on the 8 and 9," the newly promoted admiral wrote. "The *Virginia* may probably succeed in passing Old Point Comfort and the Rip Raps," Buchanan reflected. "She [is] then to be tested in a seaway…Should she encounter a gale, or a very heavy swell, I think it more than probable she would founder." In conclusion,

Buchanan counseled, "The *Virginia* [is] the most important protection to the safety of Norfolk."[11]

Since the *Virginia* was actually unfinished when she attacked the Union fleet on March 8 and had suffered significant damage during two days of combat, the ironclad was immediately placed in dry dock upon her return to Gosport Navy Yard on March 9 for modification and repair. The *Virginia*'s damage report summarized many of her required repairs:

> *The stern is twisted and the ship leaks. We lost our prow, starboard anchor and all the boats. The armor is somewhat damaged, the steam-pipes and smokestack riddled, the muzzles of two guns shot away. The colors were hoisted to the smokestack and were shot away several times...The only damage done by the* Monitor *was to the armor, the effect of shot striking obliquely on the shield, breaking the iron and sometimes displacing several feet of the outside courses and the wooden backing inside.*[12]

One shipyard worker who deserted to the Federals reported about the *Virginia*:

> *I saw her when she returned, after the first fight with the* Monitor, *and the injury done her as follows: One gun broken shot off near the trunnion, and another broken off obliquely, about eighteen inches from the muzzle. Her stern was mashed so that the wood could be strung out like a ball of thread; and they had to squeeze a whole bale of oakum into it to stop it from leaking—the planking being sprung off and gaping wide. Quite a number of the* Monitor*'s shots had plowed up the roofing so that you could lay a large watermelon in the spot where the shot had struck.*[13]

Once again, the problem of acquiring iron plate plagued the refitting project. Battle damage to the shield necessitated the replacement of several two-inch plates. Even greater attention was given to one of the *Virginia*'s primary flaws: a lack of armor below the eaves of the casemate to protect the ironclad's knuckle. Catesby Jones and Franklin Buchanan believed that "iron plating on the hull seems indispensable."[14] Time and available resources limited this effort. Eventually, only a band of 440 two-inch iron plates extending 3.5 feet below the eaves, covering 160 feet on both sides, was installed.

Mallory had anticipated the need for more iron plate for the *Virginia* as soon as he learned of the Confederate ironclad's strike against the Union

Virginia *in Dry Dock*, print, circa 1907. This view details the post-battle repairs to the *Virginia*. *Courtesy of John Moran Quarstein.*

fleet. Nevertheless, delays in producing and shipping iron to Gosport caused a controversy. The secretary of the navy advised R.M.T. Hunter, chairman of the Senate Naval Committee, "Within an hour after the telegraph announced in Richmond on the 8th of March, that the *Virginia* was engaged with the enemy's ships, I gave the order to roll and prepare iron plates to repair her assumed and anticipated damages." Mallory added on March 10, 1862, "[I] telegraphed Flag Officer Forrest that the iron was being rolled and that eight tons were on the way to Norfolk."[15] Yet as late as April 4, 1862, Mallory still struggled to acquire iron for the *Virginia*. He wrote to Sidney Smith Lee:

> *Your telegraph of this date, April 4, 1862, that you are out of iron for the* Virginia*—expected a load down last night; not arrived—want it immediately—, has just been received.*
>
> *I am surprised at this information. It was the Constructor's business to advise the department of the iron, and the quantity she required, and sufficiently in advance to guard against a delay, and the delay of the ship for a single hour, for the want of iron, is a serious responsibility.*[16]

The *Virginia*'s armament also required attention. John Mercer Brooke recognized that the *Virginia*'s ordnance required improved armor-piercing shot, and by March 10 he was already at work producing wrought iron, steel-tipped bolts for the 7- and 6.4-inch rifles. Meanwhile, Porter replaced the two damaged guns. A new ram was also required. Brooke designed a new twelve-foot-long, steel-pointed iron ram that extended from the ironclad's bow fourteen feet. A disagreement immediately arose between Brooke and Porter pertaining to the ram's design. "I suggested to the Secretary that the iron prow of the *Merrimac*," Brooke wrote in his diary, "should be square on the cutting edge to insure taking the wood of an enemy's side if struck glaring and not an acute angle; Constructor Porter proposed and was making the edge sharp. The Secretary telegraphed to make it as I suggested, Captain Lynch and Fairfax approved the change."[17]

The Brooke-Porter feud surfaced again when the *Virginia* entered dry dock on the afternoon of March 9. Stephen Mallory served as a mediator of sorts concerning design issues; however, the problems between Lieutenant John Mercer Brooke and constructor John L. Porter quickly exploded onto the pages of Southern newspapers. Porter was incensed over Brooke's continued and constant modifications to his work and was angered that Brooke received the majority of the credit for designing the *Virginia*. Most of the Confederate naval leadership recognized that the ingenuity and strength of the *Virginia* was found in Brooke's "novel plan of submerging the ends of the ship and the eaves of the casemate…[it] is the peculiar and distinctive feature of the *Virginia*." Lieutenant Robert Dabney Minor wrote to Brooke from his hospital bed on March 11:

> *You richly deserve the gratitude and thanks of the Confederacy for the plan of the now celebrated* Virginia, *and I only wish that you could have been with us to have witnessed the successful operations of this new engine of naval warfare, fostered by your care and watched over by your inventive mind.*
>
> *There will doubtless be an attempt made to transfer the great credit of planning the* Virginia *to other hands than your own. So look out for them, for to you it belongs, and the Secretary should say so in communicating his report of the victory to Congress.*[18]

Porter carried on a newspaper campaign alluding that Brooke had done little to design the ironclad. Porter believed that the credit for the *Virginia* should be given to him. The issue was never satisfactorily settled. The dispute soured the relationship between the men and had some negative impact on the refitting process.

Mallory now faced another difficult and potentially unpopular decision. The *Virginia* needed a new commander to replace the wounded Franklin Buchanan. Buchanan's wound was very painful, and it was obvious to all that he would heal slowly. Consequently, many of the *Virginia*'s officers believed that Catesby Jones should assume command of their ironclad. "By no means must any captain or commodore or even flag-officer be put over Jones," Robert Minor wrote to John Mercer Brooke. "In old Buck's sickness from his wound Jones must command the ship."[19] John Taylor Wood informed Buchanan

> *that he would be promoted to be admiral, and that, owing to his wound, he would be retired from the command of the* Virginia. *Lieutenant Jones should have been promoted, and should have succeeded him. He had fitted out the ship and armed her, and had commanded during the second day's fight. However, the department thought otherwise, and selected Commodore Josiah Tattnall; except Lieutenant Jones he was the best man.*[20]

Stephen Mallory appointed the sixty-seven-year-old Georgian Flag Officer Josiah Tattnall on March 21 to assume command of all naval forces in Virginia. Tattnall, whose father had been governor of Georgia, joined the U.S. Navy in 1812 and detailed to the USF *Constitution*. The young Tattnall fought his first engagement at Craney Island on June 22, 1813. When the British barge attack was repulsed, Tattnall waded out onto the mud flats with several volunteers and captured Admiral Sir John Borlase Warren's barge, the *Centipede*. He continued to serve with distinction throughout his career, fighting pirates in the West Indies, as well as bravely performing his duty during the Algerian and Mexican Wars. While commander of the East India Squadron, he supported British warships in their attack on Chinese forts defending the Pei-ho River in violation of U.S. neutrality. Tattnall explained his actions, stating, "Blood is thicker than water." Tattnall escorted the first Japanese ambassador to the United States. Accordingly, President James Buchanan made Tattnall his guest of honor at the White House state dinner welcoming the Japanese. Josiah Tattnall was known as the "beau ideal of a naval officer." Many fellow officers believed that Tattnall "possessed all the traits which are found in heroic characters, and with suitable opportunities, would have set his name among the great naval worthies who are historic."[21] Almost six feet tall with long arms and a protruding lower lip, Tattnall was feared in his younger days as a cutlass expert. Tattnall reluctantly resigned his U.S. Navy commission when Georgia left the Union, and he served as the commander of the Savannah Squadron until assigned to the CSS *Virginia*. His directive from Mallory was to make the Confederate ironclad "as destructive and formidable to the enemy as possible." Mallory instructed Tattnall, "Do not hesitate or wait for orders, but strike when, how, and where your judgment may dictate."[22]

Flag Officer Josiah Tattnall, CSN, photograph, circa 1855. *Courtesy of the Library of Congress.*

Tattnall took command on March 29, 1862, while the ironclad was still in dry dock. "I have had an experimental trial of wrought iron shot taken upon iron plates here with satisfactory results," Mallory informed the *Virginia*'s new commander, "and I have had some made and sent on board the *Virginia* and others are being made here and at Norfolk. These trials leave little doubt that the shot from your 7-inch guns bow and stern will penetrate the *Monitor*'s shield. But I am convinced that the *Virginia*'s new steel prow will be her most formidable weapon of defense, as well as against the *Monitor* and wooden vessels." "I suggest," Mallory also wrote, "that you converse with Buchanan as to the power and character of the ship. The heavy iron covering designed to shield her beneath the eaves of her casemate will be put on upon the first opportunity. The enemy it seems believes her to be disabled—and to be undergoing extensive repairs, and hence we may be able to surprise him, and strike him a heavy blow in the Roads."[23] Magruder was sure that once the *Virginia* was repaired and fully operational, "she would be the mistress of the roads."[24]

Repairs to the *Virginia* were still delayed by the Brooke-Porter disagreements over design modifications and lack of iron plate. Mallory fumed over the delays. Furthermore, the Confederacy had begun the construction of a second ironclad at Gosport. The CSS *Richmond* was 171 feet in length with an 11-foot draft. She would be armed with six guns, including four

7-inch Brooke rifles, and was designed as a counter to the *Monitor*. Yard commandant French Forrest argued that instead of building a new ironclad, the Confederacy should transform the sloop of war *Germantown*, raised following Gosport's capture, into an ironclad. Mallory disapproved. He wanted the work on the *Richmond* and *Virginia* immediately completed. The Confederacy needed new ironclads capable of operating in shoal-filled Southern harbors. "One such vessel," Mallory noted about the *Richmond* project, "would be worth a fleet of wooden vessels to us."[25]

Forrest's delays did not endear him to the Confederate secretary of the navy. Mallory wrote, "The work of getting out the *Virginia* and the other iron-plated gunboats in course of construction at Norfolk, ready for sea, at the earliest possible moment is the most important duty, and yet this department is ignorant of any progress made upon any of the vessels, especially the *Virginia* since it went into dock. She must be prepared for sea as soon as possible."[26] Mallory believed that Forrest had expended all of his energy preparing the *Virginia* for her first encounter with the Federal fleet and could no longer organize the repair work in a timely manner. French Forrest was unable to adequately answer the secretary's March 20 inquiry asking the yard commandant if the *Virginia* was ready for sea and, "If not, when will she be?"[27] Consequently, Captain Sidney Smith Lee, brother of General Robert E. Lee, was selected on March 24, 1862, as the new commandant of the Gosport Navy Yard.

Captain Sidney Smith Lee, CSN, photograph, 1862. *Courtesy of the Library of Congress.*

Lee was born at Stratford Plantation, Westmoreland County, Virginia, on September 3, 1802. He was appointed midshipman on December 30, 1820, and on May 17, 1828 was promoted to lieutenant. Lee fought during the siege of Vera Cruz and was promoted to commander on June 4, 1850. His next major assignment was as captain of Mathew C. Perry's flagship, the side-wheeler steamer USS *Susquehanna*, during Perry's 1853 expedition to Japan. After service as commandant of the U.S. Naval Academy and the Philadelphia Navy Yard, he resigned his commission on April 17, 1861, and was dismissed from the U.S. Navy on April 22. Captain Sidney Smith

Lee was made executive officer of the Gosport Navy Yard shortly thereafter. Lee was ordered to have the *Virginia* "ready to move at any moment" and "spare no expense" with the ironclad's refitting.[28] Workers labored around the clock on both ironclad projects, and Lee even advertised in North Carolina for additional workmen to help finish the *Richmond*.

The *Virginia*'s complement of officers also underwent some important changes following her foray into Hampton Roads. Three new officers were assigned to the ironclad after the March 8 engagement. Lieutenant John Hazelhurst Ingraham, who had previously served on the commerce raider CSS *Nashville*, was detailed to the *Virginia* on March 9, 1862, but did not participate in the engagement with the *Monitor*. Ingraham would be reassigned to the Charleston Station before the ironclad was ready for action in April. Franklin Dornin was appointed as an acting midshipman on March 8, 1862, and was transferred from the CSS *Jamestown* to the *Virginia* on March 12. Portsmouth-native First Assistant Engineer Charles Schroeder was assigned to the ironclad on March 11, 1862.

Several officers were transferred from the Savannah Squadron to continue serving under Flag Officer Josiah Tattnall. Paulding Tattnall continued in his position as the flag officer's secretary. Midshipman Barron Carter, aide to Tattnall during the Battle of Port Royal Sound, assumed this same position on the CSS *Virginia*. Since Robert Dabney Minor, Buchanan's flag lieutenant, was seriously wounded during the action with the USS *Congress*, Tattnall replaced this fine officer with Lieutenant John Pembroke Jones. Jones, a native of the Hampton Roads region, had previously served in the U.S. Navy from 1841 until resigning his commission on April 29, 1861. John Pembroke Jones was first assigned to command the Barretts Point Battery on the Elizabeth River until detailed to the Savannah Squadron. He served under Tattnall during the November 7, 1861 Battle of Port Royal Sound and as commander of the CSS *Resolute*. An outstanding and experienced officer, Jones was selected as Tattnall's flag lieutenant aboard the CSS *Virginia*. Captain Julius Ernest Meiere, CSMC, was the final transfer from Savannah. Meiere, a son-in-law of Franklin Buchanan, was assigned to the ironclad's marine guard.

The crew also underwent several changes while the ship was refitted and repaired in dry dock. All those crew members who had fought aboard the ironclad during the Battle of Hampton Roads received a fifty-dollar bounty for their distinguished service. Several of these men, like ordinary seaman Charles W. Harrison, were transferred to other warships, while others, such as Benjamin Slade, were dismissed from naval service. Replacements were needed for these men, as well as the incapacitated casualties suffered during the March 8 engagement. New seamen detailed to the ironclad before she was ready to be placed back into service included William G. Walker and John Volentine.

The *Virginia* was unfinished when she went on her March 8, 1862 "shakedown cruise." Her quarters were incomplete and were simple, temporary canvas screens. These were removed when the ship went into battle. Accordingly, during the refit, wooden partitions were installed for quarters and storage rooms. The warship's main or gun deck was located within the iron shield and was the only deck located above the waterline. The gun deck was also used as a berth deck for men assigned to gun crews. Sailors strung their hammocks between the guns and then stowed them away when not in use. Below the gun deck was the berth deck. It was mezzanine and fitted around the engine system. It contained the captain's quarters, officers' quarters, wardroom, crew berthing space, galley and sickbay. Below this was the orlop deck where the engine area, storerooms, magazine, shellroom and spiritroom were located.

As with all other Confederate ironclads, life aboard the *Virginia* was virtually intolerable. Ventilation was only available to the gun deck via the iron-grating top and through the gun ports. The berth and orlop decks received basically no outside air. The ironclad continuously leaked, and kerosene lanterns provided lighting. The spaces beneath the waterline were dark, damp and dank. Humid conditions were compounded by the heat produced by the gallery and engine system. Temperatures reaching over 130 degrees were witnessed. These abhorable conditions resulted in much sickness among the men of the *Virginia*.

The Confederacy was desperate to get the *Virginia* ready for active service. Norfolk and Richmond were under immediate threat from the Union army by early April. Major General Benjamin Huger was extremely worried about Major General Ambrose E. Burnside's progress toward the canals linking Norfolk with the North Carolina sounds. Huger also feared that the USS *Monitor* might strike against Norfolk while the *Virginia* was still in dry dock and suggested that the entrance to the Elizabeth River be blocked with obstructions. The Confederate Department of Norfolk commander received a swift reply from Richmond. "None of us are of the opinion that it would be proper to lose the vast advantages resulting from the enemy's fright at the bare idea of the *Virginia* reappearing among the wooden ship," Judah Benjamin advised Huger. "The fact of her presence guarantees you against any attempt to blockade the river." Benjamin added a cautionary note to "keep the necessary means of closing the Elizabeth River at hand for use at a moment's warning in case the *Monitor* should attempt an entrance."[29] While the CSRS *Confederate States* was prepared for rapid scuttling and the sand-filled battery CSS *Germantown* was strengthened as a gatekeeper, the B. & J. Baker Wrecking Company raised two other old hulks, the *Delaware* and *Columbus*, for use as obstructions. Unbeknownst to the Confederates, the U.S. Navy refused to let the *Monitor* strike up the Elizabeth River. "If they would only let us go up the river and get the rat in his hole," William Keeler wrote, "it would suit us exactly, much better than doing blockading duty in a diving bell."[30]

An even greater threat was now assembling at Fort Monroe: Major General George B. McClellan's huge Army of the Potomac. McClellan began moving his command, totaling 121,500 men, to the Peninsula on March 17, 1862. Magruder noted that McClellan was "straining every nerve to put a large force on the Peninsula before the *Virginia* comes out, either to operate on James River, York River, or both, whilst his troops march up."[31] McClellan, however, had no intention of using the James River and had decided to focus his campaign on the York River. McClellan wrote:

> *Important events were occurring which materially modified the designs for the subsequent campaign. The appearance of the* Merrimac *off Old Point Comfort, and the encounter with the United States Squadron on the 8th of March, threatened serious derangement of the plan for the Peninsula movement. But the engagement between the* Monitor *and* Merrimac *on the 9th of March demonstrated so satisfactory the power of the former, and the other naval preparations were so extensive and formidable, that the security of Fortress Monroe as a base of operations was placed beyond a doubt and although the James River was closed to us, the York River with its tributaries was still open as a line of water communication with the Fortress. The general plan, therefore, remained undisturbed, although less promising in its details than when the James River was in our control.*[32]

McClellan's plans were now set on the capture of Yorktown and, thence, a rapid move toward Richmond. The Union general believed that to bypass or conquer Yorktown, his army would be able to reach West Point at the confluence of the Mattaponi and Pamunkey Rivers and force the Confederates to abandon Hampton Roads and the lower James River.

Magruder looked in amazement as 389 vessels began delivering McClellan's troops, 14,592 animals, 1,224 vehicles, 44 artillery batteries,

114 seige guns "and the enormous quantity of equipage...required for an army of such magnitude."[33]

Magruder was anxious to stop the flow of Union troops and equipment to the Peninsula and wrote to General Lee:

> *It seems to me, therefore, that the* Virginia, *if she cannot get at the* Monitor*—a conflict which it will be the interest of the country to prevent—ought to so station herself outside Fort Monroe as to intercept all re-enforcements of troops and to cut off further supplies. This course, if it can be pursued at once, might prevent the advance up by land, and would also prevent the crossing of troops in large numbers on the lower James River, as far up at least as the* Virginia *could go, since, if she could pass Fort Monroe once, she could return again to the Roads, if an attempt were made to cross troops in large numbers. By taking such a position the* Virginia *would also prevent an expedition of magnitude either up York or Rappahannock Rivers.*[34] *The commander of the small Army of the Peninsula, who had previously discounted the* Virginia's *power, now sought to cooperate with the Confederate Navy to strike against the growing enemy host. Since General Joseph Eggleston Johnston had recently withdrawn his army from Manassas to the Rappahannock River, Magruder suggested to General Lee that twenty thousand men be sent to the Peninsula. "We could crush the enemy," Magruder believed, "and perhaps with the assistance of the Virginia take Fort Monroe...and vanquish or at least repel Burnside."*[35]

Yet the *Virginia* was still not out of dry dock. Magruder, fearing that McClellan would soon begin his offensive, notified the *Virginia*'s commander:

> *The enemy's transports are ready to pass his large army across lower James River, or to ascend it. In either case, the consequences may be disastrous, but if he can be prevented we have the advantage. The whole depends upon the* Virginia. *From all indications on this side, the enemy may be expected to move at any moment. He will depend upon his transports whether to cross or to ascend. If they can be destroyed, he is stopped or forced to march up the peninsula; in the latter case we can concentrate and prevent him, and thus save Norfolk and Richmond. If, on the other hand, his transports elude the* Virginia, *either by crossing over the James River at night or by ascending the river by the swash channel, or getting up the main channel so high that the* Virginia, *from her great draft of water cannot reach them, his object*

> *will be accomplished and one or both cities may fall. Should, therefore, the* Virginia *be able to take her position off Sewell's Point in the day and every night off Newport News, the fleet of transports could not pass up or across James River, and Norfolk could defend herself from other attacks.*

"If the *Monitor* does not attack you, so much the better. We have a country to save and no time for individual duels," Magruder further entreated Tattnall. "If she attacks you, I think you will destroy her, but the main object should be to destroy the enemy's transports. Should the enemy's transports once get up James River, I am sure Norfolk will fall. The Peninsula," Magruder concluded, "as a consequence, would have to be evacuated and if you got to sea with the *Virginia* you would not have a port in which you could lay without being captured sooner or later."[36]

The *Virginia* finally left dry dock on April 4, 1862. Tattnall was instructed by Mallory to attack the Union transports in Hampton Roads, which the *Virginia*'s commander hoped would provoke the *Monitor* into battle. Bad weather and mechanical problems delayed the *Virginia*'s departure for several days.

Mallory thought that the Confederate ironclad was now "in far better condition than she was before her late engagement."[37] A new funnel and steam pipes were installed, several outer iron plates were replaced, two new guns were mounted and the ship received a new anchor, flagstaff and two boats. The ironclad now featured weapons, a steel-tipped ram and wrought-iron bolts for the Brooke rifles, specifically designed to combat the *Monitor*. The *Virginia* indeed appeared as a much-improved warship; however, the refitting project was hasty and incomplete. The port shutters had been fabricated but not installed. The engines were still weak. "From my past and present experience," Ashton Ramsay concluded, "I am of the opinion that they cannot be relied upon." Ramsay advised Tattnall:

> *At the time I was ordered to the vessel I was informed that it was not the intention to take the ship where a delay, occasioned by a derangement in the machinery, would endanger her safety. The engines of this ship are not disconnected, and one cannot be worked alone. As long as the vacuum of either engine holds good, the engines might be run by working the after-engine high pressure; but as the vacuum of either engine fail, the engines would stop. Using one engine high pressure would also require a great deal of steam, which the boilers cannot generate for any length of time.*[38]

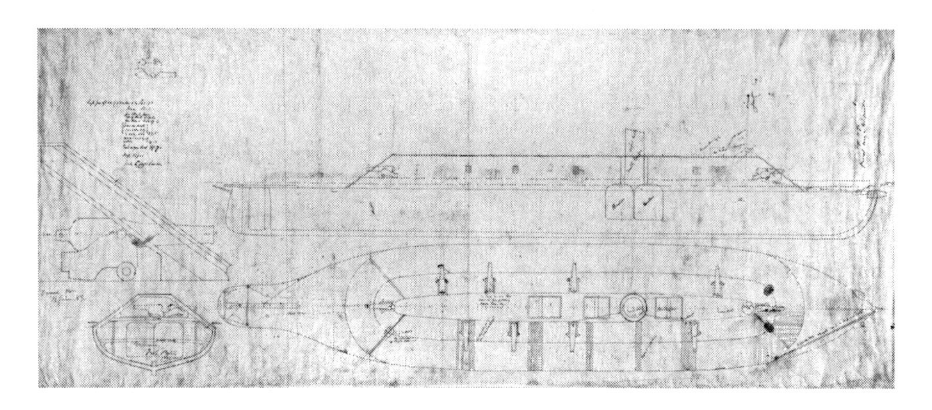

CSS *Virginia* plan, tracing, by Thomas Rowland, 1907. *Courtesy of The Mariners' Museum, Newport News, Virginia.*

An even greater problem quickly became apparent: the ironclad still did not have adequate protection below the waterline. Only two inches of iron extending 160 feet and reaching 4 feet below the shield were added to the hull, and nothing was done to protect the propeller or rudder. Ballast had to be added to lower the ironclad in the water. "These changes," John Taylor Wood noted, "with 100 tons of ballast on her fan-tails, increased her draught to 23 feet, improving her resisting powers, but correspondingly decreasing her mobility and reducing her speed to 4 knots."[39] The *Virginia*'s officers were still very dismayed with the entire repair project, citing that the improvements to the casemate, especially the iron reinforcements, were inadequately completed. Paymaster James Semple complained to Robert Minor that the project was plagued by John Porter's poor workmanship. John Taylor Wood also based the problems on "our want of resources and the difficulty of securing workmen."[40]

As the Confederacy waited for the *Virginia* to return to action, George McClellan launched his powerful army up the Peninsula. The Army of the Potomac, however, was stopped in its tracks on April 5, 1862, by Magruder's Warwick–Yorktown Line. General Barnard called the Confederate defensive system "the most extensive known to modern times" and "incapable of being carried by assault."[41] McClellan's hesitation at the Warwick River set the stage for a carefully organized Confederate ruse. Magruder began marching his soldiers "in a circle, as it were," remembered Corporal J.W. Minnish, "then on the 'Double Quick' behind the hill and woods to appear again as fresh troops arriving."[42] "It was a wonderful thing," recorded diarist Mary Chesnut about Magruder's illusion of strength, "how he played his ten thousand before McClellan like fireflies and utterly deluded him."[43] Magruder's comprehensive fortifications and clever deceptions prompted McClellan to besiege the Confederate defenses. The Army of the Potomac commander believed that he had no other choice. Since the James River was closed to Union use by the still idle CSS *Virginia*, the York River was McClellan's only riverine avenue to Richmond. Yet his access to the river was controlled by Confederate batteries at Yorktown and Gloucester Point. Flag Officer Goldsborough, fearing that the *Virginia* might attack his rear, refused to attempt any attack on the York River batteries. These circumstances forced McClellan to make Yorktown the focus of a massive siege, thus delaying the Union advance on Richmond.

Major General John Bankhead Magruder, photograph, circa 1861. *Courtesy of Virginia War Museum, Newport News, Virginia.*

Although many Confederates, such as Brigadier General Jubal Anderson Early, thought that Magruder's defense of the Peninsula was "one of the boldest exploits ever performed by a military commander," Magruder's ability to achieve success with his small force was partially due to the CSS *Virginia*.[44] The Confederate ironclad guarded his weak James River flank and threatened all U.S. Navy operations in the Hampton Roads region. The *Virginia* had influenced Federal strategic thinking and, consequently, greatly modified McClellan's campaign to capture Richmond. A new disease, "ram fever," was discovered in Virginia on March 8, 1862, and the Union command in Hampton Roads suffered greatly from its effect.

10
Equal to Five Thousand Men

General Robert E. Lee, then serving as military advisor to President Jefferson Davis, recognized the *Virginia* as a powerful weapon that should be used more than just as a deterrent but also as a tool to strike a hard blow against the enemy. Now that the Confederate ironclad was ready for sea, Lee wanted the *Virginia* to attack the Union transports in the York River. Accordingly, the Confederate general wrote to Mallory on April 8:

> *I respectfully suggest for your consideration the practicability of the* Virginia*'s passing Fort Monroe in the night to the York River. She could by destroying the enemy's gunboats and transports thwart this design. After affecting this object she could again return to Hampton Roads under cover of night. I would, however, recommend that the* Virginia*, previously to an attempt against the enemy in York River, should strike a blow at their transports and shipping in Hampton Roads and the bay outside of Forts Monroe and Calhoun, so as to prevent the possibility of an attack on Norfolk. In this manner she could so cripple their means of supplying their army as to prevent its moving against Richmond, while she could deter any movement against Norfolk. Coal could be sent by railroad and York River to Yorktown for her use.*[1]

Obviously, Lee did not understand the *Virginia*'s limitations. Mallory did and forwarded Lee's letter to Tattnall with the admonition, "I regard the *Virginia* of the first importance to the safety of Norfolk, and hence, though the suggestion of General Lee of a dash at the enemy in York River holds out a temptation to go at him at once, it should not be made if Norfolk is to be shortly exposed to capture." Mallory added, "A wholesome fear of the *Virginia* has, I think, induced him to abandon his plan of passing his troops from Newport News and Old Point to attack Norfolk, for his present more tardy operations on the peninsula. Could you destroy his transports you would scatter his army to the winds."[2]

Tattnall realized that his ironclad's engines were unreliable and did not believe that the *Virginia* could pass the Union forts without serious damage. "I presume that so long as the *Monitor* is efficient, it will not do to run up to Norfolk and destroy the yard," Tattnall advised Mallory on April 10. "Without reference to the *Monitor*, however, I think it more than doubtful whether the passage of the forts could be effected by the *Virginia* at present, for the reason I have already stated to you—the want of port covers to protect the guns and the interior of the ship," the flag officer continued his report. "In passing such formidable batteries it would be wonderful if a gun is left serviceable…If the presence of the *Virginia* at Yorktown be deemed at Richmond of such paramount importance as to call for the passage of the forts at all hazards, I will…at once attempt it." Tattnall had come to the *Virginia* seeking to enhance his career by aggressively using the *Virginia* against the Federal fleet. However, the circumstances had changed. "I have been aware from the first that my command is dangerous to my reputation," Tattnall wrote Mallory, "from the expectations of the public, founded on the success of Commodore Buchanan, and I have looked to a different field from his to satisfy them. I shall never find in Hampton Roads the opportunity my gallant friend found. I see no chance for me but to pass the forts and strike elsewhere," he concluded, "and I shall be gratified by your authority to do so as soon as the ship shall be in a suitable condition."[3] Nevertheless, Tattnall was still very willing to grapple with the *Monitor*, declaring, "I will take her! I will take her if hell's on the other side of her!"[4]

As bad weather delayed the *Virginia*'s voyage into Hampton Roads, Tattnall developed a plan of action to draw the *Monitor* into a duel. "The plan I propose is to attack the enemy's transports lying above the forts," Tattnall advised Mallory, "near Hampton Creek, in which the pilots tell me I shall be three-quarters of a mile from the transports and 1½ from the forts. These transports consist of smaller vessels of light draft." Tattnall noted that the "*Monitor* is off Hampton Creek, and will doubtless engage this ship. I shall not notice her until she closes with me, but direct my fire on the transports. There must be, however, a combat with the *Monitor*."[5]

Josiah Tattnall knew that a desperate battle would surely ensue between the two ironclads once the *Virginia* steamed out into Hampton Roads. If he could not get into position to use the new ram to pierce the *Monitor*'s hull or if the bolts for his rifled guns failed to penetrate the *Monitor*'s turret, another desperate plan was concocted to capture the Union ironclad. Information gleaned from an issue of *Scientific American*, which contained a detailed report on the Union ironclad, indicated that the *Monitor* could be boarded and captured by disabling the crew. "We had four of our small gunboats," wrote Midshipman Robert C. Foute, "ready to take the party, some of each division in each vessel. One division was provided with grappling irons and lines, another with wedges and mallets, another with tarpaulins, and the fourth with chloroform, hand grenades, etc. The idea was for all four vessels to pounce upon the *Monitor* at one time, wedge the turret, deluge the turret by breaking bottles of chloroform on the turret top, cover the pilothouse with a tarpaulin and wait for the crew to surrender." Foute added that the "plan was very simple, and seemingly entirely practical, provided we should not be blown out of the water before it could be executed."[6] Many viewed this concept with tremendous skepticism and doubted that it offered any opportunity for success. "It is just as likely," William Parker reflected, "that the *Monitor* would have towed us to Fortress Monroe if she had not sunk the whole concern before we reached her."[7]

At 6:00 a.m. on April 11, 1862, Tattnall's squadron moved down the Elizabeth River to Sewell's Point. Tattnall made a brief patriotic speech to the *Virginia*'s crew and concluded, "Now you go to your stations, and I'll go to mine."[8] Josiah Tattnall, in a daring display of Nelsonian heroism, then perched himself in an armchair on the top deck. Tattnall's operational plan was to immediately steam toward the Federal transports off Hampton Creek, which he believed would prompt the *Monitor* to attack his ironclad. The flag officer hoped that a desperate mêlée would ensue, resulting in a grand victory for the Confederacy.

When the *Virginia* entered Hampton Roads "like some huge gladiator" at 7:10 a.m., the Federals scattered to the protection of Fort Monroe "like a flock of wildfowl in the act of flight."[9] The rifled guns of the Rip Raps battery opened fire but fell short of the *Virginia* while the *Monitor* did nothing aggressive. The *Monitor*, now reinforced by the iron-hulled *Naugatuck* armed with one 100-pounder Parrott rifle, stayed in the channel between Fort Monroe and the Rip Raps. The Union ironclad had strict orders not to engage the *Virginia* unless the Confederate ironclad moved out of Hampton Roads into the open waters of the Chesapeake Bay. Tattnall refused to

The Second Trip of the Merrimac, engraving, Theodore R. Davis, 1862. *Courtesy of John Moran Quarstein.*

On the Beach.
Schooners leaving
The Octorora & Naugatuck shelling the Merrimac.

take his ironclad out of Hampton Roads, and the *Monitor* would not accept the *Virginia*'s challenge. "Each party steamed back and forth before their respective friends till dinner time," wrote William Keeler of the *Monitor*, "each waiting for the other to knock the chip off his shoulder." Keeler summarized the standoff between the two ironclads:

> *She had no desire to come under fire of the Fortress and all the gunboats, to say nothing of the rams, while engaged with us, neither did the* Monitor *with her two guns desire to trust herself to the tender mercies of the gunboats and Craney island and Sewell's point batteries while trying the iron hull of the monster. I had a fine view of her at the distance of about a mile through a good glass and I tell you she is a formidable-looking thing. I had but little idea of her size and apparent strength until now.*[10]

"Had the *Merrimack* attacked the *Monitor* where she was and still is stationed by me, I would instantly have been down before the former with all my force," Goldsborough commented. "The salvation of McClellan's army, among other things," Goldsborough continued, "greatly depends upon my holding the *Merrimack* steadily and securely in check and not allowing her to get past Fort Monroe and so before Yorktown. My game therefore is to remain firmly on the defense unless I can fight on my own terms."[11] Goldsborough's terms, as interpreted by William Keeler, were "to get the *Merrimack* in deep water where the larger steamers fitted up as rams can have a chance at her."[12] Tattnall understood the Union tactics. "The enemy's plan," the flag officer advised Mallory, "obviously, will be to get me in close conflict with the *Monitor*, and, as in that event I must occasionally lose my headway entirely, to seize the opportunity to run into me with the *Vanderbilt* and other vessels, which, for that purpose, will keep out of the melee."[13]

Goldsborough had developed his defensive posture based on the assumption that the *Virginia*'s main intent was to make a run through the Federal fleet to Yorktown. A battle in the deep water of the Chesapeake Bay presented the Union with the opportunity to use several vessels, the *Vanderbilt, Illinois* and *Arago*, which had been fitted with rams. When the *Arago* arrived in Hampton Roads, the crew learned about its mission to ram the Confederate ironclad. They mutinied and left the ship. The *Arago* was then officered by Union naval personnel and a new civilian crew, including nine contrabands. The most powerful of these rams, it was believed, was the *Vanderbilt*. The 1,700-ton side-wheel *Vanderbilt* was donated to the U.S. Navy by multimillionaire ship owner Cornelius Vanderbilt. When asked by President Abraham

Lincoln in mid-March if he could assist in preventing the Confederate ironclad from leaving Hampton Roads, Vanderbilt replied that he thought his iron-hulled steamship *Vanderbilt*, recently built at the cost of $800,000, if properly manned and altered, could do the job. The *Vanderbilt*'s bow was reinforced and fitted with a prow to ram the *Virginia*. Even the rams like the *Vanderbilt*, Goldsborough lamented, could not take the offensive unless reinforced. Goldsborough advised Gideon Welles, "It is very desirable that the new ironclad steamer at the navy yard, New York, should be got here at the earliest moment practicable."[14] The six-gun, iron-plated USS *Galena*'s absence did not spare Goldsborough from humiliation over his lack of aggression in confronting the Confederate ironclad. The infuriated Northern press lambasted Goldsborough's response to the *Virginia*. "The public," wrote a correspondent of the *New York Herald*, "are justly indignant at the conduct of our navy in Hampton Roads."[15] William Keeler expressed his frustration over the order not to engage the *Virginia* in Hampton Roads. "I believe the Department is going to build a big glass case to put us in for fear of harm coming to us."[16]

Flag Officer Louis M. Goldsborough, photograph, circa 1862. *Courtesy of The Mariners' Museum, Newport News, Virginia.*

The *Virginia* steamed around in Hampton Roads from 9:00 a.m. to 4:00 p.m., hoping that the *Monitor* would dare to attack. Commander W.N.W. Hewitt of the HMS *Rinaldo* observed that "the Confederate squadron cruised about the roads without opposition, the *Virginia* occasionally going within range of the Federal guns on the Rip Raps and Fortress Monroe as well as those of the large squadron under the guns of the fortress."[17] While the Confederate ironclad held the attention of the entire Federal fleet, the CSS *Jamestown*, commanded by Lieutenant Joseph N. Barney, captured two brigs and an Accomac schooner off Hampton Creek and towed them to Norfolk. "The capture of these vessels," an amazed Tattnall wrote to Mallory, "almost within gunshot of the *Monitor*, did not affect her movements."[18] The *Virginia*, flying the captured transport's flags upside-down under her own colors

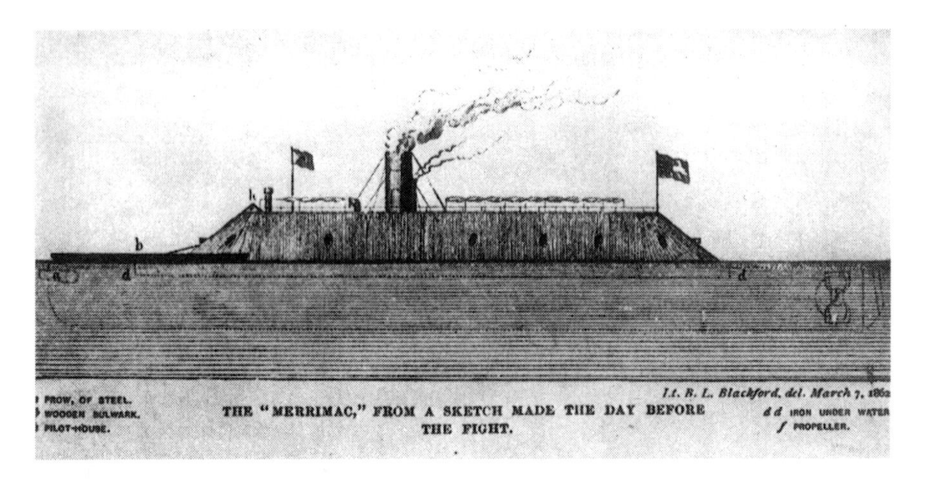

USS *Vanderbilt*, engraving, 1862. *Courtesy of U.S. Naval Historical Center, Washington, D.C.*

as an act of disdain, fired several shells at the *Naugatuck* and returned to Gosport Navy Yard. Tattnall was praised for his prudent yet gallant actions on April 11. Nevertheless, the Confederates were rather disappointed that a second battle with the *Monitor* did not develop. William Norris expressed the frustration in a faux Latin poem:

> *Supra mud-flattibus*
> *Monitoribus jugattibus*
> *Non est come—attibus*
> *Virginiabus.*[19]

Louis Goldsborough was content to wait until the ironclad USS *Galena* arrived from New York before moving into Hampton Roads. The Confederates were amazed by the Federal timidity, as John Taylor Wood wrote to his wife how "frightened they must be, with all of their forts and 3 or 400 vessels in their Navy to be afraid of our vessel."[20] William Norris capsulated the *Virginia*'s strategic victory:

> *While roaming about the* Monitor*-less Roads in triumph, the* Virginia *was a powerful support to Magruder at Yorktown. I have often heard my gallant Chief say that she was his "right wing, and was equal to five thousand men"... What were those "vast interests which the* Monitor *was protecting in her defensive role?" And how did they weigh when compared*

Capture of Two Brigs, engraving, 1862. *Courtesy of The Mariners' Museum, Newport News, Virginia.*

> *with the actual, positive tangible damage which the* Virginia *was doing their cause? If that splendid invention (as we freely admit she was, for smooth water) had been fought as she ought to have been, it might have saved them fifty thousand men.*[21]

Robert E. Lee, meanwhile, wanted the *Virginia* to achieve much more than just a strategic stalemate. "I would respectfully recommend that the *Virginia*," Lee wrote to Mallory, "after damaging the enemy's transports and destroying his means of communication in Hampton Roads, as far as practicable, turn her attention to the harbor of Yorktown, if it is considered safe for her under cover of night to pass Fort Monroe."[22] Desperate times require determined action, and Mallory finally planned an evening attack in late April. Tattnall, who was "burning to distinguish himself," made one attempt to strike at Yorktown under cover of darkness, but "his golden opportunity" for glory slipped away when the ironclad was ordered back to Norfolk by Major General Benjamin Huger. Ashton Ramsay also recounted:

> *That night we slipped down the Roads and were soon passing Fort Monroe on our way out into the Chesapeake. Presently our army signal officer began waving his lantern, communicating with our distant batteries, and then told the result to Officer Jones, who reported to Tattnall, "We have been ordered to return, sir," he said.*
>
> *Tattnall was viewing the dim outlines of the fort through his glass and pretended not to hear.*
>
> *"The order is peremptory," repeated Jones.*
>
> *Tattnall hesitated. He was of half a mind to disobey. "Old Huger has outwitted me," he muttered. "Do what you please. I leave you in command. I'm going to bed," and he went below in a high dungeon...That he did not have a chance to fight was no fault of his.*[23]

Huger canceled the sortie because he knew that the *Virginia* was the key to Norfolk's defense. There could be no attack against Yorktown until the CSS *Richmond* was ready for service. Consequently, the Confederates worked day and night to complete the ironclad, believing that the *Richmond* could handle the *Monitor* as the *Virginia* wreaked havoc amongst the Union fleet in the York River.

The *Virginia* truly needed additional support. On April 18, the *Jamestown*, *Raleigh* and *Teaser* were sent up the James River to support the Confederate right flank on Mulberry Island. Two days later, the *Patrick Henry* and *Beaufort* were assigned to Commander John Randolph Tucker's squadron stationed off the mouth of the Warwick River. This left the *Virginia* as Norfolk's only defense against an attack by the U.S. Navy. While Catesby Jones felt it was "evident that the enemy very much over-rated our power and efficiency," it was obvious to the ironclad's executive officer that the "South also had the same exaggerated idea of the vessel."[24]

Josiah Tattnall was greatly troubled by the change in command that occurred in eastern Virginia as a response to Union advances. General Joseph Eggleston Johnston had begun moving his army to the Peninsula by mid-April to help halt McClellan's advance. Accordingly, Johnston, by Special Orders No. 6, dated April 12, 1862, was placed in command of all military and naval operations in the Hampton Roads region. Tattnall was livid over this new organization and believed that he should not become "subject to the orders of General Johnston." He strongly expressed his objection, writing to Mallory:

> *This would place me, with reference to the Army, in a position never held hitherto by an officer of my rank in any naval service…*
>
> *These are times, Mr. Mallory, for frankness, and without it discord between the two arms, produced by misconception, may be fatal.*
>
> *If, therefore, I am to be placed under the command of an army officer, and, being a seaman, am to hold my action and reputation subject to the judgment of a landsman, who can know nothing of the complicated nature of naval service, I earnestly solicit to be promptly relieved from my command.*
>
> *Some younger man, whose backbone is more supple than fifty years of naval pride have made mine, can be found, I hope, for the sake of harmony, to take my place and carry out the views of the Government.*[25]

Mallory smoothed Tattnall's wounded ego, and Tattnall continued to retain his command. His ironclad also presented Tattnall with numerous problems. The engines were still unreliable and subject to constant malfunction. "Commodore Tattnall commanded the *Virginia* forty-five days," Catesby Jones noted, "of which time there were only thirteen days she was not in dock or in the hands of the Navy Yard."[26] Nothing was done to improve conditions aboard the warship for the crew.

Ashton Ramsay complained:

> *Our life on board for the weeks that followed was far from comfortable. We were within sight of the enemy, and at every movement of the opposing fleet it was "clear away for action." Steam was kept up continually. Our cabins were without air ports and no ray of light even penetrated the ward-rooms. There was no where to walk but on the upper grating—a modern prison is far more comfortable. Sometimes the sailors waded on the submerged deck, giving rise to the superstition among the darkies that they were the crew of the "dibble ship" with power to walk on the water.*[27]

Midshipman Douglas Forrest, CSN, photograph, circa 1862. *Courtesy of the Museum of the Confederacy, Richmond, Virginia.*

It is no wonder that Captain Thomas Kevill and members of the United Artillery serving on the ironclad accepted a reassignment to man a heavy battery at the Sewell's Point "Entrenched Camp" on April 19.

Joe Johnston's assumption of command on the Peninsula did not bode well for the *Virginia*'s future. Johnston believed that the Warwick–Yorktown Line should be abandoned and the capital defended on the outskirts of Richmond. Even though he lauded Magruder's "delaying tactics," the commanding general did not consider the Confederacy capable of defending both Richmond and Norfolk. "No one but McClellan could have hesitated to attack," Johnston advised. "The fight for Yorktown must be one of artillery, in which we cannot win. The result is certain; the time only doubtful."[28] Jefferson Davis, supported by Lee and Secretary of War George

Wythe Randolph, recognized that a withdrawal from the Peninsula would mean the loss of Norfolk and the CSS *Virginia*. Accordingly, Davis ordered Johnston to remain on the Peninsula. Johnston complied but believed "that events on the Peninsula would soon compel the Confederate government to adopt my method of opposing the Federal army."[29]

Johnston began orchestrating his retreat to Richmond by late April. Noting Commander John R. Tucker's "intelligence and zeal" covering the Confederate right flank, Johnston asked Tattnall if the *Virginia* could steam past Fort Monroe to Yorktown to support the Confederate withdrawal from the Peninsula. The Confederate commander concurrently advised Tattnall that the "abandonment of the peninsula will, of course, involve the loss of all our batteries on the north shore of the James River. The effect of this upon our holding Norfolk and our ships you will readily perceive."[30] Tattnall stated that the *Virginia* would do what it could for the defense of the Confederacy. The flag officer, however, critically commented that his ironclad would probably be lost in the attempt "without deriving any advantage." Such a forlorn hope strategy would cause the Confederacy, Tattnall believed, to "abandon the defense of Norfolk and the moral effect produced by the presence of the *Virginia* on the enemy's operations in the James River."[31]

11
Fire a Gun to Windward

Time was running out for the Confederate navy in Hampton Roads. On the evening of May 3, 1862, General Joseph E. Johnston ordered the evacuation of the Warwick–Yorktown Line. The retreat up the Peninsula uncovered Norfolk, forcing the Confederates to make plans to abandon the port city and naval base. When he learned of the Confederate army's retreat, Stephen Mallory immediately telegraphed Flag Officer Josiah Tattnall advising him that the Confederacy looked to the *Virginia* alone to prevent the enemy from ascending the James River. "Please endeavor to protect Norfolk," Mallory telegraphed Tattnall the very next day, "as well as the James River."[1]

Norfolk was indeed isolated, and Major General Benjamin Huger felt trapped. Huger advised Mallory that he could not defend Norfolk without the *Virginia*, as the U.S. Navy had assembled its strongest fleet to combat the Confederate ironclad. The USS *Galena* had arrived in Hampton Roads on April 27, bringing the Federal fleet's strength to three ironclads. The USRS *Naugatuck* and USS *Galena* had yet to be tested in battle. With the USS *Monitor* on hand, these ironclads composed an imposing force to counter the *Virginia*. The USS *Galena* was built in Mystic, Connecticut, and was launched on February 14, 1862. Armored with four-inch iron strips, she was 210 feet long, displaced 738 tons and had a maximum speed of eight knots. The *Galena* was armed with four IX-inch Dahlgren smoothbores and two 100-pounder Parrott rifles. The *Naugatuck* was also called the *Stevens Battery* in honor of her builder, E.A. Stevens, who had donated the iron-hulled vessel to the United States government. She displaced 192 tons and was 110 feet long. The *Naugatuck*, which was rejected by the U.S. Navy, was operated by the U.S. Revenue Marine Service, and her main armament consisted of one 100-pounder Parrott rifle.

In addition to this powerful naval force, Major General Ambrose E. Burnside was threatening Norfolk by way of the canals linking the port city with North Carolina. Brigadier General Jesse Lee Reno had already attempted to march his command through the weak Confederate forces at South Mills, North Carolina, defending the entrance to the Chesapeake & Albemarle Canal. While Reno was repulsed at South Mills on April 19, 1862, it was clear that Burnside's command was in an excellent position to move against Norfolk whenever the Union general wished to make a concentrated effort.

Mallory now recognized that Norfolk would soon be abandoned by the Confederate army despite the presence of the CSS *Virginia*. Even though the CSS *Richmond* was launched on May 1, 1862, the ironclad was far from being ready for combat. That same day, Mallory visited Portsmouth and directed Commandant S. Smith Lee to begin the evacuation of Gosport Navy Yard. All of the yard's salvageable material was dismantled and removed to Richmond, Virginia, and Charlotte, North Carolina. It was a herculean effort, made desperate by General J.E. Johnston's retreat from the Lower Peninsula. "What a terrible necessity this is," John Taylor Wood reflected in a letter to his wife, "This is war, stern, terrible war, which our sires have brought upon us."[2] Mallory, in an effort to save as much property as feasible, ordered Commander John R. Tucker's squadron to assist the evacuation. On the evening of May 5, the *Patrick Henry* and *Jamestown* steamed to Norfolk. The next evening, the two gunboats returned up the James River. The *Patrick Henry* towed the incomplete ironclad *Richmond* and the unfinished gunboat *Hampton*. The *Jamestown* pulled a brig containing heavy guns and ordnance supplies intended for use aboard the CSS *Richmond*.

The *Virginia* was now the lone naval sentinel guarding Norfolk and maintained a daily station off Sewell's Point. William Keeler of the *Monitor* remembered watching the "Big Thing" on May 7:

> [She] *again made her appearance and another just after dinner while she was in status quo under Craney Island, apparently chewing the bitter end of reflection and ruminating sorrowfully upon the future. She remained there smoking, reflecting, and ruminating till nearly sunset, when she slowly crawled off nearly concealed in a huge murky cloud of her own emission,*

Germantown *and* Merrimac *off Craney Island*, engraving, 1862. *Courtesy of the Library of Virginia, Richmond, Virginia.*

> *black and repulsive as the perjured hearts of her traitorous crew. The water hisses and boils with indignation as like some huge shiny reptile she slowly emerges from her loathsome lair with the morning light, vainly seeking with glairing* [sic] *eyes some mode of escape through the meshes of the net which she feels is daily closing her in. Behind her she already hears the hounds of the hunter and before are the ever watchful guards whom it is certain death to pass. We remain in the same position we have occupied since the fight—a sort of advance guard for the fleet.*[3]

The *Virginia*'s days of being the "cock of the walk," as John Eggleston noted, were indeed numbered as the Federals were tightening their noose around Norfolk.[4]

As the Confederates rushed to remove valuable war material from Norfolk, President Abraham Lincoln decided to go to Fort Monroe to prompt resolute action. The president had been invited to Hampton Roads by Major General John E. Wool in conjunction with Wool's plan to strike against the Confederate defenses in Norfolk. Lincoln agreed. He was disenchanted with McClellan's slow progress up the Peninsula, as well as the U.S. Navy's apparent inability to contend with the CSS *Virginia*. Lincoln arrived at Old Point Comfort on board the U.S. Revenue Cutter *Miami* during the evening of May 6, 1862. He was accompanied by Brigadier General Egbert L. Viele, Secretary of the Treasury Salmon P. Chase and Secretary of War Edwin M. Stanton.

Since McClellan was already moving up the Peninsula following the bloody Battle of Williamsburg on May 5, Lincoln's focus was on Norfolk and the CSS *Virginia*. A council of war was held with General Wool and Flag Officer Louis Goldsborough. Lincoln was still disturbed by the lack of aggressive naval action. Goldsborough had already refused McClellan's request to send the *Galena* up the James River after the occupation of Yorktown. "I dare not leave the *Merrimac* and consorts unguarded," Goldsborough had written to the Union general. "Were she out of the way everything I have here should be at work in your behalf; but as things stand you must not count upon my sending any more vessels to aid your operations than those I mentioned to you."[5] McClellan now reported to Lincoln that the Confederate batteries on Jamestown Island were vacant. Accordingly, the president wanted Union warships to immediately ascend the James River. He ordered Goldsborough to put his fleet in motion against Norfolk, as well as to open the James River in support of McClellan's advance on Richmond. Following a tour of the *Monitor* and *Galena*, Lincoln directed Goldsborough to send the *Galena*, accompanied by the gunboats *Aroostook* and *Port Royal*, "up the James River at once."[6] Goldsborough hesitated, complaining that three vessels "are too few for the work," but finally conceded to Lincoln's command.[7]

Commander John Rodgers of the USS *Galena* assumed command of the task force and, at daybreak on May 8, entered the James River. The expedition was scheduled to leave Hampton Roads at low tide to preclude the *Virginia*, which had a deep draft, from disrupting the mission once the task force had steamed into the James River. Simultaneously, Goldsborough sent another force to shell Sewell's Point. The USS *Monitor* and USRS *Naugatuck*, supported by several wooden warships, including the USS *Susquehanna* and the steamers *Dacotah* and *Seminole*, moved past the Rip Raps and began their cannonade of the Sewell's Point battery. The Union warships were circling "in front of Sewell's Point and throwing their broadsides into our works there as they passed."[8] Based on information gleaned from deserters, the Federals had caught the Confederates dismantling their guns and "pouring a continued stream of fire into the Batteries." Furthermore, the *Virginia* was at the navy yard undergoing some repairs and loading supplies.[9]

When Tattnall heard the shelling, the flag officer immediately steamed his ironclad down the Elizabeth River at full speed to contest the Union advance. As the *Virginia* neared Sewell's Point, Tattnall realized that he faced a difficult decision. He could either send his ironclad to block Rodgers's advance up the James River or he could protect the Confederate batteries defending Norfolk. Since Rodgers was already so far up the James River

CSS Virginia *Forces the Withdrawal of USS* Monitor *and other Union Warships*, lithograph, 1907. *Courtesy of John Moran Quarstein.*

where the *Virginia* could not steam due to her tremendous draft and the peremptory need to protect his base, Tattnall headed the *Virginia* toward the *Monitor* and the other Union warships. "After passing Craney Island the crew was called to quarters," Richard Curtis recalled, "buckled on their side arms already for battle, and the prospect was very good for a lively tussle."[10]

The crew was ready for a showdown with the *Monitor*; however, once again the Union ironclad retreated as the *Virginia* advanced. "At our approach they fled ignominiously and huddled for safety under the guns of Fortress Monroe,"[11] John Eggleston proudly remembered. The *Virginia* had forced "the *Monitor* and her consorts" to cease their bombardment; however, the Union ships stationed themselves just out of range of the Confederate ironclad to avoid an engagement. Then, as Catesby Jones reported,

> *Men-of-war from below the forts, and vessels expressly fitted for running us down, joined the other vessels between the forts. It looked as if the fleet was about to make a fierce onslaught upon us. But we were again to be disappointed. The* Monitor *and the other vessels did not venture to meet us, although we advanced until projectiles from the Rip Raps fell more than half a mile beyond us.*[12]

Tattnall continued to steam around in Hampton Roads for the next two hours hoping that he might induce the *Monitor* to attack but still refusing to be baited into the channel near Fort Monroe by the Union ironclad. "Finally the commodore," John Taylor Wood recounted, "in a tone of deepest disgust, gave the order: 'Mr. Jones, fire a gun to windward, and take the ship back to her buoy.'"[13] This act of disdain and defiance was considered most appropriate by the *Virginia*'s crew as, Wood noted, "It was the most cowardly exhibition I have ever seen…Goldsborough and Jeffers are two cowards."[14]

Goldsborough's actions on May 8 surprised even many of the Union naval officers. While the flag officer knew that the Confederate ironclad was now isolated in Norfolk and would eventually have to engage the Federal fleet to survive, his orders to the commanders of the ships shelling Sewell's Point were to remain on the defense.

Lieutenant David C. Constable, commander of the *Naugatuck*, noted that his instructions were to "engage the battery, and if the *Merrimack* made her appearance to fall back out of the way to induce her to come out into the roads, so that she could be attacked by the larger steamers which were then at anchor below the fortress."[15]

Members of the *Monitor*'s crew were critical of not only Goldsborough's tactics but also their own commander. Paymaster Keeler noted:

> *A good deal of fault has been found with Captain Jeffers by the officers on board for not attacking the* Merrimac *as we had her in a very favourable spot that would have given us every advantage we desired. He has always complained that he could not get permission to attack her from the Flag-Officer, but we have reason to think that he had the consent of the President to "pitch into her" if a favourable opportunity offered. Still if his orders were simply to make a reconnaissance to discover if the batteries had been strengthened or re-enforced with men or guns, he accomplished his object.*[16]

Despite Lincoln's presence, Goldsborough let slip another opportunity to engage the Confederate ironclad. Goldsborough must have been content to let events unfold as Norfolk would surely soon be captured, leaving the *Virginia* with few options.

President Lincoln was disappointed with the U.S. Navy's failure to reduce the Sewell's Point fortifications or confront the Confederate ironclad. The president viewed the entire action from the Rip Raps battery, recently renamed Fort Wool in honor of Major General John Ellis Wool. Secretary of War Edwin Stanton telegraphed Washington, D.C.:

> *President is at this moment (2 o'clock P.M.) at Fort Wool witnessing our gunboats—three of them besides the* Monitor *and* Stevens*—shelling the rebel batteries at Sewell's Point. At the same time heavy firing up the James River indicates that Rodgers and Morris are fighting the* Jamestown *and* Yorktown*... The Sawyer gun at Fort Wool has silenced one battery on Sewell's Point. The James rifle mounted on Fort Wool also does good work... The troops will be ready in an hour to move.*[17]

Though the *Virginia* stopped any landing at Sewell's Point, Lincoln had, according to one Northern correspondent, "infused new vigor in both naval and military operations here."[18] Lincoln's two-pronged naval assault worked. Rodgers was now up the James River, and Norfolk appeared ready for conquest.

When Lincoln recognized that Norfolk could not be captured by a naval attack, he began a personal reconnaissance of the coastline east of Willoughby Spit, identifying the Ocean View area as a perfect landing. On the afternoon of May 9, over six thousand troops were ferried across the Chesapeake Bay from Fort Monroe to Ocean View in canal boats. The first wave was commanded by Brigadier General Max Weber and the second by Brigadier General Joseph K.F. Mansfield. General Wool retained command of the overall operation.

While Lincoln and Wool organized the landing at Ocean View, Tattnall left the *Virginia* at her buoy off Sewell's Point to attend a council of war in Norfolk. He left the ironclad under orders not to "fight unless you are pressed."[19] Tattnall's meeting with army and navy officials was to plan the evacuation of Norfolk and decide what to do with the *Virginia* once her base was abandoned. Huger was especially worried that his command would be cut off by Union naval forces operating in the James River with the *Galena*, accompanied by the USS *Aroostook* and the USS *Port Royal*. Rodgers's squadron immediately attacked Fort Boykin, guarding Burwell's Bay. The *Galena* steamed past the fort four times, raking the colonial-era earthwork's parapets with her broadsides. The Confederates abandoned the fort. Likewise, the CSS *Jamestown* and CSS *Patrick Henry*, which were intending on steaming to Norfolk to transport more equipment to Richmond, reversed course when they noticed the Union ships had passed Newport News Point. The Confederate gunboats raced back up the James River spreading the alarm that the Union navy was in the James River. Rodgers's May 8 action against Fort Boykin threatened Suffolk, Huger's primary rail link with Richmond.

Lieutenant John Pembroke Jones, CSN, photograph, circa 1862. *Courtesy of the Hampton History Museum, Hampton, Virginia.*

The May 9, 1862 council of war was chaired by Flag Officer George N. Hollins, who had been sent to Norfolk by Secretary Mallory to resolve issues regarding the *Virginia*'s future. Huger was ill that day; however, he was represented by two army officers. Besides Hollins and Tattnall, several naval officers were present: yard commandant Sidney Smith Lee, Commander Richard L. Jones and lieutenants Catesby ap Roger Jones and John Pembroke Jones.

Josiah Tattnall wanted the *Virginia* to go out in a blaze of glory in a final encounter with the Union fleet or, perhaps, run out to sea and steam to Savannah. The voyage to Savannah was considered an impossibility as the assembled officers believed the unseaworthy ironclad would certainly founder. The other aggressive option, as John Taylor Wood later recorded, was for the *Virginia* to

> *run the blockade of the forts and done some damage to the shipping there and at the mouth of the York River, provided they did not get out of our way—for, with our great draught and low rate of speed, the enemy's transports would have gone where we could not have followed them; and the* Monitor *and other ironclads would have engaged us with every advantage, playing around us as rabbits around a sloth, and the end would have been the certain loss of the vessel.*[20]

The army pleaded that the *Virginia* should stay in Norfolk to cover its retreat. It would take ten days to remove all of the valuable equipment and supplies. Once the army had gone, the *Virginia* could then be moved up the James River. Wood remembered that

> *the pilots said repeatedly, if the ship were lightened to eighteen feet, they could take her up the James River to Harrison's Landing or City Point, where she could have been put in fighting trim again, and have been in a position to assist in the defense of Richmond.*[21]

After considering all the facts, the council's "opinion was unanimous that the *Virginia* was then employed to the best advantage, and that she should continue for the present to protect Norfolk, and thus afford time to remove the public property."[22] Tattnall returned to his ironclad moored at Sewell's Point determined to do his duty to defend Norfolk. His officers, however, did not inform the flag officer that they had observed Federal transports leaving Fort Monroe and heading out into the bay. None of the Confederates thought this unusual considering all the transports they had watched moving to and from Fort Monroe.

Meanwhile, Union troops disembarked at Ocean View at dawn on May 10 without opposition. "The landing was made without incident," Edwin Stanton reported. "[The] *Merrimack* is still off Sewell's Point." Lincoln was therefore anxious to move the Union troops toward Norfolk. The Federals had learned that morning from Confederate deserters aboard the tug *J.B. White* that the Confederates were in the process of evacuating Norfolk. Wool pushed his men forward as the Federals observed that "great volumes of smoke in the direction of Norfolk indicate that the rebels are burning the city or the navy-yard."[23] The Union troops were misinformed, but troops encountered a brief delay at Tanner's Creek. The bridge was on fire, and the Confederates fired a few artillery shells out at the approaching Federals. Nevertheless, Wool's command reached the outskirts of Norfolk by 5:00 p.m. There they were met by Mayor William W. Lamb and a select committee of the municipal council of the city of Norfolk. Lamb welcomed the Union army with a well-planned ceremony designed, according to General Egbert Viele, as "a most skillful ruse for the Confederates to secure their retreat from the city."[24] The surrender ceremony began, according to Viele, with

> *the mayor, with the formality of a medieval warden, appear*[ing] *with a bunch of rusty keys and a formidable roll of papers, which he proceeded to read with the utmost deliberation previous to delivering the "keys of the city." The reading of the documents was protracted until dark. In the meantime, the Confederates were hurrying with their artillery and stores over the ferry to Portsmouth, cutting the water-pipes, and flooding the public buildings, setting fire to the navy yard, and having their own way generally, while our General was listening in the most innocent and complacent manner to the long rigmarole so ingeniously prepared by the mayor and skillfully interlarded with fulsome personal eulogium upon himself.*[25]

Regardless of Viele's comments, Lamb's filibuster tactics enabled the destruction of the Gosport Navy Yard and other military equipment left behind by the Confederate army.

Norfolk had actually already been abandoned the day before the Union advance. Major General Benjamin Huger was seized with panic when he learned that Rodgers's squadron had already reached Jamestown Island. Huger feared that his ten-thousand-man command would soon be surrounded and left with such haste that he neglected to inform Tattnall of the evacuation.

12
Loss and Redemption

On the morning of May 10, the *Virginia* was at her mooring off Sewell's Point when Josiah Tattnall noticed that the flag was no longer flying from the Confederate fortifications. The Sewell's Point battery appeared abandoned, so the flag officer immediately dispatched his flag lieutenant, John Pembroke Jones, to Craney Island to ascertain what had happened. Jones soon learned that the Confederate army was in retreat, the navy yard in flames and the Union army was en route to Norfolk.

Tattnall was furious. He had been deceived into believing that Huger would hold Norfolk for at least another week and send the *Virginia* a prearranged signal indicating Norfolk was to be abandoned. By the time Pembroke Jones had returned to the *Virginia* at 7:00 p.m. with news that all the Confederates had left Norfolk, Craney Island had also been abandoned.

Tattnall was now faced with a difficult decision that had to be made immediately. The *Virginia* could not stay at her mooring off Sewell's Point past sunrise on May 11, as the Federals would be in control of the batteries. The flag officer once again reviewed his options. He could take his ironclad out and attack the Union fleet, perhaps destroying or significantly damaging several enemy vessels before sinking with great glory. Neither this course of action nor any effort to take the *Virginia* out to another Southern port was advisable. William Norris reported that some officers

> *suggested that we abandon her to the enemy, and after they had indulged in a sufficient amount of exultation, that Taylor Wood (our young Nelson)*

> *should slip out late some afternoon and sink her with the* Torpedo *or* Teaser*; but it was regarded as a species of ingratitude to allow the flag which she had done so much to humble, to float over her for a single moment.*[1]

Tattnall rejected all of these courses as he felt obliged by the council of war to get the huge ironclad up the James River toward Richmond. The pilots advised that this could be achieved only if the *Virginia* could reduce her draft from twenty-three feet to eighteen feet so that she could cross Harrison's Bar. Tattnall decided that the *Virginia* should follow this course. According to John Taylor Wood, Tattnall,

> *calling on hands on deck…told them what he wished done. Sharp and quick work was necessary; for, to be successful, the ship must be lightened five feet, and we must pass the batteries at Newport News and the fleet below before daylight next morning. The crew gave three cheers, and went to work with a will, throwing overboard the ballast from the fan-tails, as well as that below—all spare stores, water, indeed everything but our powder and shot. By midnight the ship had been lighted three feet, when, to our amazement, the pilots said it was useless to do more, that with the westerly wind blowing, the tide would be cut down so that the ship would not go up even to Jamestown Flats; indeed, they would not take the responsibility of taking her up the river at all. This extraordinary conduct of the pilots rendered some other plan immediately necessary.*[2]

Tattnall, who was unwell, was awakened at 1:00 a.m. and informed of the pilots' decision. He was livid. It was the second time in twenty-four hours that he had been deceived. The pilots never before mentioned that wind would be a factor navigating the shoals. Parrish, the chief pilot, calmly explained to the flag officer that since the wind was from the west rather than the east, it was blowing the water away from Harrison's Bar and making the river even shallower. Tattnall thought the pilot's plot was merely an effort to avoid fighting in a fierce battle. The flag officer lamented:

> *Had the ship not been lifted, so as to render her unfit for action, a desperate contest must have ensued with a force against us too great to justify much hope of success, and, as battle is not their occupation, they adopted this deceitful course to avoid it. I cannot imagine another motive, for I had seen no reason to mistrust their good faith to the Confederacy.*[3]

Tattnall may have cried treason, but there was little he could do. The draft reduction effort had rendered the *Virginia*, according to Ashton Ramsay, "no longer an ironclad" and unable to engage the Federal fleet.[4] Two feet of virtually unprotected hull below the shield was now exposed. The warship could only be an ironclad again by letting in water, which would flood the magazines and furnaces. "Never was a commander forced by circumstances over which he had no control into a more painful position than was Commodore Tattnall,"[5] John Taylor Wood lamented. "I had no time to lose," Tattnall wrote. Since the *Virginia* could not fight, rather than surrender to an overwhelming force, the flag officer

Flag Officer Josiah Tattnall, engraving, 1907. *Courtesy of John Moran Quarstein.*

> *determined, with the concurrence of the first and flag lieutenants, to save the crew for future service by landing them at Craney Island, the only road for retreat open to us, and to destroy the ship to prevent her falling into the hands of the enemy.*[6]

The *Virginia* slipped her cables, steamed away from her Sewell's Point mooring at about 2:00 a.m. and headed south-southwest to Craney Island. Just before heading up the Elizabeth River for the last time, Richard Curtis remembered:

> *The Boatswain Mate blew his whistle, saying all hands splice the main brace, this meant two grog tubs, one forward and one aft, the men quickly responded to this call—when I had taken my drink I went back to the forward part of the ship, very soon our ship was in motion…and was soon hard and fast aground in the light of Craney Island.*[7]

Their ironclad could not be grounded close enough to shore for the men to wade onto the beach, so the crew slowly debarked using the *Virginia*'s two cutters. The men were issued two days' rations and permitted to take some of their clothes. Midshipman Hardin Littlepage happened to see the ship's flags lying out on the gun deck as he prepared to leave the ship.

Destruction of the Virginia, engraving, 1862. *Courtesy of John Moran Quarstein.*

Littlepage quickly emptied his knapsack of all his clothes and then filled his pack with the *Virginia*'s colors. After three hours, the 350 men were safely ashore.

Ten men were selected to destroy the ironclad. Catesby Jones and John Taylor Wood spread cotton waste and other combustibles throughout the warship, while other men loaded the cannons. E.V. White "held the candle for Mr. Oliver, the gunner, to uncap the powder in the magazine to insure a quick explosion."[8] Jones then set a slow match to the powder trains, jumped into the last boat with Wood and rowed for shore "by the light of our burning ship."[9] It was daybreak, and the huge ironclad quickly became a mass of flames. "A more beautiful sight I never beheld than that great ship on fire," E.V. White noted from the shore of Craney Island, "flames issuing from the port holes, through the gratings and smokestack—the conflagration was a sight ever to be remembered."[10]

As Jones organized the crew for the twenty-two-mile march to Suffolk, many of the men must have shared similar lamentations as those so eloquently expressed by Ashton Ramsay:

> *Still unconquered we hauled down our drooping colors, their laurels all fresh and green, with mingled pride and grief, gave her to the flames, and set the imminent fires roaring against the shotted guns. The slow match, the magazine, and that last, deep, low, sullen, mournful boom told our people, now far away on the march, that their gallant ship was no more.*[11]

It was, as Richard Curtis reflected, "a sad finish for such a bright beginning."[12] As the solemn column marched along the dark, dusty road, the dejected men could hear the periodic sound of discharging guns in the distance as the flames reached the shotted cannons. Then, at precisely 4:58 a.m., they heard a terrific explosion and saw the sky turn red behind them. The men all knew that the *Virginia*, their own proud warship, was no more. The explosion marked "a bitter hour for the men," Landsman John F. Higgins sadly recalled.[13]

When the crew reached the crossroads joining the highway to Suffolk, John Higgins observed:

> *Lieutenant Jones halted and addressed a few encouraging words to his men. He then said that if any one had a family or friend in Norfolk or Portsmouth he would not blame him for returning then, "but be men," said he, meaning by the words, "be true to the South." Only two men stepped out of ranks, and the crew resumed their march to Suffolk.*[14]

The crew arrived in Suffolk by late afternoon. While they prepared to entrain for Richmond along with Huger's rearguard on the Norfolk & Petersburg Railroad, Landsman Higgins recalled that the ladies of Suffolk "prepared for us bountiful tables on both sides of the street, and dispensed gracious and patriotic hospitality to the tired and hungry men, accompanied with words of cheer for their hearts, made sad by the loss of their gallant vessel."[15]

After arriving in Suffolk, Tattnall tersely telegraphed Mallory, "The *Virginia* no longer exists."[16] Mallory was shocked when he received news of the *Virginia*'s destruction and exclaimed, "May God protect us and cure us of weakness and folly."[17] The secretary believed that Tattnall had acted prematurely and blamed the flag officer for the loss of the ironclad. John Mercer Brooke thought the fault behind the warship's loss rested on "poor leadership and lack of harmony within the Government."[18] Brooke was correct. Huger's failure to inform Tattnall of his hasty departure from Norfolk left little time to move the *Virginia* up the James River. Given sufficient time,

Destruction of the Rebel Ram Merrimac, engraving, Currier & Ives, 1862. *Courtesy of The Mariners' Museum, Newport News, Virginia.*

as well as the support of tugs and barges to carry the *Virginia*'s ordnance, the huge ironclad may have been able to reach Harrison's Bar. There, the ironclad could have been refitted to serve as a floating battery. Instead, Johnston's retreat isolated Norfolk, and Lincoln's landing on Ocean View forced Huger to prematurely evacuate Norfolk. Tattnall had few options on May 10, as his warship would soon be surrounded by land and sea. Thus, the *Virginia* was sadly scuttled by her own crew and the James River door to Richmond swung wide open for the Federals.

The Federals were overjoyed as they witnessed the *Virginia*'s demise. One young naval officer, S.R. Franklin, observed:

> *It was a beautiful sight to us in more senses than one. She had been a thorn in our side for a long time, and we were glad to have her well out of the way. I remained on deck for the rest of the night watching her burning. Gradually the casemate grew hotter and hotter, until finally it became red hot, so that we could distinctly mark its outlines, and remained in this condition for fully half an hour, when, with a tremendous explosion, the* Merrimac *went into the air and was seen no more.*[19]

A shock settled over the Union fleet as it witnessed or learned about the *Virginia*'s destruction. William Keeler remembered that "a sudden flash and a dull heavy report brought us all on deck to conjecture and surmise till the morning light should reveal the mysteries of the night." Even though Keeler was elated over the *Virginia*'s fate, he had hoped that "she would die game rather than fall by her own hand."[20]

Abraham Lincoln was pleased with the course of events and lauded General Wool for his role in the campaign, "which resulted in the surrender of Norfolk and the evacuation of strong batteries erected by the rebels on Sewell's Point and Craney Island and the destruction of the rebel iron-clad steamer *Merrimack*."[21]

Lincoln believed that the capture of Norfolk and the destruction of the Confederate ironclad were "among the most important successes of the present war."[22] McClellan was equally ecstatic and generous with his praise. The Union general telegraphed Secretary of War Edwin Stanton at Fort Monroe, "I congratulate you from the bottom of my heart upon the destruction of the *Merrimack*."[23]

By mid-morning, Goldsborough had ordered the Union fleet into Hampton Roads and the Elizabeth River. Lieutenant Thomas O. Selfridge, a survivor of the USS *Cumberland*, was the first Federal officer to land at Sewell's Point and ceremoniously raised the Stars and Stripes above the battery. The *Monitor* steamed up the Elizabeth River, stopping to take a few souvenirs from the blackened remains of the *Virginia*, and symbolically moored at the Confederate ironclad's berth at Gosport. Lincoln, along with Chase and Stanton, steamed to Norfolk aboard the USS *Baltimore* and toured the city and Gosport's smoldering ruins. The president took off his hat and bowed when his boat steamed past the USS *Monitor*.

The Federals spent little time rejoicing over their victory. Lincoln recognized that Richmond was now within the grasp of the Union navy. The president, as well as Gideon Welles and George McClellan, pressed Louis Goldsborough to strike quickly against the Confederate capital. Accordingly, Goldsborough ordered Commander John Rodgers to "push on up to Richmond if possible, without any unnecessary delay, and shell the place into a surrender...Should Richmond fall into our possession, inform me of the fact at the earliest possible moment."[24]

The Confederates had fortified several sites on the lower James River in 1861 as part of Magruder's Peninsula defensive system. Fort Boykin, Fort Huger, Mulberry Island Point, Jamestown Island and Fort Powhatan were earthen works designed to defend this approach to Richmond. Some of these forts, such as the seven-pointed, star-shaped Fort Boykin, were colonial-era fortifications rebuilt to mount modern naval ordnance. These fortifications were all designed to combat wooden ships, but not ironclads. Rodgers's squadron proved the weakness of these works when the *Galena* and other wooden gunboats entered the James River on May 8, 1862. Fort Boykin was shelled and quickly evacuated by the Confederates. Fort Huger, on Hardin's Bluff, was also shelled on May 8 and greatly reduced.

Goldsborough now ordered the *Monitor* and *Naugatuck* to reinforce Rodgers's squadron and "to reduce all the works of the enemy as they go along, spike all their guns, blow up their magazines, and then get up to Richmond."[25] On May 12, 1862, the two ironclads ascended the James River. They met token resistance at Fort Huger but soon joined Rodgers's flotilla off Jamestown Island.

Rodgers's squadron now consisted of a powerful complement of gunboats. In addition to his flagship, the six-gun ironclad USS *Galena*, Rodgers also had two additional ironclads, the USS *Monitor* and USRS *Naugatuck*, as well as the ninety-day gunboat USS *Aroostook*, commanded by Lieutenant John C. Beaumont, and the side-wheel, double-ender USS *Port Royal*. The *Port Royal* was captained by Lieutenant George Upham Morris, who had survived the sinking of the *Cumberland* on March 8, 1862. The *Aroostook*'s armament consisted of two 24-pounder smoothbores, one XI-inch Dahlgren and one 20-pounder Parrott rifle, while the *Port Royal* mounted one 100-pounder Parrott, one X-inch Dahlgren smoothbore and six 24-pounder howitzers.

Rodgers's progress upriver was slow. The squadron, due to Goldsborough's orders, stopped at each Confederate fort. The Confederates, however, had abandoned their works and spiked their guns as Johnston's army withdrew up the Peninsula. Consequently, the squadron did not reach Harrison's Landing until May 14, 1862.

Meanwhile, the Confederate capital was in an uproar over the approach of the Union fleet. The Confederate administration began preparations to abandon Richmond, and the city's government vowed to burn Richmond rather than see it fall to the Union. Richmond's lack of river defenses had been an issue for several months. Heretofore Richmond had felt secure with the *Virginia* serving as the river gatekeeper. Now that the ironclad was gone and the fortifications on the lower James were powerless to stop the Federal ironclads, all appeared to be lost. Robert E. Lee, however, was determined that Richmond "shall not be given up" and turned his attention to Drewry's Bluff.

Drewry's Bluff fortifications, photograph, 1865. *Courtesy of John Moran Quarstein.*

Drewry's Bluff was a natural defensive position on the south side of the James River, located eight miles below Richmond. The bluff, rising almost one hundred feet above the river, commanded a sharp bend in the James River and was the last place available to effectively mount a defense before reaching Richmond. Captain Alfred L. Rives of the Confederate Engineers Bureau had surveyed the James River defenses in March 1862. Rives was not confident that the lower James River batteries could stop a concentrated effort by Union ironclads to ascend the river. He advised that additional riverine defenses be constructed as part of Richmond's defensive network. A report, with comments by President Jefferson Davis, was submitted to the Confederate Congress on March 20. Rives recommended that the best place to create a defensive barrier on the river was Drewry's Bluff. The bluff's height provided an excellent position for a battery, and the river could easily be obstructed. Davis remarked that Drewry's Bluff "has intimate relations" with Richmond's defenses.[26]

Work actually began on March 17, 1862, when the Southside Heavy Artillery was assigned to the bluff. The unit's commander, Captain Augustus Hermann Drewry, was the owner of the property. Lieutenant Charles T. Mason of the Confederate Engineers Bureau designed the battery and began to place obstructions in the river with the assistance of the Confederate navy. Work on the Drewry's Bluff fortifications proceeded very slowly. A redoubt named Fort Darling had been constructed atop the bluff and armed with three heavy cannons: two 8-inch and one 10-inch Columbiads. Log cribs filled with stone had been dropped into the river and reinforced by sinking several small vessels with a one hundred-foot-wide passage.

The Union navy's entry into the James River on May 8 sent shock waves that reached all the way to Richmond. Commander John Randolph Tucker's squadron was forced to retreat up the river. When the little squadron passed Drewry's Bluff, Tucker noted the unpreparedness of the defenses. He immediately wrote to Commander Ebenezer Farrand, the recently appointed commander of Drewry's Bluff, "I feel very anxious for the fate of Richmond and would be happy to see you about the obstruction placed here—I think no time should be lost in making this point impassable."[27]

Farrand was ordered to "lose not a moment in adopting and perfecting measures to prevent the enemy's vessels from passing the river."[28] Rodgers's delays investigating each Confederate fort below City Point gave the Confederates precious time to enhance their defenses. The Confederate navy began mounting additional guns on the bluff and improving the obstructions. Reinforcements were sent to Drewry's Bluff, to include the Bedford Artillery, infantry from Brigadier General William Mahone's brigade and the crew of the CSS *Virginia*. Josiah Tattnall's decision to destroy his ironclad rather than send her on a doomed foray against the Federal fleet now paid a huge dividend to the Confederacy. The seasoned cannoneers of the *Virginia* were available to man the heavy guns at Drewry's Bluff.

Since entraining at Suffolk, the *Virginia*'s crew traveled to Richmond, arriving on May 12. The tired and depressed men marched through the city, then under martial law, and were dispatched to Drewry's Bluff. Mallory learned that the crew had been ordered to the bluff without its officers, so he detailed Lieutenant Catesby Jones to Drewry's Bluff to lead the men. Jones was joined by John Taylor Wood, Walter Raleigh Butt and Hunter

Davidson. Other officers were detailed to other special duties, such as Lieutenant John R. Eggleston, who was ordered to help build obstructions along the Warwick Bar.

Jones arrived at the bluff on the morning of May 13. He understood that "the enemy is in the river, and extraordinary exertions must be made to repel him."[29] The morale of the men was very low. Samuel Mann remembered one sailor from the *Virginia* lamented "that to attempt to defend the place would only make it a slaughter pen."[30] Although John Taylor Wood thought that the Confederate navy "for the time has been destroyed," they "must seek other ways of rendering ourselves useful."[31]

Nevertheless, Jones organized his crew members into work parties assisting Commander John R. Tucker's sailors constructing new gun emplacements. The men worked feverishly in the rain and mud for the next two days. By the morning of May 15, the sailors had mounted five guns: three 32-pounders and two 64-pounders taken from the CSS *Patrick Henry* and CSS *Jamestown*. The gun positions were dug into the brow of the bluff and revetted with logs. Tucker's men mounted a 7-inch rifle from the *Patrick Henry* in an earth-covered log casemate, located near the entrance to Fort Darling.

The Confederates made other arrangements to combat the Federals during those frantic two days. The CSS *Jamestown* was sunk, along with several other stone-laden vessels, approximately three hundred yards in front of Drewry's Bluff to enhance the obstructions. This closed the passageway to Union ships. Commander John R. Tucker held above the obstructions the remaining gunboats of the James River Squadron: *Patrick Henry*, *Raleigh*, *Teaser* and *Beaufort*. These warships were ready to engage any Union vessel that might make its way past the defenses. A Confederate marine detachment commanded by Captain John D. Simms and containing many marines from the *Virginia* dug rifle pits below the bluff. Lieutenant Wood deployed sailors as sharpshooters on the opposite bank of the river to harass the Union ships as they neared the Confederate batteries on Drewry's Bluff. Additional Confederate army units, including the Washington Artillery, Dabney's Battery and the 56th Virginia, were rushed to Chaffin's Bluff, two miles downstream from Drewry's Bluff, to help impede the Federal fleet's progress up the river.

At dawn on May 15, the Confederates at Drewry's Bluff resumed work on their fortifications. News that City Point had surrendered without opposition reached Richmond and intensified the sense of impending doom in the Confederate capital. While preparations were made to remove the government's gold supply, one Southern patriot issued this call to "Save Richmond":

> *I will be one of 100 to join any party, officered by determined and resolute officers, to board the whole fleet of gunboats and take them at all hazards, to save this beautiful city from destruction. I am not a resident of this State, but of the Confederate States, and if such a scheme can be got up, my name can be had by applying at this office.*

The editor of the *Richmond Dispatch* was so impressed by what he called "A Dashing Enterprise" that he printed the following endorsement:

> *It will be seen by an advertisement in to-day's* Dispatch *that a proposition is made to organize a party for the purpose of boarding and capturing the Yankee gunboats now endeavoring to make their way up James River to our city. That such a feat may be accomplished by bold and determined men, is not to be doubted; and surely the invaders will not be allowed to possess themselves of the capital of the Old Dominion without opposition.*[32]

There was not enough time available to test such a desperate scheme, as the Union flotilla was now ready to test the Drewry's Bluff defenses.

About 6:30 a.m. on May 15, Rodgers's command got underway from its anchorage near the mouth of Kingsland Creek, two miles below Drewry's Bluff. Rodgers decided that the *Galena*'s thin armor should be tested under fire. "I was convinced as soon as I came on board that she would be riddled with shot," Rodgers later wrote, "but the public thought differently, and I resolved to give the matter a fair trial."[33] The ironclad was constructed by Cornelius Bushnell with an experimental hull design utilizing overlapping four-inch-armor clapboard strips. The *Galena*'s sides curved from the waterline to the top deck to give the ironclad protection against shell fire from opposing ships. However, this armor-clad design would prove inadequate protection against plunging fire from Drewry's Bluff.

Despite these weaknesses, Rodgers placed the *Galena* in the lead. As his squadron passed Chaffin's Bluff, Wood's sharpshooters began to fire on the Union vessels. Wood was furious that the batteries on Chaffin's Bluff did not shell the gunboats and ordered his riflemen to open fire on the passing Union warships. The *Port Royal* and *Naugatuck* raked the shoreline with their 24-pounder howitzers as they moved toward Drewry's Bluff.

USS *Galena*, photograph depicting battle damage received at Drewry's Bluff, 1862. *Courtesy of John Moran Quarstein.*

By 7:45 a.m., Rodgers's squadron had neared the obstructions. The *Galena* steamed to within six hundred yards of Drewry's Bluff and then anchored. The river was very narrow at this point, but Rodgers gracefully swung the ironclad's broadside toward the Confederate batteries. *Virginia* boatswain Charles H. Hasker was amazed by how Rodgers placed the *Galena* into action with such "neatness and precision." Hasker, who had previously served in the Royal Navy, called the maneuver "one of the most masterly pieces of seamanship of the whole war."[34] The *Galena* received two hits while completing the maneuver and quickly became the primary target of the Confederate batteries. Once in position, however, Rodgers ordered his ironclad's port broadside battery, consisting of two IX-inch Dahlgrens and one 100-pounder Parrott, to open fire on the Confederate works atop the bluff.

The Federal fleet was at a distinct disadvantage. The obstructions effectively blocked any opportunity to run past the batteries toward Richmond. Rifle fire from along the riverbank peppered the crews of the *Port Royal*, *Naugatuck* and *Aroostook*. These vessels anchored about a half mile downriver from the obstructions and faced their bows toward the Confederate batteries. The gunboats then shelled the bluff with their 100-pounder Parrott pivot rifles. While this deployment presented more difficult targets for the Confederate artillerists atop Drewry's Bluff, Lieutenant Morris of the *Port Royal* was forced to continuously train his 24-pound howitzers on the nearby shoreline to disrupt the accurate fire of the Confederate sharpshooters in the woods. Lieutenant Beaumont of the *Aroostook* gave his men small arms to return the Confederate fire. These ships encountered other, more serious problems during the engagement. After seventeen rounds, the *Naugatuck*'s Parrott burst and forced the vessel out of action. The *Port Royal* was also disabled during the fight. Two Confederate projectiles struck the side-wheeler below her waterline, forcing the double-ender out of action.

Meanwhile, the *Galena* anchored "within point-blank range" of the Confederate batteries and the cannonade began to take effect. Ebenezer Farrand noted, "Nearly every one of our shots telling upon her iron surface."[35] Seeing that the Confederates were concentrating their fire on the *Galena*, Lieutenant Jeffers moved the *Monitor* at 9:00 a.m., virtually abreast of the *Galena*, in an effort to draw some of the Confederate shot away from the larger ironclad. The *Monitor*'s turret, however, did not permit the ironclad to elevate its two powerful XI-inch Dahlgrens to hit the Confederate batteries. The Drewry's Bluff cannoneers, knowing of the *Monitor*'s shot-proof qualities, wasted little ammunition on the ironclad. Eventually, the *Monitor* backed downstream and continued a deliberate fire from her final position below the *Galena*.

When the Union ships first steamed into sight on the morning of May 15, Midshipman Hardin Littlepage pulled the *Virginia*'s crumbled colors from his knapsack and raised the proud flag over top Drewry's Bluff. Farrand immediately ordered the three guns in the main battery to open fire. The first shot was fired by Captain Drewry. Even though the plunging Confederate shot immediately began to punish the *Galena*, the Confederates encountered numerous problems. The 10-inch Columbiad, manned by the Bedford Artillery and loaded with a double charge of powder, recoiled off its platform when the first shot was fired. This heavy naval gun was not brought back into action until the end of the engagement. The recent heavy rains caused the casemate protecting Tucker's 7-inch Brooke gun to collapse after the sailors from the *Patrick Henry* had fired just a few shots. The crew escaped and somehow was able to bring the Brooke gun back into action just before the battle concluded. Tucker commanded the five naval guns manned by sailors from the *Virginia* during the battle, while Catesby Jones was stationed

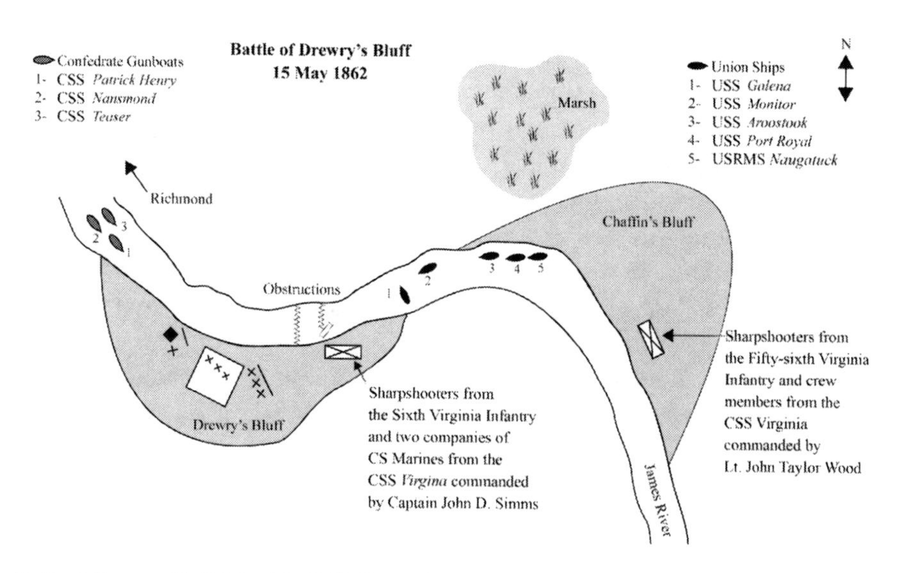

Battle of Drewry's Bluff, May 15, 1862, illustration, by Sara Kiddey. *Courtesy of The Mariners' Museum, Newport News, Virginia.*

at the Southside Artillery's position. Jones's role was to assist the volunteer artillerists to manage their heavy guns, but he was so exhausted by the efforts of the past five days that he actually dozed off while sitting on a shell box. The naval battery, despite mounting very powerful ordnance, did not play an effective role in the engagement. Positioned to the left of the fort on the brow of the bluff, the sailors continued to bang away at the Union vessels throughout the engagement and needlessly exposed themselves to Federal fire. One crew member of the *Virginia*, Landsman Michael McMore, was killed during the engagement.

Union cannon fire was rather effective early in the battle and inflicted thirteen casualties. Several men were killed by fragments from a 100-pounder shell fired by the *Galena*. Samuel Mann of the Southside Artillery remembered:

> *Shells from the* Galena *passed just over the crest of our parapets and exploded in our rear, scattering their fragments in every direction, together with the sounds of the shells from the others, which flew wide of the mark, mingled with the roar of our guns, was the most startling, terrifying and diabolical sound which I had ever heard or ever expected to hear again.*[36]

One naval officer wrote that the Federal squadron dropped "such a perfect tornado of shot and shell, right, left, front, rear, and on top of us, never seen before. It was an awful sight to see our killed and wounded, some with an arm or leg blown off, some entirely disemboweled."[37] Consequently, the Confederates retreated into bombproofs during periods of heavy Union shelling. When the Federal fire slackened, the defenders would return to their guns to return fire. This action helped to avoid greater casualties, as well as preserve the Confederates' limited supply of ammunition. Despite the Union shelling, the "batteries on the Rebel side were beautifully served," noted John Rodgers of the *Galena*, "and put shot through our sides with great precision."[38]

Some of the Confederates, like Samuel Mann, thought that the Federals "would finally overcome us";[39] however, the battle was actually decided in the Confederates' favor after almost four hours of combat. Throughout the morning, the Confederate gunners "did splendid execution with their rifled ordnance."[40] The Confederates constantly and accurately sent shot and shell into the Union vessels. They almost never missed their targets. The *Monitor* was hit three times; however, the *Galena* was struck by forty-three projectiles. The *Galena*'s weak iron plating was penetrated thirteen times during the engagement. About 11:05 a.m., the telling shot struck the *Galena*. An 8-inch shell from one of the *Patrick Henry*'s guns, manned by men from the *Virginia*, crashed through the *Galena*'s bow port and exploded. The shell ignited a cartridge, then being handled by a powder monkey, killing three men and wounding several others. The explosion sent smoke billowing out of the ironclad's gun ports. When the Confederates saw the smoke and flames rise from the *Galena*, the gunners on the bluff "gave her three hearty cheers as she slipped her cables and moved down the river."[41]

It appeared to the Confederates that the *Galena* was retreating because of the explosion. Rodgers, who had barely escaped injury when the final shell exploded on the *Galena*'s gun deck, had already decided to hoist the signal to break off action. Rodgers knew that his ironclad was running low on ammunition and had survived a damaging hail of Confederate fire. The entire squadron retreated down the river. As he watched the Union vessels steam away from Drewry's Bluff, Lieutenant John Taylor Wood hailed the *Monitor*'s pilothouse from the river bank, shouting, "Tell Captain Jeffers that is not the way to Richmond."[42]

Rodgers's squadron endured "a perfect tempest of iron raining down upon and around us" until the crippled *Galena* was forced to withdraw from action. The *Galena* suffered the greatest damage; her railings were shot away, the smokestack was riddled and she suffered twenty-four casualties. The fight demonstrated that the ironclad was "not shotproof."[43] William Keeler of the *Monitor* commented that the *Galena*'s "iron sides were pierced through and through by the heavy shot, apparently offering no more resistance than an eggshell," verifying Rodgers's opinion that "she was beneath naval criticism." When Keeler went aboard the *Galena*, he thought that the ship

> *looked like a slaughter house. The sides and ceiling overhead, the ropes and guns were spattered with blood and brains and lumps of flesh, while the decks were covered with large pools of half coagulated blood and strewn with portions of skulls, fragments of shells, arms, legs, hands, pieces of flesh and iron, splinters of wood and broken weapons were mixed in one confused, horrible mass.*[44]

The Battle of Drewry's Bluff was a dramatic Confederate victory. Richmond was under immediate threat of being captured or at least shelled by the U.S. Navy, and the cannoneers at Drewry's Bluff had saved the capital. Robert E. Lee and Jefferson Davis had both ridden to Drewry's Bluff when they heard the sound of heavy artillery. They arrived near the battle's conclusion and "seemed well pleased with the results of the engagement."[45] Commander Ebenezer Farrand was even more exuberant: "There is no doubt we struck them a hard blow." Farrand concluded his battle report, "The last that was seen of them they were steaming down the river."

Southerners were overjoyed with the victory. One patriot celebrated the Federal repulse with a short poem:

> *The* Monitor *was astonished,*
> *And the* Galena *admonished,*
> *And their efforts to ascend the stream*
> *We're mocked at.*
>
> *While the dreaded* Naugatuck,
> *With the hardest kind of luck,*
> *Was very nearly knocked*
> *Into a cocked-hat.*[46]

"The people of Richmond seemed to realize that we had saved the city from capture," Midshipman Hardin Littlepage remembered, "and early in the after-noon wagon-loads of good things came down—cakes and pies, and confections of all sorts accompanied by a delegation of Richmond ladies."[47] "O, what a feast we had!" Midshipman William F. Clayton recalled. "Surely, from the quantity and quality, the markets must have been raked clean and the dear girls sat up all night cooking."[48]

Farrand believed that "every officer and man discharged their duties with determination and coolness, and it would be doing injustice to many if I should mention or particularize any."[49] The debate over whether army or navy units deserved the greatest credit for the victory continued on into the postwar era. Many believed that it was the crew of the *Virginia*, "not Yahoos ignorant of discipline and command," who ensured victory for the Confederacy at Drewry's Bluff.[50] Augustus Drewry vehemently disagreed, stating, "I am not aware that any man connected with the navy put his hand upon any gun in the fort during the engagement."[51] Hardin Littlepage recalled serving the 10-inch Columbiad, manned by the Bedford Artillery, during the engagement. Littlepage also noted that Catesby Jones and Ebenezer Farrand assisted the Southside Artillery's Gun No. 2. A compromise was never reached, but the *Richmond Whig*, in a conciliatory fashion, wrote, "It is no empty honor to have been mistaken by either friend or foe, for the officers and men of the *Merrimac*."[52]

The role of the *Virginia*'s officers and men at Drewry's Bluff was indeed significant, even if only in a symbolic way. The crew's presence undoubtedly boosted morale, and the men manned several well-served cannons during the engagement. Their presence also gave the victory an added sense of glory. Despite the tragic loss of their ironclad, the men of the *Virginia* were willing to continue the fight against a powerful enemy during a desperate hour. Drewry's Bluff was the *Virginia*'s last opportunity to fight the *Monitor*. Even though no great drama occurred between the men of these two ironclads on May 15, the crew had redeemed itself for the destruction of its ironclad. The *Virginia*'s role at Drewry's Bluff, therefore, gave strength to a Confederacy shocked by defeat.

13
I Fought in Hampton Roads

Drewry's Bluff remained the sole gatekeeper blocking the James River approach to the Confederate capital until the commissioning of the CSS *Richmond* (often called the *Merrimac II* by the Federals) in November 1862. Even though well-positioned batteries like Drewry's Bluff, armed with high-powered rifled guns firing armor-piercing projectiles, proved to be capable of stopping an ironclad, they were vulnerable to land attack. The March 9, 1862 battle between the *Virginia* and the *Monitor* in Hampton Roads proved that only an ironclad could effectively counter another ironclad. Accordingly, the weak Confederate industrial infrastructure struggled to complete two more ironclads in Richmond shipyards before the war's end. The CSS *Fredericksburg* and CSS *Virginia II* were variations on the CSS *Virginia*. Neither of these Richmond-based ironclads, despite their design improvements, was able to match the stunning achievements of the *Virginia*.

The CSS *Virginia* was a makeshift prototype that became the most successful Confederate ironclad. Her destruction of two Union warships on March 8, 1862, proved once and for all the superiority of armored, steam-powered warships over wooden sailing vessels. It was an amazing accomplishment for the fledgling Confederate navy; however, on the next day, Southern dreams of naval superiority were dashed when the *Monitor* arrived in Hampton Roads. The inconclusive battle of March 9 made Confederate naval leaders realize that their ultimate weapon was not invincible.

The Confederacy surveyed the strengths and weaknesses of the *Virginia* prototype and initiated a major ironclad-building program. The *Merrimack*-

Virginia design, created by the unhappy team of John Luke Porter and John Mercer Brooke, became the basic model for all other Confederate ironclads. Several huge ironclads, such as the *Louisiana* and the *Arkansas*, were built as offensive weapons to challenge the blockade. The Confederates, however, quickly realized their technological and industrial limitations. While the *Virginia*, as an experimental vessel, had proven to be successful, the ironclad's very design caused her demise. The *Virginia*'s tremendous length and draft limited her operational area and precluded her ability to steam up the James River to escape the Union forces encircling Norfolk. Ironclads built thereafter needed to resolve this problem. Accordingly, Porter modified the casemate design into a smaller, lighter-draft vessel for harbor defense. The first of these new-style ironclads was the CSS *Richmond.* The *Richmond* was 180 feet in length with a 105-foot casemate and a draft of 13 feet. She was armed with four 7-inch Brooke guns. This flat-bottomed, shallow-draft ironclad design would be re-created in several lengths throughout the South. The Confederacy would attempt to build over fifty ironclads; however, only twenty-two were commissioned.

All of the Confederate ironclads were plagued by similar problems: poor propulsion systems, construction delays, limited industries producing iron and machinery, an overtaxed transportation network and a lack of sufficiently skilled workers. These factors all contrived to slow the *Virginia* construction project and weakened the development of Confederate ironclads throughout the war. Even so, the Southern ironclads were generally well armed with rifled Brooke guns and either fitted with a ram or a spar torpedo. Nevertheless, the Confederate ironclads did tremendous service for the Southern cause. The CSS *Albemarle* tipped the balance in eastern North Carolina for several months until she was destroyed by a torpedo. Ironclads in Richmond, Charleston and Savannah successfully protected these major port cities against naval attack until each city fell to Federal land-based forces. The simple casemate design proved to be the best solution for the industrially weak Confederacy.

The CSS *Virginia* was, without question, the most successful Confederate ironclad. The *Virginia* won the race for naval supremacy in Hampton Roads, thereby becoming the first ironclad to sink another warship in modern history. When the *Virginia* first appeared in Hampton Roads on March 8, 1862, she wreaked havoc amongst the Union fleet. Wooden warships appeared powerless to stop her. The *Virginia*, impervious to Federal cannon fire, rammed and sank one ship, then shelled another into submission and left her a burning hulk. Southerners rejoiced over their great naval victory, and the Northerners feared for the worst. The *Monitor*'s arrival on the evening of March 8 dampened Southern euphoria and gave hope to the Union. The next day, the two ironclads fought a memorable four-hour duel that ended inconclusively. Both sides claimed victory. Tactical success must be accorded to the *Monitor*, as the Union ironclad had effectively defended the grounded USS *Minnesota* and the rest of the Union wooden fleet. The strategic victor, however, was the *Virginia*, as the Confederate ironclad retained control of Hampton Roads.

The *Virginia*'s ability to defend Norfolk and the James River approach to Richmond significantly delayed and altered Major General George McClellan's attempt to strike at the Confederate capital by way of the Peninsula. McClellan's Peninsula Campaign was set off track from its beginning because of the Confederate ironclad's mere existence. The Federals did not understand the *Virginia*'s limitations and subsequently suffered from "ram fever" or, as William Keeler called the malaise, "*Merrimac*-on-the-brain."[1] Despite the design flaws and weak engines, the *Virginia* was a unique weapon that enabled the hard-pressed Confederacy to stop the Union advance against Richmond for over a month. Thus, the *Virginia*'s greatest victory was her service as a "fleet-in-being." She influenced every aspect of the U.S. Navy's operations during the Peninsula Campaign's first phase. Louis Goldsborough would not aggressively attack the Confederate batteries at Yorktown and Gloucester Point or make any attempt to threaten the Confederate James River flank. No Union vessel, not even the acclaimed *Monitor*, would dare challenge the *Virginia*'s supremacy as the "Mistress of the Roads." This circumstance caused McClellan to initiate time-consuming siege operations against Magruder's Warwick–Yorktown Line. The *Virginia* was the key to Magruder's vulnerable right flank and threatened to disrupt all other Union naval operations in the Hampton Roads region. The men of the *Virginia* could easily boast that they sunk two powerful Union warships, fought the *Monitor* to a standstill and cowed the entire Union fleet for over two months. "On the 8th and 9th of March, 1862," E.V. White recounted, "the Confederate States fleet successfully encountered and defeated a force equal to 2,960 men and 220 guns."[2] No other Confederate ironclad achieved such a measure of success, and all other Confederate ironclads' service would be compared to the *Virginia*'s amazing record.

Even though great accolades are heaped on John Ericsson and the *Monitor* for the technological advancement in ship design, it was the CSS *Virginia* that forged a naval revolution. The *Virginia*'s dramatic sinking of the *Congress* and *Cumberland* on March 8, 1862, proved the power of iron over wood.

The Naval Engagement between the Merrimac *and* Monitor *at Hampton Roads,* engraving, 1862. *Courtesy of John Moran Quarstein.*

When the *Cumberland* sank, so sank all of the world's wooden navies. "Not the *Monitor*," proclaimed E.V. White, "but the *Virginia* suggested the central idea of the modern cruiser and modern battleship."[3]

"She had never been the effective fighting machine that the hopes of her friends and fears of her enemies had made her,"[4] John Eggleston later admitted about the *Virginia*. Even so, the Confederacy was stunned when the *Virginia* was scuttled by her own crew on May 11, 1862. News of the *Virginia*'s destruction was devastating. A sense of doom prevailed throughout the South, and the *Virginia*'s loss was considered a symbol of the Confederacy's eventual collapse. A scapegoat needed to be found, and Josiah Tattnall was the obvious culprit. The flag officer knew when he first assumed command of the ironclad that it was "dangerous to my reputation." The *Virginia* was not equal to public expectations, and Tattnall believed he would never find in Hampton Roads the opportunity his gallant friend Buchanan had so gloriously embraced. Since Tattnall was aware that he alone would shoulder the blame for the *Virginia*'s loss, he requested an immediate court of inquiry to clear his record. The court, consisting of Flag Officer French Forrest, Flag Officer Duncan Ingraham and Captain William Lynch, was held

> *to investigate and inquire into the destruction of the steamer* Virginia, *express an opinion as to the necessity of destroying her, and state particularly whether any and what disposition could have been made of that vessel.*[5]

Although the court did not directly condemn Tattnall, the presiding officers, all of whom had sought command of the *Virginia*, found that the destruction of the *Virginia* was unnecessary and the ironclad should have been sufficiently lightened to move her up the James River to Hog Island.

A majority of naval officers were surprised with the court's opinion. Tattnall demanded a court-martial, and a lengthy defense of Tattnall appeared in the *Richmond Enquirer*. The editorial, obviously written by a crew member, stated:

> *For myself I am not only satisfied that the destruction of the* Virginia *was necessary, "At the time and place it occurred," but I assert that her destruction at the time saved the City of Richmond...The people had trusted that the existence of the* Virginia *insured our blockade of James River; and although the gallant and energetic officers of the* Patrick Henry *and* Jamestown *were working hard at Drewry's Bluff, yet the means at their command were insufficient to render the position impassable by the time the enemy's gunboats could have come up. Suddenly it bursts upon the public ear, "The* Virginia *is destroyed!" Then came "hot haste," and munitions of war and things that could assist the barricade were hurried night and day to the Bluff. The officers and crew of the* Virginia *having pushed through to Richmond, traveling unceasingly, worn out, and broken down, were sent immediately down; and ankle deep in mud, exposed to the unceasing rain for three days, without provisions or a change of clothing, they assisted, day and night, in mounting heavy guns and placing obstructions to the enemy's passage of the river. The last gun was not quite ready for action when the burst of the enemy's shell over their heads told that strife was at hand. It did come, and how gallantly the little navy maintained its reputation on that day the good citizens of Richmond may be willing to acknowledge; and perhaps they may sometimes think that some of these men were not "panic-stricken" when they destroyed the* Virginia*...I say that the destruction of the* Virginia *required the exercise of a moral courage which will outlive the late Court of Inquiry...I am proud to have been one of her crew from beginning to end, but the proudest moments, in that connection with her, were those in which I saw the flames burst from her hatches, and felt that the enemy's tread would never pollute her decks.*[6]

The court-martial was convened on July 5, 1862. The court consisted of thirteen members, including Franklin Buchanan, Sidney Smith Lee, George Hollins, Ebenezer Farrand and Matthew Fontaine Maury. Tattnall presented a lengthy defense in which he described the ironclad's weaknesses, the rapid evacuation of Norfolk and the need for the *Virginia*'s crew to aid in the defense of Drewry's Bluff. The flag officer concluded his defense, declaring:

> *Thus perished the* Virginia *and with her many high-flown hopes of naval supremacy and success. That denunciation, loud and deep, should follow in the wake of such an event might be expected from the excited mass who on occasions of vast public exigency make their wishes the measure of their expectations, and recognize in public men no criterion of merit but perfect success. But he who worthily aspires to a part in great and serious affairs must be unawed by the clamor, looking to the right-judging few for a present support, and patiently waiting for the calmer time when reflection shall assume a general sway, and by the judgment of all, full justice, though tardy, will be done to his character, motives and conduct.*[7]

The court granted Tattnall an honorable acquittal, finding that the flag officer had done his utmost as commander of the *Virginia* to engage the Union fleet until circumstances beyond his control forced him to order the *Virginia*'s destruction. Furthermore, the court acknowledged that Tattnall's decision to scuttle the *Virginia* off Craney Island on May 11, 1862, "was deliberately and wisely done" to prevent the ship and crew from being captured.[8]

Despite the judgment of the court-martial, Tattnall never recovered his prestige. He was reassigned to command of the Savannah Squadron, which contained two ironclads. While the CSS *Atlanta* was a powerful warship built from the iron hull of the blockade runner *Fingal* with excellent engines, the CSS *Georgia* contained such poor engines that she could only serve as a floating battery. Nevertheless, Tattnall sought to use the *Atlanta* against the small Union naval force blockading Savannah and capture or destroy the U.S. Navy South Atlantic Blockading Squadron station at Port Royal, South Carolina. Tattnall's failure to enact this strategy in a timely fashion prompted Mallory to relieve the flag officer of his command in late March 1863. Although Josiah Tattnall retained command of the Savannah Station until the city was evacuated in December 1864, he was broken in spirit and "very much depressed."[9] Tattnall moved to Nova Scotia when the war ended; however, he returned to Savannah in 1870 and served as inspector of ports until his death on June 14, 1871. Two U.S. Navy vessels—USS *Tattnall* (DD-125), commissioned June 26, 1919, and USS *Tattnall* (DDG-19), commissioned April 13, 1963—were named in his honor.

USS *Atlanta*, photograph, 1865. *Courtesy of The Mariners' Museum, Newport News, Virginia.*

Tattnall's replacement as commander of the *Atlanta* was Commander William Webb. Webb captained the *Teaser* during the Battle of Hampton Roads. On June 17, 1863, Webb attempted to attack two Union monitors, USS *Weehawken* and *Nahant*, guarding the approaches to Savannah in Wassau Sound, Georgia. His plan was to destroy these ironclads with his Brooke guns, ram and spar torpedo. Once the *Atlanta* achieved this goal, Webb wished to steam into Port Royal Sound, South Carolina, and destroy the Union blockading base located there. Unfortunately, the *Atlanta* ran aground on a sandbar off the tip of Cabbage Island. The two monitors then pounded the *Atlanta* into submission in fifteen minutes. Webb's audacious plan failed, even though he had modeled his assault on the daring actions of the *Virginia*'s commanders.

The *Virginia*'s other two commanders' Civil War careers were far more distinguished. Franklin Buchanan was promoted to admiral for his gallant and meritorious conduct during the March 8, 1862 engagement with the *Congress* and *Cumberland*. His wound, however, was slow to heal, and Buchanan remained on convalescence leave until August 21, 1862. Admiral

Buchanan was detailed as commander of Confederate naval forces at Mobile Bay, replacing Flag Officer Victor Randolph. Buchanan pushed forward the construction of several ironclads: the floating batteries *Tuscaloosa* and *Huntsville*, as well as the *Tennessee* and *Nashville*. As Buchanan waited for the completion of these vessels, he sought to strike at the weak Union blockading force guarding the entrance to Mobile Bay. The admiral organized his own cutting-out expedition to capture the Umadella-class ninety-day gunboat USS *Kennebec*. He planned to use the steamer CSS *Crescent*, built in Mobile in 1858, to rush the Union blockades and take her with boarders. On the evening of January 26, 1863, Buchanan, along with his son-in-law Captain Julius Meiere, CSMC, went aboard the *Crescent* and steamed toward Mobile Bay's entrance. Unfortunately, a blockader, the *Alice*, selected the very same evening to slip out of Mobile Bay. The *Alice* ran aground, alerting the *Kennebec*. With the element of surprise lost, Buchanan called off the expedition.

Despite the failure of this action, Buchanan had thoroughly rallied the Mobile Bay Squadron. He pushed the ironclad projects forward, and by the summer of 1863, both the *Tuscaloosa* and *Huntsville* were operational; however, these floating batteries were too slow, and Buchanan did not send them into Mobile Bay. His hopes for effective offensive action against the Union blockaders lay in the construction of the ironclad ram *Tennessee* and the side-wheeler ironclad *Nashville*. The *Tennessee* was commissioned one of the best Confederate-built ironclads. Armed with two 7-inch and four 6.4-inch Brooke rifles, the *Tennessee* was heavily armored. Once he had this vessel operational, Buchanan wished to wait until the *Nashville* was ready before striking out against the Union blockaders. Great things were again expected of Buchanan, and he was under considerable pressure from President Jefferson Davis to act immediately. "Everybody has taken it into their heads that *one* ship can whip a dozen and if the trial is *not made*," Buchanan wrote fellow naval officer John K. Mitchell, "we who are in her are damned for life, consequently the *trial* must be made. So goes the world."[10]

Admiral Franklin Buchanan, CSN, photograph, circa 1864. *Courtesy of the Museum of the Confederacy, Richmond, Virginia.*

USS *Atlanta*, photograph, circa 1864. A view of the former Confederate ironclad serving as a Union warship on the James River below Richmond, Virginia. *Courtesy of The Mariners' Museum, Newport News, Virginia.*

Buchanan considered a strike against New Orleans, as suggested by Stephen Mallory; however, he needed the *Nashville* and the cooperation of the Confederate army to have any chance of success. Nevertheless, Buchanan planned to attack Rear Admiral David Glasgow Farragut's fleet as it was being assembled out in the Gulf of Mexico. The *Tennessee* could dispense Farragut's force before any monitors arrived. On May 22, 1864, Buchanan prepared to attack Farragut at night; however, bad weather postponed the operation until the next evening. Unfortunately, the *Tennessee* ran aground. Buchanan reluctantly called off the attack and decided not to take any offensive action.

However, before the *Nashville* received all her armor, or any other Alabama-built ironclads were ready and able to support any offensive movement, Farragut attacked Mobile Bay Squadron on August 5. The Federal fleet was able to steam past Fort Morgan and the torpedoes guarding the entrance to the bay. While the Union ships suffered considerable damage, including the loss of the monitor *Tecumseh* when she struck a torpedo, the *Tennessee* was only able to engage several wooden vessels as they passed into the bay. Buchanan's ironclad damaged the *Kennebec*, but the Union ships were too fast and moved out into Mobile Bay.

Buchanan then steamed the *Tennessee* directly into the midst of the Union fleet. Farragut was surprised at Buchanan's tactics, declaring, "I did not think Old Buck was such a fool."[11] Buchanan was hoping to catch the Union fleet off guard and reorganizing after its bloody passage of the forts. Instead, the *Tennessee* was punished by several Union warships. The Confederate ironclad was rammed by the *Monongahela*, *Hartford* and *Lackawanna*, while heavy shot from the monitors *Chickasaw* and *Manhattan* inflicted severe damage. The *Tennessee*'s smokestack was riddled and then collapsed, her steering chains shot away and port shutters jammed. A fifteen-inch shot from the *Manhattan* penetrated the *Tennessee*'s casemate. With numerous casualties, including Buchanan wounded in the leg, the *Tennessee* surrendered.

Buchanan was held prisoner until exchanged in February 1865. The admiral did not see any further action during the war. Nevertheless, his combat record makes Buchanan one of the most daring and aggressive Confederate naval officers. In his postwar career, he served as the first president of Maryland Agricultural College (later, the University of Maryland) and as an insurance executive. His outstanding service record witnessed two U.S. Navy vessels named in his honor. The USS *Buchanan* (DD-131) was commissioned on January 20, 1919, and the USS *Buchanan* (DD-484), sponsored by his great-granddaughter Hildreth Meiere, was commissioned on March 21, 1942.

Catesby ap Roger Jones did not have a ship named in his honor; however, his Civil War record after serving as executive officer of the CSS *Virginia* was extremely distinguished. He was promoted commander for gallant and meritorious conduct aboard the *Virginia* yet never again served in combat. Jones was first assigned to command the CSS *Chattahoochee* stationed at Columbus, Georgia. His talents with ordnance soon had him detailed to supervise the Charlotte Naval Ordnance Works and then, in May 1863, to assume command of the Selma Naval Gun Foundry and Ordnance Works. Lieutenant Charles C. Simms, a shipmate aboard the *Virginia*, was his executive officer. The Selma Works rolled iron plate and produced Brooke rifles in a variety of bores. Despite the importance of his ordnance work, he sought an active command. Jones lobbied Buchanan for a ship in the Mobile Bay Squadron, and his efforts seeking sea duty perhaps prompted Robert Minor to write to Jones, "No one has yet been ordered to the *Virginia* [II] here...If you were not on more important duty, I am inclined to believe that you would have command of her."[12] Jones was finally detailed to the Mobile Bay Squadron; however, the war ended, and he was paroled aboard the USS *Stockdale*. Following the war, Jones formed a military supply company with

John Mercer Brooke and Robert Dabney Minor. The company failed, and Jones, despite surviving three fierce battles without injury, was shot and killed during an argument with a neighbor on June 17, 1877.

Jones actually served as the *Virginia*'s acting commander twice: first on March 9, 1862, following the wounding of Franklin Buchanan, and again when he led the crew to Drewry's Bluff. Once the smoke had cleared away from the bluff following the May 15 battle, the ironclad's officers and crew were assigned elsewhere. The army needed men in infantry units to help defend Richmond, while the navy needed this nucleus of trained seamen and experienced officers to fill berths at stations and aboard warships throughout the South.

Many of the *Virginia*'s former officers and men would once again appear in conspicuous leadership roles. While many crew members remained at Drewry's Bluff during the Peninsula Campaign, Robert Dabney Minor recovered from his wound received on March 8, 1862, and Thomas Kevill assisted John Mercer Brooke in building an iron-cased, 7-inch railroad gun. This mobile heavy gun was mounted on the Richmond & York River Railroad and served against the Federals during the June 29, 1862 Battle of Savage Station. Minor later became the flag lieutenant of the CSS *Virginia II* until detailed to Richmond Naval Ordnance Works. Lieutenant Hunter Davidson was assigned to assist Matthew Fontaine Maury's electric torpedo experiments. Davidson's assignment placed him in command of the CSS *Teaser*, a converted tug. The *Teaser* became the first ship to lay mines (torpedoes) and was also used to deploy the Confederate hot-air observation balloon.

On July 4, 1862, the *Teaser* encountered the USS *Monitor* and USS *Maratanza* in the James River near Turkey Point. The *Teaser*, formerly an eighty-foot-long tug, was armed with one 32-pounder rifle and one 12-pounder pivot gun. She was, however, no match for either of the Federal warships. The first shot from the *Maratanza* was high, yet the next shot of cannister prompted Davidson to order his men to abandon the *Teaser*. As soon as the crew was off the *Teaser*, a second shell struck her boiler and left her dead in the water. When the Federals inspected their prize, several torpedoes were on board the *Teaser*, as were various documents detailing where the torpedoes were to be placed in the water. Acting Assistant Paymaster William Keeler of the *Monitor* inspected the documents. Besides correspondence between Davidson and his wife, what Keeler found most interesting was:

> *The private memorandum book of Hunter Davidson who was in command. He was one of the officers of the* Merrimac *& in this book was* [sic]

Battle of Mobile Bay, lithograph, J.O. Davidson, circa 1880. *Courtesy of The Mariners' Museum, Newport News, Virginia.*

> *drafts of the* Monitor *& sketches of the mode of our capture, as they intended to attempt it. It was minute in all its details. We were to be bonded from four tugs at the same time (one of them was the* Teaser*) by men carrying turpentine, ladders, fire balls, wedges, sheets of metal, chloroform, etc. The names of the men were given, just what article each one was to carry, to what part of the* Monitor *he was to go etc., it even gave the men who were to carry matches & sand paper to rub them on.* [13]

Despite the loss of these important documents, torpedoes and the only Confederate observation balloon, Davidson was not censured for the *Teaser*'s loss. Instead, he was detailed to relieve Maury as commander of the Confederate Naval Submarine Battery Service. In conjunction with his torpedo development work, Davidson took the David class CSS *Squib* down the James River from Richmond to attack the Federal fleet in Hampton Roads. Davidson's target was the USS *Minnesota*, the very same steam frigate that had escaped destruction on March 9, 1862, because of the USS *Monitor*'s arrival in Hampton Roads. Lieutenant Davidson sought revenge and steamed the *Squib* through the entire Union squadron off Newport News Point to explode a fifty-three-pound torpedo against the *Minnesota*'s hull. Despite the heavy gunfire brought against the little torpedo boat, Davidson managed to escape. Although the *Minnesota* suffered minor damage, Stephen Mallory

CSS *Teaser* deck, photograph, 1862. *Courtesy of U.S. Naval Historical Center, Washington, D.C.*

reported that the "cool, daring, professional skill and judgment exhibited by Lieutenant Davidson in this hazardous enterprise merit high commendation and confer honor upon a service of which he is a member."[14] Davidson was promoted to the rank of commander and later commanded the blockade runner *City of Richmond* in an effort to convoy the ironclad CSS *Stonewall* from Europe to North America.

One of the most intrepid officers to serve aboard the *Virginia* was Lieutenant John Taylor Wood. Wood lamented after the Drewry's Bluff engagement that the Confederate navy had limited service opportunities afloat. Accordingly, he proposed to Stephen Mallory that a force of "horse marines" be created to strike at the U.S. Navy and Union shipping operating in the rivers of eastern Virginia. Mallory agreed with Wood's daring concept and was promoted to lieutenant commander on September 29, 1862, for his service aboard the CSS *Virginia*. He was then detailed to organize a handpicked group of men to raid Union shipping using tactics similar to Stephen Decatur's during an 1804 raid in Tripoli Harbor. Wood marched his men, with whale boats mounted on modified wagons, to the Potomac River. His first victim was the transport schooner *Frances Elmore* near Pope's Creek on October 7, 1862. Wood then moved his force to the Rappahannock River. On October 28, Wood and his horse marines captured and destroyed the 1,400-ton *Alleganian* loaded with guano off Gwynn's Island.

Wood struck again against the Union shipping during the evenings of August 22 and 23, 1863. Based at Windmill Point on the Rappahannock River, Wood's command oared out in the Chesapeake Bay and captured two Union gunboats, USS *Reliance* and USS *Satellite*, attached to the Potomac Flotilla. The *Satellite* was a steam screw tug built in 1854 and armed with two 8-inch smoothbores. On August 25, the *Satellite* captured three schooners: *Golden Rod* loaded with coal, the *Coquette* and *Two Brothers*, loaded with anchors and chains. The *Golden Rod* was stripped of all useful materials and burned. The other two schooners were taken to Urbanna. Wood realized that the Potomac Flotilla was searching for him, so he took the *Satellite* and *Reliance* to Port Royal, Virginia, where the gunboats were stripped of everything useful and then burned. All of Wood's successes prompted President Jefferson Davis to name his nephew as his aide-de-camp. Davis secured Wood's simultaneous promotion as commander, CSN, and colonel, CS Cavalry. While Wood completed several important defensive survey missions for his uncle, including an inspection of Wilmington, North Carolina's defenses, he soon was back to raiding. Wood went to Kinston, North Carolina, where he began planning a strike request against Union-controlled New Bern, North

Carolina. His goal was to capture the Union gunboat USS *Underwriter* and to use this vessel to attack the Union forts defending New Bern in conjunction with a land-based assault organized by Major General George E. Pickett to liberate the town. Wood left Kinston with 250 sailors, 25 marines and 35 officers in fourteen boats. The force was divided into two divisions, one commanded by Wood and the other by Lieutenant Benjamin P. Loyall, as well as a reserve, led by Lieutenant George Gift. Wood's plan was to take the *Underwriter* by a surprise boarding assault.

The *Underwriter* was a powerful 180-foot side-wheel steamer with an armament of one 80-pounder rifle, two 8-inch smoothbores, one 12-pounder rifle and three 12-pounder smoothbores. Once this warship was under Confederate control, she would be used to silence the Federal forts. Wood's command quietly approached the Union gunboat at 2:30 a.m. on February 2, 1864, in darkness and heavy rain. Nevertheless, the *Underwriter*'s sentry sounded the alarm as he saw the approach of the Confederate boats. The oarsmen put forth greater effort, and with a rush the Confederates were boarding the *Underwriter*. Cutlasses slashed and revolvers blazed as a desperate battle ensued. Within a few minutes, the *Underwriter* was Wood's prize. Unfortunately, the *Underwriter*'s fires were banked, and there was little steam in the boilers. Soon, the Union forts began to open fire on the captured gunboat. Consequently, Wood set the *Underwriter* ablaze and escaped upriver. This cutting-out expedition resulted in twenty-four Confederate casualties, including two of Wood's former *Virginia* shipmates. Private Perry Marry, CSMC, was slightly wounded in the cheek. Sadly, Private William Bell, CSMC—known as an excellent man, tried and faithful—was killed during the rush to sieze the *Underwriter*.

Despite these losses, Wood received unanimous thanks from the Confederate Congress for his daring commander raids against Union shipping and navy. He still wished to strike against the North. When a plan to liberate Confederate prisoners at Point Lookout, Maryland, was aborted on July 10, 1864, Wood was then detailed on July 24, 1864, to take command of the converted blockade runner *Atalanta* at Wilmington, North Carolina. The *Atalanta* was a fast cross-channel steamer with two screw propellers, twin funnels and an iron hull. She was converted into the commerce raider *Tallahassee* and armed with one 84-pounder rifle, two 24-pounder howitzers and two 32-pounder smoothbores. John Taylor Wood shipped out of New Inlet after a two-day wait on August 6, 1864. The *Tallahassee* eluded two blockades and steamed up the East Coast. The commerce raider captured her first prize on August 11, 1864, eighty miles off Sandy Hook. When

Wood arrived at Halifax, Nova Scotia, he had captured and destroyed thirty-three vessels. Unable to secure adequate coal at Halifax, Wood returned the *Tallahassee* to Wilmington on August 26, 1864.

Wood then went to Richmond to serve as aide to Jefferson Davis and was with his uncle at church when news came that Robert E. Lee could no longer hold his lines at Petersburg. John Taylor Wood escorted Davis as he fled westward on May 10, 1865. Davis was captured; however, Wood escaped. He made his way with former U.S. vice president and Confederate major general John C. Breckinridge through Florida, escaping to Cuba in a small boat.

Other officers did not attain the same high profile as Davidson and Wood or share these officers' good fortunes. John Pembroke Jones was detailed to the Wilmington Squadron. While commanding the ironclad CSS *Raleigh*, Jones attacked the blockaders guarding the Cape Fear River on May 6, 1864. The *Raleigh* was a *Richmond*-class gunboat built in Wilmington, North Carolina, and was plated with four inches of iron. She had a speed of six knots and was armed with four 6.4-inch Brooke guns. Jones, escorted by two wooden steamers, *Yadkin* and *Equator*, shipped across the New Inlet bar at 8:00 p.m. and chased the blockader *Britannia*. The *Raleigh* steamed around during the night and exchanged shots with the gunboat *Nansemond*. At daybreak, the *Raleigh* put a shot through the smokestack of the USS *Howquah*. Jones then engaged several other blockaders, including the USS *Kansas*, with a 150-pounder Parrott rifle and the USS *Mount Vernon* with a 100-pounder Parrott rifle. Fire was inaccurate, and no ship suffered significant damage. At 7:00 a.m., the *Raleigh* disengaged from this inconclusive action because her steering chains had become entangled with her propeller. While the *Raleigh* was able to safely cross the bar, she ran aground, and all efforts to refloat her were fruitless. The *Raleigh* remained abandoned in the river with "the appearance of a monstrous turtle, stranded and forlorn," until she broke apart and sank.[15] Jones was then transferred to the James River Squadron, where he stayed as commander of the CSS *Torpedo* until Richmond was evacuated.

George Washington City and Eugenius Alexander Jack went west and served as engineers aboard the CSS *Arkansas*. The *Arkansas* was even more makeshift than the *Virginia*. She was begun at Memphis but moved to Yazoo City, Mississippi, for completion. The *Arkansas* was hurriedly built with very poor engines. The ironclad was only completed thanks to the determined leadership of Lieutenant Isaac Newton Brown. Unlike all other Confederate ironclads, the sides of the *Arkansas* casemate were perpendicular, although the ends were slanted. This ersatz Confederate ironclad was plated with railroad T-iron. Iron supplies were so limited that the stern section of the casemate and the pilothouse were not plated. Armed with ten guns and manned with a crew of two hundred men, the *Arkansas* steamed down the Yazoo River on July 15 and attacked the combined Federal fleet in the Mississippi. The *Arkansas* first encountered three Union warships sent up the Yazoo River to investigate the Confederate ironclads. The USS *Carondelet* was badly damaged, and the *Arkansas* forced the Union ironclad into the riverbank and out of the fight. The rams *Queen of the West* and *Tyler* escaped downriver; however, the *Tyler* continued to shell the *Arkansas*. The pilothouse, constructed of wood, was struck, and the Yazoo River pilot was killed. The smokestack was riddled, and the breechings (connections between the stack and furnaces) were severed. This situation was serious. Steam pressure dropped, speed was greatly reduced and temperatures within the engine room soared to 130 degrees. City and Jack, despite these intolerable conditions, struggled to maintain steam pressure to keep the engines operating. Nevertheless, the *Arkansas* reached the Mississippi River and then steamed through the entire line of the combined Union fleet, anchored twelve miles above Vicksburg. Several shots from the USS *Hartford* penetrated the T-rail iron casemate and inflicted casualties. The *Arkansas* caused serious damage to several Union vessels. As the *Lancaster*, the only Union ship able to build up sufficient steam to operate, tried to ram the *Arkansas*, the Confederate ironclad somehow sent a shell through the *Lancaster*'s boiler, scalding many of the ram's crew and putting the Union vessel out of action. The *Arkansas* then made her way to Vicksburg.

Farragut vowed to destroy the *Arkansas*. That evening, the Union oceangoing vessels steamed past Vicksburg; however, they could not see the badly damaged *Arkansas* in the darkness. Another attack was planned on July 22, 1862. The ram *Queen of the West* and ironclad USS *Essex* attacked the *Arkansas* without inflicting significant damage. Both Union warships suffered heavy damage from Confederate shore battery fire.

Even though all the shot damage had yet to be repaired and the engines were so defective they could break down at any time, the *Arkansas* was ordered downriver to help Major General John C. Breckinridge's army attack Baton Rouge, Louisiana. The *Arkansas* got underway on August 3; however, she was forced to anchor for repairs. The following day, the *Arkansas* steamed to within sight of Baton Rouge when her starboard engine stopped and more repairs were required. On August 6, the *Arkansas* moved back into the channel as the USS *Essex* approached. Both engines broke down, and the *Arkansas* was scuttled by her own crew.

While officers like Jack and City sought active assignments on warships, two of the *Virginia*'s former officers played significant administrative roles for the Confederate navy. Paymaster James A. Semple, a son-in-law of former president John Tyler, joined the Office of Provisions and Clothing following the *Virginia*'s destruction. He was very zealous in the completion of his duties and eventually replaced John DeBree as chief of the bureau. Semple was noted for his ability to stockpile clothing and food in various depots throughout the South. Under Semple's guidance, in part due to supplies smuggled through the blockade, the Confederate navy was able to provide its personnel with scarce items like coffee and sugar. Semple's competence resulted in the paymaster being placed in control of the Confederate treasury during the government's retreat from Richmond in April 1865. He traveled with President Jefferson Davis. The gold, totaling $86,000, was concealed in a false carriage bottom, and Semple was to take the money to a Confederate overseas agent. The final disposition of the gold is unknown.

Henry Ashton Ramsay was sent to the Charlotte Navy Yard in North Carolina. Ramsay served as chief engineer of the yard and supervised over three hundred workers producing gun carriages, projectiles and primers. They also forged propeller shafts for ironclads throughout the South and repaired locomotives. Production lagged because of the lack of sufficiently skilled workmen. As the Confederate prospects grew dim, Ramsay organized the workers into three infantry companies. Ramsay was commissioned a lieutenant colonel in the Confederate army and took his command to serve in General Joseph E. Johnston's army in North Carolina.

Landsman William Young, CSN, photograph, circa 1861. *Courtesy of the Hampton History Museum, Hampton, Virginia.*

When Catesby Jones was detailed as commander of the CSS *Chattahoochee*, several men joined his crew in Georgia. Hardin Littlepage and Henry H. Marmaduke were both selected to serve on the gunboat by Jones. Others from the *Virginia* also shipped aboard the

Chattahoochee, including William Craig, John W. Tynan, James C. Cronin, John Dunlop, John Perry, August Lemblom and William Young. The *Chattahoochee* was a twin-screw gunboat built at Saffold, Georgia, and armed with seven guns. Following his service on the *Arkansas* and on several ships at the Savannah Station, George W. City transferred to the *Chattahoochee* as acting chief engineer on May 31, 1864. John Dunlop died on September 2, 1862, of "unknown causes" while serving on the gunboat, and William Young deserted on April 13, 1863. Young had written Robert Minor asking Lieutenant Minor to help him obtain a transfer because of the "chills and fever"[16] he suffered during his assignment on the *Chattahoochee*. John Perry volunteered for duty on the gunboat and served as master of arms until transferred to the CSS *Savannah*. James C. Cronin, Eugenius Henderson, John Joliff and William Craig were aboard the *Chattahoochee* when her boiler exploded on April 27, 1863. The *Chattahoochee* was near Blountstown, Florida, in stormy conditions preparing to try to recapture the blockade runner *Fashion*. An argument broke out in the engine room over how much water was in the boiler as the crew prepared to raise steam. A malfunctioning gauge had caused the boiler to grow red hot. When the water was poured in, it instantly vaporized and burst through piping attached to the boiler. Panicked that the magazine might explode, the crew opened the seacocks, flooding the vessel. She sank to the bottom of the river. A total of sixteen men were killed, including two former *Virginia* crew members, Joliff and Henderson; Craig was one of two wounded in the explosion.

Midshipman Hardin Littlepage, CSN, photograph, circa 1864. *Courtesy of the Museum of the Confederacy, Richmond, Virginia.*

Hardin Littlepage eventually transferred to the CSS *Atlanta* but was soon joined by two shipmates in a mission to secure warships for the Confederacy in Europe. Robert C. Foute was already serving in Savannah aboard the "mud tub" floating battery, the CSS *Georgia*. Marmaduke, Littlepage and Foute had attended the U.S. Naval Academy at Annapolis together. Foute had been the first Southern midshipman to resign from the academy, doing so the day after Abraham Lincoln's election. Franklin Buchanan noted that these midshipmen's service during the March 8, 1862 engagement with the *Cumberland* and *Congress* "gave great promise of future usefulness."[17] The young officers boarded the blockade runner *Margaret and Jessie* at Charleston, South Carolina, and ran the blockade on the evening of May 28, 1863. En route to Nassau, the blockade runner was intercepted by the USS *Rhode Island*. The *Margaret and Jessie* tried to outrun the *Rhode Island* to the Bahamas; however, even when the blockade runner entered the territorial waters of the British island of Eleuthera, the *Rhode Island* persisted in her chase. Eventually, the *Margaret and Jessie* ran aground, and the three officers had to swim to shore to avoid capture. Yet three months later, Foute, Littlepage and Marmaduke arrived in Europe and lived in Paris, awaiting a ship. Littlepage roomed with Marmaduke, while Robert Foute fell in love with the daughter of a Swedish baron, Mary Stewart deKantzow. She was the niece of Major General Philip Kearny's widow and daughter of Eloise Bullitt deKantzow, formerly of Louisville, Kentucky. Since the Confederacy could not secure enough ships from either France or Great Britain as the South appeared to be losing the war by 1864, the three officers returned to the Confederacy. Marmaduke was detailed to the CSS *Chicora* in Charleston, Littlepage to the CSS *Virginia II* in Richmond and Foute to the CSS *Savannah*. When the Confederacy collapsed in 1865, the officers somehow all joined the Semmes Naval Brigade and were captured at Saylers Creek on April 6, 1865.

Mobile Bay was another station assignment sought by the men of the *Virginia* as it enabled them to serve once again under the dynamic and aggressive leadership of Franklin Buchanan. Buchanan was not pleased with the officers in his command when he first arrived in Mobile, lamenting to Catesby Jones:

> *I have neither flag-captain nor flag-lieutenant, nor midshipman for aides; consequently, I have all the various duties to attend to from the grade of midshipman up. My office duties increase daily, which keeps me in the office until 3 o'clock, and then in the afternoon I visit the navy yard, navy store, ordnance, etc.*[18]

Although known as a strict disciplinarian, Buchanan's reputation as "a *man* and a *Commander*" beckoned men ready for active service and combat to join the Mobile Bay Squadron. Buchanan would not disappoint them.

John Eggleston, called "a clever man"[19] by Buchanan, was assigned as the squadron's flag lieutenant, and Julius Meiere commanded his father-in-law's squadron's marine detachment. Meiere was captured on August 8 following the Battle of Mobile Bay when Fort Gaines surrendered; however, he escaped from confinement at New Orleans. Benjamin Herring was an engineer aboard the CSS *Tennessee*, and Lawrence Rootes also served briefly on the ironclad. Dinwiddie Phillips and Algernon Garnett assumed surgeon billets in the Mobile Bay Squadron, Phillips aboard the *Tennessee* and Garnett on the *Baltic*, as well as the *Tuscaloosa*. Reuben Thom resigned from the CSMC due to poor health after Drewry's Bluff; however, he later returned to duty as assistant inspector general on Brigadier General Richard L. Page's staff at Mobile. Thom was later captured at Fort Morgan and spent the rest of the war as a prisoner of war. E.V. White also transferred to the Mobile Bay Squadron and served as an engineer on the CSS *Baltic*. He soon resigned from the Confederate navy and moved to Columbus, Georgia, where he invented the machinery used to manufacture nearly all buttons and buckles subsequently used by the Confederate army. White later joined the Georgia 19th Infantry Battalion (State Guards) and served until captured in the defense of Atlanta. Lieutenant Charles C. Simms was a late arrival in Mobile. Despite arriving after the Battle of Mobile Bay, he assumed command of the *Baltic* and *Nashville* until Mobile surrendered on May 5, 1865.

Several seamen also transferred to the Mobile Bay Squadron. Benjamin Allen served as an acting gunner aboard the CSS *Huntsville*, while Thomas Bennett was a seaman on the CSS *Shell Mound* until captured near Fletcher's Ferry, Alabama, on January 21, 1864. Others who served in Mobile Bay include John Carr, CSS *Selma*; Thomas Dumphrey, CSS *Baltic*; James Edwards, CSS *Morgan*; John Jones, CSS *Selma*; John McCrady, CSS *Baltic* and CSS *Tennessee*; John Riley, CSS *Tennessee*; John Williams, CSS *Tuscaloosa*; and John Wilson, CSS *Morgan*. Dumphrey, McCrady and Riley all were captured on August 5, 1864, when the *Tennessee* surrendered. McCrady, however, managed to escape from hospital barracks.

Since seamen were in such short supply in the South, many of the *Virginia*'s crew shipped aboard ironclads, blockade runners and commerce raiders throughout the South. John Barclay and Nicholas Pryde served on the commerce raider CSS *Florida*. The *Florida* was one of the most successful raiders. While under Lieutenant John Newland Moffitt's command, she captured forty-seven vessels. Then, under Lieutenant Charles N. Morris's leadership, the *Florida* captured another thirteen Union merchant ships.

Barclay was captured with the *Florida* by the USS *Wachusett* at Bahia, Brazil, on October 7, 1864. Pryde took part in the *Clarence-Tacony-Archer* expedition under the command of Lieutenant C.W. Read. Read promoted Pryde an acting master's mate; however, they were both captured on the *Clarence* off Portland, Maine. Pryde later escaped from Fort Warren on August 10, 1863. Arthur Sinclair IV was detailed to the most successful Confederate commerce raider, the CSS *Alabama*. Commander James Bullock contracted the *Alabama* to be built at the Laird Shipyard in Birkenhead, England; she possessed the necessary speed to outrun more powerful warships. Under the command of Raphael Semmes, the *Alabama* embarked on an epic two-year cruise from the Atlantic to the South China Sea and back. Until she was destroyed by the USS *Kearsage* off Cherbourgh, France, the *Alabama* sank, bonded or ransomed sixty-nine vessels. One of her victims was the USS *Hatteras*, which was the only U.S. Navy warship sunk on the high seas by a Confederate cruiser. Sinclair was promoted lieutenant during the *Alabama*'s cruise and survived the raider's sinking. He was then detailed to another commerce raider, the CSS *Rappahannock*. The *Rappahannock* was formerly the decommissioned bark-rigged, screw steamer HMS *Victor*. This ship was purchased for the Confederacy by Commander Matthew Fontaine Maury. The *Rappahannock*'s machinery failed, and she remained in Calais, France, with Sinclair on board until he was reassigned to the corvette CSS *Texas*, then under construction at Glasgow, Scotland. Sinclair served on the *Texas* until the war ended. Samuel Jones, Peter Williams and Andrew Forrest became blockade runners. Williams and Forrest were captured on the blockade runner *Greyhound* by the USS *Connecticut* and made POWs. Midshipman James C. Long was assigned to the CSS *Albemarle* and was joined on that ironclad by seven former *Virginia* shipmates, including quarter gunner William King. The *Albemarle* was one of the more successful Confederate ironclads. When she sank the USS *Southfield* on April 19, 1864, this victory liberated Plymouth; however, the *Albemarle* was damaged during her next engagement with the Federal North Carolina Squadron on May 5, 1864. The *Albemarle* was eventually sunk by a spar torpedo on October 28, 1864, during a daring attack led by Lieutenant William Cushing, USN, on Steam Picket Boat No. 1, the *Albemarle*; it was the only Confederate ironclad sunk in action by the U.S. Navy. Following the destruction of the *Albemarle* in October 1864 at Plymouth, North Carolina, Long served on the blockade runner *Owl*. King eventually deserted at Bermuda Hundred, Virginia, on April 15, 1865.

Following the Battle of Drewry's Bluff, Charles Hasker was detailed to the CSS *Richmond*. He was promoted to lieutenant on May 5, 1863, and

CSS *Albemarle*, photograph, circa 1865. *Courtesy of John Moran Quarstein.*

assigned to the CSS *Chicora*, Charleston Squadron. Hasker volunteered to serve aboard the submarine CSS *Hunley*. The *Hunley* was built in Mobile, Alabama. Local lawyer Horace L. Hunley had envisioned and underwritten the project. The crew consisted of one officer and seven sailors. She was originally named *Fish Boat*. The officer was responsible for steering the vessel and operating the weapon system while the enlisted men cranked the propeller. Following testing in Mobile Bay, the submersible was shipped to Charleston, South Carolina, in August 1863 at the request of General P.G.T. Beauregard. On August 29, 1863, Lieutenant John A. Payne, Charles Hasker and six other volunteers from the *Chicora* were preparing to make a test drive to learn the operation of the vessel when Payne accidently stepped on the lever controlling the sub's diving planes while the vessel was moving in the water. This caused the *Hunley* to dive with her hatches still open, flooding the submarine. Payne, Hasker and two unnamed volunteers escaped. The other four men drowned. Hasker never served on the submarine again as he was captured on Morris Island on September 7, 1863, and confined at Fort Warren, Massachusetts, until exchanged at Cox Wharf, Virginia, on October 18, 1864. After service aboard the CSS *Pee Dee* near Marion Court House, South Carolina, Hasker was assigned to the Semmes Naval Brigade; he surrendered at Greensboro, North Carolina, on April 28, 1865.

Edward Lakin, a former tavern keeper, was the only United Artillery officer to remain in the Confederate navy. He also served on the *Albemarle* and was then detailed as commander of the torpedo boat CSS *Scorpion.* Lakin joined the *Albemarle*'s crew after the Confederate capture of Plymouth, North Carolina, and fought during the May 5, 1864 Battle of Albemarle Sound. The Confederate ironclad had been ordered to support Major General Robert F. Hoke's effort to capture New Bern, North Carolina. The *Albemarle*, supported by the CSS *Cotton Plant* and *Bombshell*, engaged the seven-ship Union North Carolina Squadron, including the USS *Sassacus, Miami* and *Mattabessett.* The *Sassacus* and *Mattabessett* suffered severe damage, as did the *Albemarle.* The battle was a tactical draw yet a strategic defeat, as the Confederates were unable to capture New Bern. Lakin was still assigned to the ironclad when she was destroyed by Lieutenant William B. Cushing using a torpedo-tipped spar on the evening of October 24, 1864.

During the Battle of Trent's Reach on January 24, 1865, the *Scorpion* ran aground trying to help the grounded CSS *Virginia II* and CSS *Drewry* off the mud bank. The ships were under heavy bombardment from Union batteries, and the *Drewry* exploded. The *Scorpion* was badly damaged by the explosion and drifted downstream. Lakin was critically wounded during the *Drewry*'s destruction. Captain Thomas Kevill of the United Artillery returned to his unit (which was designated the 19th Virginia Battalion, Heavy Artillery) yet continued to serve alongside naval personnel. During the Battle of Savage Station, Kevill manned the railroad 7-inch Brooke gun and was stationed for the rest of the war at Drewry's Bluff. While many of the enlisted men continued to serve under Kevill in Company C, 19th Virginia Battalion, Heavy Artillery, Andrew Dalton transferred to Company D, 1st Confederate Engineer Troops, on January 2, 1864. Dalton never joined his new unit. Instead, he served with Brigadier General John Hunt Morgan and was captured near Dublin, Virginia. He was a POW at Camp Chase, Ohio, until the war's conclusion. William Francis Drake was sent on detached duty with the Confederate navy beginning on March 16, 1864. He ended up in Chunborazo Hospital #1 at Richmond, Virginia, on February 23, 1865, and was furloughed on March 23, 1865. The war ended before he could return to his unit.

Although several of the *Virginia*'s soldiers stayed in the navy after the ironclad's crew was disbanded, the majority of the soldiers returned to their old units. The service records of these men varied during the remainder of the war. John T. Moore of the 4th Georgia and Richard A. Mitchell of the 57th North Carolina were killed in action, and Robert N. Hite of the 12th Virginia was severely wounded in the arm during the Battle of Chancellorsville. He was unable to return to duty. George E. Tabb, a prewar mechanic, returned to the 34th Virginia after Drewry's Bluff; however, Lieutenant Robert Minor recommended him for a commission based on his service aboard the *Virginia*. Tabb was appointed second lieutenant in the Signal Corps and served under Lieutenant General Leonidas Polk until captured near Louisville, Kentucky. He spent the rest of the war as a POW at Johnson's Island, Ohio.

Many other former *Virginia* crewmen would be captured after they returned to infantry service. Private Benjamin Slade, originally in the 12th Virginia, joined the 16th Virginia. He deserted and was fined one month's pay. Slade endeavored to atone for his mistake. He was wounded in action at the Wilderness, captured and sent to Elmira, New York. William O. Hall was discharged after serving on the *Virginia* but was soon drafted. He went AWOL and was found guilty of desertion. On February 14, 1863, Hall was sentenced to six months' hard labor with a twelve-pound ball and three-foot chains clasped to his left ankle. He forfeited six months' pay and was given twenty-nine lashes on his bare back despite the court-martial, which noted, "This court is lenient because the accused may have intended to return to the army at some time." Hall returned to the army and was captured at Petersburg on April 2, 1865. Frederick Archer also returned to his unit, the 4th Georgia, and was captured at Gettysburg. When he was imprisoned at Fort Delaware, Archer took the oath and enlisted in the 3rd Maryland Cavalry, USA. He mustered out of the Union service on September 7, 1865, and was one of two *Virginia* crew members to serve in the Federal army during the war.

As the war moved into its final chapter in 1865, Confederate naval stations, yards and depots were captured by the rapidly advancing Union army. The Confederate navy made a vain effort to help defend major ports like Savannah and Wilmington. John F. Higgins was one of several former *Virginia* sailors who were assigned to man coastal defense positions guarding the blockade runners' haven of Wilmington, North Carolina. Higgins served on the CSS *Chickamauga* before his transfer to Battery Buchanan at Fort Fisher defending the entrance to the Cape Fear River. During the failed Federal assault on December 24, 1864, Higgins was seriously wounded, and his right leg was amputated above the knee. Just over a month later, he was captured while convalescing in General Hospital #6 in Fayetteville, North Carolina.

The evacuation of Savannah, Charleston and Wilmington left Confederate naval personnel without ships or stations. The Confederate fleet of ironclads, all of which were based on the *Virginia*'s experimental design, were destroyed

USS *Atlanta*, gun deck view, photograph, 1865. *Courtesy of The Mariners' Museum, Newport News, Virginia.*

by their crews rather than allowing them to be captured by the enemy. The sailors were transferred to Richmond and then quickly transformed into soldiers as the Confederate government prepared to abandon its capital. Rear Admiral Raphael Semmes, formerly commander of the CSS *Alabama*, assumed command of the James River Squadron. It was obvious to Semmes, as well as his men, that the Confederacy was collapsing. Semmes noted:

> *Desertion was the consequence. Sometimes an entire boat's crew would run off, leaving the officer to find his way on board the best he might...The general understanding, that the collapse of the Confederacy was at hand, had its influence with some of the more honorable of them. They reasoned that their desertion would be but an anticipation of the event by a few weeks.*[20]

Several former *Virginia* sailors decided just to go home or give themselves up to the enemy during the last months of the war. Isaac Walling deserted shortly after returning from sick leave and "obtained refuge" on the USS *Onongaga*. Both John Murphy and John Nelson deserted, took the oath and were furnished transport to New York City. John Baines, assigned to the *Virginia II*, deserted with his entire boat crew while on picket duty on the James River. John Jones also deserted from the *Virginia II*, entering Federal lines at Dutch Gap Canal. He was sent to Washington, D.C., but took the oath and was transported home to Norfolk. A former gunner on the *Virginia*, John Hunt was captured and deserted twice during the war. His final desertion was on April 5, 1865. He gave himself up at Bermuda Hundred rather than continue marching to an uncertain future away from his home in Portsmouth.

Despite the sense of impending doom, naval personnel were organized into two army commands: Semmes Naval Brigade and Tucker's Naval Brigade. Lee's lines outside Petersburg were broken on April 2, 1865, and Semmes reluctantly ordered the destruction of the James River Squadron. By the morning of April 3, 1865, the ironclads *Richmond*, *Fredericksburg* and *Virginia II* were on fire. The end of the Confederate navy in Southern waters came when

> *an explosion like a shock of an earthquake took place, and the air was filled with missiles. It was the blowing up of the* Virginia [II], *the late flag-ship. The spectacle was grand beyond description...The explosion shook the houses in Richmond and waked the echoes of the night for forty miles around.*[21]

This magnificent pyrotechnic display was a sad ending to the James River Squadron; however, the men of the Confederate navy were not ready to surrender.

Semmes's command entrained for Danville, where they were to serve as artillerists defending the new Confederate capital. Tucker's Naval Brigade, commanded by John Randolph Tucker and numbering about four hundred men, left Drewry's Bluff and joined the Army of Northern Virginia marching westward. Tucker's command formed part of Lee's rearguard. The unit, however, fought its first and last battle on April 6, 1865. The sailors fought even after the Confederates were surrounded and most infantry units had surrendered. Tucker commented, "I can't surrender"; however, the odds were against the sailors. His command was quickly overwhelmed, and they were made prisoners. Valentine Tolson and John Cunningham were two former *Virginia* men captured at Saylor's Creek. Cunningham initially refused to take the oath and was sent to the Newport News POW camp, where he remained a prisoner until June 26, 1865. John Thomas Bullock was wounded during the battle but managed to escape. He was paroled at Appomattox Court House along with a handful of former *Virginia* compatriots, including Elijah

Wilson Flake, John Flynn, Tobias Gibbons, John Steen, John Patrick Kevill, Thomas Kevill, Charles Kirby King Jr., and Robert C. Foute.

Semmes's command spent ten days in trenches outside Danville until the admiral learned of Lee's surrender. The brigade then moved into North Carolina to join General Joseph Johnston's army. Johnston also surrendered, and the brigade was paroled at Greensboro, North Carolina, on April 28, 1865. A small group of former *Virginia* crew members was included in this surrender. Walter Raleigh Butt had destroyed his ship, CSS *Nansemond*, and was detailed as the brigade's assistant adjutant general. George Walker, who had already been captured at Bermuda Hundred in October 1864 and exchanged, had served as a quartermaster during the retreat. David Stewart and Charles Hazelwood Hasker were also former POWs. Stewart had been captured by the Potomac Flotilla while delivering mail between Maryland and Virginia. Lawrence M. Rootes, aide to Franklin Buchanan during the March 8, 1862 engagement, was also paroled at Greensboro, along with George Thomas, Jeremiah Davis, James Mercer and Sterling Mathews.

The *Virginia*'s wreck off Craney Island was considered a hazard to navigation. The ironclad's destruction was very thorough. Flag Officer Charles H. Davis, Union commandant of Gosport Navy Yard, reported that portions of the *Virginia*'s casemate, some weighing as much as thirty tons, were blown into the main shipping channel. The U.S. Navy issued its first salvage contract on October 25, 1864, to Underdown & Company for the removal of the USS *Congress* and CSS *Virginia* wrecks. Underdown processed some scrap; however, the firm did not fulfill the contract, which stated that the Confederate ironclad must be completely removed and the channel cleared of her. Consequently, several other contracts were issued during the late 1860s to remove the wreck and salvage material for scrap. The two Martin boilers and some machinery were recovered by Brown, Maltby & Company in 1867. This firm also exploded two large powder changes under the stern of the *Virginia* to clear the wreck from the channel. Despite these actions, Captain Underdown continued to collect thousands of pounds of scrap from the wreck, using the tugs *Orlando* and *Pilot*. By 1868, Underdown had collected almost 290,000 pounds of cast iron, including, purportedly, the *Virginia*'s ram, drive shaft and cannons. Underdown considered his work complete that year as he exploded a powder charge under the wreck to clear the debris from the Elizabeth River. Nevertheless, the navy continued to issue contracts to remove the scuttled ironclad. Lewis Baer & Company was given a contract in 1869, and Asserson & Hebrew received a contract in 1870 to remove the hulk. The commandant of the navy yard reported on

January 17, 1871, that the wreck was completely removed when E.J. Griffith removed 102,883 pounds of iron. However, a few months later, on May 24, 1871, James Penson & Company collected 127,120 pounds of scrap. While the U.S. Navy may have considered the job complete, the *Virginia*'s hull was still in evidence and considered a navigational hazard. On October 7, 1874, B. & J. Baker & Company was hired for a second time to raise the hulk. The *Norfolk Virginian* reported:

> *The* Old Merrimac—*Operations against the wreck of the famous ironclad* Merrimac *or* Virginia, *sunk just above Craney Island, have been recommended, and the wrecking firm of B. & J. Baker & Co. has the matter in hand. There have been many bites at this cherry, but we imagine that a clean sweep will be made this time, as the firm is not in the habit of doing things by halves.*[22]

This firm recovered the draft shaft and other materials; however, the job was still incomplete when Captain William West began work on the removal. West recovered two cannons and other metal, and the project was finally considered complete when the *Norfolk Virginian* commented in a June 2, 1876 article, "The Remains of the Celebrated Ironclad *Merrimac*":

> *The remaining timbers of the once formidable floating battery, having been successfully raised by diver West, were towed up Tuesday to the Navy Yard with pontoons and lighters attached, with which she draws 24½ ft. of water. The diver has certainly been indefatigable in his efforts to raise the sunken wreck, and his efforts have been crowned with success, having cleared the navigation of the most dangerous obstacle. Although in a pecuniary point of view, it may not be so advantageous to him. The pontoons have been pumped out and the hull will be taken into dry dock for examination today. The bottom timbers are for the most part of live oak, and all the wood suitable for manufacturing into canes will be utilized for that purpose, and as but little of it is suitable, parties should not delay in sending their orders to 52 Roanoke Avenue.*
>
> *All day yesterday, the Navy Yard was crowded with parties curious to look at the remains of the once famous vessel, and a good opportunity will now be offered for doing so while the hulk is in dry dock.*[23]

Most of the material was sold for scrap; however, a few iron plates, as well as the anchor, wheel, ship's bell and drive shaft, were acquired by museums.

Wheel of the CSS *Virginia*, iron, circa 1861. *Courtesy of The Mariners' Museum, Newport News, Virginia.*

Several firms and organizations made a variety of canes and souvenirs from iron and wood portions of the vessel. Souvenirs had been cherished since the day of the battle. In fact, Acting Assistant Paymaster William Keeler of the USS *Monitor* collected mementos of the battle—shell fragments from the *Virginia*—found on the *Monitor*'s deck. Keeler sent the end of an exploded shell to President Lincoln "with the respects of the officers of the *Monitor*." When the *Virginia* was scuttled on May 11, 1862, the *Monitor* stopped to take a few souvenirs from her blackened remains.

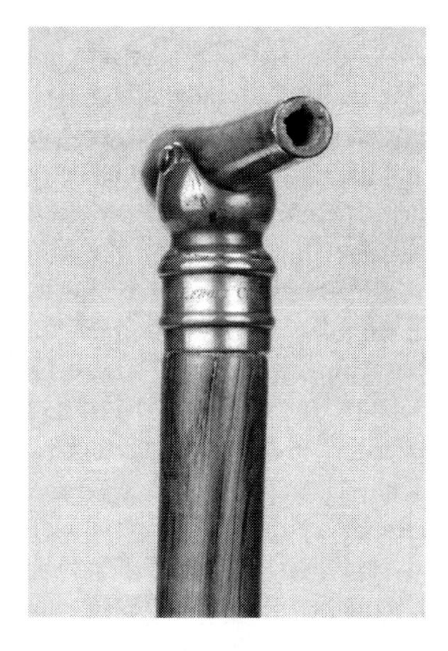

"*Merrimack* cane," iron and wood, circa 1862. *Courtesy of Stuart Mowbrey.*

Souvenirs from the *Virginia* were treated almost like holy relics or as an important connection to that famous ship. Consequently, when the ironclad was finally pulled into dry dock, the ladies of St. Mary's Catholic Church in Norfolk acquired timbers from the ship and had Mr. J.F. Freeman fashion them into canes. Canes made from the wreck were popular long before 1876, and many were even created during the war. One of the finest "*Merrimac* canes" was made of iron and wood from the Virginia's wreck. It featured a IX-inch Dahlgren model as its handle; and just beneath the pommel, a brass band engraved: "Presented by Charles F. Browne to Jesse R. Wiley/*Merrimac*." Browne was more commonly known by his pen and stage name, Artemus Ward. Ward was one of the most renowned humorists of the Civil War era. He wrote extremely popular essays, books and a syndicated newspaper column and traveled throughout the nation giving lectures to packed houses. Ward was among President Abraham Lincoln's favorite authors. One of Ward's peculiar traits was his love of oysters, and he relished eating them wherever he went. Charles Farrar Browne presented this cane to Boston, Massachusetts oyster dealer Jesse Wiley. When access to Southern oysters was cut off at the war's onset, Wiley successfully established oyster beds in the eelgrass of the South Boston flats. Obviously, Browne appreciated Wiley's ingenuity with the gift of this *Merrimac* cane. Even though many canes were made, the best market for *Virginia* relics came during the 1907 Jamestown Exposition held on Sewell's Point in Norfolk.

The Old Dominion Iron and Nail Works of Richmond produced several tokens and horseshoes noting that they were "Made from The Armour Plate The Merrimac—The First Ironclad 1862." Tredegar Iron Works also produced larger iron plate horseshoes for the 1907 event. Souvenirs—such as paperweights for the Edison Illuminating Company's 1928 annual meeting held at the new Chamberlin-Vanderbilt Hotel, Old Point Comfort, Virginia—were still in production until the 1930s.

The 1907 Jamestown Exposition included a reenactment of the battle between the *Monitor* and *Merrimack*, as well as the unveiling of a large cyclorama of the engagement. The exposition featured new articles interpreting the March 8 and 9, 1862 events and a reunion of surviving crew members from the Confederate ironclad. Most of the officers and crew returned home to their prewar occupations. The war was over, and many were anxious to renew their lives. Charles Hasker, James E. Barry Jr., Ashton Ramsay and E.V. White all went on to become very successful businessmen. Ashton Ramsay moved to Baltimore, Maryland, where he prospered as a consulting engineer until his death in 1916. White founded E.V. White & Company, General Supply Company, in Norfolk. His success led to White becoming president of Tidewater Insurance Company and founder of the Norfolk National Bank. James Barry became president of Norfolk's Bank of Commerce and spent twelve years on Norfolk's City Council. Thomas Kevill also became a prominent Norfolk official; he organized Norfolk's first professional fire department. Andrew Dalton served in a variety of political positions. Dalton was elected as a member of the Norfolk City Council, Norfolk County sheriff, Norfolk justice of the peace and Virginia state senator. Dalton even served as witness to John Patrick Kevill's application for a pension. Richard Curtis became a member of the Virginia Pilot's Association and owner of the ship chandler firm of R. Curtis Company in Norfolk. Curtis refused to sign his pension application for religious reasons; however, the pension was granted without his signature.

CSS Iron-clad Merrimac *and Her Engineers*, lithograph, 1907. *Courtesy of The Mariners' Museum, Newport News, Virginia.*

Several men left the war-torn South after the war, either fearing arrest or refusing to live in the reunified nation. Walter Raleigh Butt and Hunter Davidson went to South America. Davidson served as head of Argentina's Department of Torpedo Defense and Naval Construction. He eventually married a Paraguayan woman and lived in Paraguay until his death in 1913. He continued his interest in the Confederate navy by writing articles, letters and editorials. Walter Raleigh Butt became a captain in the Peruvian navy. He returned to Portsmouth; however, he could not find employment. Butt moved to California and served as an officer of the Pacific Mail Steamship *City of San Francisco* until his death in 1885. John Taylor Wood moved to Halifax, Nova Scotia, where he started a shipping and marine insurance firm. Wood wrote several excellent articles about the Hampton Roads battles for *Century Magazine*. Charles Schroeder joined Wood in Canada and then worked in China as an engineer for the Pacific Mail Steamship Company. Schroeder finally returned to Portsmouth and worked for E.V. White's hardware company.

Robert C. Foute endeavored to find his sweetheart from Paris, Mary deKantzow, when the war ended. Unfortunately, Foute traveled through Washington, D.C., wearing his Confederate uniform and was imprisoned for two months. He was released, then married and was ordained as an Episcopal minister. Douglas French Forrest, serving as paymaster of the CSS *Rappahannock*, was detained by French authorities until April 1865 and returned to the United States via Cuba. He was also ordained as an Episcopal minister during the postwar era.

Henry H. Marmaduke served as superintendent of Consular Bureaus of South American Republics until assuming command of the Colombian cruiser *Bogota*. He returned to the United States and worked in the U.S. Naval War Records Office with Hardin Littlepage. March 8, 1862, had been Littlepage's twenty-first birthday, and he diligently pursued the collection and publication of Confederate naval history for the rest of his life.

Not all of the crew survived the postwar era with such success. John T. Huddleston was severely wounded in the lung at the Second Battle of Manassas. His wound never properly healed. Huddleston eventually moved into the Robert E. Lee Camp Soldiers' Home in Richmond, Virginia. He lived there until he committed suicide by tying iron quoits around his waist and jumping into a pond in Byrd Park. The veteran left a suicide note revealing the location of his body and apologizing for borrowing the quoits from the Soldiers' Home.

As the years passed, several individuals sought a glorious connection with the *Virginia*. When Robert Turlington died in Phoebus, Virginia, on July 13, 1907, the local newspaper ran an obituary claiming Turlington had served as a gunner on the *Merrimac*. Turlington had actually fought in Hampton Roads aboard the CSS *Jamestown*. Edmund Curling, justice of the peace at Great Bridge, Virginia, visited his children in Norfolk in 1900 and was reported by the local newspaper as "Old Hero Visits Friends in Berkley." The article stated that Curling "has the distinction of having cut the ropes that held the *Merrimac* to the dock when she was sent out…to break the Federal blockade of Norfolk harbor."

One by one, however, the *Virginia*'s veterans passed away. John Patrick Kevill was the last survivor of the CSS *Virginia*, dying at age ninety-five on January 3, 1941. Kevill had served on the ironclad that had ushered in the age of the steel navy and was one of the men who could proudly claim, "I fought in Hampton Roads." The statement, according to William Norris, was "an open sesame to the hearts and homes of all our own countrymen. Ah! The thrilling memories of those halcyon days."[24]

Appendix I

You Say *Merrimack*, I Say *Virginia*

The March 9, 1862 Battle of Hampton Roads, similar to many other Civil War engagements, has often been called numerous other names. The Battle of the Ironclads or the *Monitor-Merrimack* are two of the frequently used titles. This confusing nomenclature begs another question: What is the proper name of the Confederate ironclad? Is it the *Merrimac*, *Merrimack* or *Virginia?*

The *Monitor* does not suffer from this type of identity crisis. The Union ironclad's inventor, John Ericsson, was asked by Assistant Secretary of the Navy Gustavus Vasa Fox to give the new ironclad, referred to as "Ericsson's Battery," a proper name. Since Ericsson believed that his innovative ironclad's "impregnable and aggressive character...will admonish the leaders of the Southern Rebellion," as well as prove to be a monitor to the Royal Navy's ironclad frigate construction program, he proposed "to name the new battery *Monitor*." The ironclad was such a success that "Monitor" became the name of an entire class and type of warship.

The Confederate ironclad's name, however, is consistently inaccurate. The most common usage is *Merrimac*. This reference, used alike by Civil War participants and historians ever since, is incorrect. The steam-powered, forty-gun frigate with a screw propeller built at Charlestown Navy Yard was named the USS *Merrimack* by John Lenthall, chief of the U.S. Bureau of Naval Construction, on September 25, 1854. Naval Constructor E.H. Delano, who designed the frigate, noted the ship's name as *Merrimack* on his plans. The frigate was the first of a class of six frigates built during the 1850s.

Each of the ships was named for an American river: *Roanoke, Wabash, Colorado, Minnesota* and *Niagara*. President Franklin Pierce was a native of Concord, New Hampshire, the county seat of Merrimack County and located on the Merrimack River. He signed the act approving the appropriation and ship names on April 6, 1854, and the frigate to be built at the Charlestown Navy Yard in Boston was spelled *Merrimack*.

Even though this evidence clearly documents that the frigate's name should always be spelled with a "k" as it was named in honor of the Merrimack River, confusion concerning the river's spelling is commonplace. The first written reference to the river dates to 1691 in the grant by the joint regents of England, William and Mary, noting the northern boundary of Massachusetts as the Merrimack River. Other references to the Merrimack spelling include Governor Thomas Hutchinson's 1764 *History of the Province of Massachusetts Bay*. The name "Merrimack" is a Native American word said to mean "swift water." By the mid-nineteenth century, many writers, Henry David Thoreau excepted, had begun to drop the "k." It appears that the spelling Merrimack is more often used at places along the river above Haverhill, New Hampshire, a town located at the head of navigation. Merrimac without the "k" is the popular spelling below Haverhill. The river formed the Merrimack Valley, which was often referred to as Merrimac Valley. This region was a major textile manufacturing area. One town in the region is named Merrimac but was not established until 1876. This circumstance and the fact that it is easier to spell Merrimack with just a "c" rather than a "k" is perhaps why so many Civil War contemporaries use the term *Merrimac* when writing about the frigate. The *Boston Evening Transcript* on June 15, 1855, referred to the frigate as the *Merrimac*.

Once the Confederates raised the burned hull of the frigate, it was reconfigured into an ironclad and christened on February 17, 1862, as the CSS *Virginia*. Confederate Secretary of the Navy Stephen R. Mallory and Flag Officer Franklin Buchanan, the ship's commander, both referred to the ironclad in all of their correspondence after this date as the *Virginia*. Consequently, from February 1862, the ironclad should always be called the *Virginia*. Unfortunately, few recognized this technicality, as even the ironclad's executive officer, Lieutenant Catesby ap Roger Jones, and chief engineer, H. Ashton Ramsay, called the vessel the *Merrimac*. Both of these men served on the frigate prior to the war, which may be the cause of their usage of the *Merrimac* name. The Southern newspapers usually referred to the vessel by its rechristened name CSS *Virginia*; however, Northern newspapers constantly used the name *Merrimac* without the "k."

William Norris perhaps expressed the best summation clarifying the ironclad's proper name when he wrote:

> *And* Virginia *was her name, not* Merrimac, *which has a nasal twang equally abhorrent to sentiment and to melody, and meanly compares with the sonorous sweetness of Virginia. She fought under Confederate colors, and her fame belongs to all of us; but there was a peculiar fitness in the name we gave her. In Virginia, of Virginia iron and wood, and by Virginians she was built, and in Virginia's waters, now made classic by her exploits, she made a record which shall live forever.*[1]

The Confederate ironclad CSS *Virginia* will forevermore be indiscriminately called *Merrimac, Merrimack* and *Virginia.* Accordingly, the battle, too, will be known by several titles, but the ironclad should always be remembered as the vessel that proved once and for all that iron ships would rule the waves.

Appendix II
CSS *Virginia* Designers

At the onset of the war, the Confederate secretary of the navy recognized that the South needed a novel weapon to contest the North's overwhelming naval superiority. "I regard the possession of an iron-plated ship as a matter of the first necessity," Mallory advised the Confederate Congress. "Such a vessel at this time could traverse the entire coast of the United States," he added, "prevent all blockades, and encounter, with a fair prospect of success, their entire Navy."[1] Mallory realized that if the Confederacy could only purchase or build a fleet of ironclads, the Union wooden warships would be powerless to stop the South from gaining control of the sea.

While Mallory sent agents to Europe to secure ironclads, he also sought to begin an ironclad construction program in the South. However, building ironclads would prove to be a very difficult task. The Confederacy lacked adequate industrial resources to effectively construct ironclads. Nevertheless, Mallory was determined to initiate an ironclad program and believed that the recently captured Gosport Navy Yard provided the South with the opportunity. He accordingly instructed the brilliant naval scientist John Mercer Brooke to begin designing a warship. On June 21, 1861, Mallory organized a committee consisting of Brooke, Naval Constructor John L. Porter and Chief Engineer William P. Williamson to further develop plans. At a meeting in Richmond on June 23, 1861, Porter brought a model of a harbor-defense vessel he had previously devised (1846 Pittsburgh model). Brooke modified the design with submerged ends to enhance seaworthiness, since Mallory wanted an oceangoing warship, and increased the casemate's slope to improve the shot-proof qualities. Meanwhile, Williamson searched for a power plant. Since none could be quickly produced, Williamson suggested that the engines of the burned USS *Merrimack* be used. Porter concurred and believed that it would be best to reuse the *Merrimack*'s hull. Despite misgivings about the ship's draft, the men adapted the design to the *Merrimack*. Mallory approved the concept and detailed the men to begin work on the project. Mallory divided project responsibilities as follows:

Conversion Supervision: John L. Porter
Engines Revitalization: William P. Williamson
Armament and Armor: John Mercer Brooke

Since the *Merrimack*'s conversion was an experiment, many disagreements arose between Brooke and Porter during the conversion. Responsibilities overlapped, and Brooke was continually suggesting critical modifications. Mallory and the ironclad's executive officer, Lieutenant Catesby ap Roger Jones, generally supported Brooke's recommendations, which only increased the rancor between the two men.

The disputes between Brooke and Porter came to the foreground following the *Virginia*'s success in combat on March 8 and 9, 1862. The two-day engagement, particularly the dramatic sinking of the USS *Cumberland*, brought great acclaim to the *Merrimack-Virginia* project. Both Brooke and Porter claimed to be the ship's designer.

Several newspapers immediately following the battle gave the design honors to Porter, while Mallory accorded credit to Brooke in his report to the Confederate Congress on March 29, 1862. This prompted Porter to publish a letter in the *Richmond Examiner* a few days later claiming credit. Brooke sought to end the argument. He applied for and was granted Patent 100 from the Confederate Patent Office on July 29, 1862. The patent award did not settle the issue, and the two men carried on a war of words during the postwar era.

Who then deserves the honor of being named as the *Virginia*'s designer? William Williamson must be granted the acclaim of first recommending that the *Merrimack*'s hull and engines be used for the project. Porter immediately supported the concept. While the constructor advocated that the committee's design be adapted to the *Merrimack*, the design was not solely based on Porter's 1848 model. Porter's 1848 armored floating battery design was greatly modified by John Mercer Brooke. It was Brooke who suggested the submerged ends, the ram and valuable alterations to the casemate's configuration. Even though Porter played a critical role in the project as the on-site supervisor and having drawn all the plans, Jefferson Davis, Stephen Mallory, Catesby Jones and numerous other naval officers all agreed that Brooke, not Porter, deserved the distinction of being named the *Virginia*'s designer.

John Mercer Brooke

John Mercer Brooke was perhaps one of the most inventive and practical minds in the Confederate navy. Born at Fort Brooke near Tampa, Florida, on December 18, 1825, Brooke was the son of Brigadier General George Mercer Brooke and Lucy Thomas. He was appointed by Virginia as a midshipman in the U.S. Navy in 1841 and graduated from the U.S. Naval Academy in 1847. Brooke was assigned to oceanographic survey work, during which time he invented a deep-sea sounding apparatus for topographic mapping of the ocean floor. His fathometer won the Gold Medal of Science from the Academy of Berlin. He was promoted to lieutenant in 1855. Brooke was mapping the North Pacific when the Civil War began and resigned his commission when Virginia left the Union. He was officially dismissed from the U.S. Navy on April 20, 1861.

John Mercer Brooke immediately joined the Virginia State Navy and served as naval aide to General Robert E. Lee assisting the construction of water batteries along the James River. Brooke was appointed a lieutenant in the Confederate navy on May 2, 1861, and worked on several special projects for Secretary of the Navy Stephen Mallory, including uniform design. Brooke designed the buttons eventually worn by all Confederate naval personnel, as well as developing the initial specifications for Mallory's ironclad construction program. Brooke is credited with devising the casemate's slope to effectively deflect shot and the concept of submerged ends to enhance seaworthiness. Once the conceptual work on the South's first ironclad was accomplished, Brooke was detailed to the Bureau of Ordnance and Hydrography on June 24, 1861. While he continued to supervise the production of iron plate for the *Merrimack-Virginia* conversion, as well as other project details, Brooke invented a rifled gun for use by the Confederate navy. The Brooke gun, primarily produced in 6.4- and 7-inch models, became the standard rifled gun for seacoast fortifications, as well as aboard warships. Brooke's rifle was one of the most successful weapons produced during the Civil War.

John Mercer Brooke was promoted to the rank of commander on September 13, 1862, and was detailed as chief of the Bureau of Ordnance and Hydrography in March 1863. Brooke escaped from Richmond on April 2, 1865, and traveled to Greensboro, North Carolina, where he was paroled. Brooke was "one of the few men in the Confederate States Navy who showed genius during the Civil War," and Stephen Mallory noted that "whatever success attended the efforts of the Confederate Navy was, in no small degree, due to your skill and ability."[2]

Brooke was appointed professor of physics and astronomy at the Virginia Military Institute (VMI) in October 1865. While serving as a professor, he also sought to market his technological skills. He formed a partnership with Robert Dabney Minor and Catesby ap Roger Jones in 1866 to sell military technology to foreign governments, particularly Japan. The "Civil Bureau of Supply" failed and dissolved in 1869. A vain effort to revitalize the company in 1870 during the Franco-Prussian War did not achieve any success. Brooke then rejected an offer to serve as colonel of ordnance in the Egyptian army and remained at VMI. He died following forty years of service at the Institute on December 14, 1906.

JOHN LUKE PORTER

John Luke Porter was born on September 19, 1813, in Portsmouth, Virginia. He worked in his father's shipyard as a carpenter until obtaining a position at Gosport Navy Yard following his father's death. Porter, thanks to the patronage of Lieutenant William W. Hunter, was ordered to Pittsburgh, Pennsylvania, to work on the production of two iron-hulled warships, *Allegheny* and *Water Witch*. While superintending the construction of the *Allegheny*, Porter "conceived the idea of an iron-clad, and made a model with the exact shield which I placed on the *Merrimac*."[3] The 1846 Porter plan was a floating battery with the casemate running the entire length of the vessel. Porter submitted the plan to the Navy Department; however, no official action was taken.

Porter suffered several other disappointments at this time. The *Allegheny* proved to be a dismal failure due to mechanical problems, and in 1847, Porter failed his examination for appointment as a naval constructor. He returned to Gosport Navy Yard and, in 1852, was elected president of the Portsmouth town council. Porter retrieved his naval reputation while supervising the refitting of the USF *Constellation*. Named as Gosport's acting naval constructor, he completed the *Constellation* project and supervised the construction of the steam screw frigate *Colorado*. In 1856, he was assigned to the Warrington Naval Yard at Pensacola, Florida. Porter passed the naval constructor examination in 1857 and supervised the construction of the USS *Seminole*. Several construction defects found in the *Seminole* resulted in Porter's court-martial in 1860. He was exonerated of all charges and, in 1861, was reappointed to Gosport Navy Yard.

When Virginia seceded, Porter resigned his appointment as naval constructor and joined the Virginia State Navy. He was appointed as naval constructor in the Confederate navy and immediately began work on the *Merrimack-Virginia* conversion project. Porter worked on the initial concept with John Mercer Brooke and William P. Williamson and then was placed in charge of the ironclad's overall construction. While supervising the *Merrimack*'s transformation, Porter completed work on numerous other Confederate shipbuilding projects. Porter wrote to a friend:

> *I never was so busy in all my life. I have all the work in the navy yard to direct, and all the duties of the Bureau of Construction...I have all the planning of the various gunboats to do which are being built all over the South...The Secretary refers most of the matter concerning the building of vessels, buying of materials, etc. to me.*[4]

Porter organized repairs to the *Virginia* following the March 8 and 9 engagements and began work on the keel-up ironclad project known as the CSS *Richmond*.

Reassigned to Richmond following the loss of Gosport Navy Yard, Porter completed the CSS *Richmond* and supervised the construction of two other ironclads, CSS *Fredericksburg* and CSS *Virginia II*, at Rockets Navy Yard. He also designed and completed plans for most of the ironclads constructed thereafter in the South, including the CSS *Albemarle* and CSS *Tennessee*. Porter was responsible for hull design, decks, fittings, steering systems and overall supervision of construction for all the ironclads produced in the South. He was named chief naval constructor on January 7, 1864, and served at the Wilmington Station until paroled at Greensboro, North Carolina, on April 28, 1865.

After the war, Porter returned to Portsmouth and worked for two local shipyards until named superintendent of Norfolk County Ferries in 1883. John L. Porter died on December 14, 1893.

William Price Williamson

William Price Williamson was born in Norfolk, Virginia, on July 26, 1810. He was the son of Thomas and Anne Williamson. His father was clerk of the Bank of Virginia, Norfolk branch, and served one term (1829) as mayor of Norfolk. William Williamson studied mechanical engineering in New York and was appointed by Virginia to the U.S. Navy as an assistant engineer. Assigned to Gosport Navy Yard in 1842, he supervised the yard's machine works and by October was appointed a chief engineer. Williamson refused to take the oath of allegiance to the United States and was dismissed from the U.S. Navy on March 25, 1861. He was imprisoned from April 21 to May 22, 1861. Upon his release, he joined the Confederate navy and was appointed chief engineer on June 10, 1861.

Williamson served at Gosport Navy Yard and was responsible for refitting the *Merrimack*'s engines and mechanical systems. His outstanding engineering talents resulted in his appointment as the engineer in chief of the Confederate navy in April 1862. In this position, he supervised the acquisition, installation and operation of all Confederate warships' engines and boilers. Williamson conducted all engine system trials and inspections, as well as supervising all chief engineers in the Confederate navy. When Richmond was abandoned, Williamson fled to Greensboro, North Carolina, where he was paroled on May 1, 1865. William Williamson returned to Norfolk and died there on October 20, 1870.

Appendix III

The Commanders

The CSS *Virginia* never formally had a commanding officer. Confederate Secretary of the Navy Stephen Mallory knew the man he wanted for the command: Franklin Buchanan. Yet he hesitated to name him the ship's captain because of the time-honored seniority system, which placed two men, French Forrest and Victor Randolph, senior to Buchanan. Forrest actively sought the position, and Lieutenant Catesby ap Roger Jones later wrote that "having had so much to do with her I was actually oppressed with the undue expectation" of being named the ironclad's commander.[1] Both men would be disappointed.

Stephen Mallory had chaired the Senate Committee on Naval Affairs from 1853 to 1861. During his tenure as committee chairman, Mallory sponsored legislation creating the Naval Retiring Board to remove overage and inefficient officers from the service. Over two hundred officers were identified for discharge or retirement as a result of the board's conclusions. Mallory did not have the time in 1862 to proceed with a similar process, so he sidestepped the issue by naming Flag Officer Franklin Buchanan commander of the James River defenses. His flagship, of course, would be the CSS *Virginia*.

Three very able and experienced officers would use the *Virgini*a as their flagship during the ironclad's brief career. Buchanan's wounding on March 8 while destroying the USS *Congress*, promoted Jones to acting commander. Jones proved himself during the March 9 fight with the USS *Monitor*. Since Buchanan's wound was slow to heal, a permanent commander of the

James River Defenses was needed. Jones was once again overlooked. John Taylor Wood, who favored Jones as best suited to lead the ironclad into her next battle, noted that "Jones is not old enough, this is our system." Once again, Mallory was somewhat trapped by the seniority system; however, his selection brought another aggressive officer onto the *Virginia*'s deck. Flag Officer Josiah Tattnall, who had served with distinction in the U.S. Navy since 1812, assumed command on March 29, 1862. While Tattnall reflected that he would "never find in Hampton Roads the opportunity my gallant friend found," he steamed the *Virginia* into the harbor on several occasions to tempt the *Monitor* into battle.[2] Tattnall retained the *Virginia* as his flagship until he was forced to order the ironclad's destruction on May 11, 1862.

Of the three commanders, Buchanan and Tattnall were both seasoned veterans of the old navy, trained in the heroic mold of commanders like Oliver Hazard Perry. Jones, on the other hand, was an experienced officer who worked with John Dahlgren in the development of naval ordnance. Jones probably knew the *Virginia* better than any other officer and perhaps should have been named the ironclad's commander from the very beginning.

FRANKLIN BUCHANAN

Able, courageous and experienced, Franklin Buchanan was perhaps the most aggressive senior officer to join the Confederate navy. His strategic flair, discipline and heroic qualities made him respected and admired by all those around him. "A typical product of the old-time quarter deck, as indomitably courageous as Nelson, and as arbitrary," Lieutenant John Eggleston described Buchanan. "I don't think," he added, "the junior officer or sailor ever lived with nerve sufficient to disobey an order given by the old man in person."[3] Ashton Ramsay thought Buchanan to be "one of grandest men who ever drew a breath of salt air."[4] Consequently, Buchanan was Stephen Mallory's only choice to take the Confederate ironclad into battle.

Franklin Buchanan was born in the prominent Druid Hill section of Baltimore, Maryland, at his family estate, Auchentorlie, on September 17, 1800. He was the son of a prominent Maryland physician and a founder of the Medical Society of Baltimore, George Buchanan. His mother, Laetitia, was the daughter of former Pennsylvania governor and signer of the Declaration of Independence Thomas McKeen. He enjoyed a very comfortable life as a child and spent much of his youth with his grandfather in Pennsylvania. On July 12, 1815, he was appointed by Pennsylvania as a midshipman and served initially under Oliver Hazard Perry on board the USS *Java*. During the next five years, he served aboard the *Constitution* and several other ships in the Mediterranean, China and the West Indies. Buchanan was assigned as the second officer of the *Dorothea* in February 1821. He then requested and received permission to become a mate on a merchant ship destined for China. After fifteen months at sea, he returned to the U.S. Navy and was promoted lieutenant on January 13, 1825. His first assignment was to deliver the USS *Baltimore* to Emperor Dom Pedro I of Brazil at Rio de Janeiro. Buchanan was detailed to the USS *Constellation* until transferred to the USS *Delaware*. He was aboard the *Delaware* when the seventy-four-gun ship of the line transported U.S. Minister Edward Livingston to France in 1833 and was invited, along with the *Delaware*'s other officers, to attend a state dinner with King Louis Philippe.

Assigned to shore duty in 1834, Buchanan tested guns at the Philadelphia Navy Yard and commanded the Baltimore rendezvous. Buchanan served on the USS *Constitution* and *Falmouth* during a Pacific cruise from April 1839 to June 1840. He was promoted commander on September 8, 1841, and was detailed as captain of the USS *Mississippi* in 1842. While commanding the USS *Vincennes* in 1843, his ship assisted two British merchant ships in peril in Galveston Harbor, and Buchanan received the official thanks of the British government.

In 1845, Buchanan submitted a plan to establish a naval academy. The concept was approved, and Buchanan was selected by Secretary of the Navy George Bancroft as the first superintendent of the new U.S. Naval Academy at Annapolis, Maryland. While he only served as superintendent for two years, he rapidly transformed old Fort Stevens into a viable midshipman's school. He established the academy's curriculum and management and interjected a stern sense of discipline, which fostered the academy's growth.

Buchanan immediately sought an active sea command when the Mexican War erupted. Named captain of the twenty-two-gun sloop *Germantown* in March 1847, Buchanan served under commodores David Connor and Matthew Calbraith Perry. He participated in the siege of Vera Cruz and helped to coordinate the landing of General Winfield Scott's troops. His role in the capture of San Juan d'Uuoa was particularly distinguished. Buchanan was honored by the Maryland legislature and awarded 160 acres in Iowa by the U.S. Congress. Following the war, Buchanan served on several naval boards, including the Lighthouse Committee, as well as recruiting duty in Baltimore, Maryland. Buchanan was detailed as captain of the flagship USS *Susquehanna* during Commodore Matthew C. Perry's expeditions to Japan in

1852 and 1853. He was purportedly the first senior officer to step foot on Japanese soil. Promoted to captain on September 14, 1855, he commanded the Washington Navy Yard until April 20, 1861.

Like so many other Southern officers serving in the prewar U.S. Navy, Buchanan was torn between his loyalty to the Union and his devotion to his home state during the secession crisis. "I am as strong a Union man as any in the country," Buchanan wrote. "Union under the Constitution and laws and as to the stars and stripes I have as strong a loyal feeling for them as anyone who was ever born; I have fought my country's enemies under the glorious stripes, and will do again when occasion calls for my services, but as to fight my own countrymen and relatives under it I never can. I am no secessionist," Buchanan added, "and did not admit to the right of secession, but at the same time I admit the right to revolution."[5] Buchanan proudly displayed his fidelity to the Stars and Stripes during the wedding of his daughter, Nannie, to Lieutenant Julius Ernest Meiere, USMC, on April 3, 1861. The reception was held in the Washington Navy Yard headquarters. "The house was everywhere festooned with the American flag, even to the bridal bed," remembered David Dixon Porter. Even though a number of suspected disloyal officers were present, President Abraham Lincoln attended the ceremony and even cut the wedding cake. The turmoil of secession touched the gala when "Buchanan's daughter, Elizabeth Taylor, then fifteen years old, refused to shake hands with Mr. Lincoln, at first who called her a little rebel and finally won her over with bonbons and the charm of his personality." A little more than two weeks later, David Porter reflected, "the Commandant, including his new son-in-law, resigned their commissions and left the Washington Navy Yard to take care of itself."[6]

Buchanan's Southern connections immediately placed him under suspicion by the Lincoln administration. "I was, by reliable friends, put on my guard as respected...Buchanan," Gideon Welles wrote, as he was "being courted and caressed by the Secessionists." Buchanan, however, remained loyal to the Union until the April 19, 1861 Baltimore Riots. Believing that his native state of Maryland would soon secede, Buchanan resigned his commission on April 20, 1861. When Maryland did not leave the Union, he sought reinstatement. "I never was an advocate for secession," Buchanan reflected to a friend. "I am a strong Union man...I have had a horror of fighting against the 'stars and stripes.'" He wrote to Gideon Welles, "I am ready for service," noting that he hoped to be assigned overseas so that he would not have to fight against the South. "Your name," Welles replied, "has been stricken from the rolls of the Navy."[7]

Even though he was courted by the Confederacy, Buchanan retired to his estate near Easton, Maryland, until September 1861. Then, after transferring all his property to his family, he went south and was commissioned a captain in the Confederate navy on September 15, 1861. Mallory, despite some questions by other naval officers as to Buchanan's fidelity to the South, appointed Buchanan as head of the Office of Orders and Detail. Buchanan administered this important office with tremendous imagination and supervised the publication of the Confederacy's *Navy Regulations*. He developed a system of merit promotions to replace the seniority system. Franklin Buchanan's work and commitment to the Confederacy made him one of Mallory's most trusted advisors.

On February 24, 1862, Buchanan was named flag officer in command of the James River naval defenses. This appointment was Mallory's effort to circumvent the old seniority system by placing Buchanan in command of the CSS *Virginia*. The new ironclad became Buchanan's flagship, and he was determined to take the *Virginia* as quickly as possible.

He endeavored to gain the cooperation of Major General John Bankhead Magruder to attack Camp Butler at Newport News Point. Magruder, however, rejected the concept, and Buchanan proceeded with his own plan. On March 8, 1862, Buchanan rammed and sank the USS *Cumberland*, destroyed the USS *Congress* and sank three other smaller steamers. It was the Confederacy's greatest naval victory; however, Buchanan was unable to fully enjoy his success. While supervising the destruction of the USS *Congress*, Buchanan climbed atop the *Virginia* and exchanged rifle fire with Union troops on the shore. Buchanan was wounded in the hip by a bullet that grazed his femoral artery. Consequently, he turned the command of the *Virginia* over to his executive officer, Lieutenant Catesby ap Roger Jones.

Buchanan's wound was very painful and healed very slowly. Two months passed before he could walk with crutches, and the wound continued to bother him throughout the summer. He convalesced first in the Portsmouth Naval Hospital and then completed his recovery at Greensboro, North Carolina.

On August 21, 1862, Buchanan's appointment as the first admiral in the Confederate navy was confirmed by the Confederate Congress. As the ranking officer in the Confederate navy, Mallory assigned Buchanan to command the naval defenses in Mobile Bay, Alabama. Mallory hoped that the Confederate admiral would be able to use the ironclads under construction at Selma, Alabama, to break the blockade and, perhaps, liberate New Orleans and the lower Mississippi River.

Only the CSS *Tennessee* was completed by early 1864. Buchanan added this ironclad to his fleet of three wooden gunboats and made plans to attack the blockading squadron once additional ironclads were completed. However, Admiral David Glasgow Farragut struck first. On August 5, 1864, Farragut's fleet of wooden frigates and monitors steamed past the forts guarding the entrance to Mobile Bay. Admiral Buchanan, using the *Tennessee* as his flagship, then attacked the entire Union fleet with his ironclad. Buchanan was wounded in his leg during the engagement, and the *Tennessee*, disabled by Union shot and shell, surrendered. The Confederate admiral was imprisoned at Fort Lafayette, New York, until exchanged in February 1865. He was reassigned to Mobile, but the city surrendered and the Confederacy collapsed before he could return to duty.

After the war, Buchanan retired to Talbot County, Maryland. He became president of the Maryland Agricultural College (later the University of Maryland) and then moved back to Mobile, Alabama, as secretary of the Alabama Life Association. He died on April 11, 1874, at his estate, The Rest, near Easton, Maryland.

CATESBY AP ROGER JONES

Admiral David Dixon Porter remarked at the outbreak of the Civil War that he regretted the loss of only two officers to the Union service. John Mercer Brooke was one of these officers who joined the Confederate navy in 1861; the other was ordnance expert Catesby ap Roger Jones.

Tall, well proportioned and always immaculately dressed, Jones maintained a commanding presence despite walking with a slight limp caused by a gunshot wound to the hip. Jones was born in Fairfield, Virginia, on April 15, 1821. His father, Brigadier General Roger Jones, was adjutant general of the U.S. Army. His uncle, Commodore Thomas ap Catesby Jones, was a hero of the Battle of New Orleans during the War of 1812 and received national acclaim in 1842 for his capture of Monterey, California. The unusual Welsh idiom "ap" in Jones's name means "son of." Thus, Catesby was the son of Roger Jones.

Jones entered the U.S. Navy as a midshipman on June 18, 1836. He was promoted to passed midshipman in 1842 and served under his uncle, Thomas ap Catesby Jones, during the "Exploring Expedition." Promoted to master on September 14, 1848, and lieutenant on May 12, 1849, Jones was then assigned to the Washington Navy Yard. During the 1850s, Jones worked on artillery experiments with Captain John Dahlgren, chief of the U.S. Navy Ordnance Bureau. Jones went on the first voyage of the USS *Merrimack* to test the frigate's cannons. All of the new guns mounted in the *Merrimack*'s broadside were of Dahlgren's design.

After Virginia's secession, Jones resigned from the U.S. Navy and was appointed a captain in the Virginia State Navy. Jones was immediately dispatched to Norfolk with Captain Robert B. Pegram to assume command of the naval station and enlist seamen. He coordinated the capture of the Norfolk Arsenal at Fort Norfolk on April 19, 1861, which prompted the destruction of the Gosport Navy Yard the next evening by retreating Union naval forces.

Jones was commissioned lieutenant in the Confederate navy on June 11, 1861, and assigned to command the batteries on Jamestown Island. While serving at Jamestown, Jones worked with John Mercer Brooke testing the iron plate and rifled cannons for the *Merrimack* conversion project. Because of his experience and ordnance expertise, Stephen Mallory ordered Jones to assume the position of executive officer of the new ironclad in November 1861. Jones prepared the ironclad's battery, began recruiting a crew and sought to resolve the differences between the ironclad's designers Brooke and Porter.

While Jones was extremely dismayed that he was not named the *Virginia*'s commander, he continued to serve as the ironclad's executive officer under Flag Officer Franklin Buchanan. Jones became acting commander of the *Virginia* when Buchanan was wounded during the fight with the USS *Congress* and ably commanded the ironclad in its engagement with the USS *Monitor*. He was once again disappointed when Flag Officer Josiah Tattnall replaced Buchanan. Jones remained as the ironclad's executive officer until the *Virginia* was destroyed to prevent its capture on May 11, 1862. He then led the crew to Drewry's Bluff, where he supervised the construction of a battery to help defend this approach to Richmond. He commanded the battery during the May 15, 1862 fight with Commander John Rodgers's squadron, including the ironclads *Monitor*, *Galena* and *Naugatuck*.

Jones was then assigned to command the CSS *Chattahoochee* at Columbus, Georgia; however, in 1863, he was sent to supervise the Charlotte Naval Ordnance Works. He was finally recognized for his gallant and meritorious conduct aboard the *Virginia* when promoted to the rank of commander on April 29, 1863. By this time, he had assumed command of the Selma Iron Works. Soon renamed the Confederate Naval Gun Foundry, Jones's brilliant mind and exceptional engineering talents quickly transformed the foundry into a major manufacturer of rifled cannon. He personally supervised each large casting and, on one occasion, narrowly escaped death or injury from an explosion. The Selma Confederate Naval Gun Foundry produced 143 heavy guns, many of which were used on the CSS *Tennessee* and in the forts defending Mobile, Alabama. As the war neared its end, Jones was assigned to the Mobile Bay Squadron. He was paroled on May 9, 1865, on board the USS *Stockdale.*

After the war, Jones established a business partnership with former naval associates John Mercer Brooke and Robert D. Minor. Together they purchased military supplies in the United States and sold them to foreign governments. The business failed. Jones died on June 20, 1877, after being shot by a Selma neighbor during a quarrel between two of their children.

JOSIAH TATTNALL

Dashing, daring and devoted to his nation, Josiah Tattnall was known throughout the U.S. Navy as the "beau ideal of a naval officer." His imposing stature and dedicated service made Tattnall a legendary figure. Almost six feet tall with a large head, sunken blue eyes and a protruding lower lip, Tattnall's great strength and rather long arms made him feared in his younger days as a cutlass expert. "He possessed all of the trait's which are found in heroic characters," William Parker wrote, "and with suitable opportunities, would have set his name among the great naval worthies who are historic."

Josiah Tattnall was born at his family estate, Bonaventure, near Savannah, Georgia, on November 9, 1795. His father was Josiah Tattnall, governor of Georgia. Orphaned at the age of ten, he was educated in England under the supervision of his maternal grandfather. Tattnall returned to Savannah six years later, and on January 1, 1812, he was appointed midshipman in the U.S. Navy. During the War of 1812, he was assigned to the eighteen-gun brig *Epevier* at Savannah and then the thirty-eight-gun frigate *Constellation.* While serving on the USS *Constellation* trapped in Norfolk by the British blockade, he fought with distinction in the repulse of the British attack on Craney Island on June 23, 1813. Tattnall led a small detachment at the end of the battle onto the mud flats off Craney Island to capture Admiral Sir John Borlase Warren's barge, the *Centipede.* Following the war, Tattnall served on the *Epevier*, *Constellation* and *Ontario* in operations against the Barbary States.

He was promoted to lieutenant on April 1, 1818, and was assigned to the USS *Macedonian* in the Pacific. In 1821, Tattnall married Henrietta Fenwick Jackson, his mother's cousin, and took a leave of absence to attend Patridge's

Military School. He then served in Commodore David Porter's squadron suppressing piracy in the West Indies aboard the USS *Jackal* in 1823. After two years' service in the Mediterranean Squadron on the USS *Constitution* and USS *Brandywine*, Tattnall returned to the West Indies. During 1828, he completed surveys of the fortifications on the Tortugas Reef while assigned to the USS *Erie*. Shortly after he assumed command of the ten-gun schooner *Grampus* in 1831, he captured the Mexican pirate brig *Montezuma*, which had recently attacked the American schooner *William A. Turner*. As commander of the barque *Pioneer* in 1836, Tattnall became a national hero when he returned President Antonio Lopez de Santa Anna of Mexico to Vera Cruz after his capture by Texans during their war for independence. Tattnall was lauded for his thoughtful handling of Santa Anna. Many people credited Tattnall with saving the Mexican president's life.

Josiah Tattnall's forthright and dynamic leadership led to his promotion to the rank of commander in 1838. He spent most of the 1840s as commandant of the Boston Navy Yard and commanded the eighteen-gun corvettes *Fairfield* and *Saratoga* in the Mediterranean Squadron. When the Mexican War erupted, Tattnall was captain of the gunboat *Spitfire* in the Gulf of Mexico. He was severely wounded in the arm during the siege of Vera Cruz and was once again acclaimed as a national hero. Tattnall was honored by the Georgia legislature and presented with a gold sword. He was then reassigned to command the Charlestown Navy Yard near Boston.

Promoted to the rank of captain on February 5, 1850, he took command of the USS *Saranac*, USS *Independence* and the Sacketts Harbor Naval Station. In 1857, Tattnall was named commodore of the East India Squadron. He ignored the U.S. neutrality concerning the Taiping Rebellion in China when he aided British warships trapped by Chinese forts on the Pei-ho River. Tattnall exclaimed, "Blood is thicker than water!" when asked to justify his actions. When Tattnall returned, escorting Japan's first ambassador to the United States in 1860, he was feted as President James Buchanan's guest of honor at the White House.

When Georgia left the Union, Tattnall regretfully resigned his U.S. Navy commission on February 20, 1861. He had opposed secession, noting to one junior officer about his grandfather, "He was a Tory, sir, stood by his King, sir." Nevertheless, Tattnall was named the senior flag officer in the fledgling Georgia State Navy on February 28, 1861. One month later, he was commissioned a captain in the Confederate navy and assigned command of the naval defenses of Georgia and South Carolina. Tattnall, with only one riverboat and three tugs, harassed the Union blockade but could do little to stop Flag Officer Samuel Francis DuPont's capture of Port Royal Sound, South Carolina, in November 1861.

On March 21, 1862, Josiah Tattnall received the prestigious assignment of replacing the severely wounded Franklin Buchanan as flag officer of the James River Defenses with the CSS *Virginia* as his flagship. Despite Tattnall's age, Mallory believed that the veteran flag officer was an excellent selection for the post. Even though Tattnall's new command brought him back to the scene of his first battle at Craney Island, he recognized that his new assignment placed his fifty-year naval reputation at risk. He wrote to Mallory, "I have been aware from the first that my command is dangerous to my reputation, from the expectations of the public, founded on the success of Commodore Buchanan, and I have looked to a different field from his to satisfy them. I shall never find in Hampton Roads the opportunity my gallant friend found."

During the next six weeks, Tattnall endeavored to lure the *Monitor* into battle within the confines of Hampton Roads. While the public clamored for more, the Confederate retreat from the Lower Peninsula forced the evacuation of Norfolk and Gosport Navy Yard. Tattnall was faced with two options: send the *Virginia* out to attack the Union fleet or take the ironclad up the James River toward Richmond. Unfortunately, there was not adequate time to lighten the deep-draft ironclad sufficiently to navigate the James River. Rather than surrender his ship, Tattnall ordered the *Virginia*'s destruction on May 11, 1862. Tattnall was blamed for the ironclad's loss, yet his actions were upheld by a court-martial. Instead of chastisement, Tattnall was exonerated and complimented for his fidelity.

Tattnall returned to Savannah as commander of the naval defenses of Georgia and South Carolina. Delays in sending the ironclad, CSS *Atlanta*, into combat against the Union blockading squadron prompted Mallory to relieve Tattnall. Tattnall was reassigned to command the Savannah shore station and directed the construction of two ironclads, the *Savannah* and *Milledgeville*. The warships were destroyed by Tattnall to prevent their capture when Major General William T. Sherman occupied Savannah.

When the war ended, Tattnall was paroled on May 9, 1865, and moved his family to Nova Scotia. Penniless, he returned to Savannah in 1869 to serve as the inspector of the port. He died seventeen months later. Buried at Bonaventure, Savannah's church bells tolled for ninety minutes, the flags all flew at half mast and businesses closed in his honor.

Appendix IV

CSS *Virginia* Officers' Assignment Dates

Catesby ap Roger Jones, Nov. 11, 1861
Algernon S. Garnett, Nov. 18, 1861
John Taylor Wood, Nov. 25, 1861
Charles Carroll Simms, Nov. 30, 1861
Benjamin S. Herring, Dec. 1, 1861
John Randolph Eggleston, Dec. 3, 1861
Henry Hungerford Marmaduke, Dec. 3, 1861
Marshall P. Jordan, Dec. 3, 1861
Eugenius Alexander Jack, Dec. 3, 1861
George Washington City, Dec. 4, 1861
Hunter Davidson, Dec. 8, 1861
John W. Tynan, Dec. 9, 1861
James Allen Semple, Jan. 1, 1862
Henry Ashton Ramsay, Jan. 1, 1862
Hardin Beverly Littlepage, Jan. 1, 1862
Lawrence M. Rootes, Jan. 1, 1862
William Parrish, Jan. 1, 1862
Loudon Campbell, Jan. 1, 1862
Elsberry Valentine White, Jan. 18, 1862
Walter Raleigh Butt, Jan. 22, 1862
Dinwiddie Brazier Phillips, Jan. 27, 1862
Arthur Sinclair IV, Feb. 1, 1862
Robert Chester Foute, Feb. 12, 1862
William James Craig, Feb. 19, 1862

Robert Dabney Minor, Feb. 26, 1862
James Crosby Long, Mar. 1, 1862
John Hazelhurst Ingraham, Mar. 9, 1862
Charles Schroeder, Mar. 11, 1862
Franklin B. Dornin, Mar. 12, 1862
Josiah Tattnall, Apr. 1, 1862
Paulding Tattnall, Apr. 1, 1862
John Pembroke Jones, Apr. 1, 1862
Barron Carter, Apr. 1, 1862

Appendix V

Confederate Navy Volunteers Aboard the CSS *Virginia*

The *Virginia*'s muster rolls list 431 men serving aboard the ironclad from March 8, 1862, through May 11, 1862. Of this number, only 75 originally enlisted in the Confederate navy. "There was a sprinkling of old man-of-war's men," Lieutenant John R. Eggleston later wrote about the ironclad's crew, "whose value at the time could not be overestimated." These U.S. Navy veterans and merchant sailors included:

Albright, E., paymaster's clerk
Armstrong, Alex, landsman
Ault, Charles, seaman
Ball, Lemuel, seaman
Baxter, Henry, coal heaver
Benthall Robert, ordinary seaman
Brogan, John, coal heaver
Brower, Emsey H., ship's corporal
Cahill, James, captain of forecastle
Canning, William, quartermaster
Carey, Thomas, quartermaster
Clarke, William T., pilot
Collins, James, first class fireman
Collins, Phillipp, seaman
Cunningham, John, ordinary corporal
Cunningham, Thomas, pilot

Dinning, Cornelius, fireman
Divers, David, seaman
Domat, Moses, landsman
Donnovant, Michael, landsman
Duncan, Thomas M., quartermaster
Fisher, Joseph, seaman
Fisher, William, seaman
Gaskill, Silas, ordinary seaman
Gillan, James, seaman
Goff, John, first class fireman
Gray, James H., captain of top
Hannon, James, ordinary seaman
Harrison, Charles W., ordinary seaman
Harvey, Lawrence, landsman
Hinds, Laurence, quartermaster
Hoyt, Henry, ordinary seaman
Hunt, John, seaman
Ives, Emerson H., seaman
Jarvis, William R., carpenter's mate
Johnson, Elisha R., gunner's mate
Joice, John, second class fireman
Jones, John, landsman
Jones, Samuel, first class fireman
King, Alfred, ordinary seaman
Lackie, Pierre, officer's cook
Leary, James, ordinary seaman
Lilles, Michael, second class fireman
Lindsay, Hugh, carpenter
Loyd, William, coal heaver
McCrady, John, coal heaver
McCubbins, John, coal heaver
Meads, Charles C., yeoman
Mercer, James E., seaman
Mercer, Thomas P., seaman
Moore, Robert W., landsman
Mulroy, John, first class fireman
Murphy, Patrick, ordinary seaman
Nelson, John, seaman
Ollsen, Jacob, seaman

Patrick, James A., landsman
Peterson, Andrew G., boatswain's mate
Riley, Owen, coal heaver
Ryan, James, ship's carpenter
Saunders, Thomas, seaman
Scott, James R., first class fireman
Sheriff, Benjamin, quarter gunner
Shever, James, officer's cook
Spence, Robert, seaman
Stewart, David, captain of hold
Tolson, Valentine, seaman
Waters, John, captain of top
Waters, Robert, coal heaver
Webb, James, officer's steward
Wilkins, Willis A., carpenter's mate
Williams, Hezekiah, pilot
Wright, George, pilot

Appendix VI
Confederate Marines Aboard the CSS *Virginia*

Captain Reuben T. Thom, the first officer appointed to the Confederate States Marine Corps (CSMC), began recruiting Company C, CSMC, on March 25, 1861. Even though Thom was a distinguished Mexican War veteran, he had no prior Marine Corps experience. Thom was fortunately able to enlist as his first recruit a twelve-year United States Marine Corps (USMC) veteran, Jacob Scholls. Scholls immediately was promoted to first sergeant and assisted Thom with the organization of Company C. Thom's company was initially assigned to the Warrington Navy Yard at Pensacola, Florida.

While stationed at Warrington Navy Yard, Company C formed part of a battalion of three hundred marines commanded by Lieutenant Colonel Henry B. Tyler. However, Captain Thom had already been sent to the New Orleans Naval Station to take command of fifty-five recruits. On July 6, 1861, Thom commanded this detachment of marines when Ship Island was occupied by the Confederates. Thom and his men engaged the USS *Massachusetts,* an iron-hulled screw steamer armed with one 42-pounder and four 8-inch guns. The Confederate artillery out-ranged those of the *Massachusetts* and forced the Union gunboat to disengage. Lieutenant Alexander F. Warley, the senior naval officer of the expedition, lauded the efforts of Captain Thom, noting, "Where work was to be done, there was the captain to be found and his men working as I never saw raw recruits work before."[1]

Thom received orders on November 29, 1861, to take Company C to Gosport Navy Yard. The company had just been issued Enfield rifled muskets (November 28, 1861) and arrived at Gosport on December 7, 1861. The company was initially stationed aboard the CSRS *Confederate States*. Marine detachments were assigned as guards aboard the CSS *Jamestown* (Second Lieutenant James R.Y. Fendall with twenty men) and the CSS *Yorktown* (First Lieutenant Richard H. Henderson with twenty-four men). Thom retained command of the rest of the company while these marines served as guards at the navy yard. Thom took fifty-four marines as guards on the CSS *Virginia* the day the ironclad was commissioned, February 17, 1862.

When the CSS *Virginia* steamed down the Elizabeth River to attack the Union warships in Hampton Roads on March 8, 1862, Captain Reuben Thom commanded broadside Guns #8 and #9. Captain Thomas Kevill of the United Artillery aided the management of Gun #9. During the engagement with the USS *Cumberland* and USS *Congress*, Flag Officer Franklin Buchanan noted that the "Marine Corps was well represented by Captain Thom, whose tranquil mien gave evidence that the hottest fire was no novelty to him."

Aboard the *Virginia* with Thom were fifty-four marines from Company C:

Aenchbacker, Samuel N., private
Aird, Hugh, private
Baines, John, musician
Barclay, John, private
Bell, William, private
Bessant, Abraham W., private
Brennan, Thomas, private
Brewster, William H., private
Briggs, William, corporal
Campbell, James, private
Charlesworth, Joshua, sergeant
Coleman, Thomas, private
Coyle, Robert, private
Creilly, Thomas, private
Cunningham, James, private
Curtis, Charles Samuel, private
Davis, Theodore, private
Davis, William Henry, private
Dewey, William H., private

Dohran, Thomas, private
Driscoll, Owen, private
Egan, Daniel, private
Faley, John, private
Gibbons, Tobias, private
Hampton, Cornelius, private
Hart, Charles J., corporal
Hickey, John, private
Howell, John, private
Huddleston, William C., private
Jones, Thomas, private
Kelly, Michael, private
Langtree, John W., private
Leahy, James, private
Litchfield, Orson, private
Marcy, Michael, private
Marry, Peter, private
McGinnis, Patrick, private
McLaughlin, Peter, private
Moriarty, Daniel, private
Phillips, Jerome B., private
Purcell, Thomas, private
Sailor, John, private
Scholls, Jacob S., first sergeant
Schwartz, Henry, private
Shavor, Jacob, private
Smith, James F., private
Smyth, Samuel Bell, private
Sparks, Charles M., private
Stack, Garrett N., private
Turner, Robinson, private
Watson, Oliver, Private
Wenzel, Joseph, private
Whitten, James, private

It is believed that many of these marines served on cannon crews during the March 8 and 9 engagements.

Following the battle with the USS *Monitor*, command of the *Virginia*'s Marine Guard fell on Captain Julius Ernest Meiere, CSMC. Meiere, a former member of the USMC and a son-in-law of Franklin Buchanan, commanded the detachment through the Battle of Drewry's Bluff, when Company C members served as sharpshooters along the banks of the James River. Captain John D. Simms was in overall command of the marine battalion from all the Confederate gunboats stationed at Drewry's Bluff.

Appendix VII

Confederate Army Volunteers Aboard the CSS *Virginia* by Unit Designation

The vast majority of the CSS *Virginia*'s crew were Confederate soldiers who "transferred to steamer *Merrimac*" as a result of intensive recruitment organized by Catesby Jones and John Taylor Wood. Despite their lack of naval experience, these soldiers-turned-sailors "proved themselves to be as gallant and trusty a body of men as anyone wished to command," remembered John Taylor Wood.

Lieutenant John R. Eggleston asserted that the crew was "made up mostly of volunteers from the various regiments stationed about Norfolk at the time." Of the 300 soldiers serving on the ironclad, 124 had previously enlisted in Virginia regiments. The United Artillery (41st Virginia) accounted for 41 of these volunteers. North Carolina provided the next largest number of men, 58, followed closely by Louisiana, 42. All but two Confederate states, Tennessee and Florida, provided volunteers from army units. One border state, Maryland, provided transfers to the ironclad.

Alabama

3RD Alabama Infantry

Co. D: McDevett, Charles, private

Co. G: Black, John G., private

Claysing (Claysung), Ernest R., private

Roach, James I., corporal

Scott, William, private

Co. H: Harvey, James M., private

8th Alabama Infantry

Co. E: Allen, Benjamin S., private
Gallager, Charles, private
Hughes, Patrick, private
Lawler, William F., private
Lembler, August W., private
Martin, Patrick, private
Merriam, James, private
Williams, Peter, private

12th Alabama Infantry

Co. A: Bennett, Thomas, private
Edwards, James, private
Hansell, Thomas, private
Martin, Thomas, private
Perry, John, private
Pettit, Edmond, private

Arkansas

1st Arkansas

(Colquitt's) Infantry Reardon, John W., private
Co. F: Coughlan, Patrick, private
Co. H: Rudd, James, private
Co. I: Durand, Julius, private
Co. K: Carr, James, private

Georgia

3rd Georgia Infantry

Co. D: Jones, William T., private
Co. G: Davis, Richard L., private
Co. H: Jones, William H.H., private
Russell, George, private
Tinsley, Jefferson M., private

4th Georgia Infantry

Co. A: Archer, Frederick, private
Co. C: Lamb, Andrew J., private

Moore, George T., private
Moore, John T., private
Nelson, Rambling W., private
Southall, Benjamin F., private
Tharpe, Marcellas A., private

5TH GEORGIA INFANTRY
Co. B: Manly, Richard, private

22ND GEORGIA INFANTRY
Co. A: O'Halloran, Dixon R., private
Cobb's Legion Cavalry
Co. B: Sprague, Chauncey A., private
Cobb's Legion
Co. C: Brown, Andrew, private.
Co. G: McHenry, William, private

LOUISIANA

1ST LOUISIANA
(Nelligan's) Infantry
Co. B: Herd, William C., private
Co. C: Allman, William D., private
Logan, Patrick H., private
Morton, Edwin, private
Co. E: Dunlop, John, private
Co. F: Maxwell, John, private
McBride, William E., private
McGowan, John, private
Co. K: Brown, William H., private

2ND LOUISIANA INFANTRY
Co. A: Muirhead, Philip T., private
Co. C: Volentine, John C., third corporal
Co. E: Scott, John, private
Co. F: Barnard, George W., private
Co. I: Henderson, Eugene T., private
Sharpe, Andrew Jackson, private
Co. K: Mathews, Sterling N., corporal

5th Louisiana Infantry
Co. F: Hayward, William H., second corporal

8th Louisiana Infantry
Co. D: Ahern, David, Private

10th Louisiana Infantry
Co. A: Ryall, Peter, private
Co. C: Harrington, Edward, private
Noon, Patrick, private
Co. D: Murphy, John, private
Co. G: May, George, private
McCoy, Thomas, private
Williams, George, private
Witz, William, private

14th Louisiana Infantry
Co. F: Hoar, John, private

15th Louisiana Infantry
Co. A: McClure, Michael, private
McNamee, Christopher, private
McQuinn, John, private
Co. D: Dunbar, Charles (John), private
Genzmer, William, private
Jones, David, private
Pryde, Nicholas P., private
Ross, James, private
Co. E: McAdams, Francis, third sergeant
Co. F: Williams, John, private
Co. H: Fitzpatrick, Dennis, private
Levy, Benjamin, private

Louisiana Zouave Battalion
Co. F: Doret (Dorrey), Benjamin, private

Confederate Army Volunteers Aboard the CSS *Virginia* by Unit Designation

Maryland

1st Maryland Artillery
Battery McLaughlin, Ephraim K., private

Mississippi

12th Mississippi Infantry
Co. C: Gardell, John, third corporal

North Carolina

1st North Carolina Artillery
Co. K: Rosler, John A., private
Wilson, Charles, private.

1st North Carolina Infantry
Co. A: Carmine, Joseph, private
Leonard, John H., private
Litchfield, Spencer, private
Whitaker, William C.M., private
Co. G: Ayers (Airs), Robert E., private
Allen, Gabriel, private
Barnes, Leonard, private
Cullifer, Joseph, private
Cullington, James, private
Douglas, William W., private
Hodder, Humphrey, private
Lee, Thomas J., private
Mills, Benjamin B., private
Morris, William, private
Myers, Joseph H., private
Oliver, Joseph L., private
Pritchett, Adam, private
Runnels, John, private
Co. H: Belanger, William, private
Tetterton, William Ropheus, private

2nd North Carolina
Infantry Co. G: Thomas, George, private

3rd North Carolina Infantry
Co. C: Riley, John G., private
Co. E: Wilson, John A., private
Co. I: Parish, Hillsman, private

6th North Carolina Infantry
Co. F: Wood, Levin H., private

8th North Carolina Infantry
Co. A: Walker, George W., private

12th North Carolina Infantry
Co. D: Young, William H., private

13th North Carolina Infantry
Co. A: Harralson, Brice, private
Wood, Levin, private
Co. B: Sheffield, James M., private
Co. D: Wright, Sidney R., private
Co. E: Mitchell, Richard A., private
Ward, William H., private
Co. G: Cogins (Coggins), Thomas, private
Hedgepeth, Joseph S., private
Price, William M., private
Co. K: Lyon, William W., private

14th North Carolina Infantry
Co. D: Stroup, Alford W., private
Co. E: Harrison, Howell W., private
Co. F: Craig, William Pleasant, private
Co. K: Powers, William R., private
Flake, Elijah Wilson, private
Little, William C., private

17th North Carolina Infantry
Co. I: Joliff, John R., private
Co. L: Salyer, Samuel, private

Confederate Army Volunteers Aboard the CSS *Virginia* by Unit Designation

South Carolina

1st South Carolina (McCreary's) Infantry
Co. I: Callahan, John C., private
Deary, Thomas, private
Hopkins, Joseph, private
Riddock, Joseph, private
Saunders, George, private
Skerrit, James, private
Truesdale, Stephen P., private
Whalen, Edward, private
Whelin, Edward, private
Co. K: Hickey, David, private
Co. L: Anderson, Frank, private
Waldeck, Louis, private

Texas

4th Texas Infantry
Co. B: Cronin, James C., private
Delley, John, private

Virginia

2nd Virginia Infantry
Co. F: Burke, William G., private

5th Virginia Battalion
Mason, George, first corporal

6th Virginia Infantry
Co. D: Harris, William R., private
Co. G: Sinclair, Arthur, IV, private
Co. H: Ferris, James J., Jr., private
Porter, Christopher, private
Turner, Robert G., private
Walker, William G., private
Young, Ephraim, private
Co. I: Ryan, John T., private

9th Virginia Infantry
Co. A: Agnew, John W., private
Bowers, George F., private
Leonard, Jacob K., private
Leonard, Samuel, private
Williams, Pleasant H., private
Co. B: Beveridge, David, private
Forrest, Andrew H., private
Miller, Lafayette, private
Co. D: Dobbs, Thomas E., private
Gray, William, private
Hall, Carey J., private
Pitt, Lorenzo D., private
Walton, John W., private
Co. E: Halstead, Alexander, private
Co. F: Higgins, John F., private
Sturges, John J., private
Co. G: Williams, John Q.A., private
Co. H: Davis, Jeremiah A., private
Co. I: Bunting, William H., private
Henry, William C., private
Co. K: Creekmur, Charles J., private

12th Virginia Infantry
Co. K: Cooper, Astley A., private
Eanes, Charles Peter, private
Gilmore, Allen, private
Hite, Robert N., private
Huddleston, John T., private
Rainey, Theophilus, private
Slade, Benjamin, private
Thayer, Martin G., private
Traylor, Thomas A., private

14th Virginia Infantry
Co. I: Dumphrey, Thomas, private
McKue, John, private
Naughton, John, private

Confederate Army Volunteers Aboard the CSS *Virginia* by Unit Designation

32ND VIRGINIA INFANTRY
Co. A: Curtis, Richard, private
Co. E: Davis, John C., private
Co. I: Messick, Zadock Wesley, private
Co. K: Baines, Mathew, private
Palmer, Charles K., private

34TH VIRGINIA INFANTRY
Co. A: Tabb, George E., Jr., third sergeant

41ST VIRGINIA INFANTRY
Co. C: Carlin, John A., private
Cross, Thomas W., private
Hall, William O., private
Jones, Albert A., private
Perkinson, Edward, private
Satchfield, Francis, private
(United Artillery)
Co. E: Applewhite, Edward, private
Barrett, Henry C., private
Barry, James E., Jr., first lieutenant
Bell, Miles K., private
Belote, John Krendal, private
Bowers, George, private
Bullock, John Thomas, sergeant
Burns, William A., private
Capps, John C., private
Carr, John, private
Carstarphen, Richard W., private
Colonna, William B., private
Crosby, William H., private
Dalton, Andrew J., private
Drake, William Francis, private
Dudley, William, private
Duncan, James Marshall, private
Duncan, William C., private
Flynn, John, private
Gillis, John, private
Griswold, Albert C., private
Howard, Charles H., private
Johnson, William, private

Kevill, John Patrick, private
Kevill, Thomas, captain
Knight, George, private
Knowles, Daniel, private
Lakin, Edward, second lieutenant
McCarthy, Neil, private
Mowle, Jacob R., private
Nance, Robert, private
Richardson, Benjamin A., private
Scultatus, George, private
Sharp, Charles, private
Smith, John David, private
Spence, Alexander, private
Spence, Charles H., private
Steen, John, private
Stillman, Eleazor, private
Walling, Isaac H., private
Co. G: Stevens, James H., private
Co. K: Harrell, Wilson, private

53RD Virginia Infantry
Co. H: McGraw, Patrick, private

57TH Virginia Infantry
Co. I: Barker, George W., Jr., private
Barker, John H., private

115TH Virginia Militia
Co. C: Wainwright, John William, private

Lynchburg Artillery
Rice, Robert J., private

Magruder Light Artillery
Smith, George N., private

Richmond Fayette Artillery
Crump, Charles D., private

Salem Flying Artillery
Fogler, Charles J., private

Appendix VIII

CSS *Virginia* Casualties, March 8, 1862

During her two days of combat with the Union fleet, the *Virginia* only suffered severe casualties during the engagement with the USS *Cumberland*, USS *Congress* and Union shore batteries at Newport News Point. The *Cumberland* caused the major structural damage to the ironclad. Furthermore, the stricken sloop's broadsides inflicted the most serious casualties, killing two men and wounding at least one other. The other serious casualties occurred while negotiating the surrender of the USS *Congress*.

Eyewitness accounts vary as to the exact number of men injured and how they received their wounds. Surgeon Dinwiddie Phillips's report indicates that two men were killed and eight wounded. The surgeon's report included:

KILLED IN ACTION

Dunbar, Charles
Waldeck, Louis

WOUNDED IN ACTION

Officers
Buchanan, Franklin
Marmaduke, Henry H.
Minor, Robert Dabney

Seamen
Burke, William G. (Listed as William Burkes)
Ives, Emsley H. (Listed as Emerson Ivas)
Leonard, John H.
Capps, John C.
Dalton, Andrew Joseph

In a postwar article entitled "The Career of the *Merrimac*," Phillips wrote:

> *Total losses on this day (there were none on the 9th) was two men killed, two officers and one man seriously wounded, and fourteen men so slightly injured as to be able to return to duty the next day.*

This is the first of many discrepancies concerning casualties on March 8. Both Catesby Jones and John Taylor Wood wrote that the *Virginia* suffered twenty-one casualties. John Eggleston reported fourteen in his narrative, and Ashton Ramsay recorded nineteen.

These variations are most likely due to the confusion of battle on the smoke-filled gun deck. Furthermore, many of these accounts were written years later. Other than the obvious injuries to those men killed and the wounding of key officers, it was difficult to trace all the minor concussions or injuries received by sailors during the engagement.

The official report indicates that "our loss is two men killed and 19 wounded." It is impossible to identify all of the wounded; however, several casualties are well documented in battle narratives. Two men were killed. Charles Dunbar was killed by shell fragments just before the *Virginia* rammed the *Cumberland*. The shell struck the sill of the bow port when the 7-inch Brooke gun was being reloaded. Richard Curtis served on this gun crew and wrote that Charles Dunbar was killed "when captain of the gun cried 'sponge,' and he jumped over the breechin and was instantly killed by being shot in the head from the deck of the *Cumberland*." Dunbar died at Curtis's feet. Either this same broadside or a subsequent one from the sloop struck and knocked off the muzzle of Gun #2. The 6.4-inch Brooke rifle's muzzle was cut off at the trunnions. This incident resulted in the death of Louis Waldeck and the wounding of Midshipman Henry H. Marmaduke. Marmaduke was painfully wounded in the arm yet remained at his post throughout the battle.

A massive broadside from the *Cumberland* also struck Gun #9, commanded by Captain Thomas Kevill. Even though only a small section of the IX-inch Dahlgren was knocked off, the shot hit the smoothbore causing the gun to discharge. The two members of the United Artillery are listed on the surgeon's report; however, they were not wounded by this incident. The report notes that John C. Capps and Andrew J. Dalton were wounded by "musket balls coming thru gun port."

Another United Artillery member, Daniel Knowles, was reportedly wounded in the head during the battle. Knowles is probably one of approximately ten men who suffered minor injuries during the battle but were not included in the surgeon's report. This wound could have occurred when a broadside from the *Cumberland* struck and parted the hog chain, driving it into the ship and wounding several men. David Knowles probably received his injury during the loss of Gun #9's muzzle or, more likely, from standing too close to the casemate's shield when it was struck by shot. He could have been one of the several men who, according to Ashton Ramsay, "leaned against the shield [and] were stunned and carried below, bleeding at the ears."

Franklin Buchanan's wounding is perhaps the best documented of all injuries sustained on the ironclad on March 8, 1862. Buchanan was standing atop the ironclad while engaged with the *Congress* and U.S. Army units from Camp Butler when he was shot in the thigh. The bullet grazed his femoral artery and caused a very painful wound. He was taken below for treatment. The flag officer had taken such an exposed position because he was enraged by Union rifle fire from Newport News Point after the *Congress* had surrendered. Bullets had peppered the gunboats CSS *Raleigh* and CSS *Beaufort*, causing several casualties on those ships and forcing them to back away from the frigate. Buchanan then sent his flag lieutenant in a cutter to burn the *Congress*. En route to the frigate, Minor's boat was hit by a fuselage of bullets. Two of his men were knocked down by shot, and then Minor was hit in the ribs. The bullet glanced into his chest to within an inch of his heart. There is no record of which sailors were with Minor in the boat; however, Eggleston wrote that one of the men "had an eye shot out." This sailor could be the seriously wounded man mentioned by Dinwiddie Phillips. William G. Burke, Emsley H. Ives and John H. Leonard were all listed as wounded by Phillips. Unfortunately, there is no record of how they were injured or the cause of their wounds.

The *Virginia* did not suffer any significant casualties during her engagement with the USS *Monitor* on March 9, 1862.

Appendix IX

CSS *Virginia* Personnel Paroled at Appomattox, Virginia, and Greensboro, North Carolina

Appomattox, Virginia, April 9, 1865

Agnew, Jonathan W.
Aird, Hugh
Brown, William H.
Bullock, John Thomas
Colonna, William Bramwell
Dudley, William G.
Duncan, James Marshall
Flake, Elijah Wilson
Flynn, John
Folger, Charles J.
Foute, Robert Chester
Gibbons, Tobias
Gillis, John
Harrell, Wilson
Kevill, John Patrick
Kevill, Thomas
King, Charles Kirby, Jr.
Rudd, James
Scultatus, George
Steen, John

Greensboro, North Carolina, April 26–28, 1865

Butt, Walter Raleigh
Davis, Jeremiah A.
Hasker, Charles Hazelwood
Hedgepeth, Joseph S.
Mathews, Sterling N.
Rootes, Lawrence M.
Stewart, David
Tharpe, Marcellus Augustus
Thomas, George
Walker, George W.
Wood, Levin H.

Appendix X

CSS *Virginia* Officer Assignments, March 8, 1862

COMMANDER, JAMES RIVER DEFENSES
Flag Officer Franklin Buchanan

CAPTAIN'S STAFF
Flag Lieutenant: Lieutenant Robert Dabney Minor
Aides: Acting Midshipman Lawrence Rootes
Lieutenant Douglas Forrest, CSA
Clerk: Acting Midshipman Arthur Sinclair IV

EXECUTIVE OFFICER
Lieutenant Catesby ap Roger Jones

MEDICAL STAFF
Surgeon Dinwiddie B. Phillips
Assistant Surgeon Algernon S. Garnett

GUN #1: BOW PIVOT 7-INCH BROOKE RIFLE
Lieutenant Charles C. Simms, commanding

GUN #2: 6.4-INCH BROOKE RIFLE
Lieutenant Hunter Davidson, commanding
Midshipman Henry H. Marmaduke

Gun #3: 6.4-Inch Brooke Rifle
Lieutenant Hunter Davidson, commanding

Gun #4: IX-Inch Dahlgren "Hot Shot" Smoothbore
Lieutenant John Randolph Eggleston, commanding
Midshipman Hardin Littlepage

Gun #5: IX-Inch Dahlgren "Hot Shot" Smoothbore
Lieutenant John Randolph Eggleston, commanding

Gun #6: IX-Inch Dahlgren Smoothbore
Lieutenant Walter Raleigh Butt, commanding
Midshipman Robert Chester Foute

Gun #7: IX-Inch Dahlgren Smoothbore
Lieutenant Walter Raleigh Butt, commanding

Gun #8: IX-Inch Dahlgren Smoothbore
Captain Reuben Thom, CSMC, commanding
Midshipman William James Craig

Gun #9: IX-Inch Dahlgren Smoothbore
Captain Reuben Thom, CSMC, commanding
Captain Thomas Kevill, CSA

Gun #10: 7-Inch Brooke Rifle
Lieutenant John Taylor Wood, commanding

Powder Division Commander
Paymaster James A. Semple
Midshipman James C. Long

Engineering Staff
Chief Engineer H. Ashton Ramsay
Engine Room: First Assistant Engineer John Tynan
First Assistant Engineer Loudon Campbell
Third Assistant Engineer George W. City

FIRE ROOM
Second Assistant Engineer Benjamin Herring
Third Assistant Engineer E.A. Jack

SIGNAL BELL, GUN DECK
Third Assistant Engineer E.V. White

BOATSWAIN
Charles Hasker

MASTER AT ARMS
William Harris

GUNNER
Charles B. Oliver

CARPENTER
Hugh Lindsey

PILOTS
Acting Master William Parrish
Civilian Pilot George Wright
Civilian Pilot William T. Clarke
Civilian Pilot Thomas Cunningham
Civilian Pilot Hezehiah Williams

Appendix XI

CSS *Virginia* Dimensions and Statistics

When Third Assistant Engineer E.A. Jack first viewed the *Virginia*, he pronounced that the ironclad "was no mean machine." The *Virginia* had just left dry dock, and Jack thought:

> *She was a curious looking craft. The sloping sides of the shield sinking below the water about two feet and inclining upward pierced with gun ports looked like a sloping Mansard roof of a house, but a sight of the black mouthed guns peeping from the ports gave altogether different impressions and awakened hopes that ere long she would be belching fire and death from those ports, to the enemy and crashing into their wooden vessels with that formidable ram sending them to the bottom. But one thing appeared against her effectiveness, her immense draught of water—over twenty feet—would keep her to deeper channels and let the lighter vessels escape her.*

The *Virginia* looked fearsome indeed, but she was a makeshift prototype, primarily designed to fight against wooden ships. While her armament was outstanding and her armor adequate, she still suffered from a poor power plant and a deep draft. Nevertheless, the CSS *Virginia* was truly a transitional warship. Built atop the hull of a wooden frigate with weak engines, the ironclad was an experiment in ship design and construction for the Confederacy. Despite her limitations, the *Virginia* won the greatest Confederate naval victory during the Civil War and served as the prototype for all other Confederate ironclads.

Battery:
two 7-inch Brooke rifle pivots
two 6.4-inch Brooke rifles in broadside
four IX-inch Dahlgren smoothbores in broadside
two IX-inch Dahlgren smoothbores in broadside modified as hot shot guns
two 12-pounder howitzers on deck
one 1,500-pound cast-iron projecting the stern by 2 feet

Gun Ports: fourteen Elliptical, four to a side, three to an end

Armor: two courses of 2-inch plate, each plate 8 feet long and 3.5 feet wide

First Course: horizontal

Second Course: vertical

Engines: two horizontal, back acting; two cylinders; 72 inches diameter, 3-foot stroke

Boilers: four Martin type; average steam pressure 18 pounds

Speed: 8.89 knots (The pilots believed the ironclad reached a speed of 8 knots on her trial trip down the Elizabeth River. Most observers rated the ironclad's speed to be between 4 and 5 knots.)

Draft: Loaded: 22 feet; after repairs and modifications completed by April 1862, draft is listed at 23 feet
Unloaded: 19 feet, 6 inches
Beam: 51 feet, 2 inches
Depth: 27 feet, 6 inches
Tonnage: 3,200

Length: 275 feet (USS *Merrimack* prior to conversion)
262 feet, 9 inches (after conversion to CSS *Virginia*)

Casemate: Length, 170 feet (at base)
167 feet, 7 inches (on gun deck)
66 feet from edge of shield to rudder
29 feet, 6 inches from stern to shield
Height: 7 feet from gun deck to grating

Appendix XII
The Crew of the CSS *Virginia*

Introduction to the Roster

The following pages chronicle the service records of the 431 men assigned to the CSS *Virginia* during her short career. The starting point for this roster was the *Virginia*'s muster roll found in the *Official Records of the Union and Confederate Navies in the War of Rebellion* (commonly referred to as the *ORN*). Published by the U.S. Navy Department from 1894 to 1922, this thirty-one-volume series can be found in most large public libraries. While the *ORN* lists the names of the men who served on the *Virginia*, it provides no information on their lives, either before or after their time on this historic vessel.

Unfortunately, most Confederate navy records did not survive the Civil War. Many original documents were destroyed when vessels were scuttled or burned to prevent their capture by Union forces. While the Confederate Navy Department did maintain duplicate copies of records such as pay and muster rolls, relatively few of these escaped the fires that ravaged Richmond when the Confederate government evacuated in April 1865. Much of the surviving information is found in the National Archives' Record Group 45 (Naval Records Collection of the Office of Naval Records and Library). This resource was used extensively to prepare the roster that follows.

To obtain able-bodied men for naval service, the Confederate navy established recruiting stations or naval rendezvous at Macon, Mobile, New Orleans, Raleigh, Richmond and Savannah. Recruits enlisted or "shipped" (hence the phrase "shipping out") for a period of one to three years and were then assigned to a specific vessel or navy station. Recruiting, however, seldom met the Confederate navy's manpower needs. As a result, the navy was frequently forced to request volunteers from the Confederate army. Such was the case with the CSS *Virginia*. While the majority of her officers had served in the United States Navy before the war, most of the crew had previously served in the Confederate army.

Armed with this knowledge, Compiled Service Records files (found in the National Archives' Record Group 109, War Department Collection of Confederate Records) were reviewed to attempt to identify soldiers who transferred to the *Virginia*. The service records consist of cards with the individual's name, rank and service data extracted from a variety of sources, such as original muster rolls, payrolls and hospital registers. Occasionally, the records provide interesting data such as where and when a soldier was born, his occupation and physical characteristics such as height, complexion and eye and hair color.

Since many of the names listed in the *ORN* were common, identifying which soldiers transferred to the *Virginia* was an extremely time-consuming process. For example, hundreds of men named David Jones served in the Confederate army. Another complicating factor is that surnames were frequently misspelled, and men are sometimes listed by their first names in one record and by their middle names in another. The quality of a given record depends on the educational level of the officer or clerk and his dedication to the task. As the fortunes of the Confederacy waned, so too did the accuracy and completeness of records.

In the roster that follows, some men are listed as "reduced in rank." This routinely occurred when an enlisted man was detailed to a special assignment outside the unit. There was generally little stigma attached to such an action, and it was merely an indication of reduced responsibilities, not ineptitude.

Men listed as "away without leave" (AWOL) or "deserted" also bears further discussion. Sometimes a man was listed as AWOL or deserted when he had been killed, hospitalized or captured. Some soldiers and sailors did go AWOL, especially in the early stages of the war. The reason was usually a family member's illness or other problem at home and not cowardice. After a short period of time, the soldier or sailor commonly returned to his original unit. During his absence, he was listed as AWOL or deserted depending on the whim of the unit's clerk. Some men became disenchanted with their commander and enlisted in a different unit or branch of service. While still serving "the cause," the soldier or sailor would eventually be listed as either AWOL or deserted in his original unit's records.

Each biographical sketch is arranged chronologically to the maximum extent possible. Space requirements have made it necessary to use phrases rather than complete sentences and to make liberal use of the abbreviations that follow this introduction. The term "no further record" (NFR) has been used to show when a man's service ended without explanation.

Data from Record Groups 45 and 109 were supplemented by postwar rosters, state pension applications and marriage, death and census records. Inevitably, errors have crept into the biographical sketches in this volume due to errors made by clerks long since dead. Descendants with additional information concerning their ancestors are urged to contact the publisher or author.

Abbreviations

A&IGO: Adjutant and Inspector General's Office
ANV: Army of Northern Virginia
b.: born
Brig.: Brigadier
btn.: battalion
Btry.: battery
bur.: buried
ca.: circa
capt.: captain
Cav.: Cavalry
Cem.: Cemetery
CH: Court House
cmdr.: Commander
Co.: Company or county
CO: commanding officer
comp.: complexion
Conf.: Confederate
corp.: corporal
CRR: clothing receipt roll
CS: Confederate States
CSA: Confederate States of America
CSMC: Confederate States Marine Corps
CSN: Confederate States Navy
CSRS: Confederate States Receiving Ship

CSS: Confederate States Ship
d.: died
Dept.: Department
det.: detachment
Ft.: Fort
Gen.: General
GO: General Order
Hosp.: Hospital
ind.: independent
int.: intermittent
KIA: killed in action
Lt.: Lieutenant
MWIA: mortally wounded in action
NFR: no further record
occ.: occupation
OR: Official Records of the Union and Confederate Armies
ORN: Official Records of the Union and Confederate Navies
para.: paragraph
POW: prisoner of war
pvt.: private
QM: quartermaster
regt.: regiment
RR: railroad
RTD: returned to duty
sgt.: sergeant
SO: Special Order
St.: Street
thru: through
UCV: United Confederate Veterans
USN: United States Navy
USNA: United States Naval Academy
USRMS: United States Revenue Marine Service
USRS: United States Receiving Ship
USF: United States Frigate
USS: United States Ship
w/: with
w/o: without
WIA: wounded in action

The Crew of the CSS *Virginia*

Aenchbacker, Samuel N.: b. 1839, Switzerland; occ. tailor; resided Columbus, Georgia; enlisted at Montgomery, Alabama as a pvt. in Co. C, CSMC, 3/29/61; seventh man to enlist in CSMC; assigned to Marine Guard, CSS *Virginia*, ca. 11/61–ca. 5/62; in charge of squad of marines onboard *Virginia* when she was launched, 2/62; helped man Gun #6 during Battle of Hampton Roads, Virginia, 3/8–9/62; in Bruton Parish Episcopal Church Hosp., Williamsburg, Virginia w/ syphilis, 5/12–14/62; sgt., 4th qtr. '62; on CRR CSS *Drewry*, 1st qtr. '63; on CRR, Rocketts Navy Yard, Richmond, Virginia, 7/20/63; reduced to pvt., ca. 9/20/63; extra duty as carpenter, Camp Beall, Drewry's Bluff, Virginia, 12/9–23/63; assigned to Marine Guard, CSS *Fredericksburg*, James River Squadron, 5/14/64; on same duty thru 12/31/64; postwar, occ. carpenter; resided, Columbus, Georgia; d. of apoplexy (stroke), 9/10/1901; bur. Linwood Cem., Columbus, Georgia, 9/11/1901; on *Virginia* muster roll in *ORN* as "Samuel Archbacker."[1]

Agnew, Jonathan W.: b. ca. 1835, Bedford Co., Virginia; occ. clerk/merchant; resided Salem, Virginia; enlisted at Norfolk, Virginia, for 1 year as a pvt. in 1st Co. A, 9th Virginia Infantry, 5/14/61; described as 5'9¾", dark comp., dark hair; blue eyes; present until discharged per SO '62, Dept. of Norfolk, 3/21/62; served as a landsman, CSS *Virginia*; appears to have left crew of *Virginia* before 4/1/62; officially discharged from CSN, 10/62; may have later served as 3rd Lt., Co. D, 5th Virginia Cavalry; paid at Orange CH, Virginia, 4/15/63; 2nd Lt., 5/13/63; POW, Aldie, Virginia, 6/17/63; sent to Old Capital Prison, Washington, D.C.; Baltimore, Maryland; Johnson's Island, Ohio; and finally Point Lookout, Maryland; exchanged, 3/10/64; WIA (right thigh flesh wound) Yellow Tavern, Virginia, 5/11/64; in hosp., Richmond, Virginia, 5/13/64; transferred to Gen. Hosp., Liberty, Virginia, 5/27/64; RTD, 7/7/64; company commander, 7/22–10/23/64; ordered arrested as a deserter, 10/26/64; absent sick, Gen. Hosp., Liberty, 1/31–2/12/65; paroled, Appomattox CH, 4/9/65; postwar: resided Bedford Co., Virginia.[2]

Ahern, David: b. ca. 1836, Ireland; occ. laborer; single; resided New Orleans, Louisiana; enlisted at Camp Moore, Louisiana, for 1 year as a pvt. in Co. D, 8th Louisiana Infantry, 6/19/61; present thru 9/61; absent sick, in Gen. Hosp., Richmond, Virginia, 10/14/61; discharged, Richmond, Virginia, 10/31/61; shipped for 1 year as an ordinary seaman on receiving ship, CSS *Confederate States*, 1/7/62; served as an ordinary seaman, CSS

Virginia, 2/18/62; reenlisted for the war, received $50 bounty as member of crew, *Virginia*, 3/25/62; appears to have left crew of *Virginia* before 4/1/62; apparently RTD w/ Co. D, 8th Louisiana Infantry, date unknown; MWIA, 2/63; NFR.[3]

Aird, Hugh: enlisted at New Orleans, Louisiana, as a pvt. in Co. C, CSMC, 6/4/61; assigned to Marine Guard, CSS *Virginia*, ca. 11/61–ca. 5/62; sent to Charleston, South Carolina, as part of Monitor Boarding Party under Capt. Wilson, ca. 3/63; on duty with Co. C until 5/14/64; assigned to Marine Guard, CSS *Fredericksburg*, James River Squadron, 5/14–12/31/64; paroled, Appomattox CH, 4/9/65; NFR.[4]

Albright, E.: married; appointed as a paymaster's clerk, ca. 6/29/61; served on CSS *McRae* and *Pontchartrain*, New Orleans Naval Station, '61; assigned to CSS *Virginia*, 1/1/62; served at Richmond Naval Station, attached to Drewry's Bluff, Virginia and CSS *Roanoke* thru 3/31/64; NFR.[5]

Alexander, Joseph W.: b. North Carolina; appointed by North Carolina to USNA as acting midshipman, 9/21/53; midshipman, 1/10/57; passed midshipman, 6/25/60; master, 10/24/60; Lt., 3/11/61, tendered resignation, 7/4/61; dismissed from USN, 7/5/61; appointed 1st Lt., CSN, 10/23/61; served on CSS *Virginia*, '61; commanded CSS *Raleigh*, '61–4/15/62; took part in Battle of Hampton Roads, Virginia, 3/8–9/62; served on CSS *Atlanta*, '62–'63; captured by USS *Weehawken*, 6/17/63; escaped, 8/19/63; recaptured and returned to Ft. Warren, Massachusetts, 9/7/63; 1st Lt., Provisional Navy, 6/2/64; paroled, 9/28/64; exchanged, 10/18/64; served on CSS *Virginia II*, James River Squadron, '64; commanded CSS *Beaufort*, James River Squadron, '64–'65; NFR; not on *Virginia* muster roll in *ORN*.[6]

Allen, Benjamin S.: b. ca. 1837, Boston, Massachusetts; occ. riverman; resided Mobile, Alabama; single; enlisted at Mobile for the war as a pvt. in Co. E, 8th Alabama Infantry, 5/6/61; described as 5'5", florid comp., dark hair, gray eyes; present thru 8/61; "transferred to C.S. Navy by order of Secretary of War," 2/10/62; shipped for 3 years as a seaman, Naval Rendezvous, Richmond, Virginia, 2/16/62; served as a seaman, CSS *Virginia*; reenlisted for the war, received $50 bounty as member of crew, *Virginia*, 3/25/62; paid for service on the *Virginia*, 4/1–5/12/62; superannuated for rations, Drewry's Bluff, Virginia, 5/12–24/62; on *Virginia* CRR, Drewry's Bluff, 7/20/62; quarter gunner, 1/2/63; paid, Drewry's Bluff, 5/13/62–6/30/63 and 10/1/63–3/26/64; served on CSS *Huntsville*, Mobile Squadron, '64; acting gunner, Provisional Navy, date unknown; surrendered, 5/4/65; paroled, Nanna Hubba, Alabama, 5/10/65.[7]

Allen, Gabriel: b. ca. 1832, Washington Co., North Carolina; resided Washington Co.; enlisted at Plymouth, North Carolina, for the war as a pvt. in Co. G, 1st North Carolina Infantry, 6/28/61; present thru 10/61; absent, on 12 day furlough, 11–12/61 roll; present until transferred to CSN per SO 17, A&IGO 62, para 23, 2/3/62; served as an ordinary seaman, CSS *Virginia*; superannuated for rations, Drewry's Bluff, Virginia, 5/12–24/62; NFR.[8]

Allman, William D.: b. ca. 1839, Holly Springs, Mississippi; occ. sailor; enlisted at New Orleans, Louisiana, as a pvt. in Co. C, 1st (Nelligan's) Louisiana Infantry, 5/22/61; described as 5'6", dark comp., black hair, black eyes; present, 5/31/61; ordered to Richmond, Virginia, per dispatch of secretary of war; present, 9–12/61; "discharged by S.O. of Gen. Huger to join the *Merrimac*," per 1–2/62 muster roll; served as a landsman, CSS *Virginia*; paid for service on the *Virginia* and Drewry's Bluff, Virginia, 4/1–7/23/62; superannuated for rations, Drewry's Bluff, 5/12–24/62; listed as having deserted on final *Virginia* payroll, dated 9/30/62; NFR; may have later enlisted at Richmond for the war as a 3rd sgt. in Co. B, 39th Btn., Virginia Cavalry, 8/12/62; present, as 4th sgt., 11–12/62; absent, on leave as 3rd sgt., 2/18–3/3/63; absent, recruiting horses, 3–4/63; listed as "captured in Pennsylvania" on 7–12/63 rolls; paroled, Richmond, 4/21/65; on *Virginia* muster roll in *ORN* as "W.D. Allmond."[9]

Anderson, Frank: b. ca. 1838, Glasgow, Scotland; resided New York; enlisted at Light Knot Springs, South Carolina, for 1 year as a pvt. in Co. L, 1st (McCreary's) South Carolina Infantry, 9/4/61; described as 5'9", dark comp., dark hair, dark eyes; present, 11–12/61; transferred to CSS *Virginia* per SO 12, Dept. of Norfolk, 1/18/62; reenlisted for the war, received $50 bounty as member of crew, *Virginia*, 3/25/62; paid for service as a seaman, CSS *Beaufort*, James River Squadron, 4/1–15/62; superannuated for rations, Drewry's Bluff, Virginia, 5/12–24/62; paid for service as an officers' steward, Charleston Naval Station, 10/1–31/63; served as a painter on receiving ship CSS *Indian Chief*, Charleston Naval Station, '65; deserted, Charleston, South Carolina, 2/23/65; took oath and released; on *Virginia* muster roll in *ORN* as "James Anderson."[10]

Applewhite, Edward F.: b. ca. 1824, Virginia; occ. tavern keeper; single; resided and enlisted at Norfolk, Virginia, for 1 year as a pvt. in the United Artillery (1st Co. E, 41st Virginia Infantry), 4/19/61; present, 7/1/61–3/10/62; "volunteered to go onboard the Ironclad Steamer *Virginia*," 3/6/62; manned gun on CSS *Virginia* during Battle of Hampton Roads, Virginia, 3/8–9/62; reenlisted for the war, received $50 bounty; mustered out, 3/10/62; listed as AWOL on 9/62 roll; on CRRs, 12/22/62, 6/2/63, 9/14/63, 11/1/63, and 3/18/64; paid, 5/1–6/30/64; in Chimborazo Hosp. #1, Richmond, Virginia, w/ secondary syphilis, 8/13–24/64; detailed as a blacksmith, Drewry's Bluff, Virginia, 10/10–12/31/64; deserted, Bermuda Hundred, Virginia, 4/5/65; took oath, sent to Washington, D.C., furnished transport to Norfolk, 4/10/65; postwar: occ. boatkeeper; single; resided Norfolk, '70; also found as "Applewaite."[11]

Archer, Frederick: b. ca. 1843, Talbot Co., Georgia; occ. farmer; enlisted at Talbotton, Georgia, for 1 year as a pvt. in Co. A, 4th Georgia Infantry, 4/26/61; described as 5'6", fair comp., dark hair, dark eyes; present thru 9/61; in regt. hosp., Camp Jackson, Virginia, w/ int. fever, 9/27–29, 10/4–7, 10/18–31, 11/1–5, 11/6–16, 11/17–19, 12/25–26/61, 1/6–19, and 1/23–25/62; transferred to CSS *Virginia* per SO 28, Dept. of Norfolk, 2/10/62; served as a landsman, CSS *Virginia*; reenlisted for the war, received $50 bounty as member of crew, *Virginia*, 3/25/62; paid for service on the *Virginia* and Drewry's Bluff, Virginia, 4/1–5/15/62; on *Virginia* CRR, Drewry's Bluff, 7/20/62; listed as having deserted on final *Virginia* payroll, dated 9/30/62; RTD w/Co. A, 4th Georgia Infantry; POW, Gettysburg, Pennsylvania, 7/4/63; sent to Ft. Delaware, Delaware, via Harrisburg and Philadelphia, 7/9/63; enlisted at Ft. Delaware as a pvt. in Co. D, 3rd Maryland Cavalry, U.S. Army, 9/4/63; mustered in at Baltimore, Maryland, 9/15/63; transferred to Co. E, 12/64; mustered out, Vicksburg, Mississippi, 9/7/65; Georgia commissioner of pensions requested verification of service, 1916.[12]

Armstrong, Alex: served as a landsman, CSS *Virginia*; reenlisted for the war, received $50 bounty as member of crew, *Virginia*, 3/25/62; paid for service on the *Virginia*, 4/1-5/12/62; on *Virginia* CRR, Drewry's Bluff, Virginia, 7/20/62; paid, Drewry's Bluff, 5/13/62–6/30/63 and 10/1/63–3/31/64; POW near Richmond, Virginia, 4/3/65; in Jackson Gen. Hosp., Richmond w/ chronic diarrhea, 4/8/65; escaped 4/19/65; NFR.[13]

Ault, Charles: shipped for the war as a seaman, CSS *Forrest*, North Carolina Squadron, 7/25/61; served as a seaman, CSS *Virginia*; appears to have left crew of *Virginia* before 4/1/62; NFR; on *Virginia* muster roll in *ORN* as "Charles Alt."[14]

Ayers, Robert E.: b. ca. 1836, Washington, North Carolina; occ. waterman; married; resided Washington Co.; enlisted at Plymouth, North Carolina, for the war as a pvt. in Co. G, 1st North Carolina Infantry, 6/24/61; present until transferred to CSN per SO 17, A&IGO 62, para 23, 2/3/62; served as an ordinary seaman, CSS *Virginia*; reenlisted for the war, received $50 bounty as member of crew, *Virginia*, 3/25/62; appears to have left crew of *Virginia* before 4/1/62; reportedly "deserted and joined the enemy;" NFR; also found as "Ayres" and "Airs."[15]

Baines, John: enlisted at Mobile, Alabama, as a musician in Co. C, CSMC, 8/26/61; probably served at Pensacola Navy Yard until Co. C transferred to Gosport Navy Yard, Portsmouth, Virginia, 11/29/62; assigned to Marine Guard, CSS *Virginia*, ca. 11/61–ca. 5/62; on duty with Co. C until arrested by civil authorities, 2/23/64; RTD, 10/7/64; assigned to CSS *Virginia II*, James River Squadron, 12/31/64; deserted with boat's crew on James River, 10/65; took oath, Bermuda Hundred, Virginia, 3/11/65; sent to Washington, D.C.; furnished transport to Norfolk, Virginia.[16]

Baines, Matthew: b. ca. 1830, Virginia; occ. laborer; single; resided Fox Hill Parish, Elizabeth City Co., Virginia; enlisted at Elizabeth City Co. for 1 year as a pvt. in Co. K, 32nd Virginia Infantry, 5/14/61; joined Co. F in Williamsburg, Virginia, when Co. K disbanded, 6/61; present, 9–12/61; listed as "transferred to the *Merrimac*" on 3/62 roll; served as a landsman, CSS *Virginia* before 4/1/62; listed as having deserted on final *Virginia* payroll, dated 9/30/62; RTD w/ Co. K, 32nd Virginia Infantry, 8/3/62; NFR; also found as "Bains."[17]

Ball, Lemuel: shipped at Norfolk, Virginia, as an ordinary seaman, CSS *Ellis*, North Carolina Squadron, 8/2/61; paid, CSS *Ellis*, 8/2/61–2/7/62; paid for service as a seaman, CSS *Virginia*, 4/1–5/12/62; superannuated for rations, Drewry's Bluff, Virginia, 5/12–24/62, on *Virginia* CRR, Drewry's Bluff, 7/20/62; paid, Drewry's Bluff, 5/13–9/13/62; listed as discharged on final *Virginia* payroll, dated 9/30/62; NFR.[18]

Barclay, John: enlisted at Mobile, Alabama, as a pvt. in Co. C, CSMC, 8/27/61; probably served at Pensacola Navy Yard until Co. C transferred to Gosport Navy Yard, Portsmouth, Virginia, 11/29/62; received transport from Weldon, North Carolina, to Portsmouth, Virginia, 1/5/62; detailed to Machine Shop, Gosport Navy Yard, 1/29/62; paid for service, assigned to Marine Guard, CSS *Virginia*, 2/11–5/12/62; on Co. C CRR, 3rd qtr. '62; listed as having deserted on final *Virginia* payroll, dated 9/30/62; NFR; may have later served as a seaman on the cruiser CSS *Florida*, '64; captured by USS *Wachusett* off Bahia, Brazil, 10/7/64; POW, Point Lookout, Maryland, 11/15/64; sent to Ft. Warren, Massachusetts, 11/26/64; took oath and paroled, 2/1/65; also found as "John Bartley."[19]

Barker, George Washington, Jr.: b. near Danville, Virginia; enlisted at Bachelor's Hill, Virginia, for 1 year as a pvt. in Co. I, 57th Virginia Infantry, 8/1/61; absent, sick in Masonic Hosp., 10/61 roll; present, 11–12/61;

volunteered for duty in CSN; billeted on receiving ship, CSS *Confederate States*; served as a landsman, CSS *Virginia*; paid, Richmond Naval Station, 4/1–5/12/62; superannuated for rations, Drewry's Bluff, Virginia, 5/12–24/62; on *Virginia* CRR, Drewry's Bluff, 7/20/62; paid, Drewry's Bluff, 5/13/62–6/30/63 and 10/1/63–3/31/64; reportedly discharged due to sickness, returned home; NFR; postwar: resided Charleston, West Virginia, before moving west.[20]

BARKER, JOHN W.: enlisted at Bachelor's Hill, Virginia, for 1 year as a pvt. in Co. I, 57th Virginia Infantry, 8/1/61; absent, sick in Masonic Hosp., 10/61 roll; present, in confinement, 11–12/61; absent, on 34-day furlough in Henry Co., Virginia, 1/25/62; volunteered for duty in CSN; served as a landsman, CSS *Virginia*; superannuated for rations, Drewry's Bluff, Virginia, 5/12–24/62; on *Virginia* CRR, Drewry's Bluff, 7/20/62; paid, Drewry's Bluff, 5/12–12/23/62; in Howard's Grove Hosp., Richmond, Virginia, w/ varioloid, 12/24/62–3/5/63; paid, Drewry's Bluff, 3/5–6/30/63 and 10/1/63–3/31/64; took oath, Burkeville, Virginia, 4/26/65; NFR; also found as "John H. Barker."[21]

BARNARD, GEORGE W.: enlisted at Camp Walker, Louisiana, for 1 year as a pvt. in Co. F, 2nd Louisiana Infantry, 5/9/61; present until "enlisted on the *Merrimac* for 3 years," 3/19/62; billeted on receiving ship, CSS *Confederate States*; served as a landsman, CSS *Virginia*; reenlisted for 3 years on CSS *Virginia*, 3/19/62; superannuated for rations, Drewry's Bluff, Virginia, 5/12–24/62; reenlisted at Winchester, Virginia, for the war as a pvt. in Co. F, 2nd Louisiana Infantry, 11/1/62; present until listed as AWOL, 9/25/63; NFR.[22]

BARNES, LEONARD: b. ca. 1840; resided Washington Co., North Carolina; enlisted at Plymouth, North Carolina, for the war as a pvt. in Co. G, 1st North Carolina Infantry, 6/26/61; present, 7–8/61; present, sick, 11–12/61; transferred to CSN per SO 17, A&IGO 62, para 23, 2/3/62; served as a landsman, CSS *Virginia*; reenlisted for the war, received $50 bounty as member of crew, *Virginia*, 3/25/62; appears to have left crew of *Virginia* before 4/1/62; NFR; not on *Virginia* muster roll in *ORN*.[23]

BARRETT, HENRY C.: b. 1837, England; parents, Antony and Ann Barrett; occ. mariner; enlisted at Norfolk, Virginia, for 1 year as a pvt. in the United Artillery (1st Co. E, 41st Virginia Infantry), 4/19/61; described as 5'8", black hair, brown eyes; present until "detailed to go on the *Merrimac*" per SO 9, Dept. of Norfolk, 1/19/62; billeted on receiving ship CSS *Confederate States*; served as a ship's carpenter, CSS *Virginia*, 1/19–5/10/62; listed as having deserted on final *Virginia* payroll, dated 9/30/62; NFR.[24]

Barry, James E., Jr.: b. 1818, Savannah, Georgia; occ. wealthy Norfolk, Virginia wholesale crockery merchant; enlisted at Ft. Norfolk, Virginia, for 1 year as a Lt. in the United Artillery (1st Co. E, 41st Virginia Infantry), 4/19/61; briefly served in Norfolk Light Artillery Blues (Grandy's Battery); 1 of 31 men from the United Artillery detailed to man guns on CSS *Virginia*; took part in Battle of Hampton Roads, Virginia, 3/8–9/62; unit became Co. C, 19th Btn., Virginia Heavy Arty., 4/19/62; manned heavy btry. at Entrenched Camp on evacuation of Norfolk, 5/10/62; led RR artillery at Savage Station, Virginia, 6/62; ordered to report w/ co. to S.S. Lee, capt., CSN at Drewry's Bluff, Virginia; present, Btry #8, 8–9/62; company became an independent btry., 10/1/62; present, 11/62–1/63; took part in naval raids led by Lt. John Taylor Wood, CSN, '62–'63; in Montgomery Hosp., White Sulphur Springs, West Virginia, 8/12–19/64; present, 12/31/64; in Stuart Gen. Hosp., Richmond, Virginia, w/ intermittent fever, 1/28–3/6/65; resigned due to poor health, 2/10/65; postwar: occ. Norfolk businessman; 12 years on city council; president, Bank of Commerce; member, Pickett-Buchanan Camp, UCV.[25]

Baxter, Henry: billeted on receiving ship CSS *Confederate States*; served as a coal heaver on CSS *Virginia*; paid for service on the *Virginia* and Drewry's Bluff, Virginia, 4/1/62–6/30/63; discharged, 4/7/63; may have later served as an ordinary seaman on CSS *Virginia II*, James River Squadron, '64–'65; NFR.[26]

Belanger, William: b. ca. 1834, Martin Co., North Carolina; resided Martin Co.; enlisted at Williamston, North Carolina, for the war as a pvt. in Co. H, 1st North Carolina Infantry, 6/24/61; present until discharged, Richmond, Virginia, and transferred to CSN per SO 17, A&IGO 62, para 23, 2/3/62; served as an ordinary seaman, CSS *Virginia*; reenlisted for 3 years, received $50 bounty, 4/21/62; paid for service on the *Virginia*, 4/1–5/12/62; on *Virginia* CRR, Drewry's Bluff, Virginia, 7/20/62; paid for service, Drewry's Bluff, 5/13/62–9/30/63; paid for service, naval det. on special expedition from Drewry's Bluff to Charleston Naval Station, 10/1–31/63; NFR.[27]

Bell, Miles K.: b. 1828, Norfolk, Virginia; resided and enlisted at Norfolk for 1 year as a pvt. in the United Artillery (1st Co. E, 41st Virginia Infantry), 4/19/61; "volunteered to go onboard the Ironclad Steamer *Virginia*," 3/6/62; manned gun on CSS *Virginia* during Battle of Hampton Roads, Virginia, 3/8–9/62; reenlisted for war, received $50 bounty, 3/11/62; on CRRs, 6/2/63, 9/14/63, 10/31/63 and 3/18/64; in Gen. Hosp. #9, Richmond, Virginia, 8/6/64; in Chimborazo Hosp. #5, Richmond, w/

debility, 8/7/64; in Chimborazo Hosp. w/ remittent fever, 8/9–10/5/64; in Chimborazo Hosp., 10/10/64; on CRR, Howard's Grove Hosp., Richmond, 10/12/64; present, 11–12/64; NFR; postwar: no occ. listed; single; resided Norfolk, '70.[28]

Bell, William: enlisted at New Orleans, Louisiana, as a pvt. in Co. C, CSMC, 5/1/61; on Capt. Holmes's CRR, ca. 5/10/61; probably served at Pensacola Navy Yard until Co. C transferred to Gosport Navy Yard, Portsmouth, Virginia, 11/29/62; assigned to Marine Guard, CSS *Virginia*, ca. 11/61–ca. 5/62; sent to Charleston, South Carolina, as part of *Monitor* Boarding Party under Capt. Wilson, ca. 3/63; on duty with Co. C until KIA during the cutting out of USS *Underwriter*, 2/2/64; mentioned as "an excellent man, tried and faithful."[29]

Belote, John Krendal: b. ca. 1823, Virginia; occ. oysterman; married; resided Gloucester Co., Virginia; enlisted at Norfolk, Virginia, for 1 year as a pvt. in the United Artillery (1st Co. E, 41st Virginia Infantry), 4/19/61; "volunteered to go onboard the Ironclad Steamer *Virginia*," 3/6/62; manned gun on CSS *Virginia* during Battle of Hampton Roads, Virginia, 3/8–9/62; reenlisted for war, received $50 bounty, 3/11/62; on CRRs, 12/22/62, 6/2/63, 9/14/63, 11/31/63 and 3/18/64; absent, on detached duty w/ CSN, per SO 63, A&IGO 64, dated 3/16/64; absent, sick, 12/31/64; POW, Gen. Hosp. #13, Raleigh, North Carolina, 4/13/65; paroled, Raleigh, 5/6/65.[30]

Bennett, Thomas: resided Rutherford Co., Tennessee; enlisted at Mobile, Alabama, for the war as a pvt. in Co. A, 12th Alabama Infantry, 6/4/61; present on 7–8/61 rolls; transferred to CSN per SO 89, A&IGO 62, para 10, 3/27/62; shipped for 1 year as an ordinary seaman, Naval Rendezvous, Richmond, Virginia, 3/31/62; served as an ordinary seaman, CSS *Virginia*; reenlisted for 3 years, received $50 bounty, 4/21/62; paid for service on the *Virginia*, 4/1–5/12/6; superannuated for rations, Drewry's Bluff, Virginia, 5/12–24/62; on *Virginia* CRR, Drewry's Bluff, 7/20/62; paid for service, Drewry's Bluff, 5/13/62–6/30/63; listed as having deserted on payroll, dated 9/30/63; may have later served as a seaman, CSS *Shell Mound*, Mobile Squadron; captured, Fletcher's Ferry, Alabama, 1/21/64; POW, Nashville, Tennessee; sent to Louisville, Kentucky, and Rock Island, Illinois; apparently escaped/ paroled; surrendered, 5/20/65; took oath and paroled, Nashville, Tennessee, 5/27/65; described then as 5'7", dark comp., dark hair, hazel eyes; possibly "Thomas A. Bennett" who was b. ca. 1822, North Carolina; married; resided Rutherford Co., Tennessee, '50.[31]

Benthall, Robert: b. ca. 1841, Virginia; parents, Robert & Teabilla Benthall; occ. mariner; single; resided Baltimore, Maryland; served as an ordinary seaman on CSS *Virginia*; paid, Drewry's Bluff, Virginia, 5/12/62–1/1/63; appointed by Maryland as an acting master's mate, date unknown; acting master, 3/14/64; served on CSS *Missouri*, Red River Squadron, '63–'65; paroled, Alexandria, Louisiana, 6/3/65; listed destination as Baltimore.[32]

Bessant, Abraham W.: enlisted at Montgomery, Alabama, as a pvt. in Co. C, CSMC, 4/1/61; probably served at Pensacola Navy Yard until Co. C transferred to Gosport Navy Yard, Portsmouth, Virginia, 11/29/62; assigned to Marine Guard, CSS *Virginia*, ca.11/61–c.5/62; with Co. C at Camp Beall, Drewry's Bluff, Virginia; sent to Charleston, South Carolina, as part of Monitor Boarding Party under Capt. Wilson, ca. 3/63; paid extra as a carpenter, 12/10/63–1/17/64; on expedition to Wilmington, North Carolina, 7/64; retained by Lt. Crenshaw as part of his guard for the CSS *Tallahassee*; detailed to Capt. Alfred C. Van Benthuysen, CSMC, Wilmington, 7/30/64; on extra duty as an orderly to post commander, 7–8/64; served as a cook, 9–10/64; served as an orderly to btn. commander, 11–12/64; NFR; possibly "Abraham Bessant" who was b. ca. 1804, South Carolina; occ. farmer; married; resided Duval Co. Florida; also found as "Bessent."[33]

Beveridge, David: b. ca. 1834, Scotland; occ. waterman; resided, Baltimore, Maryland; enlisted at Norfolk, Virginia, for 1 year as a pvt. in Co. B, 9th Virginia Infantry, 6/5/61; present until discharged per SO 18, Dept. of Norfolk, 1/29/62; billeted on receiving ship, CSS *Confederate States*; served as a landsman, CSS *Virginia*; reenlisted for the war, received $50 bounty; on *Virginia* CRR, Drewry's Bluff, Virginia, 7/20/62; listed as transferred to CSS *Patrick Henry* on *Virginia* muster roll dated 7/31/62, paid for service on the *Virginia* and Drewry's Bluff, 4/1–11/30/62; paid as a quarter master, CSS *Albemarle*, North Carolina Squadron, 7/1–9/30/64; deserted, Bermuda Hundred, Virginia, 4/5/65; sent to Washington, D.C., from City Point, Virginia, 4/12/65; took oath and provided transport to Philadelphia, Pennsylvania.[34]

Black, John G.: b. ca. 1837, England; enlisted at Montgomery, Alabama, as a pvt. in Co. G, 3rd Alabama Infantry, 6/17/61; described as 5'8", dark hair; discharged to serve on CSS *Virginia* per SO 28, Dept. of Norfolk, 2/8/62; billeted on receiving ship, CSS *Confederate States*; served as a landsman, CSS *Virginia*; superannuated for rations, Drewry's Bluff, Virginia, 5/12–24/62; paid, Drewry's Bluff, 5/12–11/8/62; NFR.[35]

BLACK, WILLIAM: billeted on receiving ship CSS *Confederate States*; served as a 2nd class fireman, CSS *Virginia*; reenlisted for the war, received $50 bounty as member of crew, *Virginia*, 3/25/62; appears to have left crew of *Virginia* before 4/1/62; may have later shipped for the war, Naval Rendezvous, Raleigh, North Carolina, 6/11/64; NFR; possibly "William Black" who enlisted at New Orleans, Louisiana, as a pvt. in Co. K, 1st Louisiana Heavy Artillery, 5/31/61; present on all rolls to 2/62; NFR.[36]

BOWERS, GEORGE F.: b. ca. 1833, Virginia; occ. tinsmith; single; resided and enlisted at Salem, Virginia, for 1 year as a sgt. in 1st Co. A, 9th Virginia Infantry, 5/14/61; mustered in at Lynchburg, Virginia, 5/16/61; reduced to pvt., 12/10/61; present until "volunteered to go onboard the Ironclad Steamer *Virginia*," 3/6/62; manned gun on CSS *Virginia* during Battle of Hampton Roads, Virginia, 3/8–9/62; enlisted at Norfolk, Virginia, for the war as a pvt. in the United Artillery (1st Co. E, 41st Virginia Infantry), 3/10/62; received $50 bounty; on CRR, 6/2/63; absent, on detached duty w/ CSN per SO 63, A&IGO 64, dated 3/16/64; deserted, 12/22/64; entered Federal lines, 12/26/64; took oath, 1/3/65; furnished transport to Norfolk, Virginia.[37]

BRENNAN, THOMAS: enlisted at Memphis, Tennessee, as a pvt. in CSMC, 10/10/61; probably transferred from Capt. Hays's det., Co. D, to Co. C, Pensacola Navy Yard, ca. 11/27/61; transferred with Co. C to Gosport Navy Yard, Portsmouth, Virginia, 11/29/61; assigned to Marine Guard, CSS *Virginia*, ca. 11/61–ca. 5/62; in Bruton Parish Episcopal Church Hosp., Williamsburg, Virginia, 5/10–14/62; corp., 4th qtr. '62; listed as a pvt. on 5/9/63 roll; NFR; also found as "Thomas Brannan."[38]

BREWSTER, WILLIAM H.: b. ca. 1834; enlisted at Memphis, Tennessee, as a pvt. in CSMC, 9/13/61; probably transferred from Capt. Hays's det., Co. D, to Co. C, Pensacola Navy Yard, ca. 11/27/61; transferred with Co. C to Gosport Navy Yard, Portsmouth, Virginia, 11/29/61; assigned to Marine Guard, CSS *Virginia*, ca. 11/61–ca. 5/62; advertised as a deserter, 6/12/62; described then as 5'6½"; dark comp., light brown hair, hazel eyes; still listed as having deserted on final *Virginia* payroll, dated 9/30/62; NFR.[39]

BRIGGS, WILLIAM: b. ca. 1810, Baltimore, Maryland; occ. seaman, married; resided at Baltimore, Maryland; enlisted at Mobile, Alabama, as a pvt. in Co. C, CSMC, 8/15/61; corp., '61; probably served at Pensacola Navy Yard until Co. C transferred to Gosport Navy Yard, Portsmouth, Virginia, 11/29/61; assigned to Marine Guard, CSS *Virginia*, ca. 11/61–ca. 5/62; d. 6/22/62, Camp Beall, Drewry's Bluff, Virginia; bur. Hollywood Cem., Richmond, Virginia.[40]

Brogan, John: shipped for the war as a coal heaver, Naval Rendezvous, New Orleans, Louisiana, 1/20/62; served as a coal heaver, CSS *Virginia*; appears to have left crew before 4/1/62; listed as having deserted on final *Virginia* payroll, dated 9/30/62; NFR.[41]

Brower, Emsey H.: b. ca. 1834, Randolph Co., North Carolina; billeted on receiving ship, CSS *Confederate States*; served as the ship's corp., CSS *Virginia*; reenlisted for the war, received $50 bounty as member of crew, *Virginia*, 3/25/62; appears to have left crew of *Virginia* before 4/1/62; NFR; possibly "E.H. Brower" who was b. ca. 1830, North Carolina; occ. farmer; married; resided Bladen Co., North Carolina, in '60.[42]

Brown, Andrew: served as a landsman, CSS *Virginia*; reenlisted for the war, received $50 bounty as member of crew, *Virginia*, 3/25/62; superannuated for rations, Drewry's Bluff, Virginia, 5/12–24/62; shipped for the war as a landsman, Naval Rendezvous, Richmond, Virginia, 9/16/62; served as a landsman, CSS *Palmetto State*, Charleston Squadron, '63–'64; NFR; possibly "Andrew Brown," who was b. ca. 9/1834; enlisted at Decatur, Georgia, for the war as a pvt. in Co. C, Cobb's Legion, Georgia Infantry, 8/1/61; mustered in, 8/5/61; present, 11–12/61; NFR.[43]

Brown, John: enlisted at Memphis, Tennessee, as a pvt. in CSMC, 10/29/61; assigned to Georgia/South Carolina Naval Stations, 10–11/61; probably transferred from Capt. Hays's det., Co. D, to Co. C, Pensacola Navy Yard, ca. 11/27/61; transferred with Co. C to Gosport Navy Yard, Portsmouth, Virginia, 11/29/61; assigned to Marine Guard, CSS *Virginia*, ca. 11/61–ca. 5/62; on Co. C CRR, 3rd qtr. '62; listed as having deserted on final *Virginia* payroll, dated 9/30/62; NFR.[44]

Brown, William H.: b. England; occ. seaman; single; resided and enlisted at New Orleans, Louisiana, for 1 year as a pvt. in Co. K, 1st (Nelligan's) Louisiana Infantry, 4/25/61; present thru 12/61; "discharged by order of Gen. Huger to join the *Merrimac*," 1/31/62; served as a landsman, CSS *Virginia*; paid for service on the *Virginia* and Drewry's Bluff, Virginia, 4/1–5/15/62, superannuated for rations, Drewry's Bluff, 5/12–24/62; listed as having deserted on final *Virginia* payroll, dated 9/30/62; served as an ordinary seaman, CSS *Chattahoochee*, '64; may have later enlisted in Amelia Co., Virginia, for the war as a pvt. in Co. E, 18th Virginia Infantry, 10/12/64; transferred to Johnston Artillery (Epes's Co., Virginia Heavy Artillery), per SO 313, HQ ANV, 12/21/64; present, 12/31/64; paroled, Appomattox CH, 4/9/65.[45]

Buchanan, Franklin: b. 9/17/1800, "Auchentorlie," Baltimore, Maryland; father, Dr. George Buchanan, one of the founders of the Medical Society of

Baltimore; mother, Laetitia McKean, daughter of former Gov. McKean of Pennsylvania; resided in Pennsylvania as youth; appointed by Pennsylvania as a midshipman, USN, 6/12/15; initially served on USS *Java* under Commodore Oliver Hazard Perry; served for five years on various ships, chiefly in the Mediterranean; received permission from Navy Dept. to serve for fifteen months as a mate on merchant vessel going to China; Lt., 1/13/25; delivered USS *Baltimore* to Emperor of Brazil at Rio de Janeiro, 7/26; served on USS *Constellation* in Mediterranean; served on USS *Delaware* which carried U.S. minister Edward Livingston to France, '33; invited with other officers to dine w/ King Louis Philippe; ordered to shore duty, tested guns at Philadelphia Navy Yard; commanded receiving ship at Baltimore; served on USS *Constitution* and *Falmouth* during Pacific cruise, 4/39–6/40; Comdr., 9/8/41; commanded USS *Mississippi*, '42; commanded USS *Vincennes*, '42–'43; assisted two British merchant-men in peril in Galveston Harbor and received official thanks from British government; first superintendent of USNA, Annapolis, Maryland, 10/10/45–3/2/47; commanded USS *Germantown*, 3/47; participated in landing of Gen. Scott's troops at Vera Cruz, and took part in the capture of Juan d'Ulloa during the Mexican War; shore duty primarily in Baltimore, Maryland; took command of USS *Susquehanna*, Commodore Matthew Perry's flagship, during voyage to Japan, '52; Captain, 9/14/55; took commanded of the Washington Navy Yard, '59; tendered his resignation after 4/19/61 Baltimore riots, was dismissed from USN, 4/22/61; tried to revoke his resignation when Maryland failed to leave the Union; decided to join the Confederacy and traveled to Richmond, Virginia, to offer his services, 8/61; commissioned captain, CSN, 9/5/61; chief, Bureau of Orders and Detail, Richmond, '61–'62; assigned to James River Squadron, '62; assumed command of CSS *Virginia*, 2/24/62; WIA during engagement with USS *Cumberland* and *Congress*, 3/8/62; on convalescence leave, Norfolk, Virginia, and Greensboro, North Carolina, until 8/15/62; admiral, for gallant and meritorious conduct during the Battle of Hampton Roads, 8/21/62; commanded naval forces at Mobile Bay, '62–'64; admiral, Provisional Navy, 6/2/64; supervised construction of CSS *Tennessee* and was on board her when she was defeated by Adm. David Farragut's fleet during Battle of Mobile Bay, 8/6/64; WIA and taken POW; held at Ft. Lafayette, New York, until exchanged, 2/65; reassigned to Mobile, but did not reach the city before its surrender; paroled 5/17/65; postwar: president, Maryland Agricultural College (later the University of Maryland); returned to Mobile, Alabama, and worked as an insurance executive; d. 5/11/74 at his res. "*The Rest*," Davis, Talbot Co., Maryland; U.S.S. *Buchanan* (DD-131), commissioned, 1/20/1919, and USS *Buchanan* (DD-484), sponsored by his great-granddaughter, Hildreth Meiere, commissioned, 3/21/1942, were named in his honor.[46]

Bullock, John Thomas: b. ca. 1834, Virginia; occ. ship's carpenter; married; resided and enlisted at Norfolk, Virginia, for 1 year as a sgt. in the United Artillery (1st Co. E, 41st Virginia Infantry), 4/19/61; on special duty as a carpenter, constructing coastal batteries near Norfolk, 11/61; "volunteered to go onboard the Ironclad Steamer *Virginia*," 3/6/62; manned gun on CSS *Virginia* during Battle of Hampton Roads, Virginia, 3/8–9/62; reenlisted for war, received $50 bounty, 3/10/62; on CRRs, 12/22/62, 6/2/63, 9/14/63, and 11/31/63; absent, on 30-day medical furlough, 12/7/63; on CRR, 3/18/64; absent, on furlough as pvt., 10/10/64; present, 11–12/64; WIA, Saylor's Creek, Virginia, 4/6/65; promoted 1st sgt.; paroled, Appomattox CH, 4/9/65.[47]

Bunting, William H.: b. ca. 1843, Northampton Co., Virginia; occ. farmer; enlisted at Churchland, Virginia, for 1 year as a pvt. in Co. I, 9th Virginia Infantry, 5/15/61; mustered in at Churchland, 6/12/61; described as 5'7", light comp., light hair, blue eyes; present until discharged "by reason for service on the *Merrimac*," 1/18/62; served as an ordinary seaman, CSS *Virginia*; paid for service on the *Virginia* and Drewry's Bluff, Virginia, 4/1–10/15/62; listed as having deserted, 10/15/62; NFR.[48]

Burke, William G.: b. ca. 1827; occ. comedian; enlisted at Harpers Ferry, Virginia, for 1 year as a pvt. in Co. F, 2nd Virginia Infantry, 5/1/61; mustered in 5/11/61; 2nd corp., 8/2/61; present until discharged and "transferred to another government branch" per SO 146, A&IGO 61, para 9, 9/7/61; "Navy Dept. asks his transfer to Norfolk Navy Yard"; appointed as a master's mate, CSN, Gosport Navy Yard, Portsmouth, Virginia, 9/23/61; served at Richmond Naval Station, '61; discharged, 11/25/61; served as a seaman, CSS *Virginia*; WIA, 3/8/62; appears to have left crew of *Virginia* before 4/1/62; NFR; listed as "William Burkes" in surgeon's report; also found as "Burk."[49]

Burns, William A.: b. ca. 1839, Isle of Wight, Virginia; occ. seaman; single; resided and enlisted at Norfolk, Virginia, for 1 year as a pvt. in the United Artillery (1st Co. E, 41st Virginia Infantry), 4/19/61; "volunteered to go onboard the Ironclad Steamer *Virginia*," 3/6/62; manned gun on CSS *Virginia* during Battle of Hampton Roads, Virginia, 3/8–9/62; reenlisted for war, received $50 bounty, 3/10/62; on CRRs, 12/22/62, 6/2/63, 9/14/63, 11/31/63 and 3/18/64; absent, on detached duty w/ CSN per SO 63, A&IGO 64, dated 3/16/64; present, 12/31/64; POW; sent to Point

Lookout, Maryland; released on oath, 6/23/65; described then as 5'4½", dark comp., brown hair, blue eyes; post-war: occ. waterman; married; resided Norfolk.[50]

Butt, Walter Raleigh: b. 12/10/39, Portsmouth, Virginia; resided Washington Territory; appointed by Washington Territory to USNA, 9/21/55; midshipman, 6/9/59; Lt., 8/31/61; refused to take oath of allegiance to U.S.; dismissed from USN, 10/5/61; imprisoned for two months at Ft. Lafayette, New York, Ft. Warren, Massachusetts, and on USS *Congress* at Newport News, Virginia; paroled, 12/21/61; exchanged, 1/62; 1st Lt., CSN, 1/8/62; reported to CSS *Virginia*, 1/24/62; commanded Guns #6 & #7 during Battle of Hampton Roads, Virginia, 3/8–9/62; served at Drewry's Bluff, Virginia, '62–'63; in Europe trying to obtain command of commerce cruiser, 3/63–7/64; took part in Battle of Trent's Reach, 1/24/65; commanded CSS *Nansemond*, James River Squadron, until it was destroyed, 4/3/65; assistant adjutant general, Semmes naval brigade during retreat from Richmond, Virginia; surrendered with Gen. Johnston's army; paroled, Greensboro, North Carolina, 4/28/65; postwar: captain, Peruvian navy; returned to Portsmouth; member, Stonewall Camp, UCV, Portsmouth; unable to find productive work; moved to California; joined former comrade William Harwar Parker as officer of Pacific Mail Steamship *City of San Francisco*; d. California, 4/26/85; bur. Cedar Grove Cem., Portsmouth, Virginia.[51]

Cahill, James H.: resided New Orleans, Louisiana; paid as a seaman, Richmond Naval Station, 1/1–2/15/62; served as captain of forecastle, CSS *Virginia*; capt. of gun #1; reenlisted for the war, received $50 bounty as member of crew, *Virginia*, 3/25/62; paid for service on the *Virginia*, 4/1–5/12/62; in Bruton Parish Episcopal Church Hosp., Williamsburg, Virginia, w/ catarrh, 4/10–27/62; on *Virginia* CRR, Drewry's Bluff, Virginia, 7/20/62; gunner's mate, 8/28/62; paid, Drewry's Bluff, 5/13–9/30/62; paid, Richmond Naval Station, 10/1–12/31/62; paid, Drewry's Bluff, 1/1–6/30/63 and 10/1/63–3/26/64; gunner, CSN, date unknown; served on CSS *Fredericksburg*, James River Squadron, '64–'65; NFR.[52]

Callahan, John C.: enlisted at Richmond, Virginia, for the war as a pvt. in Co. I, 1st (McCreary's) South Carolina Infantry, 8/19/61; transferred to CSN, 1/17/62; served as a landsman/ordinary seaman, CSS *Virginia*; superannuated for rations, Drewry's Bluff, Virginia, 5/12–24/62; paid, Drewry's Bluff, 6/1–12/31/62 and 7/1–12/31/63; paid for service as a landsman and master at arms, CSS *Palmetto State*, Charleston Squadron, South Carolina, 1/1–12/31/64; NFR.[53]

Campbell, James: enlisted at New Orleans, Louisiana, as a pvt. in Co. C, CSMC, 5/23/61; assigned to Georgia/South Carolina Naval Stations, 10–11/61, assigned New Orleans Naval Station, 1/62; transferred to Gosport Navy Yard, Portsmouth, Virginia; assigned to Marine Guard, CSS *Virginia*, 5/12/62; on Co. C CRR, 3rd qtr. '62; listed as having deserted on final *Virginia* payroll, dated 9/30/62; NFR.[54]

Campbell, Loudon: b. 1826, Alexandria, Virginia; parents, William F. Campbell and Elizabeth Smedley; appointed by Virginia to USN as 3rd assistant engineer, 7/21/58; 2nd assistant engineer; tendered resignation and dismissed from USN, 5/6/61; appointed as a 2nd assistant engineer, CSN, 6/17/61; 1st assistant engineer, 3/24/62; served on CSS *Patrick Henry*, 7/1–9/30/61; assigned to CSS *Virginia*, 1/1/62; served in the *Virginia*'s engine room during Battle of Hampton Roads, Virginia, 3/8–9/62; served at Columbus Naval Ironworks, '62; served on CSS *Georgia*, Savannah River Squadron, '62; served on CSS *Palmetto State*, Charleston Squadron, '62–'63; assigned to Savannah Naval Station, '63–'64, 1st assistant engineer, Provisional Navy, 6/2/64; chief engineer, Provisional Navy (nominated) to rank from 10/26/64; served on CSS *Chattahoochee*, '64; part of boarding party led by Lt. G.W. Gift in abortive attempt to capture USS *Somerset* and *Adela* on the Appalachicola River, spring '64; NFR.[55]

Canning, William: served as a quartermaster, CSS *Virginia*; reenlisted for the war, received $50 bounty as member of crew, *Virginia*, 3/25/62; appears to have left crew of *Virginia* before 4/1/62; NFR.[56]

Capps, John C.: b. ca. 1842, Virginia; occ. apprentice boat builder; single; resided and enlisted at Norfolk, Virginia, for 1 year as a pvt. in the United Artillery (1st Co. E, 41st Virginia Infantry), 4/19/61; "volunteered to go onboard the Ironclad Steamer *Virginia*," 3/6/62; slightly WIA by Minié ball from USS *Cumberland* or *Congress* coming in gun port, 3/8/62; reenlisted for war, received $50 bounty, 4/62; on guard duty, Battery #8, 9/62; on CRRs, 12/22/62, 6/2/63, 9/14/63, 11/31/63 and 3/18/64; shipped for the war, Naval Rendezvous, Richmond, Virginia, 5/64; served as a landsman, CSS *Virginia II*, James River Squadron, '64–'65; deserted, Bermuda Hundred, Virginia, 3/22/65; sent to City Point, Virginia; arrived, Washington, D.C., 3/27/65; took oath, 3/29/65; furnished transport to Norfolk; postwar: occ. driver; married; resided Norfolk, '70.[57]

Carey, Thomas: served as a quartermaster, CSS *Virginia*; reenlisted for the war, received $50 bounty as member of crew, *Virginia*, 3/25/62; appears to have left crew of *Virginia* before 4/1/62; NFR.[58]

Carlin, John A.: b. ca. 1842, Virginia; occ. foundry apprentice; single; resided Old St., east of Cross St., Petersburg, Virginia; enlisted at Petersburg for 1 year as a pvt. in Co. C, 41st Virginia Infantry, 5/4/61; described as 5'8", fair comp., light hair, blue eyes; corp., 9/61; present until 3/62; served as a landsman, CSS *Virginia*, 3/27–4/10/62; discharged per SO 72, Dept. of Norfolk, 4/10/62; worked as machinist thru out war; postwar: occ. locomotive engineer for 30 years on Norfolk and Western RR; resided Petersburg; d. Crewe, Virginia, 1907; bur. St. Joseph's Roman Catholic Cem., Petersburg.[59]

Carmine, Joseph: b. ca. 1841; resided Chowan Co., North Carolina; enlisted at Edenton, North Carolina, for the war as a pvt. in Co. A, 1st North Carolina Infantry, 5/18/61; present until transferred to CSN per SO 17, A&IGO 62, para 23, 2/1/62; served as an ordinary seaman, CSS *Virginia*; paid, Drewry's Bluff, Virginia, 5/13/62–6/30/63; paid for service, naval det. on special expedition from Drewry's Bluff to Charleston Naval Station, 10/1–31/63; NFR; on *Virginia* muster roll in *ORN* as "Joseph Carmines."[60]

Carr, James: b. ca. 1833; occ. laborer; enlisted at Little Rock, Arkansas, for 1 year as a pvt. in Co. K, 1st (Colquitt's) Arkansas Infantry; mustered in at Lynchburg, Virginia, 5/21/61; present, 5–6/61; reenlisted for 2 years, received $50 bounty, 1/24/62; transferred to CSN per SO 17, A&IGO 62, para 23; served as a 2nd class fireman, CSS *Virginia*; reenlisted for the war, received $50 bounty as member of crew, *Virginia*, 3/25/62; paid for service on the *Virginia* and Drewry's Bluff, Virginia, 4/1/62–7/5/63; listed as having deserted, 7/5/63; NFR.[61]

Carr, John: b. ca. 1830, Ireland; occ. mariner; enlisted at Norfolk, Virginia, for 1 year as a pvt. in the United Artillery (1st Co. E, 41st Virginia Infantry), 5/8/61; described as 5'9", light hair, gray eyes; present until discharged and transferred to CSN per SO 9, Dept. of Norfolk, "to go on *Merrimac*," 1/16/62; served as a 1st class fireman, CSS *Virginia*; paid for service on the *Virginia*, 4/1–5/12/62; on *Virginia* CRR, Drewry's Bluff, Virginia, 7/20/62; shipped for the war, Naval Rendezvous, Richmond, Virginia, received $50 bounty, 7/24/62; paid, Drewry's Bluff, 5/13–8/21/62; served as a landsman, CSS *Selma*, Mobile Squadron, '62–'63; in Jackson Gen. Hosp., Richmond, w/ chronic diarrhea, 4/65; d. 4/28/65; bur. Hollywood Cem., Richmond.[62]

Carstarphen, Richard W.: b. ca. 1834, Virginia; resided and enlisted at Norfolk, Virginia, for the war as a corp. in the United Artillery (1st Co. E, 41st Virginia Infantry) and received $50 bounty, 3/6/62; "volunteered to go onboard the Ironclad Steamer *Virginia*," 3/6/62; manned gun on CSS

Virginia during Battle of Hampton Roads, Virginia, 3/8–9/62; on CRRs, 12/22/62, 6/2/63, 9/14/63, 11/31/63 and 3/18/64; absent, on detached duty w/ CSN, per SO 63, A&IGO 64, dated 3/16/64; present, 11–12/64; deserted, 4/5/65; sent to City Point, Virginia, 4/18/65; sent to Washington, D.C.; took oath, 5/10/65; furnished transport to Norfolk.[63]

Carter, Barron: b. ca. 1844, Richmond, Georgia; father, John Carter; resided Augusta, Georgia; appointed by Georgia to USNA, 9/28/60; tendered resignation as acting midshipman, USN, 1/21/61; resignation accepted, 1/25/61; appointed master's mate, CSN, 6/19/61; acting midshipman, 7/8/61; served at Savannah Naval Station, '61–'62; aide to Flag Officer Tattnall during Battle of Port Royal, South Carolina, 11/7/61; served as acting midshipman, CSS *Virginia*, 4/1–5/13/62; served on CSS *Georgia*, *Savannah* and *Isondiga*, Savannah River Squadron, '62–'63; served on CSS *Patrick Henry*, James River Squadron, '63–'64; midshipman, Provisional Navy, 6/2/64; assigned to Wilmington Naval Station, '64; NFR.[64]

Charlesworth, Joshua: b. ca. 1836, Baltimore, Maryland; occ. machinist; enlisted at Philadelphia, Pennsylvania, for 4 years as a pvt. in USMC, 9/28/57; described as 5'8½"; fair comp., brown hair, blue eyes; resided Portsmouth, Virginia; deserted USMC at Gosport Navy Yard, Portsmouth, 4/20/61; enlisted at Montgomery, Alabama, as a sgt. in Co. C, CSMC, 4/27/61; probably served at Pensacola Navy Yard until Co. C transferred to Gosport Navy Yard, Portsmouth, Virginia, 11/29/61; assigned to Marine Guard, CSS *Virginia*, ca. 11/61–ca.5/62; on Co. C CRR, 3rd qtr. '62; listed as having deserted on final *Virginia* payroll, dated 9/30/62; NFR.[65]

Churges, Charles: served as an ordinary seaman, CSS *Virginia*; reenlisted for the war, received $50 bounty as member of crew, *Virginia*, 3/25/62; superannuated for rations, Drewry's Bluff, Virginia, 5/12–24/62; NFR; also found as "Schurgs."[66]

City, George Washington: b. Washington, D.C.; appointed by Virginia to USN as 3rd assistant engineer, 1/12/54; 2nd assistant engineer, 5/9/57; 1st assistant engineer, 8/2/59; tendered resignation from USN 7/27/61; dismissed 8/1/61; appointed 1st assistant engineer, CSN, 8/29/61; served on CSS *Richmond*, Richmond Naval Station, '61; paid for service, Richmond Naval Station, 10/1–12/3/61; assigned to CSS *Virginia*, 12/4/61–2/26/62; assigned to Gosport Navy Yard, '62; served on CSS *Arkansas*, Mississippi River Squadron, '62; took part in Battle of Vicksburg; hospitalized due to exhaustion when *Arkansas* destroyed by crew; assigned to Jackson Naval Station, '62; assigned to Savannah Naval Station, '62–'63; assigned to Charleston Station, '63; served on CSS *Isondiga*, *Savannah* and

Macon, Savannah River Squadron, '63–'64; served on CSS *Chattahoochee*, '64; acting chief engineer, 5/31/64; 1st assistant engineer, Provisional Navy, 6/2/64; paroled, Augusta, Georgia, 5/3/65; postwar: entered Maryland Line Confederate Soldiers Home, Pikesville, Maryland, sometime after '88; NFR.[67]

Clarke, William T.: b. ca. 1829, Virginia; occ. pilot; married; resided Hampton, Virginia; net worth $1,800 in '60; served as a civilian pilot, CSS *Virginia*; took part in Battle of Hampton Roads, Virginia, 3/8–9/62; served at Richmond Naval Station, '62; NFR; cousin of Richard Curtis, who also served on the *Virginia*.[68]

Claysing, Ernest R.: b. ca. 1838, Kingdom of Hanover (Germany); enlisted at Montgomery, Alabama, as a pvt. in Co. G, 3rd Alabama Infantry, 4/26/61; described as 5'7"; discharged to serve on CSS *Virginia* per SO 28, Dept. of Norfolk, 2/8/62; served as a landsman, CSS *Virginia*; appears to have left crew of *Virginia* before 4/1/62; NFR; on *Virginia* muster roll in *ORN* as "Ernest R. Claysung."[69]

Cogins, Thomas: b. ca. 1840, Edgecombe Co., North Carolina; occ. farmer; resided Edgecombe Co.; enlisted at Tarboro, North Carolina, for 1 year as a pvt. in Co. G, 13th North Carolina Infantry, 5/8/61; mustered in at Garysburg, North Carolina, 5/16/61; described as 5'11", light comp., light hair, blue eyes; present until transferred to CSN, 2/19/62; billeted on receiving ship, CSS *Confederate States*; served as a landsman, CSS *Virginia*; reenlisted for the war, received $50 bounty; paid for service on the *Virginia*, 4/1–5/12/62; superannuated for rations, Drewry's Bluff, Virginia, 5/12–24/62; on *Virginia* CRR, Drewry's Bluff, 7/20/62; paid, Drewry's Bluff, 5/13/62–6/30/63; paid for service, naval det. on special expedition from Drewry's Bluff to Charleston Naval Station, 10/1–31/63; reenlisted at Turkey Ridge, Virginia, for the war as a pvt. in Co. G, 13th North Carolina Infantry, 6/8/64; present thru 10/31/64; NFR; on *Virginia* muster roll in *ORN* as "Thomas Coggins."[70]

Coleman, Thomas: enlisted at New Orleans, Louisiana, as a pvt. in Co. C, CSMC, 5/31/61; on New Orleans Naval Station, 1/62; assigned to Marine Guard, CSS *Virginia*, 1/62–5/12/62; on Co. C CRR, 3rd qtr. '62; listed as having deserted on final *Virginia* payroll, dated 9/30/62; NFR.[71]

Collins, James: b. ca. 1843, Ireland; resided New Orleans, Louisiana; served as a 1st class fireman, CSS *Virginia*; reenlisted for the war, received $50 bounty as member of crew, *Virginia*, 3/25/62; paid for service on the *Virginia* and Drewry's Bluff, Virginia, 4/1/62–7/30/63; on *Virginia* CRR, Drewry's Bluff, 7/20/62; served on CSS *Patrick Henry*, James River Squadron; served as a fireman, CSS *Charleston*, Charleston Squadron; entered Federal lines, 2/8/65; POW, Charleston, South Carolina, 3/2/65; described then as 5'5", dark comp., auburn hair, gray eyes; sent to Point Lookout, Maryland, and then to Washington, D.C.; furnished transport to New Orleans; possibly "James Collins," who enlisted at New Orleans, Louisiana, for the war as a 2nd corp. in Co. A, 15th Louisiana Infantry, 6/3/61; present, 9–10/61; 3rd sgt., 9/25/61; NFR.[72]

Collins, Phillipp: served as a seaman, CSS *Virginia*; appears to have left crew of *Virginia* before 4/1/62; NFR.[73]

Colonna, William Bramwell: b. ca. 1832, Northampton, Virginia; occ. farmer; married; resided Accomac Co., Virginia; enlisted in Virginia for 1 year as a pvt. in Co. C, 1st (Butler's) South Carolina Infantry, 2/21/61; mustered in at Sullivan's Island, South Carolina, 5/17/61; present until enlistment ended, 2/21/62; enlisted at Norfolk, Virginia, for the war as a pvt. in the United Artillery (1st Co. E, 41st Virginia Infantry) and received $50 bounty, 3/6/62; "volunteered to go onboard the Ironclad Steamer *Virginia*," 3/6/62; manned gun on CSS *Virginia* during Battle of Hampton Roads, Virginia, 3/8–9/62; on CRRs, 12/22/62, 6/2/63, 9/14/63, 11/31/63 and 3/18/64; absent, on detached duty w/ CSN, per SO 63, A&IGO 64, dated 3/16/64; present, 11–12/64; paroled, Appomattox CH, 4/9/65; postwar: resided Warwick Co., Virginia; married '80; member, Pickett-Buchanan Camp, UCV, Norfolk, Virginia; entered R.E. Lee Camp Soldiers' Home, Richmond, 2/24/94; d. of paralysis, 4/28/1903; buried Hollywood Cem., Richmond; widow, Daisy W. Colonna, received pension 7/1/1924; also found as "Colona."[74]

Cooper, Astley A.: b. ca. 1835, Petersburg, Virginia; occ. coach trimmer; resided and enlisted at Petersburg, Virginia, for 1 year as a pvt. in Co. K, 12th Virginia Infantry, 5/4/16; described as 5'7", dark comp., dark hair, hazel eyes; corp., 8/21/61; 5th sgt., 10/1/61; reenlisted for war, received $50 bounty, 2/11/62; "discharged for naval service" per SO 69, Dept. of Norfolk, 3/28/62; served as a landsman, CSS *Virginia*; paid, Drewry's Bluff, Virginia, 5/12/62–6/30/63 and 10/1/63–3/64; d. of chronic diarrhea, South Carolina Hosp., Petersburg, Virginia, 3/30/64.[75]

Cooper, James R.: reenlisted for the war, received $50 bounty as member of crew, CSS *Virginia*, 3/25/62; shipped for the war at Orange CH, Virginia, 3/31/64; NFR; not on *Virginia* muster roll in *ORN*.[76]

Coughlan, Patrick: b. ca. 1837; occ. laborer; enlisted at Little Rock, Arkansas, for 1 year as a pvt. in Co. F, 1st (Colquitt's) Arkansas Infantry, 5/8/61; mustered in at Lynchburg, Virginia, 5/20/61; listed as AWOL,

12/28/61; present, 2/3/62; transferred to CSN per SO 22, A&IGO 62, para 11, 2/3/62; served as a landsman, CSS *Virginia*; superannuated for rations, Drewry's Bluff, Virginia, 5/12–24/62; on *Virginia* CRR, Drewry's Bluff, 7/20/62; paid, Drewry's Bluff, 5/12–12/31/62; NFR; also found as "Caughlin;" on *Virginia* muster roll in *ORN* as "Patrick Coghlan."[77]

COYLE, ROBERT: b. ca. 1822, Ireland; occ. dairyman; married; resided and enlisted at New Orleans, Louisiana, as a pvt. in Co. C, CSMC, 6/17/61; assigned New Orleans Naval Station, 1/62; assigned to Marine Guard, CSS *Virginia*, 1–5/12/62; on Co. C CRR, 3rd qtr. '62; listed as having deserted on final *Virginia* payroll, dated 9/30/62; NFR.[78]

CRAIG, WILLIAM JAMES: b. Kentucky; appointed by Kentucky to USNA, 11/30/58; tendered resignation as an acting midshipman, 8/6/61; resignation accepted, 8/12/61; appointed as an acting midshipman, CSN, 8/28/61; served on CSS *Patrick Henry*, 12/20–31/61; left North Carolina Squadron for CSS *Virginia*, 12/31/61; assigned to receiving ship, CSS *Confederate States*, '62, assigned to CSS *Virginia*, 2/19/62; served as captain of Gun #8 during Battle of Hampton Roads, Virginia; paid for service, Charleston Station, 5/18–6/30/62; served on CSS *Huntress*, Charleston Squadron, '62; served on CSS *Chattahoochee*, 8/1/62–'63; injured in boiler explosion, 5/27/63; served on CSS *Patrick Henry*, James River Squadron, '63; served on CSS *Georgia* and *Sampson*, Savannah River Squadron, '63–'64; passed midshipman, 1/8/64; served on CSS *Virginia II*, James River Squadron, '64; master in line of promotion, Provisional Navy, 6/2/64; served on the cruiser, CSS *Tallahassee*, '64; in Gen. Hosp. #4, Wilmington, North Carolina, w/ diarrhea, 10/11/64; assigned Mobile Squadron, '64–'65; surrendered, 5/4/65; paroled, 5/10/65; NFR.[79]

CRAIG, WILLIAM PLEASANT: b. ca. 1840, Buncombe Co., North Carolina; occ. shoemaker; resided Buncombe Co.; enlisted at Asheville, North Carolina, for 1 year as a pvt. in Co. F, 14th North Carolina Infantry, 5/3/61; mustered in at Raleigh, North Carolina, 5/25/61; described as 5'10", light comp., sandy hair, hazel eyes; present until "discharged by order of Gen. Huger to go onboard the *Merrimac*," 2/18/62; served as a landsman, CSS *Virginia*; paid for service on the *Virginia*, 4/1–5/12/62; on *Virginia* CRR, Drewry's Bluff, Virginia, 7/20/62; reenlisted for the war, received $50 bounty; paid, Drewry's Bluff, 5/13/62–1/5/63; listed as having deserted on payroll, dated 3/31/63; enlisted for the war as a pvt. in Co. D, 6th North Carolina Cavalry; listed as AWOL on 7–8/63 rolls; court-martialed, 9/9/64; sentence suspended by order of secretary of war, 9/12/64; on CRR, 11/14/64; NFR.[80]

Creekmur, Charles J.: b. 9/14/28, Norfolk Co., Virginia; occ. clerk; married; resided Portsmouth, Virginia; served in Mexican War with 1st Virginia Volunteers; enlisted at Portsmouth in Co. K, 9th Virginia Infantry, 4/20/61; received numerous recommendations for commission in Confederate army; discharged at Pinner's Point, Norfolk Co. and transferred to CSN per SO 17, A&IGO 62, para 16; served as purser's/paymaster steward, CSS *Virginia*; on *Virginia* CRR, Drewry's Bluff, Virginia, 7/20/62; attached to Drewry's Bluff and CSS *Roanoke* thru 12/17/63; NFR; postwar, member, Pickett-Buchanan Camp UCV, Norfolk; d. 12/6/1901; bur. Oak Grove Cem., Portsmouth; also found as "Creekmore."[81]

Creilly, Thomas: b. ca. 1829; enlisted at New Orleans, Louisiana, as a pvt. in Co. C, CSMC, 6/4/61; assigned New Orleans Naval Station, 1/62; assigned to Marine Guard, CSS *Virginia*, 1–5/12/62; advertised as a deserter, 6/12/62; described then as 5'5½", dark comp., auburn hair, blue eyes; RTD w/ Co. C by 3rd qtr. '62; sent to Charleston, South Carolina, as part of Monitor Boarding Party under Capt. Wilson, ca. 3/63; assigned to Marine Guard, CSS *Virginia II*, James River Squadron, 5/14/64–2/18/65; confined, Libby Prison, Richmond, Virginia, 4/10/65; NFR; also found as "Cribly," "Crilly," "Crilley" and "Crielly"; also found as "Charles Crilley"; on *Virginia* muster roll in *ORN* as "Thomas Crilley."[82]

Cronin, James C.: enlisted at Camp Clark, Texas, for the war as a pvt. in Co. B, 4th Texas Infantry, 7/11/61; present, 9–10/61; discharged to "work in another government branch" per SO 192, A&IGO 61, para 12; 10/28/61; paid for service as boatswain's mate on receiving ship, CSS *Confederate States*, 1/1–9/62; paid for service on CSS *Sea Bird*, North Carolina Squadron, 1/10–2/9/62; served as a boatswain's mate, CSS *Virginia;* manned Gun #3 during Battle of Hampton Roads, Virginia, 3/8–9/62; reenlisted for the war, received $50 bounty as member of crew, *Virginia*, 3/25/62; paid for service on the *Virginia* and Drewry's Bluff, Virginia, 4/1–6/30/62; on *Virginia* CRR, Drewry's Bluff, 7/20/62; volunteered for duty on CSS *Chattahoochee*, 8/1/62; paid for service, CSS *Chattahoochee*, 8/1–12/31/62 and 4/1–6/12/63; on board when boiler exploded, 5/27/63; transferred to CSS *Savannah*, Savannah River Squadron, 6/13/63; appointed by Georgia as a boatswain, 7/11/63; paid for service, Charleston Naval Station, 10/1–31/63; served on CSS *Savannah*, '64; boatswain, Provisional Navy, 6/2/64; NFR.[83]

Crosby, William H.: b. ca. 1830, Maine, occ. barkeeper; married, Norfolk, Virginia, 3/4/60; resided and enlisted at Norfolk for 1 year as a pvt. in the United Artillery (1st Co. E, 41st Virginia Infantry), 4/19/61; corp., 8/25/61;

present thru 3/6/62; "volunteered to go onboard the Ironclad Steamer *Virginia*," 3/6/62; manned gun on CSS *Virginia* during Battle of Hampton Roads, Virginia, 3/8–9/62; reenlisted for war, received $50 bounty, 3/10/62; 2nd sgt.; on CRRs, 12/22/62, 6/2/63, 9/14/63, 11/31/63 and 3/18/64; in Chimborazo Hosp. #9, Richmond, Virginia, 9/16/64; in Chimborazo Hosp. #4, w/ int. fever, 9/17–22/64; absent, sick, in Chimborazo Hosp., 10/10/64 and 12/31/64 rolls; deserted, 4/5/65; sent to City Point, Virginia, 4/8/65; sent to Washington, D.C., 4/12/65; took oath, furnished transport to Norfolk; postwar: occ. policeman; resided Norfolk next to William Crosby, formerly of the United Artillery and CSS *Virginia*; d. of tuberculosis, 2/2/97; widow, Eliza A. Crosby, received pension, witnessed by former *Virginia* shipmate Andrew J. Dalton, Norfolk justice of peace, 5/7/1900.[84]

CROSS, THOMAS W.: b. 1841; Dinwiddie Co., Virginia; occ. farmer; enlisted at Petersburg, Virginia, for 1 year as a pvt. in Co. C, 41st Virginia Infantry, 5/9/61; mustered in, 5/11/61; described as 5'11", fair comp., light hair, blue eyes; present, 7–8/61; discharged by SO 69, Dept. of Norfolk, 3/28/62; served as a landsman, CSS *Virginia*; appears to have left crew of *Virginia* before 4/1/62; may have later enlisted at Henrico Co., Virginia, for the war as a pvt. in Co. E, 5th Btn. Virginia Infantry, 6/1/62; absent sick, 5–6/63; NFR.[85]

CRUMP, JOHN D.: b. ca. 1831; occ. sailor; enlisted as a pvt. in the Richmond Fayette Artillery (Co. B, 38th Btn. Virginia Light Artillery), 4/25/61; described then as 5'7", light comp., dark hair, dark eyes; discharged and transferred to CSN; appointed as an acting master's mate, CSN; paid for service on CSS *Jamestown*, James River Squadron, 10/1–31/61 and 12/1–31/61; shipped for 3 years as a seaman, Naval Rendezvous, Richmond, Virginia, 2/27/62; served as a coxswain, CSS *Virginia*; reenlisted for the war, received $50 bounty as member of crew, *Virginia*, 3/25/62; paid for service on the *Virginia* and Drewry's Bluff, Virginia, 4/1–9/30/62; superannuated for rations, Drewry's Bluff, 5/12–24/62; paid, Richmond Naval Station, 10/1–11/30/62; paid, Drewry's Bluff, 12/1/62–6/30/63 and 10/31/63–3/31/64; paid as a quarter master starting 1/2/63; paid as a seaman starting 1/17/63; NFR; on *Virginia* muster roll in *ORN* as "John C. Crump."[86]

CULLIFER, JOSEPH: b. ca. 1834, Washington Co., North Carolina; occ. farmer; resided Washington Co.; enlisted at Plymouth, North Carolina, for the war as a pvt. in Co. G, 1st North Carolina Infantry, 6/24/61; present until transferred to CSN per SO 17, A&IGO 62, para 23, 2/3/62; served as an ordinary seaman, CSS *Virginia*; reenlisted for the war, received $50 bounty as member of crew, *Virginia*, 3/25/62; paid for service on the *Virginia*, 4/1–5/12/62; superannuated for rations, Drewry's Bluff, Virginia, 5/12–24/62;

on *Virginia* CRR, Drewry's Bluff, 7/20/62; paid, Drewry's Bluff, 5/13/62–3/31/64; NFR; on *Virginia* muster roll in *ORN* as "Joseph Cullipher."[87]

Cullington, James: b. ca. 1841, Washington Co., North Carolina; resided Washington Co.; enlisted at Plymouth, North Carolina, for the war as a corp. in Co. G, 1st North Carolina Infantry, 6/24/61; reduced to pvt., 11/15/61; present until transferred to CSN per SO 17, A&IGO 62, para 23, 2/3/62; served as a seaman on CSS *Virginia*; reenlisted for the war, received $50 bounty as member of crew, *Virginia*, 3/25/62; paid for service on the *Virginia*, 4/1–5/12/62; superannuated for rations, Drewry's Bluff, Virginia, 5/12–24/62; on *Virginia* CRR, Drewry's Bluff, 7/20/62; paid, Drewry's Bluff, 5/13/62–9/30/63; paid for service, naval det. on special expedition from Drewry's Bluff to Charleston Naval Station, 10/1–31/63; paid for service as a fireman, CSS *Albemarle*, North Carolina Squadron, 7/1–9/30/64; NFR.[88]

Cunningham, James: enlisted at Memphis, Tennessee, as a pvt. in CSMC, 9/7/61; probably transferred from Capt. Hays's det., Co. D, to Co. C, Pensacola Navy Yard ca. 11/27/61; transferred with Co. C to Gosport Navy Yard, Portsmouth, Virginia, 11/29/61; assigned to Marine Guard, CSS *Virginia*, ca. 11/61–ca. 5/62; on Co. C CRR, 3rd qtr. '62; NFR.[89]

Cunningham, John: b. ca. 1816, Scotland; occ. grocer/ship's chandler; married; resided Chatham Co., Georgia; served as a seaman, CSS *Petrel*; POW, Charleston, South Carolina, 7/28/61; sent to Ft. Delaware, Delaware; escaped or exchanged; served as an ordinary seaman, CSS *Virginia*; reenlisted for the war, received $50 bounty as member of crew, *Virginia*, 3/25/62; appears to have left crew of *Virginia* before 4/1/62; shipped for the war as an ordinary seaman, Naval Rendezvous, Richmond, Virginia, 8/7/62; paid, Drewry's Bluff, Virginia, 8/13–11/11/62; served as an ordinary seaman, CSS *Savannah*, Savannah River Squadron, '62; paid, Drewry's Bluff, 4/1–9/30/63; paid for service, naval det. on special expedition from Drewry's Bluff to Charleston Naval Station, 10/1–31/63; paid as a seaman, CSS *Yadkin*, North Carolina Squadron, 2/1–4/25/64; served as a seaman, CSS *Arctic*, North Carolina Squadron, '64; served as an ordinary seaman, CSS *Tennessee*, Mobile Squadron; POW, Mobile Bay, 8/5/64; escaped 9/27/64; served as a seaman, CSS *Virginia II*, James River Squadron, '64–'65; pvt., Co. D, 4th Btn. Virginia Infantry, Local Defense (Naval Btn.); POW, Farmville, Virginia, 4/6/65; sent to Newport News, Virginia, 4/16/65; took oath, 6/26/65; described then as 5'4", light comp., dark hair, blue eyes.[90]

Cunningham, Thomas: b. Virginia; occ. pilot; married; resided Hampton, Virginia; served as a civilian pilot, CSS *Virginia*; took part in the Battle of Hampton Roads, Virginia, 3/8–9/62; NFR.[91]

Curtis, Charles Samuel: b. ca. 1825, Virginia; single; resided Macon Co., Alabama; enlisted at Mobile, Alabama, as a pvt. in Co. C, CSMC, 8/7/61; probably served at Pensacola Navy Yard until Co. C transferred to Gosport Navy Yard, Portsmouth, Virginia, 11/29/61; assigned to Marine Guard, CSS *Virginia*, 4/1–5/12/62; sent to Charleston, South Carolina, as part of Monitor Boarding Party under Capt. Wilson, ca. 3/63; extra duty, carpenter, Camp Beall, Virginia, 12/9/63–1/29/64; assigned to Marine Guard, CSS *Virginia II*, James River Squadron, 5/14/64; RTD w/ Co. C, Camp Beall, Virginia, 9/18/64; on daily duty as a carpenter until 12/31/64; confined in Libby Prison, Richmond, Virginia, 4/10/65; also found as "Samuel C. Curtis."[92]

Curtis, Richard: b. ca. 1839; Hampton, Virginia; occ. boatman; enlisted at Hampton, Virginia, for 1 year as a pvt. in Co. A, 32nd Virginia Infantry, 5/13/61; fought in Battle of Big Bethel, Virginia, 6/10/61; present thru 8/31/61; "transferred to *Merrimack* by order of Gen. Magruder," 1/3/62; served as an ordinary seaman, CSS *Virginia*; assigned to 1st division, stationed at bow gun, 3/8/62; listed as a POW on *Virginia* muster roll, 7/31/62, but paid for service as an ordinary seaman on the *Virginia* and Drewry's Bluff, Virginia, 4/1–7/24/62 and as a landsman, 7/25–8/10/62; listed as discharged on final *Virginia* payroll, dated 9/30/62; NFR; cousin of William Clarke, civilian pilot, CSS *Virginia*; postwar: resided 1123 Highland Ave., Norfolk, Virginia; refused to sign pension application for religious reasons; pension approved anyway, 9/12/24.[93]

Dalton, Andrew Joseph: b. ca. 1843, Dublin, Ireland; resided Norfolk, Virginia; occ. apprentice printer to T.G. Broughton, ed. *Norfolk Herald*, '54–'59; master printer, '59–'60; enlisted at Charleston, South Carolina, for 1 year as a pvt. in Co. C, 1st South Carolina Arty, 2/10/61; fired on Ft. Sumter; present, 5–10/61; corp., 9/1/61; discharged 2/10/62; enlisted as pvt. in the United Artillery (1st Co. E, 41st Virginia Infantry), 3/4/62; "volunteered to go onboard the Ironclad Steamer *Virginia*," 3/6/62; took part in Battle of Hampton Roads, Virginia, 3/8–9/62; WIA by musket balls coming thru gun port, 3/8/62; unit became 1st Co. C, 19th Btn. Virginia Heavy Artillery; WIA twice in war; unit became an independent btry., 10/1/62; transferred to Co. D, 1st Confederate Engineer Troops, 1/2/64; never joined unit; with Brig. Gen. John Hunt Morgan's Cavalry; POW, Dublin, Virginia '64; sent to Camp Chase, Ohio; postwar: occ. printer; member, Norfolk City Council; Norfolk street inspector; Norfolk Co. sheriff; Norfolk justice of the peace; Virginia state senator; member, Pickett-Buchanan Camp, UCV, Norfolk; alive 1914.[94]

Daniels, Austin: served as a landsman, CSS *Virginia*; paid for service on the *Virginia* 5/1–12/62; superannuated for rations, Drewry's Bluff, Virginia, 5/12–24/62; on *Virginia* CRR, Drewry's Bluff, 7/20/62; paid, Drewry's Bluff, 5/13/62–6/30/63; served as an officer's steward, Drewry's Bluff, 10/8/63–3/31/64; NFR.[95]

Davidson, Hunter: b. 9/20/26, Georgetown, Washington, D.C.; appointed to the USN by Virginia as a midshipman, 10/29/41; passed midshipman, 8/10/47; married Mary S. Ray of Anne Arundel Co., Maryland, 7/20/52; master, 9/14/55; Lt., 9/15/55; instructor, USNA at start of Civil War; resided, Stevensville, Maryland; tendered resignation and dismissed from USN, 4/23/61; appointed 1st Lt., CSN, 6/10/61; served on CSS *Patrick Henry*, 8/1–9/30/61; left North Carolina Squadron for CSS *Virginia*, 12/7/61; assigned to CSS *Virginia*, 12/8/61; commanded first division and Guns #2 & #3 during the Battle of Hampton Roads, Virginia; paid for service, Richmond Naval Station, 5/12–6/30/62; aided Comdr. Matthew F. Maury's development of electric torpedoes, which defended Confederate rivers/harbors; commanded CSS *Teaser*, James River Squadron; chief, CSN Torpedo Bureau, 9/62; responsible for destruction of USS *Commodore Barney* and *Commodore Jones*; commanded CSS *Squib* when it torpedoed USS *Minnesota* off Newport News, Virginia, 4/9/64; 1st Lt., Provisional Navy, 6/2/64; cmdr., Provisional Navy, for gallant and meritorious conduct during attack on *Commodore Barney* and *Commodore Jones*, 6/10/64; commanded blockade runner, CSS *City of Richmond*; delivered men/supplies to the blockade runner, CSS *Stonewall*, Quebon Bay, France, 1/65; postwar, invited by to Argentina by President Sarmiento, '75; directed Argentine navy's Dept. of Torpedo Defense and Naval Construction for 12 years; resided Paraguay, '87; d. Pirayu, Paraguay, 2/16/1913.[96]

Davis, Jeremiah A.: b. ca. 1824, Virginia; occ. farm laborer; married; resided Dumfries, Virginia; enlisted at Fletcher's Chapel, Lunenburg Co., Virginia, for 1 year as a pvt. in 1st Co. H, 9th Virginia Infantry, 6/11/61; present until "discharged by order of Gen. Huger to go on *Merrimack*," 1/31/62; billeted on receiving ship, CSS *Confederate States*; served as quarter master, CSS *Virginia*; paid for service on the *Virginia* and Drewry's Bluff, Virginia, 4/1–9/30/62; listed as discharged on final *Virginia* payroll, dated 9/30/62; may have later enlisted at Richmond, Virginia, as a pvt. in Co. F, 4th Btn. Virginia Infantry, Local Defense (Naval Btn.), 6/27/63; present, 8/2/64; served as a landsman/ordinary seaman, Co. K, Naval Brigade; paroled Greensboro, North Carolina, 4/26/65.[97]

DAVIS, JOHN COLBERT: resided Baltimore, Maryland; enlisted at Hampton, Virginia, as a pvt. in Co. E, 32nd Virginia Infantry, 5/19/61; transferred to CSN per SO 262, A&IGO 61, para 16, 12/20/61; shipped for 1 year as an ordinary seaman, CSS *Jamestown*, James River Squadron, 12/21/61; billeted on receiving ship, CSS *Confederate States*; served as a landsman, CSS *Virginia*; reenlisted for the war, received $50 bounty as member of crew, *Virginia*, 3/25/62; paid for service on the *Virginia* and Drewry's Bluff, Virginia, 4/1–11/11/62; served as an ordinary seaman, CSS *Jamestown*, James River Squadron, '62; served as a landsman, CSS *Baltic*, Mobile Squadron, '62–'63; NFR; postwar: entered Maryland Line Confederate Soldiers Home, Pikesville, Maryland, sometime after '88; NFR; on *Virginia* muster roll in *ORN* as "James C. Davis."[98]

DAVIS, RICHARD L.: enlisted at Augusta, Georgia, for 1 year as a pvt. in Co. G, 3rd Georgia Infantry, 4/26/61; transferred to CSN; billeted on receiving ship, CSS *Confederate States*; served as a 1st class fireman, CSS *Virginia*; reenlisted for the war, received $50 bounty as member of crew, *Virginia*, 3/25/62; superannuated for rations, Drewry's Bluff, Virginia, 5/12–24/62; appears to have left crew of *Virginia* before 4/1/62; reportedly served with Savannah River Squadron, 11/64; NFR; Georgia commissioner of pensions requested verification of service, 1916.[99]

DAVIS, THEODORE: b. ca. 1837; enlisted at Memphis, Tennessee, as a pvt. in CSMC, 8/10/61; probably transferred from Capt. Hays's det., Co. D, to Co. C, Pensacola Navy Yard, ca. 11/27/61; probably served at Pensacola Navy Yard until Co. C transferred to Gosport Navy Yard, Portsmouth, Virginia, 11/29/61; assigned to Marine Guard, CSS *Virginia*, ca. 11/61–ca. 5/62; advertised as a deserter, 6/12/62; described then as 5'9", florid comp., black hair, blue eyes; sent to Charleston, South Carolina, as part of Monitor Boarding Party under Capt. Wilson, ca. 3/63; corp. on rolls thru 9/20/63; NFR; on *Virginia* muster roll in *ORN* as "Theodor Davis."[100]

DAVIS, WILLIAM HENRY: b. ca. 1834, Madison Co., Alabama; parents, A. Charles and Cynthia Davis; enlisted at Montgomery, Alabama, as a pvt. in Co. C, CSMC, 4/20/61; probably served at Pensacola Navy Yard until Co. C transferred to Gosport Navy Yard, Portsmouth, Virginia, 11/29/61; assigned to Marine Guard, CSS *Virginia*, ca. 11/61–ca. 5/62; on Co. C rolls thru 1/26/63; NFR.[101]

DEARY, THOMAS: b. ca. 1828; enlisted at Charleston, South Carolina, for 6 months as a pvt. in Co. M, 1st South Carolina Infantry, 4/22/61; mustered out at Richmond, Virginia, 7/9/61; enlisted at Richmond for the war as a pvt. in Co. I, 1st (McCreary's) South Carolina Infantry, 7/20/61;

transferred to CSN, 1/17/62; billeted on receiving ship, CSS *Confederate States*; served as a boatswain's mate, CSS *Virginia*; reenlisted for the war, received $50 bounty as member of crew, *Virginia*, 3/25/62; paid for service as a boatswain's mate on the *Virginia* and Drewry's Bluff, Virginia, 4/1–6/14/62; superannuated for rations, Drewry's Bluff, 5/12–24/62; paid as a seaman, 6/15/62–3/31/63; listed as having deserted on 12/31/62–3/31/63 payrolls; may have later enlisted at Charleston, South Carolina, for the war as a pvt. in Co. H, 27th South Carolina Infantry, 4/7/63; present, 8/63; absent, on detached service, 9–12/63; absent, detailed to Brig. Gen. Roswell S. Ripley's boat, 12/29/63–8/31/64; NFR; also found as "Thomas Deery."[102]

Delley, John: enlisted at Camp Clark, Texas, for the war as a pvt. in Co. B, 4th Texas Infantry; present, 9–10/61; "discharged to work in another government branch" per SO 192, A&IGO 61, para 12, 10/28/61; paid for service on receiving ship, CSS *Confederate States*, 1/1–9/62; paid, North Carolina Squadron, 1/10–2/9/62; served as a coxswain, CSS *Virginia*; reenlisted for the war, received $50 bounty as member of crew, *Virginia*, 3/25/62; paid for service on the *Virginia*, 4/1–5/12/62; superannuated for rations, Drewry's Bluff, Virginia, 5/12–24/62; paid, Drewry's Bluff, 5/13/62–6/30/63; served as a quarter master, Drewry's Bluff, 12/6–31/63; served as a seaman, CSS *Virginia II*, James River Squadron, '64–'65; in Jackson Gen. Hosp., Richmond, Virginia, 4/8/65; listed as having deserted, 4/19/65; NFR; also found as "Dellay"; on *Virginia* muster roll in *ORN* as "John Delly."[103]

Dewey, William H.: enlisted at Mobile, Alabama, as a pvt. in CSMC, 8/19/61; on Co. C CRRs, 4th qtr. '61 & 3rd qtr. '62; assigned to Marine Guard, CSS *Virginia*, ca. 11/61–ca. 5/62; listed as having deserted before 9/30/62; NFR.[104]

Dinning, Cornelius: served as a 2nd class fireman, CSS *Virginia*; paid for service on the *Virginia*, 4/1–5/12/62; superannuated for rations, Drewry's Bluff, Virginia, 5/12–24/62; on *Virginia* CRR, Drewry's Bluff, 7/20/62; paid, Drewry's Bluff, 5/13–12/31/62; NFR.[105]

Divers, David: served as a seaman, CSS *Virginia*; reenlisted for the war, received $50 bounty as member of crew, *Virginia*, 3/25/62; paid for service on the *Virginia* and Drewry's Bluff, Virginia, 4/1–5/15/62; listed as having deserted on final *Virginia* payroll, dated 9/30/62; NFR; also found as "David Devers"; possibly "David Devers" who enlisted at New Orleans, Louisiana, for the war as a pvt. in Co. A, 15th Louisiana Infantry, 6/30/61; present, 9–10/61; NFR.[106]

Dixon, Sylvester: b. ca. 1834, North Carolina; occ. mariner; resided Carteret Co., North Carolina; shipped for the war as a landsman, CSS *Forrest*, North Carolina Squadron, 8/16/61; served as a landsman, CSS *Virginia*; paid for service on the *Virginia* and Drewry's Bluff, Virginia, 4/1–7/31/62; superannuated for rations, Drewry's Bluff, 5/12–24/62; listed as discharged on final *Virginia* payroll, dated 9/30/62; enlisted at Beaufort Co., North Carolina, for the war as a pvt. in Co. K, 3rd North Carolina Cavalry, 5–6/62; present until transferred to 1st North Carolina Artillery; enlisted at Craven Co., North Carolina, for the war as a pvt. in 1st Co. I, 1st North Carolina Artillery, 1/1/63; on CRR, 5/9/64; transferred to Co. A, 1st Btn. North Carolina Local Defense Troops, 4/16/64; present until listed as having deserted on 12/64 roll; NFR.[107]

Dobbs, Thomas E.: b. ca. 1841, Portsmouth, Virginia; occ. carpenter; enlisted at Naval Hosp., Portsmouth, Virginia, for 1 year as a pvt. in Co. D, 9th Virginia Infantry, 4/27/61; mustered in, Naval Hospital, Portsmouth, 5/5/61; described as 5'6½", dark comp., black hair, black eyes; discharged from regt., 4/1/62; served as a landsman, CSS *Virginia*; paid for service on the *Virginia* and Drewry's Bluff, Virginia, 4/1–5/15/62; superannuated for rations, Drewry's Bluff, 5/12–24/62; listed as having deserted on final *Virginia* payroll, dated 9/30/62; enlisted at Richmond, Virginia, for the war as 1st sgt., 24th Btn. Virginia Partisan Rangers, 7/20/62; unit disbanded by order of the decretary of ear, 1/3/63; enlisted at Warm Springs, Virginia, for the war as a pvt. in Co. K, 20th Virginia Cavalry, 4/1/63; listed as AWOL, 7/1/63; had enlisted at Richmond for the war as 5th sgt. In the Confederate Engineer Troops, 6/30/63; assigned to Co. F, 1st Confederate Engineer Troops, 10/3/63; absent, on 30-day furlough, 2/24/64; listed as AWOL, 3/25/64; reduced to pvt., listed as a deserter, 8/31/64; NFR.[108]

Dohran, Thomas: b. ca. 1826, Ireland; occ. clerk; single; resided and enlisted at New Orleans, Louisiana, as a pvt. in CSMC, 5/6/61; on Capt. Holmes's CRR, 5/10/61 & 7/11/61; on Co. C CRR, 4th qtr. '61; probably with Co. C when it transferred to Gosport Navy Yard, Portsmouth, Virginia, 11/29/61; assigned to Marine Guard, CSS *Virginia*, ca. 11/61–5/12/62; on Co. C CRRs, 3rd & 4th qtr. '62 and 1/23/63; NFR; on *Virginia* muster roll in *ORN* as "Thomas Dohan."[109]

Domat, Moses: b. ca. 1846; resided Orleans Parish, Louisiana; served as a landsman, CSS *Virginia*; reenlisted for the war, received $50 bounty as member of crew, *Virginia*, 3/25/62; paid for service on the *Virginia* and Drewry's Bluff, Virginia, 4/1/62–3/31/64; superannuated for rations, Drewry's Bluff, 5/12–24/62; on *Virginia* CRR, Drewry's Bluff, 7/20/62; served as a landsman, CSS *Virginia II*, James River Squadron, '64–'65; volunteered for special service under command of Lt. Wharton, 1/13/65; captured during armed boat expedition to destroy boats/bridges near Kingston, Tennessee, 2/26/65; POW, Knoxville, Tennessee; sent to Nashville, Tennessee, 3/14/65; sent to Camp Chase, Ohio; took oath, 6/13/65; described then as 5'9", dark comp., dark hair, gray eyes.[110]

Donnovant, Michael: shipped for 1 year as a landsman, Naval Rendezvous, Richmond, Virginia, 7/1/61; served on CSS *Patrick Henry*, James River Squadron, 7/1–8/4/61; paid for service on receiving ship, CSS *Confederate States*, 1/1–9/62; assigned to CSS *Jamestown*, North Carolina Squadron, 1/10–2/24/62; served as a landsman, CSS *Virginia*; reenlisted for 3 years, received $50 bounty, 4/21/62; paid for service on the *Virginia*, 4/1–5/12/62; on *Virginia* CRR, Drewry's Bluff, Virginia, 7/20/62; paid, Drewry's Bluff, 5/13/62–7/2/63; listed as having deserted on 9/30/63 payroll; served as a landsman, CSS *Patrick Henry*, James River Squadron, '64; NFR; also found as "Donnvant" and "Dunavent."[111]

Doret, Benjamin: enlisted at New Orleans, Louisiana, for 1 year as a pvt. in Co. F, C.S. Zouave Btn. Louisiana Infantry, 4/61; present, 5–8/61; discharged, 1/31/62; listed as transferred to CSN on regt. return, 1/62; served as an ordinary seaman, CSS *Virginia*; reenlisted for the war, received $50 bounty as member of crew, *Virginia*, 3/25/62; superannuated for rations, Drewry's Bluff, Virginia, 5/12–24/62; NFR; also found as "Dorrey;" on *Virginia* muster roll in *ORN* as "Benjamin Dorry."[112]

Dornin, Franklin B.: b. Maryland; resided, Baltimore, Maryland; appointed as a captain's clerk, CSN, 11/8/61; paid for service, Richmond Naval Station, 11/8/61–3/12/62; appointed by Maryland as an acting midshipman, CSN, 3/8/62; assigned to CSS *Virginia*, 3/12/62; served on CSS *Jamestown*, James River Squadron, '62; served at Drewry's Bluff, Virginia, 5/12–8/28/62; served on CSS *Baltic*, Mobile Squadron, '62–'63; served on CSS *Patrick Henry* and *Roanoke*, James River Squadron, '64; midshipman, Provisional Navy, 6/2/64; assigned to Mobile Squadron, '64; surrendered, 5/4/65; paroled, Nanna Hubba, Alabama, 5/10/65; listed destination as 31 Cathedral St., Baltimore.[113]

Douglas, William Wilson: b. ca. 1835, North Carolina; occ. shinglemaker; married; resided Washington Co., North Carolina; enlisted at Plymouth, North Carolina, for the war as a pvt. in Co. G, 1st North Carolina Infantry, 6/24/61; present until transferred to CSN per SO 17, A&IGO 62, para 23, 2/3/62; served as a landsman, CSS *Virginia*; reenlisted for the war, received $50 bounty as member of crew, *Virginia*, 3/25/62; appears to have left crew of *Virginia* before 4/1/62; served as an ordinary seaman, CSS *Arctic*, North Carolina Squadron, '64; NFR; on *Virginia* muster roll in *ORN* as "William W. Douglass."[114]

Drake, William Francis: resided Northampton Co., Virginia; enlisted at Norfolk, Virginia, for 1 year as a pvt. in the United Artillery (1st Co. E, 41st Virginia Infantry), 4/19/61; present thru 3/6/62; "volunteered to go onboard the Ironclad Steamer *Virginia*," 3/6/62; stationed at port bow gun of CSS *Virginia* during Battle of Hampton Roads, Virginia, 3/8–9/62; reenlisted for war, received $50 bounty, 3/10/62; on CRRs, 12/22/62, 6/3/63, 9/14/63, 11/2/63 and 3/18/64; absent, on detached duty w/ CSN, per SO 63, A&IGO 64, dated 3/16/64; present, 12/31/64; in Chimborazo Hosp. #1, Richmond, Virginia, 2/23/65; furloughed, 3/23/65; NFR; postwar: occ. teacher; entered Confederate Home, Raleigh, North Carolina, 11/1902; d. 8/13/1930; bur. Oakwood Cem., Richmond.[115]

Driscoll, Owen: enlisted at New Orleans, Louisiana, as a pvt. in Co. C, CSMC, 6/6/61; probably served at Pensacola Navy Yard until Co. C transferred to Gosport Navy Yard, Portsmouth, Virginia, 11/29/61; assigned to Marine Guard, CSS *Virginia*, ca. 11/61–ca. 5/62; on duty with Co. C thru 5/14/64; assigned to Marine Guard, CSS *Fredericksburg*, James River Squadron, thru 12/31/64; NFR.[116]

Dudley, William G.: b. ca. 1824, Maryland; occ. shipwright; married; resided and enlisted at Norfolk, Virginia, for 1 year as a pvt. in the United Artillery (1st Co. E, 41st Virginia Infantry), 4/19/61; present thru 3/10/612; "volunteered to go onboard the Ironclad Steamer *Virginia*," 3/6/62; manned gun on CSS *Virginia* during Battle of Hampton Roads, Virginia, 3/8–9/62; reenlisted for war, received $50 bounty, 3/10/62; on CRRs, 12/22/62, 6/3/63, 9/14/63, 11/2/63 and 3/18/64; absent, on detached service w/ CSN, per SO 63, A&IGO 64, dated 3/16/64; present, 12/31/64; paroled, Appomattox CH, 4/9/65.[117]

Dumphrey, Thomas: b. ca. 1828; occ. miner; enlisted at Chester, Virginia, as a pvt. In Co. I, 14th Virginia Infantry, 5/11/61; present 7–8/61; absent, under arrest on charges of insubordination while on guard duty, 9–10/61; sentenced to 20 days' hard labor by order of court-martial, 12/20/61; transferred to CSN by order of Gen. Magruder to serve on ironclad CSS *Virginia*, 4/2/62; listed as having deserted on final *Virginia* payroll, dated 9/30/62; NFR; also found as "Dunphey," "Dunphy" and Dumphry."[118]

Dunbar, Charles: b. ca. 1830, New Brunswick, Canada; occ. sailor; resided and enlisted at New Orleans, Louisiana, for 1 year as a pvt. in Co. D, 15th Louisiana Infantry, 8/16/61; transferred to CSN, 2/62; manned bow gun on CSS *Virginia*; KIA when capt. of the gun cried "sponge," and he jumped over the breechin, threw his head partly out of the port and was instantly killed by being shot in the head from the deck of the USS *Cumberland*"; first man killed during the Battle of Hampton Roads.[119]

Duncan, James Marshall: b. 4/7/42, Pearthshire, Scotland; enlisted at Norfolk, Virginia, as a pvt. in the United Artillery (1st Co. C, 19th Btn. Virginia Heavy Artillery), 5/8/62, unit became an independent btry., 10/1/62; absent, on detached duty aboard the CSS *Fredericksburg*, James River Squadron, per SO 63, A&IGO 64, dated 3/16/64; present, 12/31/64; paroled, Appomattox CH, 4/9/65; postwar: returned to Scotland; returned to U.S. five years later; worked in steel mills, Butler Co., Pennsylvania; d. 1/5/1911, Wolf Trap Farm, Fairfax Co., Virginia; bur. St. Andrew's Chapel Cem., Vienna, Virginia.[120]

Duncan, Thomas M.: paid as a seaman, CSS *Winslow*, North Carolina Squadron, 7/30–11/30/61; paid, Gosport Navy Yard, Portsmouth, Virginia, 12/13/61; paid as a seaman, North Carolina Squadron, 12/13/61–3/31/62; served as a quarter master, CSS *Virginia*; reenlisted for the war, received $50 bounty as member of crew, *Virginia*, 3/25/62; paid for service on the *Virginia* and Drewry's Bluff, Virginia, 4/1/62–6/30/63 and 10/1/63–1/64; POW, New Bern, North Carolina, 2/2/64; sent to Ft. Norfolk, Virginia; arrived Point Lookout, Maryland, 2/27/64; sent to Ft. Warren, Massachusetts, 9/20/64; NFR; on *Virginia* muster roll in *ORN* as "Thomas W. Duncan."[121]

Duncan, William C.: b. ca. 1832, Maryland; occ. mariner; enlisted at Norfolk, Virginia, for 1 year as a pvt. in the United Artillery (1st Co. E, 41st Virginia Infantry), 4/19/61; described as 5'10", light comp., light hair, blue eyes; AWOL, summer '61, joined crew of merchant ship attempting to run blockade of Hampton Roads; returned to regt. w/o punishment by order of Gen. Huger, 9/2/61; present until "detailed to go on *Merrimac*" by SO 9, Dept. of Norfolk, 1/16/62; served as a quarter gunner, CSS *Virginia*; reenlisted for the war, received $50 bounty as member of crew, *Virginia*, 3/25/62; paid for service on the *Virginia*, 4/1–5/12/62; on *Virginia* CRR, Drewry's Bluff, Virginia, 7/20/62; paid, Drewry's Bluff, 5/13/62–3/31/63; NFR; postwar: entered Maryland Line Confederate Soldiers Home, Pikesville, Maryland, sometime after '88.[122]

Dunlop, John: b. 3/8/37, Portland, Maine; parents, John and Lois Dunlop; occ. sailor; enlisted at New Orleans, Louisiana, for 1 year as a pvt. in Co. E, 1st (Nelligan's) Louisiana Infantry, 4/28/61; described as 5'3½", dark comp., dark hair, gray eyes; present thru 12/61; "discharged by order of Gen. Huger to join CSN," 1/18/62; served as a seaman, CSS *Virginia*; reenlisted for the war, received $50 bounty as member of crew, *Virginia*, 3/25/62; paid for service on the *Virginia* and Drewry's Bluff, Virginia, 4/1–9/15/62; in Bruton Parish Episcopal Church Hosp., Williamsburg, Virginia, w/ scalp wound, 5/10–13/62; on *Virginia* CRR, Drewry's Bluff, 7/20/62; volunteered for duty on CSS *Chattahoochee*, 8/1/62; d. Soldiers' Home, Columbus, Georgia, 9/2/62; bur. Linwood Cem., Columbus, Georgia; on *Virginia* muster roll in *ORN* as "John Dunlap."[123]

Durand, Julius: b. 11/7/38, Elizabethtown, New York; parents, Merari Durand and Ann Ferguson; occ. bookmaker; enlisted at Monticello, Arkansas, for 1 year as a pvt. in Co. I, 1st (Colquitt's) Arkansas Infantry, 5/8/61; mustered in at Lynchburg, Virginia, 5/19/61; reenlisted for 2 years, received $50 bounty, 1/10/62; transferred to CSN per SO 17, A&IGO 62, para 23, 2/25/62; served as an officer's steward, CSS *Virginia*; reenlisted for the war, received $50 bounty as member of crew, *Virginia*, 3/25/62; paid as a seaman, CSS *Virginia* 4/1–5/12/62; superannuated for rations, Drewry's Bluff, Virginia, 5/12–24/62; on *Virginia* CRR, Drewry's Bluff, 7/20/62; paid as a seaman, Drewry's Bluff, 5/13–10/9/62 and as a landsman, 10/10–11/11/62; appointed gunner, CSN; served on the cruiser, CSS *Chickamauga*, North Carolina Squadron, '64; NFR; d. 9/2/1918.[124]

Eanes, Charles Peter: b. ca. 1843, Chesterfield Co., Virginia; father, German Eanes, occ. laborer; enlisted at Petersburg, Virginia, as a pvt. in Co. K, 12th Virginia Infantry, 5/4/61; described as 6'1", light comp., light hair, blue eyes; discharged to join CSN per SO 18, Dept. of Norfolk, 1/29/62; part of Capt. William F. Lynch's squadron during Battle of Roanoke Island, North Carolina, 2/62; served as a landsman, CSS *Virginia*; reenlisted at Petersburg as a pvt. in Co. K, 12th Virginia Infantry, 5/12/62; absent, sick, 8–9/62; in Gen. Hosp. #12, Richmond, Virginia, 11/26–12/31/62; in C.S. Hosp., Petersburg, Virginia, 2/63; present, 3–8/63; in Gen. Hosp. #9, Richmond, 9/9/63; in Chimborazo Hosp., 9/10–10/63; listed as AWOL, 11–12/63; absent, sick, 1–2/64; in Bruton Parish Episcopal Church Hosp., Williamsburg, Virginia, w/ conjunctivitis, 2/22–3/14/64; detailed for special duty with Petersburg & Weldon RR per SO 80, A&IGO 64, para 19, 4/5/64; in Gen. Hosp. #9, Richmond, 4/11–12/64 and 5/16–17/64; NFR; postwar: occ. foreman, bridge maintenance crew, Petersburg & Weldon RR; later manager, Petersburg lumber mill/box factory; elected Petersburg commissioner of revenue, '94; member, A.P. Hill Camp, U.C.V., Petersburg.[125]

Edwards, James: b. Alabama; enlisted at Mobile, Alabama, for the war as a pvt. in Co. A, 12th Alabama Infantry, 6/4/61; present on 7–8/61 rolls;

transferred to CSN per SO 89, A&IGO 62, para 10, 3/27/62; shipped for 1 year as a seaman, Naval Rendezvous, Richmond, Virginia, 3/31/62; served as a seaman, CSS *Virginia*; reenlisted for 3 years, received $50 bounty, 4/21/62; superannuated for rations, Drewry's Bluff, Virginia, 5/12–24/62; may have later served as a seaman, CSS *Arctic*, North Carolina Squadron, '62–'63; appointed by Alabama as 3rd assistant engineer, 6/10/63; served on CSS *Morgan*, Mobile Squadron, '63; NFR.[126]

Egan, Daniel: enlisted at New Orleans, Louisiana, as a pvt. in Co. C, CSMC, 5/30/61; probably served at Pensacola Navy Yard until Co. C transferred to Gosport Navy Yard, Portsmouth, Virginia, 11/29/61; assigned to Marine Guard, CSS *Virginia*, ca. 11/61–ca. 5/62; served with Co. C, Camp Beall, Drewry's Bluff, Virginia; returned from Wilmington Expedition, 7/23/64; surrendered to officer of USS *Monadnock* on James River, Virginia; NFR; also found as "Eagan"; possibly "Daniel Egan," who was b. ca. 1827, Ireland; occ. laborer; single; resided New Orleans.[127]

Eggleston, John Randolph: b. 1831, Amelia Co., Virginia; parents, William Eggleston and Frances Peyton Archer; appointed by Mississippi to USNA as a midshipman, 8/2/47; passed midshipman, 6/10/53; master, 9/15/55; Lt., 9/16/55; assigned to USS *Wyandotte* at start of Civil War; tendered resignation from USN, 1/13/61; resignation accepted, 1/22/61; appointed 1st Lt., CSN, 4/5/61; married Sarah Dabney of Hinds, Mississippi, 5/29/61; special duty, New Orleans, Louisiana, '61; served on CSS *McRae*, New Orleans Naval Station, '61; assigned to CSS *Virginia*, 12/3/61; commanded Guns #4 & #5 during Battle of Hampton Roads, Virginia, 3/8–9/62; assigned Drewry's Bluff, Virginia, '62; Flag Lt., Mobile Squadron, '62–'64; assigned Mobile Naval Station, '64–'65; 1st Lt., Provisional Navy, 6/2/64; paroled, Jackson, Mississippi, 4/14/65; d. 1915.[128]

Faley, John: enlisted at New Orleans, Louisiana, as a pvt. in Co. C, CSMC, 6/21/61; assigned to Marine Guard, CSS *Virginia*, ca. 11/61–ca. 5/62; on Co. C CRR, 3rd qtr. '62; listed as having deserted on final *Virginia* payroll, dated 9/30/62; NFR; also found as "Faaley" and "Feely"; on *Virginia* muster roll in *ORN* as "John Fusley"; possibly "John J. Fealey" who was b. ca. 1830, Ireland; occ. watchman; married; resided New Orleans.[129]

Ferris, James J., Jr.: b. ca. 1842, Brantford, Ontario, Canada; parents, James and Sarah Ferris; occ. house carpenter; single; resided and enlisted at Norfolk, Virginia, for 1 year as a pvt. in Co. H, 6th Virginia Infantry, 4/19/61; mustered in at Craney Island, Virginia, 6/30/61; described as 5'9", dark comp., black hair, gray eyes; present thru 12/61; extra duty, Engineering Dept., 1/62; discharged per SO 69, Dept. of Norfolk, 3/30/62; served as a landsman, CSS *Virginia*; appears to have left crew of *Virginia* before 4/1/62; reenlisted for the war as a pvt. in Co. H, 6th Virginia Infantry, 5/12/62; present thru 7/63; POW, Falling Waters, Maryland, 7/14/63; sent to Baltimore, Maryland, and then Point Lookout, Maryland, 8/17/63; took oath, 4/12/64; postwar: occ. house carpenter; resided Norfolk, '69; also found as "Ferriss."[130]

Fisher, Joseph: shipped for the war as a seaman, CSS *Forrest*, North Carolina Squadron, 7/25/61; served as a seaman, CSS *Virginia*; reenlisted for war, received $50 bounty; paid for service on the *Virginia*, 4/1–5/12/62; superannuated for rations, Drewry's Bluff, Virginia, 5/12–24/62; on *Virginia* CRR, Drewry's Bluff, 7/20/62; paid, Drewry's Bluff, 5/13/62–6/30/63; paid for service, naval det. on special expedition from Drewry's Bluff to Charleston Naval Station, 10/1–31/63; served as a seaman, CSS *Virginia II*, James River Squadron, '64; NFR.[131]

Fisher, William: paid for service as a seaman, CSS *Ellis*, North Carolina Squadron, 8/2–10/3/61; served as a seaman, CSS *Virginia*; appears to have left crew of *Virginia* before 4/1/62; NFR.[132]

Fitzpatrick, Dennis: served as a 2nd class fireman, CSS *Virginia*; reenlisted for the war, received $50 bounty as member of crew, *Virginia*, 3/25/62; paid for service on the *Virginia* 4/1–5/12/62; superannuated for rations, Drewry's Bluff, Virginia, 5/12–24/62; on *Virginia* CRR, Drewry's Bluff, 7/20/62; paid, Drewry's Bluff, 5/13/62–7/31/63; NFR; possibly "Dennis Fitzpatrick," who enlisted at New Orleans, Louisiana, for the war as a pvt. in Co. H, 15th Louisiana Infantry, 7/2/61; NFR.[133]

Flake, Elijah Wilson: b. 1/15/41, Anson Co., North Carolina; parents, William Cameron Flake and Martha Emaline Huntley; occ. farmer; resided Anson Co.; enlisted at Camp Bee, Virginia, as a pvt. in Co. K, 14th North Carolina Infantry, 9/5/61; described as 5'6", dark comp., dark hair, hazel eyes; present until "transferred to Steamer *Merrimac* by order of Gen. Huger," 2/15/62; served as landsman, CSS *Virginia*; paid for service on the *Virginia*, 2/16–5/12/62; on *Virginia* CRR, Drewry's Bluff, Virginia, 7/20/62; paid, Drewry's Bluff, 5/13–9/13/62; discharged, 9/13/62; enlisted at Goldsboro, North Carolina, for the war as a pvt. in Co. K, 26th North Carolina Infantry, 2/1/63; present or accounted for until WIA, Gettysburg, Pennsylvania, 7/3/63; absent, sick at home on furlough, 1–2/64; absent, on detailed duty, Winder Gen. Hosp.; 5/64–2/65; RTD w/ Co. K, 26th North Carolina Infantry; paroled, Appomattox CH, Virginia, 4/9/65; postwar: occ. farmer; married; resided Wadesboro, North Carolina; d. 1918; on *Virginia* muster roll in *ORN* as "Elisha W. Flake."[134]

Flynn, John: b. ca. 1831, Virginia; occ. cooper; single; resided and enlisted at Norfolk, Virginia, for 1 year as a pvt. in the United Artillery (1st Co. E, 41st Virginia Infantry), 4/19/61; corp. between 4–8/61; 5th sgt., 8/25/61; present thru 3/6/62; "volunteered to go onboard the Ironclad Steamer *Virginia*," 3/6/62; manned gun on CSS *Virginia* during Battle of Hampton Roads, Virginia, 3/8–9/62; reenlisted for war, received $50 bounty, 3/10/62; on CRRs, 12/22/62, 6/3/63, 9/14/63, 11/2/63, and 3/18/64; in Gen. Hosp. #9, Richmond, Virginia, 9/8–9/64; in Chimborazo Hosp., Richmond, w/debility, 9/7–10/14/64; in Chimborazo Hosp., 10/10/64; in Chimborazo Hosp. #2 w/debility, 10/10–13/64; absent, detailed to work at Richmond Naval Ordnance Works per SO 288, A&IGO 64, 12/5/64; paroled, Appomattox CH, 4/9/65.[135]

Folger, Charles J.: b. ca. 1830, Minster, Prussia; occ. coppersmith; resided and enlisted at Salem, Virginia, for 1 year as a pvt. in 1st Co. A, 9th Virginia Infantry, 5/14/61; mustered in at Lynchburg, Virginia, 5/16/61; described as 5'4½", ruddy comp., light hair, blue eyes; transferred to the Salem Flying Artillery (Griffin's Co., Virginia Light Artillery), 2/2/62; present until discharged per SO 69, Dept. of Norfolk, 3/29/62; served as a landsman, CSS *Virginia*; received $50 bounty; on *Virginia* CRR, Drewry's Bluff, Virginia, 7/20/62; paid, Drewry's Bluff, 5/12–12/14/62; listed as having deserted, 12/14/62; reenlisted at Merrimans, Virginia, for the war as a pvt. in the Salem Flying Artillery, 12/62; corp., 9/20/63; present thru 12/64; paroled, Appomattox CH, 4/9/65; d. 1903, bur. East Hill Cem., Roanoke, Virginia; on *Virginia* muster roll in *ORN* as "Charles I. Folger."[136]

Forrest, Andrew H.: b. ca. 1841, Matthews Co., Virginia; occ. waterman; resided Elizabeth River Parish, Norfolk Co., Virginia; enlisted at Norfolk as a pvt. in Co. B, 9th Virginia Infantry, 6/5/61; described as 5'10", light comp., gray eyes; present until discharged per SO 18, Dept. of Norfolk, 1/29/62; served as an ordinary seaman, CSS *Virginia*; reenlisted for the war, received $50 bounty; paid for service on the *Virginia*, 4/1–5/12/62; on *Virginia* CRR, Drewry's Bluff, Virginia, 7/20/62; paid, Drewry's Bluff, 5/13/62–3/31/64; captured on blockade runner, CSS *Greyhound* by USS *Connecticut*, 5/12/64; POW, Camp Hamilton, Virginia, 5/16/64; released and sent to Baltimore, Maryland, 6/14/64; captured, 4/3/65; POW, Jackson Gen. Hosp., Richmond, Virginia, 4/8/65; left hosp. w/o permission, 4/18/65; postwar: occ. oysterman; resided Crittenden, Virginia; married, Norfolk, 4/24/72; received pension, 1910; d. of tuberculosis, Crittenden, 4/14/1921; widow, Georgia Forrest, received pension, 1922.[137]

Forrest, Douglas French: b. 8/17/37, Fairfax, Virginia; parents, Commodore French Forrest, USN/CSN and Emily Douglas Simms; attended University of Virginia; occ. lawyer; resided Ellicott City, Maryland; enrolled at Alexandria, Virginia, for 1 year as a 2nd Lt., 17th Virginia Infantry, 4/17/61; mustered in at Manassas Junction, 5/28/61; present, 7–8/61; detached as aide to Gen. Isaac R. Trimble, 9/14/61; present, 1–2/62; served on CSS *Virginia* as CSA volunteer aide to Flag Officer Buchanan, 3/8/62; took flag of USS *Congress* to Richmond, Virginia, 3/9/62; appointed by Virginia as an assistant paymaster, CSN, Yorktown, Virginia, 4/27/62; resigned from CSA, 5/6/62; paid, Wilmington Naval Station, 5/6–10/31/62; served on CSS *Arctic*, North Carolina Squadron, '62; assigned Richmond Naval Station, '62–'63; served abroad, '63–'64; assistant paymaster, Provisional Navy, 6/2/64; served on the cruiser, CSS *Rappahannock*, Calais, France, '64; detained with her crew by French authorities until 4/65; returned to U.S. via Cuba; paroled, Houston Texas, 6/26/65; postwar: joined law firm of Joseph Packard, Baltimore, Maryland, 12/65; attended Virginia theological Seminary, Alexandria, Virginia, '70–'73; married Sallie Winston of Richmond, Virginia, 1/9/73; ordained as Episcopal minister, '74; rector of churches in Wytheville, Virginia; Howard Co., Maryland; Washington, D.C.; Clarksburg, West Virginia; Coronado Beach, California; and Jacksonville, Florida; d. of heart attack while visiting sister's home, Ashland, Virginia, 5/3/1902.[138]

Foute, Robert Chester: b. 4/14/41, Greenville Co., Tennessee; parents, Dr. George Washington Foute and Mary Amanda Broyles; appointed by Tennessee to USNA, 9/20/58; tendered resignation as an acting midshipman, USN, 12/3/60; resignation accepted, 12/4/60; appointed as an acting midshipman, CSN, 6/11/61; assigned to Savannah Naval Station, '61–'62; assigned to CSS *Virginia*, 2/12/62; served as captain of Gun #6 during Battle of Hampton Roads, Virginia, 3/8–9/62; master in line of promotion, 10/4/62; served on CSS *Georgia*, Savannah River Squadron, '62–'63; passenger on steamer *Margaret and Jessie* when chased ashore by USS *Rhode Island*, 5/30/63; special service abroad in unsuccessful attempt to secure ironclad contracts, 8/63–9/64; 2nd Lt., 1/7/64; 1st Lt., Provisional Navy, 6/2/64; ordered to CSS *Savannah*, Savannah River Squadron, 10/28/64; served on CSS *Columbia* and *Chicora*, Charleston Squadron, '64–'65; assigned to Tucker's Naval Brigade, Drewry's Bluff; '65; major, CSA artillery; paroled, Appomattox CH, 4/10/65; arrested in Washington after Lincoln's assassination, 4/14/65; held in Washington Street Federal Prison, Alexandria, Virginia, for two months; postwar: married Mary Stewart deKantzow, daughter of Swedish baron Frederich deKantzow and Eloise Bullitt, 9/19/66; ordained as a minister, Christ Church, Savannah, Georgia, '74; rector, Grace Church, San Francisco, California, '83; d. San Francisco, California, 7/28/1903; bur. Cypress Lawn Cem., Coloma, California.[139]

GALLAGHER, CHARLES: b. Dublin, Ireland; occ. laborer; enlisted at Mobile, Alabama, as a pvt. in Co. E, 8th Alabama Infantry, 5/6/61; transferred to CSN by order of secretary of war, 2/12/62; shipped for 3 years as a seaman, Naval Rendezvous, Richmond, Virginia, 2/14/62; served as a seaman, CSS *Virginia*; reenlisted for the war, received $50 bounty as member of crew, *Virginia*, 3/25/62; paid for service on the *Virginia*, 4/1–5/12/62; on *Virginia* CRR, Drewry's Bluff, Virginia, 7/20/62; paid, Drewry's Bluff, 5/13/62–3/31/63; NFR.[140]

GARDELL, JOHN: b. ca. 1830; enlisted at Brookhaven, Mississippi, for 1 year as a 3rd corp. in Co. G, 12th Mississippi Infantry, 5/12/61; absent, wounded, in Gen. Hosp., Warrenton, Virginia, 6/22/61; in Gen. Hosp. #1, Danville, Virginia, 1/22/62; in Gen. Hosp. #21, Richmond, Virginia, 3/9–27/62; shipped for 1 year as a seaman, Naval Rendezvous, Richmond, 3/25/62; served as a seaman, CSS *Virginia*; appears to have left crew of *Virginia* before 4/1/62; NFR; possibly "John Gardell" who was b. ca. 1830, Pennsylvania; occ. carpenter, single; resided Bibb Co., Georgia, '60.[141]

GARNETT, ALGERNON S.: b. 1834, "Wakefield," Westmoreland Co., Virginia; father, Col. Henry T. Garnett; attended George Washington College, Washington, D.C.; graduated, University of Virginia, '53; graduated, Jefferson Medical College, Philadelphia, Pennsylvania, '56; appointed by Virginia to USN as an assistant surgeon, 5/16/57; assigned to USS *Wyandotte* at start of Civil War; tendered resignation from USN, 4/22/61; dismissed, 5/10/61; appointed as an assistant surgeon, CSN, 6/24/61; took part in capture of steamer *St. Nicholas* in Chesapeake Bay, 6/29/61; served on CSS *Patrick Henry*, 6/26–9/6/61; transferred to CSS *Fredericksburg*, 9/7/61; assigned Aquia Creek, Virginia, '61; married Alice E. Scott, Richmond, Virginia, 10/30/61; assigned to CSS *Virginia*, 11/18/61; took part in Battle of Hampton Roads, Virginia, 3/8–9/62; detailed to receiving ship, CSS *Confederate States*, 4/10/62; surgeon, 8/22/62; assigned Drewry's Bluff, Virginia, '62–'63; served on CSS *Baltic* and *Tuscaloosa*, Mobile Squadron, '63–'64; surgeon, Provisional Navy, 6/2/64; postwar: professor, University of Alabama, Tuscaloosa for 4 years; moved to Hot Springs, Arkansas; member, American Association for the Advancement of Science; medical director of Dept. of Arkansas, Medical Corps, UCV, 9/6/93; d. 1913, Hot Springs, Arkansas.[142]

Gaskill, Silas: b. ca. 1842, Sea Level, Carteret Co., North Carolina; parents, Joel Gaskill and Annis Lupton; shipped for the war as a seaman, CSS *Forrest*, North Carolina Squadron, 8/10/61; served as an ordinary seaman, CSS *Virginia*; reenlisted for the war, received $50 bounty; paid for service on the *Virginia*, 4/1–5/12/62; superannuated for rations, Drewry's Bluff, Virginia, 5/12–24/62; on *Virginia* CRR, Drewry's Bluff, 7/20/62; paid, Drewry's Bluff, 5/13/62–6/30/63; paid for service, naval det. on special expedition from Drewry's Bluff to Charleston Naval Station, 10/1–31/63; served as a landsman, CSS *Fredericksburg*, James River Squadron, '64–'65; ordered to Naval Ordnance Works, Richmond, Virginia, 1/13/65; NFR; postwar: occ. farmer; resided Beaufort Co. North Carolina; married Mary T. Salter of Carteret Co., North Carolina, '67.[143]

Genzmer, William: enlisted at Amite City, Louisiana, for the war as a pvt. in Co. D, 15th Louisiana Infantry, 8/7/61; present thru 10/31/61; shipped for the war as a seaman, Naval Rendezvous, New Orleans, Louisiana; served as a seaman, CSS *Virginia*; reenlisted for the war, received $50 bounty as member of crew, *Virginia*, 3/25/62; appears to have left crew of *Virginia* before 4/1/62; NFR; also found as "Gansmere"; on *Virginia* muster roll in *ORN* as "William Genzmar."[144]

Gibbons, Tobias: b. ca. 1825, Ireland; occ. laborer; single; resided and enlisted at New Orleans, Louisiana, as a pvt. in Co. C, CSMC, 5/30/61; probably served at Pensacola Navy Yard until Co. C transferred to Gosport Navy Yard, Portsmouth, Virginia, 11/29/61; assigned to Marine Guard, CSS *Virginia*, 4/1–5/12/62; court-martialed, 8/62, reason unknown; on CSS *Drewry* CRR, 1st qtr. '63; assigned to Marine Guard, Rocketts Navy Yard, Richmond, Virginia, 6/20–7/20/63; assigned to Marine Guard, CSS *Richmond*, James River Squadron, 1/19–12/31/64; paroled, Appomattox CH, 4/10/65; listed residence as St. Louis, Missouri; on *Virginia* muster roll in *ORN* as "Tobias Givens."[145]

Gillan, James: shipped for the war as a seaman, CSS *Forrest*, North Carolina Squadron, 7/25/61; served as a seaman, CSS *Virginia*; paid for service on the *Virginia*, 4/1–5/12/62; superannuated for rations, Drewry's Bluff, Virginia, 5/12–24/62; on *Virginia* CRR, Drewry's Bluff, 7/20/62; paid, Drewry's Bluff, 5/13–8/15/62; listed as discharged on final *Virginia* payroll, dated 9/30/62; may have later served as a seaman, CSS *Chicora*, Charleston Squadron, 10/1–31/63; transferred to unknown vessel; NFR; also found as "Gilliam" and "Gillen"; on *Virginia* muster roll in *ORN* as "James Gilland."[146]

Gillis, John: b. ca. 1826, New York; enlisted at Norfolk, Virginia, for 1 year as a pvt. in the United Artillery (1st Co. E, 41st Virginia Infantry), 4/19/61; left co. to join a privateer, 11/61; ordered back to co. by Gen. Huger, 12/18/61; present thru 3/6/62; "volunteered to go onboard the Ironclad Steamer *Virginia*," 3/6/62; manned a gun on CSS *Virginia* during Battle of Hampton Roads, Virginia, 3/8–9/62; reenlisted for war, received $50 bounty, 3/10/62; on CRRs, 12/22/62, 6/3/63, 9/14/63 and 11/2/63; listed as a 4th corp. on CRR, 3/18/64; absent, on detached duty w/ CSN, per SO 63, A&IGO 64, dated 3/16/64; present, 12/31/64; paroled as 4th sgt., Appomattox CH, 4/9/65; postwar: occ. rigger; married; resided Norfolk; member, Pickett-Buchanan Camp, UCV, Norfolk; d. 9/11/96, Robert E. Lee Camp Soldier's Home, Richmond.[147]

Gilmore, Allen: b. ca. 1840, Glasgow, Scotland; occ. paperhanger; enlisted at Petersburg, Virginia, for 1 year as a pvt. in Co. K, 12th Virginia Infantry, 5/4/61; discharged having supplied a substitute, 9/17/61; rejoined Co. K as substitute for J.B. Shackleford, 1/19/62; present until discharged at Craney Island, Virginia, for naval service per SO 69, Dept. of Norfolk, 3/28/62; served as landsman, CSS *Virginia*; appears to have left crew of *Virginia* before 4/1/62; RTD w/ Co. K, 5/15/62; listed as AWOL, 8/62–1/63; present, 3–4/63; deserted near Hagerstown, Maryland, 7/10/63; sent to Harrisburg, Pennsylvania, then Ft. Mifflin, Pennsylvania; took oath and released, 12/17/63; described then as 5'10", dark comp., black hair, black eyes; listed next residence as New York City.[148]

Goff, John: b. Ireland; served as a 1st class fireman, CSS *Virginia*; in Bruton Parish Episcopal Church Hosp., Williamsburg, Virginia, w/ hernia, 5/10/62; transferred to unknown hosp., 6/2/62; may have later served on the blockade runner, CSS *R.E. Lee*, '63; in Gen. Hosp. #4, Wilmington, North Carolina, w/ debility, 5/9/63; NFR.[149]

Grant, Randolph: served as a landsman, CSS *Virginia*; paid for service on the *Virginia* and Drewry's Bluff, Virginia, 4/1–8/25/62; listed as transferred on final *Virginia* payroll, dated 9/30/62; NFR; not on *Virginia* muster roll in *ORN*.[150]

Gray, James H.: served as captain of top, CSS *Virginia* before 4/1/62; listed as having deserted on final *Virginia* payroll, dated 9/30/62; NFR.[151]

Gray, William: b. ca. 1828, New York, New York; occ. sailor; enlisted at Naval Hosp., Portsmouth, Virginia, for 1 year as a pvt. in Co. D, 9th Virginia Infantry, 4/27/61; mustered in at Naval Hosp., Portsmouth, 5/5/61; described as 5'7½", dark comp., dark eyes; detached as a quarter gunner, 6/8–10/31/61; present, on daily duty as acting ordnance sgt., 11–12/61;

discharged by SO 18, Dept. of Norfolk, 1/29/62; served as captain of top, CSS *Virginia*; appears to have left crew of *Virginia* before 4/1/62; may have later served as a seaman, CSS *Spray* and paroled at St. Marks, Florida, 5/12/65.[152]

Griswold, Albert C.: b. 12/25/37, Wales; father, master tailor; immigrated to New York City, '52; served in USN as ship's tailor on USS *Macedonian* and *Cumberland*, '57–'61; deserted USN, 4/61; enlisted at Norfolk, Virginia, for 1 year as a pvt. in the United Artillery (1st Co. E, 41st Virginia Infantry), 4/19/61; present thru 3/6/62; "volunteered to go onboard the Ironclad Steamer *Virginia*," 3/6/62; manned gun on CSS *Virginia* during Battle of Hampton Roads, Virginia, 3/8–9/62; reenlisted for war; received $50 bounty, 3/10/62; on CRRs, 12/22/62 and 6/3/63; court-martialed per GO 37, para 8, 8/19/63; on CRRs, 9/14/63, 11/2/63 and 3/18/64; POW, Howlett's Farm, Virginia, 7/7/64; sent to City Point, Virginia, and then confined Ft. Hamilton, Virginia, 7/11/64; sent to Ft. Monroe, Virginia, to Point Lookout, Maryland, 7/15/64; sent to Elmira, New York, 8/3/64; exchanged, Aiken's Landing, Virginia, 2/20/65; deserted, entered Federal lines, Bermuda Hundred, Virginia, 4/5/65; sent to Aiken's Landing, Virginia; arrived Washington, D.C., 4/10/65; took oath, furnished transport to Norfolk; postwar: occ. merchant-tailor; resided Norfolk; member, Pickett-Buchanan Camp, UCV; IOOF; vestryman, St. Peter's Episcopal Church, Norfolk; Knight of Pythias; alive '99.[153]

Guy, Samuel: served as a landsman, CSS *Virginia*; superannuated for rations, Drewry's Bluff, Virginia, 5/12–24/62; NFR.[154]

Hall, Carey J.: b. 4/14/42, Norfolk, Virginia; occ. wheelwright; single; resided Portsmouth, Virginia; enlisted at Naval Hosp., Portsmouth, Virginia, for 1 year as a pvt. in Co. D, 9th Virginia Infantry, 4/27/61; mustered in at Naval Hosp., 5/5/61; described as 5'7½", light comp., light hair, blue eyes; absent sick, Gen. Hosp., 7–8/61; present until discharged per SO 72, Dept. of Norfolk, 4/1/62; served as a landsman, CSS *Virginia*; reenlisted for the war, received $50 bounty; paid, Drewry's Bluff, Virginia, 5/12/62–6/30/63; paid for service, naval det. on special expedition from Drewry's Bluff to Charleston Naval Station, 10/1–31/63; served at New Bern, North Carolina; '64; served in James River Squadron, '65; paroled, Liddy Prison, Richmond, Virginia, 4/22/65; postwar: keeper of Portsmouth Cem.; resided 623 County St., Portsmouth; member, Stonewall Camp, UCV, Portsmouth; d. 2/12/1916; bur. Oak Grove Cem., Portsmouth.[155]

Hall, James: served as an officer's steward, CSS *Virginia*; paid for service on the *Virginia* and Drewry's Bluff, Virginia, 4/1–6/15/62; superannuated for rations, Drewry's Bluff, 5/12–24/62; on *Virginia* CRR, Drewry's Bluff, 7/20/62; listed as transferred on final *Virginia* payroll, dated 9/30/62; NFR; not on *Virginia* muster roll in *ORN*.[156]

Hall, William H.: resided Maryland; acting 3rd assistant engineer, 4/2/62; assigned to CSS *Virginia*; 4/21/62; served at Drewry's Bluff, Virginia, 5/13–8/26/62; resigned, 8/27/62; NFR.[157]

Hall, William O.: b. ca. 1837, Chesterfield Co., Virginia; occ. laborer; resided Chesterfield Co.; enlisted at Petersburg, Virginia, for 1 year as a pvt. in Co. C, 41st Virginia Infantry, 5/9/61; mustered in, 5/11/61; described as 5'6", light comp., brown hair, blue eyes; present until discharged per SO 69, Dept. of Norfolk, 3/26/62; served as a landsman, CSS *Virginia*; discharged revoked, drafted back into 41st Virginia Infantry for the war, received $50 bounty; present, 4/30–10/31/62; AWOL, 12/10–29/62; found guilty of desertion, 2/14/63; sentenced to 6 months hard labor, 12 lb. ball & 3' chain to left ankle, to be closely confined between labor in place designated by commanding general, forfeited 6 months' pay, 29 lashes on bare back, his name, crime and punishment to be published in Petersburg newspaper and *Richmond Enquirer*, "this court is lenient because accused may have intended to return to army at some time"; in Howard's Grove Hosp., Richmond, Virginia, w/ smallpox, 4/22–5/13/63; in CS Hosp., Petersburg, Virginia, 7/30–9/30/64; granted 60-day sick furlough, 9/30/64; POW, Petersburg, 4/2/65; held at Hart's Island, New York, until 6/20/65; postwar: resided Swannsboro, Virginia, 4/1900; claimed he had bullet wound in hip and shell wound in hip and thigh (possibly wounded at Crater); received pension, 4/3/1900.[158]

Halstead, Alexander: b. ca. 1843, Princess Anne Co., Virginia; occ. farmer; enlisted at Gosport Navy Yard, Portsmouth, Virginia, for 1 year as a pvt. in 1st Co. E, 6th Virginia Infantry, 5/17/61; mustered in at Entrenched Camp, Norfolk, Virginia, 6/30/61; described as 5'6", dark comp., dark hair, dark eyes; present until discharged, 9/1/61; enlisted as a substitute for John Anderson at Craney Island, Virginia, for 1 year as a pvt. in Co. D, 9th Virginia Infantry, 9/7/61; present until discharged per SO 18, Dept. of Norfolk, 1/29/62; served as a landsman, CSS *Virginia*; reenlisted for the war, received $50 bounty; paid for service on the *Virginia*, 4/1–5/12/62; superannuated for rations, Drewry's Bluff, Virginia, 5/12–24/62; on *Virginia* CRR, Drewry's Bluff, 7/20/62; paid, Drewry's Bluff, 5/13/62–9/30/63; paid for service, naval det. on special expedition from Drewry's Bluff to Charleston Naval Station, 10/1–31/63; NFR; postwar: occ. farmer; married; resided Norfolk Co., Virginia, '70.[159]

Hampton, Cornelius Sidney: b. 2/6/40, Franklin Co., Kentucky; parents, Thomas Hampton and Emily Pemberton; enlisted at Mobile, Alabama, as a pvt. in Co. C, CSMC, ca. 10/23/61; sgt. on recruiting duty in Mobile, 10/23–11/21/61; paid for boarding self; on Co. C CRRs, 4th qtr. '61 and 3rd qtr. '62; assigned to Marine Guard, CSS *Virginia*; listed as having deserted on final *Virginia* payroll, dated 9/30/62; NFR; postwar: occ. farm laborer; resided Frankfort Co., Kentucky.[160]

Hannon, James: served as an ordinary seaman, CSS *Virginia*; paid for service as an ordinary seaman on the *Virginia*, 4/1–5/12/62; superannuated for rations, Drewry's Bluff, Virginia, 5/12–24/62; on *Virginia* CRR, Drewry's Bluff, 7/20/62; paid as an ordinary seaman, Drewry's Bluff, 5/13/62–1/16/63 and as a landsman, 1/17–12/31/63; served as an ordinary seaman, CSS *Fredericksburg*, James River Squadron, '64–'65; ordered to report to Naval Ordnance Works, Richmond, Virginia, 1/13/65; deserted; POW, Ft. Monroe, Virginia; took oath and paroled, 4/14/65; also found as "Hannan."[161]

Hansell, Thomas: enlisted at Mobile, Alabama, for the war as a pvt. in Co. A, 12th Alabama Infantry, 6/4/61; present on 7–8/61 rolls; in Chimborazo Hosp. #3, Richmond, Virginia, w/ irritatus spinalis, 3/7–4/12/62; transferred to CSN per SO 89, A&IGO 62, para 10, 3/27/62; shipped for 1 year as a seaman, Naval Rendezvous, Richmond, Virginia, 3/31/62; paid for service as an ordinary seaman, CSS *Virginia*, 4/12–5/12/62; reenlisted for 3 years, received $50 bounty, 4/21/62; in Bruton Parish Episcopal Church Hosp., Williamsburg, Virginia, w/ int. fever, 5/10–17/62; paid, Drewry's Bluff, Virginia, 5/13–11/7/62; NFR.[162]

Harralson, Brice: b. ca. 1831, Caswell Co., North Carolina; occ. merchant; resided Caswell Co.; enlisted at Yanceyville, North Carolina, for 1 year as a pvt. in Co. A, 13th North Carolina Infantry; mustered in at Raleigh, North Carolina, 6/24/61; described as 5'9", dark comp., light hair, gray eyes; present until absent on furlough, 10/31–11/10/61; present until "entered naval service on *Merrimack*," 2/14/62; served as a landsman, CSS *Virginia*; paid for service on the *Virginia*, 4/1–5/12/62; superannuated for rations, Drewry's Bluff, Virginia, 5/12–24/62; on *Virginia* CRR, Drewry's Bluff, 7/20/62; paid, Drewry's Bluff, 5/13–8/20/62; discharged, 8/20/62; enlisted at Ft. Lee, North Carolina, for the war as a pvt. in Co. K, 1st North Carolina Artillery, 3/17/64; present thru 10/64; WIA and POW, Ft. Fisher, North Carolina, 1/15/65; arrived Point Lookout, Maryland, via Ft. Monroe, Virginia, 1/31/65; transferred to U.S. Gen. Hosp. w/ fractured left elbow, 2/2/65; took oath and released, 6/28/65; postwar: occ. dry goods retail merchant; single; resided Yanceyville, North Carolina; on *Virginia* muster roll in *ORN* as "Brice Harrolson"; also found as "Harrelson."[163]

Harrell, Wilson: b. ca. 1838, Nansemond Co., Virginia; occ. farmer; married; resided Suffolk, Virginia; enlisted at South Quay, Virginia, for 1 year as a pvt. in Co. K, 41st Virginia Infantry, 6/6/61; described as 5'9", sandy hair; discharged "to join *Merrimac*, 2/7/62"; paid for service as a landsman, CSS *Virginia*, 2/7–5/15/62; listed as having deserted 5/15/62 on final *Virginia* payroll, dated 9/30/62; reenlisted at Petersburg for the war as a pvt. in 41st Virginia Infantry, 5/7/63; WIA by Minié ball, left thigh, Cold Harbor, 6/64; in Chimborazo Hosp. No. 4, Richmond, Virginia, 6/30–7/27/64; granted 60-day sick furlough, 7/27/64; POW, Burgess' Mill, Virginia, 10/27/64; Point Lookout, Maryland, 10/31/64–3/65; exchanged, Aiken's Landing, Virginia, 3/28/65; paroled, Appomattox CH, 4/9/65; NFR.[164]

Harrington, Edward: b. ca. 1823, Ireland; occ. laborer; married; resided New Orleans, Louisiana; enlisted at Camp Moore, Louisiana, for the war as a pvt. in Co. C, 10th Louisiana Infantry, 7/22/61; present thru 12/61; "transferred to Steamer *Virginia*," 1/3/62; officially transferred to CSN, 2/62; served as a coal heaver, CSS *Virginia*; paid for service on the *Virginia*, 4/1–5/12/62; on *Virginia* CRR, Drewry's Bluff, Virginia, 7/20/62; paid, Drewry's Bluff, 5/13–8/12/62; discharged, 8/12/62; NFR.[165]

Harris, William R.: b. ca. 1815, North Carolina; occ. seaman; married; resided and enlisted in Norfolk, Virginia, for 1 year as a pvt. in Co. D, 6th Virginia Infantry, 5/8/61; present until discharged per SO 12, Dept. of Norfolk, "for duty on board *Virginia*," 1/18/62; served as a quarter gunner, CSS *Virginia*, '62; reenlisted for the war, received $50 bounty as member of crew, *Virginia*, 3/25/62; appears to have left crew of *Virginia* before 4/1/62; served as a carpenter, CSS *Virginia II*, James River Squadron, '64–'65; NFR; postwar: occ. laborer; married; resided Norfolk Co., Virginia.[166]

Harrison, Charles W.: paid for service as a quarter master, CSS *Jamestown*, James River Squadron, 10/1–31/61 and 12/1–31/61; shipped for 1 year as a seaman, CSS *Jamestown*, 3/22/62; served as an ordinary seaman, CSS *Virginia*; reenlisted for 3 years, received $50 bounty, 4/21/62; paid for service on the *Virginia* and Drewry's Bluff, Virginia, 4/1–5/15/62; listed as having deserted on final *Virginia* payroll, dated 9/30/62; paid as a seaman, Drewry's Bluff, 1/31–3/31/63 and 7/31–9/30/63; NFR.[167]

Harrison, Howell W.: b. ca. 1841, Johnston Co., North Carolina; occ. cooper; resided Johnston Co.; enlisted at Raleigh, North Carolina, for 1 year as a pvt. in Co. E, 14th North Carolina Infantry, 5/1/61; mustered in at

Raleigh, 5/13/61; described as 6', dark comp., brown hair, hazel eyes; present until "transferred to the C.S. War Steamer *Virginia*," 2/15/62; served as a landsman, CSS *Virginia*; paid for service on the *Virginia* and Drewry's Bluff, Virginia, 2/18–7/1/62; superannuated for rations, Drewry's Bluff, 5/12–24/62; discharged, 7/1/62; may have later enlisted at Camp Holmes, North Carolina, for the war as a pvt. in Co. A, 10th Btn. North Carolina Heavy Artillery, 9/1/62; on daily duty w/ Signal Corps as a guard, Ft. Caswell, North Carolina, 11–12/63; present thru 9/30/64; NFR; on *Virginia* muster roll in *ORN* as "Howell W. Harrisson."[168]

Hart, Charles J.: b. ca. 1836, New York; occ. carpenter; single; resided Mobile Co., Alabama; enlisted at Montgomery, Alabama, as a pvt. in Co. C, CSMC, 3/26/61; probably served at Pensacola Navy Yard until Co. C transferred to Gosport Navy Yard, Portsmouth, Virginia, 11/29/61; assigned to Marine Guard, CSS *Virginia*; corp., date unknown; listed as having deserted before 6/30/62, but apparently RTD by 3rd qtr. '62; assigned to Marine Guard, Rocketts Navy Yard, Richmond, Virginia, 10/8/63; on daily duty guarding naval flour at mill, 7–8/64; served at Camp Beall, Drewry's Bluff, Virginia, 12/31/64; NFR; also found as "Heart."[169]

Harvey, James Madison: b. 3/16/42, Selma, Alabama; occ. farm laborer; single; resided Talladego Co., Alabama; enlisted as a pvt. in Co. H, 3rd Alabama Infantry, 4/13/61; described as 5'8"; discharged to served on CSS *Virginia* per SO 28, Dept. of Norfolk, 2/10/62; billeted on receiving ship, *CSS Confederate States*; served as a landsman, CSS *Virginia*; reenlisted for the war, received $50 bounty; paid for service on the *Virginia* and Drewry's Bluff, Virginia, 4/1/62–9/1/63; on *Virginia* CRR, Drewry's Bluff, 7/20/62; paid for service as a landsman, CSS *Patrick Henry*, James River Squadron, 10/1–31/64; NFR; postwar: married Martha Jane Paralee Kite, Texas, '66; d. 4/20/94; bur. Sand Springs Cem., Mineola, Texas.[170]

Harvey, Lawrence: served as a landsman, CSS *Virginia*; paid for service on the *Virginia* and Drewry's Bluff, Virginia, 4/1/62–6/30/63; on *Virginia* CRR, Drewry's Bluff, 7/20/62; listed as a POW on *Virginia* muster roll, 7/31/62; listed as discharged on payroll, dated 9/30/63; NFR.[171]

Hasker, Charles Hazelwood: b. 1831, London, England; served in Royal Navy as youth; came to U.S. at age 17; resided Portsmouth, Virginia; appointed by Virginia as a boatswain, USN, 3/28/57; tendered resignation and dismissed from USN, 6/4/61; appointed boatswain, CSN, 6/11/61; paid for service, New Orleans Naval Station, 7/1–8/14/61; served on CSS *Fanny*, North Carolina Squadron, '61; served on CSS *Virginia* during Battle of Hampton Roads, Virginia, 3/8–9/62; assigned to Drewry's Bluff,

Virginia until 8/2/62; served on CSS *Richmond*, James River Squadron, '62–'63; Lt. for the war, 5/5/63; served on CSS *Chicora*, Charleston Squadron, '63; served on CSS *Hunley*; POW, Morris Island, South Carolina, 9/7/63; confined, Ft. Warren, Massachusetts; 1st Lt., Provisional Navy, 6/2/64; paroled, 9/28/64; exchanged, Cox Wharf, Virginia, 10/18/64; served on CSS *Peedee*, Marion CH, South Carolina, '64; assigned to Semmes naval brigade, '65; paroled, Greensboro, North Carolina, 4/28/65; postwar: occ. partner, Hasker & Marcuse business firm; resided Richmond, Virginia; entered R.E. Lee Camp Soldiers' Home, Richmond, 4/19/89; d. 7/8/98; bur. Oakwood Cem., Richmond.[172]

HAYWARD, WILLIAM H.: served as a landsman, CSS *Virginia*; reenlisted for the war, received $50 bounty as member of crew, *Virginia*, 3/25/62; shipped for the war, Naval Rendezvous, Richmond, Virginia, 5/6/62; listed as having deserted on final *Virginia* payroll, dated 9/30/62; NFR; also found as "Haywood"; possibly "William H. Hayward," who was b. ca. 1838, New Orleans, Louisiana; occ. clerk; resided New Orleans, Louisiana; single; enlisted at Camp Moore, Louisiana, for the war as a 2nd corp. in Co. F, 5th Louisiana Infantry, 5/10/61; 1st corp., 8/61; present on 9–10/61 rolls; 5th sgt., 11/22/61; present on 1–2/61 rolls; listed as AWOL on 7–8/62 rolls; present on 9–12/62 rolls; in Gen. Hosp. #9, Richmond, Virginia, 1/11/63; transferred to Gen. Hosp. #1, Richmond, 3/4/63; in Louisiana Hosp., Richmond, 5–6/63; discharged 6/15/63; NFR.[173]

HEDGEPETH, JOSEPH S.: b. 1842, Edgecombe Co., North Carolina; occ. farmer; resided Edgecombe Co.; enlisted at Tarboro, North Carolina, for 1 year as a pvt. in Co. G, 13th North Carolina Infantry, 5/8/61; sick in hosp., 5–6/61 roll; present until "discharged for naval service," 2/20/62; served as a landsman, CSS *Virginia*; reenlisted for the war, received $50 bounty as member of crew, *Virginia*, 3/25/62; paid for service on the *Virginia*, 4/1–5/12/62; superannuated for rations, Drewry's Bluff, Virginia, 5/12–24/62; on *Virginia* CRR, Drewry's Bluff, 7/20/62; paid, Drewry's Bluff, 5/13/62–3/31/64; paroled, Greensboro, North Carolina, 4/28/65; on *Virginia* muster roll in *ORN* as "Joseph Hedgepath."[174]

HENDERSON, EUGENIUS TURNER: b. ca. 1841, Macon Co., Alabama; parents, John Carter Henderson and Mary Moreland; attended Auburn University; occ. merchant; resided Tuskegee, Alabama; enlisted at New Orleans, Louisiana, for 1 year as a pvt. in Co. I, 2nd Louisiana Infantry, 7/9/61; described as 5'6", dark comp., dark hair, gray eyes; present thru 10/61; absent, sick in hosp., Williamsburg, Virginia, 11–12/61; "discharged by reason of transfer to the Steam Ship *Virginia*," 3/15/62; served as a landsman, CSS *Virginia*; paid for service on the *Virginia* and Drewry's Bluff, Virginia, 3/18–9/30/62; on *Virginia* CRR, Drewry's Bluff, 7/20/62; listed as discharged on final *Virginia* payroll, dated 9/30/62; served as a paymaster's clerk, CSS *Chattahoochee*, '63; paid for service, CSS *Chattahoochee*, 12/1–31/62 and 4/1–5/27/63; killed in boiler explosion onboard the *Chattahoochee*, 5/27/63; bur. First Methodist Church Cem., Chattahoochee, Florida; also found as "E.V. Henderson."[175]

Henry, William C.: b. ca. 1843, Fairfax Co., Virginia; occ. sailor; enlisted at Churchland, Virginia, for the war as a pvt. in Co. I, 9th Virginia Infantry, 5/15/61; mustered in at Churchland, 6/12/61; described as 5'11", light comp., light hair, blue eyes; present until discharged "by reason for service on the *Merrimac*," 1/18/62; served as a landsman, CSS *Virginia*; in Bruton Parish Episcopal Church Hosp., Williamsburg, Virginia, w/ debility, 5/10–27/62; may have later enlisted at Lynchburg, Virginia, as a pvt. in Coakley's Co., 5th Virginia Cavalry, 6/24/62; present thru 6/30/62; NFR.[176]

Herd, William C.: b. ca. 1837, Monroe Co., Ohio; resided Alexandria, Louisiana; occ. machinist; single; resided Alexandria, Louisiana; enlisted at New Orleans, Louisiana, for 1 year as a pvt. in Co. B., 1st (Nelligan's) Louisiana Infantry, 8/2/61; described as 5'8", florid comp., dark hair, gray eyes; present until "discharged by S.O. 18, Dept. of Norfolk, the party having entered the navy," 1/29/62; served as a 1st class fireman, CSS *Virginia*; superannuated for rations, Drewry's Bluff, Virginia, 5/12–24/62; apparently RTD w/ 1st (Nelligan's) Louisiana Infantry; listed as having deserted, Bunkers Hill, Virginia, '62; POW, Malvern Hill, Virginia, 7/1/62; sent to Harpers Ferry, West Virginia; NFR; on *Virginia* muster roll in *ORN* as "William C. Hurd."[177]

Herring, Benjamin Simms: b. 3/4/37, Duplin Co., North Carolina; parents, Bryan Whitfield Herring and Penelope Simms; attended the University of Mississippi; appointed by North Carolina to USN as a 3rd assistant engineer, 8/11/60; assigned to USS *Richmond* at start of Civil War; tendered resignation from USN, 7/5/61; dismissed, 7/8/61; appointed as an acting 3rd assistant engineer, CSN, 7/23/61; served on CSS *Jamestown*, James River Squadron, '61; acting 2nd assistant engineer, 11/25/61; assigned to CSS *Virginia*, 12/1/61; assigned to her fireroom in charge of auxiliary steam pumps during Battle of Hampton Roads, Virginia, 3/8–9/62; assigned to Columbus Naval Ironworks, '62–'63; served on CSS *Gaines*, Mobile Squadron, '63; 2nd assistant engineer, 8/13/63; 1st assistant engineer, 8/20/63; served on CSS *Tennessee*, Mobile Squadron, '63–'64; served on CSS *Webb*, Red River Squadron, '64; 1st assistant engineer, Provisional

Navy, 6/2/64; assigned to Mobile Squadron, '64–'65; surrendered, 5/4/65; paroled, 5/10/65; postwar: occ. farmer; resided Tallahassee, Florida; married Rosa Reynolds of Alabama; d. at home, 9/17/1915; wife received Florida pension, 1924.[178]

Hickey, Daniel: b. ca. 1835; occ. laborer; enlisted at Charleston, South Carolina, for the war as a pvt. in Co. K, 1st (McCreary's) South Carolina Infantry, 6/28/61; described as 5'9", light comp., light brown hair, blue eyes; absent, in hosp., 12/31/61; transferred to CSN per 2/28/62 muster roll; served as a coal heaver, CSS *Virginia*; appears to have left crew of *Virginia* before 4/1/62; NFR.[179]

Hickey, John: b. Ireland; occ. laborer; resided and enlisted at New Orleans, Louisiana, as a pvt. in Co. C, CSMC, 6/11/61; probably served at Pensacola Navy Yard until Co. C transferred to Gosport Navy Yard, Portsmouth, Virginia, 11/29/61; assigned to Marine Guard, CSS *Virginia*, ca. 11/61–5/12/62; served with Co. C, Drewry's Bluff, Virginia, until transferred to Wilmington Naval Station, 1/20/64; assigned to Marine Guard, CSS *North Carolina*, North Carolina Squadron, 1/22–9/30/64; transferred to CSS *Arctic*, North Carolina Squadron, 9/30/64; POW, Ft. Fisher, North Carolina, 1/15/65; released on oath from Point Lookout, Maryland, 5/13/65; NFR.[180]

Higgins, John Francis: b. ca. 1842, Baltimore, Maryland; single; resided Baltimore, Maryland; enlisted at Chuckatuck, Virginia, for 1 year as a pvt. in Co. F, 9th Virginia Infantry, 5/18/61; mustered in at Chuckatuck, 5/22/61; described as 5'8", light comp., light hair, blue eyes; present until discharged by SO 30, Dept. of Norfolk, for "service on *Merrimac*," 2/12/62; served as a landsman, CSS *Virginia*; took part in Battle of Hampton Roads, Virginia, 3/8–9/62; reenlisted for the war, received $50 bounty as member of crew, *Virginia*, 3/25/62; paid for service on the *Virginia*, 4/1–5/12/62; superannuated for rations, Drewry's Bluff, Virginia, 5/12–24/62; on *Virginia* CRR, Drewry's Bluff, 7/20/62; paid, Drewry's Bluff, 5/13/62–9/30/63; paid for service, naval det. on special expedition from Drewry's Bluff to Charleston Naval Station, 10/1–31/63; served on the cruiser, CSS *Chickamauga*; North Carolina Squadron; WIA, Battery Buchanan, Ft. Fisher, North Carolina, 12/24/64; right leg amputated above knee; in hosp., Wilmington, North Carolina; in Gen. Hosp. #6, Fayetteville, North Carolina, 1/28/65; took oath and paroled, 5/24/65; postwar: resided Crittenden, Virginia; received $60 from Virginia to purchase artificial leg, '86; married, Chestnut Hill, Maryland, 1/1/1907; d. of heart trouble, Crittenden, 12/19/1924; widow, Mary S. Higgins, received pension, 9/28/34.[181]

HINDS, LAURENCE: enlisted for 3 years, received $50 bounty, 4/21/62; served as a quarter master, CSS *Virginia*; paid for service on the *Virginia*, 4/1–5/12/62; on *Virginia* CRR, Drewry's Bluff, Virginia, 7/20/62; paid, Drewry's Bluff, 5/13/62–6/30/63; shipped for the war at Orange CH, Virginia, 3/31/64; served as a seaman, CSS *Fredericksburg*, James River Squadron, '64; WIA, left hand and forearm, 10/22/64; deserted, POW, Libby Prison, Richmond, Virginia, 4/10/65; also found as "Hines" and "Hindes."[182]

HITE, ROBERT NICHOLAS: b. ca. 1842, Prince George Co., Virginia; occ. farm laborer; single; enlisted at Petersburg, Virginia, for 1 year as a pvt. in Co. K, 12th Virginia Infantry, 5/4/61; described as 6'1", light comp., light hair, gray eyes; present until discharged for naval service per SO 69, Dept. of Norfolk, 3/28/62; served as a landsman, CSS *Virginia*; superannuated for rations, Drewry's Bluff, Virginia, 5/12–24/62; RTD w/Co. K, 12th Virginia Infantry, 5/62; absent sick, 2/21–3/18/63, C.S. Hosp., Petersburg, Virginia; 1–2/63; present, 3–4/63; severely WIA left wrist & hand by shell fragment, Chancellorsville, Virginia, 5/1/63; in Chimborazo Hosp. #1, Richmond, Virginia, 5/6/63; granted 40-day furlough, 6/18/63; in C.S. Hosp., 7/18/63; in Hen. Hosp #9, Richmond, 8/17/63; detailed to Tredegar Ironworks, 9/63; left arm almost completely paralyzed; little finger on right hand amputated, Chimborazo Hosp. #9, 11/19/64; listed as having deserted, Ft. Powhatan, James River, Virginia, 3/10/65; postwar: resided Petersburg; married, 3/15/78; received pension, 6/19/91; d. of chronic bronchitis, Petersburg, 2/4/1920; widow, Sarah J. Hite, received pension, 1920.[183]

HOAR, JOHN: served as a coal heaver, CSS *Virginia*; reenlisted for the war, received $50 bounty as member of crew, *Virginia*, 3/25/62; appears to have left crew of *Virginia* before 4/1/62; NFR; possibly "John Hoar," who was b. ca. 1837, Massachusetts; occ. sailor; single; resided New Orleans, Louisiana; enlisted at New Orleans as a pvt. in Co. F, 14th Louisiana Infantry, 6/19/61; mustered in at Camp Pulaski, Georgia, 7/2/61; present, 7–8/61; transferred to Co. D, 11–12/61; NFR.[184]

HODDER, HUMPHREY: b. ca. 1838, Washington, North Carolina; resided Washington Co.; enlisted at Plymouth, North Carolina, for the war as a pvt. in Co. G, 1st North Carolina Infantry, 6/24/61; present until transferred to CSN per SO 17, A&IGO 62, para 23, 2/3/62; served as an ordinary seaman, CSS *Virginia*; reenlisted for the war, received $50 bounty as member of crew, *Virginia*, 3/25/62; paid for service on the *Virginia*, 4/1–5/12/62; superannuated for rations, Drewry's Bluff, Virginia, 5/12–24/62; on *Virginia*

CRR, Drewry's Bluff, 7/20/62; paid, Drewry's Bluff, 5/13/62–3/31/63; listed as died in the line duty on 3/31/63 payroll; on *Virginia* muster roll in *ORN* as "Humphrey Hardder."[185]

Hopkins, Joseph: enlisted at Richmond, Virginia, for the war as a pvt. in Co. I, 1st (McCreary's) South Carolina Infantry, 9/1/61; transferred to CSN, 1/17/62; paid for service as a landsman on CSS *Virginia* and Drewry's Bluff, Virginia, 4/1–11/11/62; shipped for the war as a seaman, received $50 bounty, Naval Rendezvous, Richmond, 7/21/62; may have later served as a seaman, CSS *Arctic*, North Carolina Squadron, '63; NFR.[186]

Howard, Charles H.: b. ca. 1841, USA; occ. mariner; enlisted at Norfolk, Virginia, for 1 year as a pvt. in the United Artillery (1st Co. E, 41st Virginia Infantry), 4/19/61; described as 5'7", dark comp., brown hair, brown eyes; "present until detailed to go on *Merrimac*" per SO 9, Dept. of Norfolk, 1/16/62; enlisted in CSN, 1/19/62; paid, North Carolina Squadron, 1/10–2/9/62; paid as a landsman, CSS *Virginia*, 3/20–5/12/62; paid, Drewry's Bluff, Virginia, 6/1/62–6/30/63 and 10/31/63–3/31/64; NFR; on *Virginia* muster roll in *ORN* as "Henry H. Howard."[187]

Howell, John: enlisted at New Orleans, Louisiana, as a pvt. in Co. A, CSMC, 5/6/61; on Capt. Holmes's CRR, ca. 5/10/61; on Co. C CRR, 4th qtr. '61; assigned to Marine Guard, CSS *Virginia*, ca. 11/61–ca. 5/62; on Co. C CRRs, 3rd qtr. '62; listed as having deserted before 9/30/62; NFR.[188]

Hoyt, Henry: served as an ordinary seaman, CSS *Virginia*; reenlisted for the war, received $50 bounty as member of crew, *Virginia*, 3/25/62; appears to have left crew of *Virginia* before 4/1/62; NFR; also found as "Hoyst."[189]

Huddleston, John T.: b. ca. 1836, Petersburg, Virginia; occ. iron molder; enlisted at Petersburg for 1 year as a pvt. in Co. C, 41st Virginia Infantry, 5/9/61; mustered in at Petersburg, 5/11/61; described as 5'7", light comp., brown hair, light eyes; present until transferred to Co., K, 12th Virginia Infantry, 11/4/61; present until discharged per SO 69, Dept. of Norfolk, 3/29/62; served as a landsman, CSS *Virginia*; appears to have left crew of *Virginia* before 4/1/62; reenlisted at Falling Creek, Virginia, for the war as a pvt. in Co. K, 12th Virginia Infantry, 8/15/62; WIA, right lung, Battle of Second Manassas, 8/30/62; Minié ball not extracted, wound remained open 2 years later; in hosp., 9/62–4/63; present, 4–8/63; absent, sick w/ remittent fever, 11/63; detailed light duty in Richmond hosp., then Petersburg; deserted and took oath, Natchez, Mississippi, 4/28/65; furnished transport to City Point, Virginia; postwar: occ. barman/iron molder; resided Natchez; returned to Virginia and resided in Chesterfield Co.; received pension, 6/11/88; entered R.E. Lee Camp Soldiers' Home,

Richmond, Virginia, 9/17/96; committed suicide by tying iron quoits from Soldiers' Home around waist and jumping in Byrd Park pond, 6/30/1911; left suicide note revealing location of body and apologizing for borrowing the quoits; buried, Hollywood Cem., Richmond.[190]

HUDDLESTON, WILLIAM G.: b. 3/8/43, Montgomery, Alabama; occ. overseer, single; resided Macon Co., Alabama; enlisted at Montgomery as a pvt. in Co. C, CSMC, 3/28/61; served at Pensacola Navy Yard until Co. C transferred to Gosport Navy Yard, Portsmouth, Virginia, 11/29/61; assigned to Marine Guard, CSS *Virginia*, ca. 11/61–5/12/62; served in Battle of Drewry's Bluff, Virginia; corp. on CRR, 4th qtr. '62; pvt. on CRR, 8/2/63; on CRRs thru 11/1/63; served in ANV; POW, Wheeling, West Virginia, date unknown; NFR; postwar: resided Montgomery, Alabama, then moved to Texas; married Ella B. Chase, 8/66; married Selena A. Williams after '79; member, Camp #1555, UCV, Jacksonville, Texas; Texas commissioner of pensions requested verification of service, 10/25/1909; d. Jacksonville, Texas, 10/19/29.[191]

HUGHES, PATRICK: b. ca. 1833, Dublin, Ireland; occ. riverman; enlisted at Mobile, Alabama, as a pvt. in Co. E, 8th Alabama Infantry, 5/6/61; transferred to CSN by order of secretary of war, 2/12/62; shipped for 3 years as a seaman, Naval Rendezvous, Richmond, Virginia, 2/14/62; served as a seaman, CSS *Virginia*; reenlisted for the war, received $50 bounty as member of crew, *Virginia*, 3/25/62; appears to have left crew of *Virginia* before 4/1/62; may have been captured in City Hosp., Mobile, Alabama, 4/12/65; paroled, Mobile, 5/11/65.[192]

HUNT, JOHN: served as a seaman, CSS *Virginia*; manned gun #4 during Battle of Hampton Roads, Virginia, 3/8–9/62; reenlisted for the war, received $50 bounty as member of crew, *Virginia*, 3/25/62; superannuated for rations, Drewry's Bluff, Virginia, 5/12–24/62; listed as POW, 6/7–7/31/62; apparently RTD since he was paid for service on the *Virginia* and Drewry's Bluff, 4/1/62–3/31/64; may have later served on CSS *Fredericksburg*, James River Squadron, '64; POW, Bermuda Hundred, Virginia, 10/9/64; sent to Washington, D.C., 10/11/64; took oath and reportedly furnished transport to New York City; apparently escaped and RTD w/ Naval Btn.; deserted, Bermuda Hundred, 4/5/65; took oath and sent to City Point, Virginia, 4/8/65; sent to Washington, D.C., 4/12/65; furnished transport to New York City; NFR; possibly "John E. Hunt" who was b. ca. 1822, New York; occ. oysterman; married; resided, Portsmouth, Virginia.[193]

INGRAHAM, JOHN HAZELHURST: b. South Carolina; appointed by South Carolina to USNA as an acting midshipman; tendered resignation, 2/10/61;

resignation accepted, 2/14/61; appointed midshipman, CSN, 4/23/61; served on Savannah Naval Station, '61; acting master, 9/24/61; served on CSS *Nashville*, '61–'62; 2nd Lt. and Lt. for war, 2/8/62; boarding officer of *Nashville* when it captured *Robert Gilfillan*, 2/26/62; assigned to CSS *Virginia*, 3/9–31/62; served on receiving ship, CSS *Indian Chief*, and CSS *Chicora* and *Huntress*, Charleston Squadron, '62; service abroad, '63–'64; served on CSS *Georgia*, Savannah River Squadron, '63–'64; 1st Lt., 1/6/64; served on CSS *Chicora* and commanded batteries, Charleston harbor, '64; 1st Lt., Provisional Navy, 6/2/64; assigned to Battery Wood, James River, 10/64; NFR.[194]

Ives, Emsley H.: paid for service as a quarter master, CSS *Ellis*, North Carolina Squadron, 8/2–10/3/61; served as a seaman, CSS *Virginia*; WIA 3/8/62; reenlisted for the war, received $50 bounty; paid for service on the *Virginia*, 4/1/–5/12/62; superannuated for rations, Drewry's Bluff, Virginia, 5/12–24/62; on *Virginia* CRR, Drewry's Bluff, 7/20/62; paid, Drewry's Bluff, 5/13/62–6/30/63 and 10/1/63–3/31/64; in Jackson Gen. Hosp., Richmond, Virginia, w/ chronic diarrhea, 4/18/65; listed as having deserted, 4/19/65; NFR; listed as "Emerson Ivas" in surgeon's report; possibly "E.W. Ives" who was b. ca. 1820, North Carolina; occ. tradesman; married; resided Beaufort Co., North Carolina.[195]

Jack, Eugenius Alexander: b. 7/17/40, Portsmouth, Virginia; occ. machinist/chief engineer; resided and enrolled at Portsmouth, Virginia, as a pvt. in Co. K, 9th Virginia Infantry, 4/20/61; mustered in 4/30/61; described as 5'6", dark comp., dark hair, dark eyes; on det. duty, Gosport Navy Yard per SO 197, Dept. of Norfolk, 7/25/61; discharged, Pinner's Point, Virginia, per SO 405, Dept. of Norfolk; transferred to CSN; appointed by Virginia as acting 3rd assistant engineer, CSN, 11/29/61; assigned to CSS *Virginia*, 12/3/61; served in her fireroom during Battle of Hampton Roads, Virginia, 3/8–9/62; served on CSS *Arkansas*, Mississippi River Squadron, '62; 1st assistant engineer, 8/16/62; served on CSS *Arctic*, *North Carolina* and *Tallahassee*, North Carolina Squadron, '62–'64; paid, Wilmington Station, 8/16–3/31/64 and 9/22–10/19/64; served on CSS *Columbia* and *Palmetto State*, Charleston Squadron, '64–'65; acting chief engineer, CSS *Richmond*, James River Squadron; captured, Saylor's Creek, Virginia, 4/6/65; sent to Old Capital Prison, Washington, D.C., then Johnson's Island, Ohio; released on oath, 6/18/65; described then as age 24, dark comp., dark hair, hazel eyes; postwar: married Ella Brown Ege of Richmond, Virginia, 6/8/69; occ. engineer, U.S. Revenue Cutter Service; resided 302 Dinwiddie St., Portsmouth, Virginia; member, Stonewall Camp, UCV, Portsmouth; d. Alton, Illinois, 12/18/1911; bur. Cedar Grove Cem., Portsmouth.[196]

JARVIS, WILLIAM R.: b. ca. 1836, Virginia; occ. ship joiner; married; resided Norfolk, Virginia; shipped for the war as a seaman, Naval Rendezvous, Richmond, Virginia, 10/23/61; on special duty, St. Marks, Florida, reported for duty as a carpenter's mate, CSS *Virginia*, 3/4/62; paid for service on the *Virginia*, 4/1–5/12/62; superannuated for rations, Drewry's Bluff, Virginia, 5/12–24/62; on *Virginia* CRR, Drewry's Bluff, 7/20/62; paid, Drewry's Bluff, 5/13–12/31/62; carpenter, Provisional Navy, 6/2/64; served on CSS *Richmond* and *Virginia II*, James River Squadron, '64–'65; paroled, Burkeville Junction, Virginia, 4/65; postwar: member, Pickett-Buchanan Camp, UCV, Norfolk, Virginia; alive '86.[197]

JOHNSON, ELISHA R.: b. ca. 1816, Virginia; occ. butcher, married; resided Elizabeth River Parish, Norfolk Co., Virginia; served as gunner's mate, CSS *Virginia;* paid for service on the *Virginia* and Drewry's Bluff, Virginia, 4/1–8/27/62; superannuated for rations, Drewry's Bluff, 5/12–24/62; appointed by Virginia as acting gunner, CSN, 8/29/62; served on CSS *Chicora*, Charleston Squadron, '62–'63; gunnery instructor, C.S. Naval Academy, Richmond, fall '63; paid for service, Charleston Naval Station, 10/6–11/30/63; served on CSS *Chicora*, '64; gunner, Provisional Navy, 6/2/64; paid for service on CSS *Patrick Henry*, James River Squadron, 11/1–12/31/64; NFR.[198]

JOHNSON, WILLIAM: b. ca. 1829, Norway; occ. mariner; enlisted at Norfolk, Virginia, for 1 year as a pvt. in the United Artillery (1st Co. E, 41st Virginia Infantry), 4/19/61; described as 5'8", sandy hair, blue eyes; deserted and joined crew of privateer fitting out of Norfolk, Virginia, summer '61; returned to regt. by order of Gen. Huger, 9/3/61; present until "detailed to go on the *Merrimac*" per SO 9, Dept. of Norfolk, 1/16/62; served as a gunner's mate, CSS *Virginia*, paid for service on the *Virginia* and Drewry's Bluff, Virginia, 4/1–7/23/62; superannuated for rations, Drewry's Bluff, 5/12–24/62; WIA during Battle of Drewry's Bluff, 5/15/62; on *Virginia* CRR, Drewry's Bluff, 7/20/62; listed as in hosp. on *Virginia* muster roll, dated 7/31/62; listed as discharged on final *Virginia* payroll, dated 9/30/62; transferred to CSN per SO 125, A&IGO 63, para 13, 5/26/63; paid for service, Charleston Naval Station, 10/1–31/63; may have later served as a landsman, CSS *Huntsville*, Mobile Squadron, '63; may have later enlisted at Navy Dept., Richmond, Virginia, as a pvt. in Co. D, 4th Btn. Virginia Infantry, Local Defense (Naval Btn.), 6/23/63; present, 8/2/64; on CRR, 4th qtr. '64; served as a seaman, CSS *Palmetto State*, Charleston Squadron, South Carolina, '64; deserted, Bermuda Hundred, Virginia, 4/5/65; took oath and sent to City Point, Virginia; sent to Washington, D.C., 4/12/65; furnished transport to New York City.[199]

Joice, John: served as a 2nd class fireman, CSS *Virginia*; reenlisted for the war, received $50 bounty as member of crew, *Virginia*, 3/25/62; paid for service on the *Virginia* and Drewry's Bluff, Virginia, 4/1/62–7/30/63; paid for service as a 1st class fireman, Charleston Naval Station, 10/6–31/63; served as a seaman, CSS *Chicora*, Charleston Squadron, '63–'64; POW, 4/3/65; in Jackson Gen. Hosp., Richmond, Virginia, w/ chronic diarrhea, 4/8/65; left hosp. w/o permission, 4/17/65; NFR; also found as "Joyce."[200]

Joliff, John R.: b. ca. 1839, Perquimans Co., North Carolina; occ. farmer; married; enlisted as a pvt. in Co. I, 17th North Carolina Infantry (1st Org.); present until discharged before 7/30/61; served as a seaman on CSS *Virginia*; reenlisted for the war, received $50 bounty as member of crew, *Virginia*, 3/25/62; paid for service on the *Virginia* and Drewry's Bluff, Virginia, 4/1–6/30/62; superannuated for rations, Drewry's Bluff, 5/12–24/62; on *Virginia* CRR, Drewry's Bluff, 7/20/62; volunteered for duty on CSS *Chattahoochee*, 8/1/62; paid for service, CSS *Chattahoochee*, 4/1–5/27/63; killed in boiler explosion onboard the *Chattahoochee*, 5/27/63; bur. First Methodist Church Cem., Chattahoochee, Florida; on *Virginia* muster roll in *ORN* as "J.R. Joliff."[201]

Jones, Albert A.: b. ca. 1829, Chesterfield Co., Virginia; occ. stonemason; married; resided Chesterfield Co.; enlisted at Petersburg, Virginia, for 1 year as a pvt. in Co. C, 41st Virginia Infantry, 5/9/61; mustered in, 5/11/61; described as 5'8", dark comp., brown hair, brown eyes; discharged to work at Gosport Navy Yard per SO 18, Dept. of Norfolk, 1/29/62; served as a landsman, CSS *Virginia*; appears to have left crew of *Virginia* before 4/1/62; enlisted at Petersburg as a pvt. in Petersburg Horse Artillery (Graham's Co., Virginia Horse Artillery), 6/25/62; listed as a bugler, 9–10/64 rolls; present thru 2/28/65; deserted, took oath, and sent to City Point, Virginia, 3/4/65; furnished transport to Newark, New Jersey.[202]

Jones, Catesby ap Roger: b. 4/15/21, Fairfield, Virginia; parents, Roger Jones, adjutant general, U.S. Army, and Mary Anne Mason Page; appointed by Virginia to USN as a midshipman, 6/18/36; passed midshipman, 7/1/42; served under uncle, Commodore Thomas ap Catesby Jones, during "Exploring Expedition"; master, 9/14/48; Lt., 5/12/49; served at Washington Navy Yard, where he assisted John Dahlgren with his ordnance experiments; took part in the USS *Merrimack*'s maiden voyage as ordnance officer in charge of testing its Dahlgren guns; tendered resignation from USN, 4/17/61; appointed Capt., Virginia State Navy, 4/18/61;

resignation from USN finally accepted, 5/13/61; commissioned as a Lt., CSN, 6/11/61; commanded Jamestown, Virginia defenses, '61; conducted ordnance experiments and tested CSS *Virginia* armor plating with Lt. John Mercer Brooke; paid for service, Richmond Naval Station, 10/1–11/10/61; assigned as executive officer, *Virginia*, 11/11/61; commanded *Virginia* against the USS *Monitor* on 3/9/62 after Capt. Buchanan WIA on previous day; executive officer under Flag Officer Josiah Tattnall until *Virginia*'s destruction; served with Southside Artillery during Battle of Drewry's Bluff, Virginia, 5/15/62; 1st Lt., 10/23/62; commanded CSS *Chattahoochee*, '62–'63; supervised Charlotte Naval Works, '63; assumed command of Selma Naval Gun Foundry and Ordnance Works, which cast heavy artillery for fortifications/vessels at Mobile, Alabama, 5/9/63; commander for gallant and meritorious conduct aboard the *Virginia*, 5/29/63; commander, Provisional Navy, 6/2/64; assigned to Mobile Squadron, '65; paroled on board USS *Stockdale* off Mobile, Alabama, 5/9/65; postwar: established military supply company with John Mercer Brooke and Robert Dabney Minor but company failed; d. Selma, Alabama, killed in argument with neighbor, 6/17/77; bur. Live Oak Cem., Selma.[203]

Jones, David: b. ca. 1833, Wales; occ. sailor; single; resided and enlisted at New Orleans, Louisiana, as a pvt. in Co. D, 15th Louisiana Infantry, 7/18/61; transferred to CSN, 2/62; served as a seaman, CSS *Virginia*; paid for service on the *Virginia*, 4/1–5/12/62; reenlisted for 3 years, received $50 bounty, 4/21/62; superannuated for rations, Drewry's Bluff, Virginia, 5/12–24/62; on *Virginia* CRR, Drewry's Bluff, 7/20/62; paid, Drewry's Bluff, 5/13/62–3/31/64; quarter gunner, 1–3/64; may have later served as a seaman, CSS *Spray*, '65; surrendered, Tallahassee, Florida, 5/10/65; paroled, St. Marks, Florida, 5/12/65; described then as 5'11½", dark comp., dark hair, blue eyes.[204]

Jones, John: shipped for the war as a landsman, CSS *Forrest*, North Carolina Squadron, 7/25/61; served as a landsman, CSS *Virginia*; reenlisted for the war, received $50 bounty as member of crew, *Virginia*, 3/25/62; paid for service as an ordinary seaman on the *Virginia*, 4/1–5/12/62; superannuated for rations, Drewry's Bluff, Virginia, 5/12–24/62; on *Virginia* CRR, Drewry's Bluff, 7/20/62; paid, Drewry's Bluff, 5/13–8/31/62; listed as discharged on final *Virginia* payroll, dated 9/30/62; may have later served as a seaman, CSS *Arctic*, North Carolina Squadron, '63 and as captain of aft gun, CSS *Virginia II*, James River Squadron, '64; NFR.[205]

Jones, John Pembroke: b. 1825, Pembroke Farm, near Hampton, Virginia; parents, John and Mary Booker Jones; resided Elizabeth City Co., Virginia;

attended John Baytop Carey's school, Hampton; attended College of William and Mary, '40; appointed by Virginia to USN as a midshipman, 10/19/41; took part in Siege of Buena Vista during Mexican War; passed midshipman, 8/10/47; master, 3/1/55; Lt., 9/14/55; performed coastal surveys of North Carolina/Virginia; married Jane Vance London from Wilmington, North Carolina area; assigned to USS *Richmond* at start of Civil War; tendered resignation and dismissed from USN, 4/29/61; appointed as a 1st Lt., CSN, 5/2/61; assigned James River naval batteries, '61; commanded naval battery, Barrett's Point, Elizabeth River, Virginia, '61; took part in Battle of Port Royal, South Carolina, 11/7/61; commanded CSS *Resolute*, Savannah River Squadron, '61–'62; assigned as Flag Lt., CSS *Virginia*, 4/1/62; commanded CSS *Georgia* and *Savannah*, Savannah River Squadron, '62–'63; served on CSS *Nansemond*, James River Squadron, '63–'64; commanded CSS *Raleigh*, *Arctic* and *North Carolina*, North Carolina Squadron, '64; 1st Lt., Provisional Navy, 6/2/64; married Mary Willis of Savannah, Georgia, '64; commanded submarine defenses, Chaffin's Bluff, Virginia, 9/64; assigned CSS *Torpedo*, James River Squadron, '64–'65; postwar: resided Airlie Farm, Fauquier Co., Virginia, for several years; married Georgia Newton of Norfolk; moved to California; d. Pasadena, California, 5/25/1910.[206]

Jones, Samuel: served as a 1st class fireman, CSS *Virginia;* reenlisted for the war, received $50 bounty as member of crew, *Virginia*, 3/25/62; paid for service on the *Virginia*, 4/1–5/12/62; on *Virginia* CRR, Drewry's Bluff, Virginia, Virginia, 7/20/62; listed as having transferred as a 1st class fireman, CSS *Beaufort*, on *Virginia* muster roll, 7/31/62; paid, Drewry's Bluff, 5/13/62–9/30/63; may have later been captured on the blockade runner, CSS *Greyhound* by USS *Connecticut*, 5/12/64; POW, Camp Hamilton, Virginia, 5/16/64; released and sent to Baltimore, Maryland, 6/6/64; NFR; possibly "Samuel Jones," who enlisted at New Orleans, Louisiana, for the war as a pvt. in Co. A, 15th Louisiana Infantry, 6/30/61; present, 9–10/61; NFR.[207]

Jones, Thomas: enlisted at Mobile, Alabama, as a pvt. in Co. C, CSMC, 8/28/61; probably served at Pensacola Navy Yard until Co. C transferred to Gosport Navy Yard, Portsmouth, Virginia, 11/29/61; assigned to Marine Guard, CSS *Virginia*, ca. 11/61–5/12/62; assigned to CSS *Richmond*, James River Squadron, 12/22/62; on same duty, 12/31/64; NFR.[208]

Jones, William Henry Harrison: b. ca. 1840, Georgia; occ. day laborer; single; resided Newton Co., Georgia; enlisted at Covington, Georgia, for 1 year as a pvt. in Co. H, 3rd Georgia Infantry, 4/25/61; "transferred to CSS *Merrimac*," 1/13/62; served as a 2nd class fireman, CSS *Virginia*;

superannuated for rations, Drewry's Bluff, Virginia, 5/12–24/62; listed as having transferred to CSS *Hampton*, James River Squadron, on *Virginia* muster roll, 7/31/62; listed as having deserted on final *Virginia* payroll, dated 9/30/62; RTD w/ Co. H, 3rd Georgia Infantry; KIA, Chancellorsville, Virginia, 5/3/63; Georgia commissioner of pensions requested verification of service, 1916.[209]

Jones, William T.: enlisted at Madison, Georgia, for 1 year as a pvt. in Co. D, 3rd Georgia Infantry, 4/24/61; mustered in at Augusta, Georgia, 5/2/61; transferred to CSN, 1/13/62; served as a 1st class fireman, CSS *Virginia;* paid for service on the *Virginia* and Drewry's Bluff, Virginia, 4/1–5/18/62; listed as having deserted on 5/18/62 on final *Virginia* payroll, dated 9/30/62; reportedly served as a seaman, vessel unknown, '64; NFR; Georgia Commissioner of Pensions requested verification of service, 1916.[210]

Jordan, Marshall P.: b. ca. 1833, Virginia; appointed by Virginia to USN as a 3rd assistant engineer, 12/24/53; 2nd assistant engineer, 5/9/57; married; resided Norfolk, Virginia, assigned to USS *Brooklyn* at start of Civil War; tendered resignation, 4/25/61; dismissed from USN, 5/30/61; appointed as 1st assistant engineer, CSN, 7/6/61; served on CSS *Rappahannock*, '61; served on CSS *Patrick Henry*, 7/20–12/2/61; left North Carolina Squadron for CSS *Virginia*, 12/2/61; assigned, CSS *Virginia*, 12/3/61–1/13/62; assigned, Richmond Naval Station, 1/14–6/21/62; assigned, Charleston Station, 10/1/62–1/63; served on CSS *Palmetto State*, Charleston Squadron, '62–'64; 1st assistant engineer, Provisional Navy, 6/2/64; NFR.[211]

Kelly, Michael: enlisted at Mobile, Alabama, as a pvt. in Co. C, CSMC, 7/15/61; probably served at Pensacola Navy Yard 7–11/61 until Co. C transferred to Gosport Navy Yard, Portsmouth, Virginia, 11/29/61; assigned to Marine Guard, CSS *Virginia*, ca. 11/61–5/12/62; on Co. C CRR, 3rd qtr. '62; probably deserted before 9/30/62; NFR; also found as "Kelley."[212]

Kevill, John Patrick: b. 10/5/44, Charlestown, Massachusetts; father, Patrick Kevill; orphaned when very young; moved to Norfolk, Virginia, to live with uncle, Thomas Kevill; occ. store clerk; single; resided and enlisted at Norfolk, Virginia for 1 year as a pvt. in the United Artillery (1st Co. E, 41st Virginia Infantry), 8/19/61; present thru 3/6/62; "volunteered to go onboard the Ironclad Steamer *Virginia*," 3/6/62; served as a powder boy, CSS *Virginia* during Battle of Hampton Roads, Virginia, 3/8–9/62; reenlisted for war, received $50 bounty, 3/10/62; on CRRs, 12/22/62, 6/3/63, 9/14/63 and 11/2/63; in Chimborazo Hosp. #2, Richmond, Virginia w/ debility, 8/6/64; in Chimborazo Hosp. #1, 8/11–9/13/64; granted 30-day furlough; absent, det. as quartermaster sgt., 11–12/64 muster roll; paroled, Appomattox CH, 4/9/65; postwar: occ. manifest clerk, Old Dominion Steamship Line; 1st Lt., Norfolk Light Artillery Blues, Virginia State Militia; member, Pickett-Buchanan Camp UCV; received pension, witnessed by former *Virginia* shipmate, Andrew J. Dalton, Norfolk justice of peace, 7/14/1914; resided Olney Road, Norfolk; d. 1/3/1941; bur. Cedar Grove Cem., Norfolk; last survivor of the CSS *Virginia*.[213]

Kevill, Thomas: b. 4/5/26, Sligo, Ireland; immigrated to Canada, then Boston, Massachusetts; occ. clothing merchant; resided Norfolk, Virginia, '48; married Augustine L. Shield, '50; formed and headed United Volunteer Fire Co., Norfolk, '60; organized United Artillery Co. from ranks of fireman as a militia unit before Ft. Sumter; enlisted at Norfolk for 1 year as a capt. in the United Artillery (1st Co. E, 41st Virginia Infantry), 4/19/61; occupied Ft. Norfolk, Elizabeth River, Virginia defenses, 4/19/61–3/62; capt., CSA, 7/12/61; "volunteered to go onboard the Ironclad Steamer *Virginia*," 3/6/62; commanded 9-inch broadside gun #9 on CSS *Virginia* during Battle of Hampton Roads, Virginia, 3/8–9/62; unit became Co. C, 19th Btn. Virginia Heavy Arty., 4/19/62; manned heavy btry. at Entrenched Camp on evacuation of Norfolk, 5/10/62; led RR arty. at Savage Station, 6/62; on court-martial duty, Richmond, Virginia, 8–9/62; company became an independent btry., 10/1/62; present, 12/62–1/63; stationed much of war at Drewry's Bluff, Virginia; absent, sick on 12/31/64 roll; paroled, Appomattox CH, 4/9/65; postwar: occ. clothing merchant; organized and chief of Norfolk's first professional fire dept.; retired '92; member, Pickett-Buchanan Camp UCV; d. 1896; bur. St. Mary's Catholic Cem., Norfolk.[214]

King, Alfred William: b. ca. 1826, England; occ. pressman; married; resided Mobile, Alabama; shipped for the war as an seaman, Naval Rendezvous, Mobile, 10/30/61; served as an ordinary seaman, CSS *Virginia*; paid for service on the *Virginia*, 4/1–5/12/62; superannuated for rations, Drewry's Bluff, Virginia, 5/12–24/62; on *Virginia* CRR, Drewry's Bluff, 7/20/62; paid, Drewry's Bluff, 5/13/62–9/30/63; paid for service, naval det. on special expedition from Drewry's Bluff to Charleston Naval Station, 10/1–31/63; paid for service as a quarter gunner, CSS *Albemarle*, North Carolina Squadron, 7/1–9/30/64; deserted, Bermuda Hundred, Virginia, 4/15/65; sent to City Point, Virginia; NFR.[215]

King, Charles Kirby, Jr.: b. Virginia; appointed by D.C. to USNA, 9/25/56; resigned as an acting midshipman, USN, 5/16/61; appointed as a midshipman, CSN, 6/15/61; acting master, 9/24/61; assigned to CSN Office of Provisions and Clothing, Richmond Naval Station, 9/24–10/4/61; assigned to receiving ship, CSS *Confederate States*, '61; 2nd Lt. and Lt. for war,

2/8/62; paid for service, Richmond Naval Station, 10/4/61–2/27/62; assigned to CSS *Virginia*, 4/30–5/13/62; paid for service, Richmond Naval Station, 5/13–6/30/62; 1st Lt., 10/2/62; served on CSS *Georgia*, Savannah River Squadron, '62–'63; service abroad, '63–'64; passenger on steamer *Margaret and Jesse* when it was chased ashore by USS *Rhode Island*, 5/30/63; served on the cruiser, CSS *Georgia*, '63–'64; served on CSS *Charleston*, Charleston Squadron, '64; 1st Lt., Provisional Navy, 6/2/64; paroled, Appomattox CH, 4/9/65.[216]

KNIGHT, GEORGE A.: b. ca. 1841, Virginia; single; resided Portsmouth, Virginia; enlisted at Norfolk, Virginia, as a pvt. in the United Artillery (1st Co. E, 41st Virginia Infantry), 6/2/61; present thru 3/6/62; "volunteered to go onboard the Ironclad Steamer *Virginia*," 3/6/62; manned gun on CSS *Virginia* during Battle of Hampton Roads, Virginia, 3/8–9/62; reenlisted for war, received $50 bounty, 3/10/62; on CRRs, 12/22/62, 6/3/63, 9/14/63, 11/2/63 and 3/18/64; in Gen. Hosp. #9, Richmond, Virginia, 9/8–9/64; in Chimborazo Hosp. #3, Richmond, w/ int. fever, 9/9–16/64; present, 10/10/64; deserted, 12/22/64, entered Federal lines, 12/26/64; sent to Washington, D.C.; took oath, furnished transport to Norfolk, 1/3/65; NFR.[217]

KNOWLES, DANIEL: b. ca. 1830, Pennsylvania; served as a classboy, USS *Albany* during Mexican War, '47; WIA twice; drew pension for service; occ. waggoner, net worth $25,000, '60; married; resided and enlisted at Norfolk, Virginia, for 1 year as a pvt. in the United Artillery (1st Co. E, 41st Virginia Infantry), 4/19/61; 1st sgt., summer '61; present thru 3/6/62; "volunteered to go onboard the Ironclad Steamer *Virginia*," 3/6/62; member of gun crew, CSS *Virginia* during Battle of Hampton Roads, Virginia; WIA in head; reenlisted for war, received $50 bounty, 3/10/62; on duty as a guard, Btry #8, 9/62; promoted 2nd Lt.; on CRR, 12/22/62; in Chimborazo Hosp. #1, Richmond, Virginia, w/ remittent fever, 8/11–26/64; furloughed to Wilmington, North Carolina, 8/29/64; RTD, 11/3/64; signed roll as commanding co., 12/31/64; resigned due to poor health and desire to "get off James River," 3/23/65; listed next residence as Wilmington, North Carolina; NFR; postwar: occ. plumber and as a rigger at Norfolk Naval Yard; d. Norfolk Protestant Hosp., 4/5/1905; bur. Elmwood Cem, Norfolk.[218]

LACKIE, PIERRE: served as officer's cook, CSS *Virginia*; paid for service on the *Virginia* and Drewry's Bluff, Virginia, 4/1–7/26/62; superannuated for rations, Drewry's Bluff, 5/12–24/62; discharged, 7/26/62; NFR.[219]

LAKIN, EDWARD: b. ca. 1834, New York; occ. tavern keeper; resided and enlisted at Norfolk, Virginia, for 1 year as a 2nd Lt. in the United Artillery (1st

Co. E, 41st Virginia Infantry), 4/19/61; appointed, 2nd Lt., CSA, 7/1/61; assigned to Ft. Norfolk, Virginia, 7/61–5/62; served as a gun capt., CSS *Virginia*, 5/10–11/62; present, Btry #8, 8/62; present, commanding co. at Camp Rhett, Virginia, 9/62; present, Ft. Drewry, Virginia, 11/62–1/63; ordered to Wilmington, North Carolina, by SO 22, para 31, A&IGO 64, 1/27/64; transferred to CSN, 6/29/64; 1st Lt., Provisional Navy, 7/26/64; commanded CSS *Drewry*, James River Squadron, '64; paid for service on CSS *Albemarle*, North Carolina Squadron, 6/27–9/30/64; commanded the torpedo boat, CSS *Scorpion*, James River Squadron, '65; wounded by explosion of CSS *Drewry*, 1/24/65; NFR.[220]

Lamb, Andrew J.: b. ca. 1839, Twiggs Co., Georgia; enlisted at Twiggs Co. for 1 year as a pvt. in Co. C, 4th Georgia Infantry, 4/25/61; described as 6', light comp., light hair, gray eyes; in regt. hosp., Camp Jackson, Virginia, w/ int. fever, 9/19–29/61; in regt. hosp. w/cholera, 10/5–16 and 10/18–21/61; in regt. hosp. w/rheumatism, 1/18–19/62; in regt. hosp. w/sciatica, 1/22–28 and 2/2–6/62; "transferred to *Merrimac*" per SO 28, Dept. of Norfolk, 2/10/62; paid for service as a landsman, CSS *Virginia*, 2/4–18/62; reenlisted for the war, received $50 bounty; paid for service on the *Virginia*, 4/1–5/12/62; on *Virginia* CRR, Drewry's Bluff, Virginia, 7/20/62; paid, Drewry's Bluff, 5/13/62–9/30/63; paid for service, naval det. on special expedition from Drewry's Bluff to Charleston Naval Station, 10/1–31/63; reportedly served at Drewry's Bluff, '64; NFR; Georgia commissioner of pensions requested verification of service, 1916.[221]

Langtree, John W.: enlisted at Memphis, Tennessee, as a pvt. in CSMC, 8/7/61; probably transferred from Capt. Hays's det., Co. D, to Co. C, Pensacola Navy Yard, ca. 11/27/61; on Co. C CRR, 4th qtr. '61; assigned to Marine Guard, CSS *Virginia*, ca. 11/61–5/12/62; sgt., 4th qtr. '62; reduced to pvt. by 5/9/63; on Co. C CRR, 6/30/63; NFR; also found as "Langtry."[222]

Lawler, William F.: b. ca. 1841, Waterford, Ireland; occ. seaman; enlisted at Mobile, Alabama, as a pvt. in Co. E, 8th Alabama Infantry, 5/6/61; "transferred to CSN by order of Secretary of War," 2/12/62; shipped for 3 years as a seaman, Naval Rendezvous, Richmond, Virginia, 2/14/62; served as a seaman, CSS *Virginia*; reenlisted for the war, received $50 bounty as member of crew, *Virginia*, 3/25/62; superannuated for rations, Drewry's Bluff, Virginia, 5/12–24/62; NFR; may have later enrolled as a 2nd Lt., Co. F, 35th Alabama Infantry prior to 9/24/62; WIA, left leg, Battle of Corinth, Mississippi, 10/3/62; paid 7/24–10/24/63; unable to perform duties in field due to wound; resigned, Camden, Alabama, 11/24/63; resignation accepted 1/11/64; NFR.[223]

Leahy, James: enlisted at Memphis, Tennessee, as a pvt. in CSMC, 8/9/61; on CRR, Co. C, 4th qtr. '61; probably transferred from Capt. Hays's det., Co. D, to Co. C, Pensacola Navy Yard; probably served at Pensacola until Co. C transferred to Gosport Navy Yard, Portsmouth, Virginia, 11/29/61; assigned to Marine Guard, CSS *Virginia*, ca. 11/61–5/12/62; assigned to Wilmington Naval Station, 2/17–12/31/64; paid for service on CSS *Arctic*, North Carolina Squadron, 7/1–9/30/64; surrendered to officer of USS *Monadnock*, James River, 4/5/65; sent to Washington, D.C.; furnished transport to Nashville, Tennessee, 4/12/65; also found as "Lehy" and 'Leighiegh."[224]

Leary, James: served as an ordinary seaman, CSS *Virginia;* reenlisted for 3 years, received $50 bounty, 4/21/62; paid for service on the *Virginia* and Drewry's Bluff, Virginia, 4/1/62–6/30/63; superannuated for rations, Drewry's Bluff, 5/12–24/62; on *Virginia* CRR, Drewry's Bluff, 7/20/62; served on CSS *Patrick Henry*, James River Squadron; paid for service, Drewry's Bluff, 4/1–9/30/63; listed as having deserted on payroll, dated 3/31/63; paid for service as an ordinary seaman, CSS *Chicora*, Charleston Squadron, 7/1/63–3/31/64; NFR.[225]

Lee, Thomas J.: b. ca. 1839, Washington, North Carolina; occ. seaman; single; resided Currituck Co.; enlisted at Plymouth, North Carolina, for the war as a pvt. Co. G, 1st North Carolina Infantry, 7/3/61; present until transferred to CSN per SO 17, A&IGO 62, para 23, 2/3/62; served as an ordinary seaman, CSS *Virginia*; appears to have left crew of *Virginia* before 4/1/62; NFR.[226]

Lembler, August W.: b. ca. 1825, Russia; occ. riverman; enlisted at Mobile, Alabama, as a pvt. in Co. E, 8th Alabama Infantry, 5/6/61; "transferred to CSN by order of Secretary of War," 2/12/62; shipped for 3 years as a seaman, Naval Rendezvous, Richmond, Virginia, 2/16/62; served as a seaman, CSS *Virginia*; reenlisted for the war, received $50 bounty as member of crew, *Virginia*, 3/25/62; paid for service on the *Virginia* and Drewry's Bluff, Virginia, 4/1–6/30/62; superannuated for rations, Drewry's Bluff, 5/12–24/62; on *Virginia* CRR, Drewry's Bluff, 7/20/62; volunteered for duty on CSS *Chattahoochee*, 8/1/62; served as captain of forecastle, CSS *Chattahoochee* thru 6/12/63; transferred to CSS *Savannah*, Savannah River Squadron, 6/13/63; served as a pvt., Naval Btn.; POW, Burkeville, Virginia, 4/6/65; sent to City Point, Virginia, and then Point Lookout, Maryland, 4/14/65; took oath and released, 6/14/65; on *Virginia* muster roll in *ORN* as "August W. Lemblom."[227]

Leonard, Jacob K.: b. ca. 1838, Roanoke Co., Virginia; occ. farmer; resided and enlisted at Salem, Virginia, for 1 year as a pvt. in 1st Co. A, 9th

Virginia Infantry, 5/14/61; mustered in at Lynchburg, Virginia, 5/16/61; described as 5'8", dark comp., dark hair, dark eyes; reenlisted for 2 years, $50 bounty due, 2/13/62; present until discharged per SO 69, Dept. of Norfolk, 3/29/62; served as a landsman, CSS *Virginia*; superannuated for rations, Drewry's Bluff, Virginia, 5/12–24/62; NFR; on *Virginia* muster roll in *ORN* as "Jacob K.Lenard."[228]

LEONARD, JOHN H.: b. ca. 1836; resided Maryland; enlisted at Edenton, North Carolina, for the war as a pvt. in Co. A, 1st North Carolina Infantry, 5/18/61; present until confined, sentenced to 30 days' labor and loss of $10 pay, 12/22/61; present until transferred to CSN per SO 17, A&IGO 62, para 23, 2/1/62; served as a seaman, CSS *Virginia*; WIA, 3/8/62; reenlisted for 3 years, received $50 bounty, 4/21/62; NFR.[229]

LEONARD, SAMUEL: b. ca. 1840, Roanoke Co., Virginia; occ. farmer; resided and enlisted at Salem, Virginia, for 1 year as a pvt. in 1st Co. A, 9th Virginia Infantry, 5/14/61; mustered in at Lynchburg, Virginia, 5/16/61; described as 5'6", ruddy comp., light hair, gray eyes; reenlisted for 2 years, $50 bounty due, 2/13/62; present until discharged per SO 69, Dept. of Norfolk, 3/29/62; served as a landsman, CSS *Virginia*; paid for service on the *Virginia* and Drewry's Bluff, Virginia, 4/1–5/15/62; superannuated for rations, Drewry's Bluff, 5/12–24/62; listed as discharged on final *Virginia* payroll, dated 9/30/62; NFR.[230]

LEVY, BENJAMIN A.: served as an ordinary seaman, CSS *Virginia*; reenlisted for the war, received $50 bounty as member of crew, *Virginia*, 3/25/62; superannuated for rations, Drewry's Bluff, Virginia, 5/12–24/62; NFR; possibly "Benjamin Levy," who enlisted at New Orleans, Louisiana, for the war as a pvt. in Co. H, 15th Louisiana Infantry, 7/21/61; NFR.[231]

LILLES, MICHAEL: served as a 2nd class fireman, CSS *Virginia*; superannuated for rations, Drewry's Bluff, Virginia, 5/12–24/62; listed as having deserted on final *Virginia* payroll, dated 9/30/62; NFR.[232]

LINDSAY, HUGH: b. ca. 1815, Ireland; carpenter, USN, 11/4/40; dismissed from USN, 12/9/53; occ. carpenter; single; resided Portsmouth, Virginia; appointed carpenter, CSN, 8/14/61; billeted on receiving ship, CSS *Confederate States*, '61; served on CSS *Virginia* during Battle of Hampton Roads, Virginia, 3/8–9/62; assigned Drewry's Bluff, Virginia, until 12/31/62; served on CSS *Patrick Henry*, James River Squadron, '63; assigned Savannah Naval Station, '63; served on CSS *Raleigh*, North Carolina Squadron, '63; resigned 6/22/63; NFR; also found as "Lindsey."[233]

LINDSAY, JAMES E.: b. 1829, Guilford, North Carolina; parents, Jeduthan Harper Lindsay and Miss Strange; appointed by North Carolina as an assistant surgeon, USN; tendered resignation, 7/16/61; dismissed from USN, 10/10/61; prisoner, Ft. Warren, Massachusetts, 10/15/61; appointed as an assistant surgeon, CSN, 1/23/62; exchanged, 1/24/62; served on CSS *Virginia*, '62; assigned Drewry's Bluff, Virginia, '62–'63; passed assistant surgeon, 4/20/63; service abroad, '63–'65; passed assistant surgeon, Provisional Navy, 6/2/64; NFR.[234]

LITCHFIELD, ORSON: b. ca. 1838, Missouri; parents, Israel Litchfield and Chloe Keith; occ. overseer; enlisted at Memphis, Tennessee, as a pvt. in CSMC, 8/6/61; on Co. C CRR, 4th qtr. '61; probably transferred from Capt. Hays's det., Co. D, to Co. C, Pensacola Navy Yard, ca. 11/27/61; assigned to Marine Guard, CSS *Virginia*, ca. 11/61–5/12/62; paid for service, CSS *Virginia*, 1/1–5/12/62; on CRR, Co. C, 3rd qtr. '62; honorably discharged and paid transport. from Richmond, Virginia, to Memphis, 6/7/62; may have later enlisted at New Hanover Co., North Carolina, as a pvt. in Co. G, 61st North Carolina Infantry, 8/23/62; listed as AWOL, date not specified; NFR; on *Virginia* muster roll in *ORN* as "Orison Litchfield."[235]

LITCHFIELD, SPENCER: b. ca. 1837, North Carolina; occ. mariner; single; resided Elizabeth City, North Carolina; enlisted at Edenton, North Carolina, for the war as a pvt. in Co. A, 1st North Carolina Infantry, 5/18/61; present on all rolls until transferred to CSN per SO 17, A&IGO 62, para 23, 2/1/62; served as an ordinary seaman, CSS *Virginia*; reenlisted for the war, received $50 bounty as member of crew, *Virginia*, 3/25/62; paid for service on the *Virginia*, 4/1–5/12/62; on *Virginia* CRR, Drewry's Bluff, Virginia, 7/20/62; paid, Drewry's Bluff, 5/13/62–3/31/64; NFR; on *Virginia* muster roll in *ORN* as "Spence Lichfield."[236]

LITTLE, WILLIAM C.: b. ca. 1840, Wake Co., North Carolina; occ. laborer; resided Wake Co.; enlisted at Raleigh, North Carolina, as a pvt. in Co. K, 14th North Carolina Infantry, 5/21/61; described as 5'10", dark comp., dark hair, hazel eyes; present until "transferred to Steamer *Merrimack*" per SO 36, Dept. of Norfolk, 2/15/62; served as a landsman, CSS *Virginia*; reenlisted for the war, received $50 bounty as member of crew, *Virginia*, 3/25/62; paid for service on the *Virginia*, 4/1–5/12/62; in Bruton Parish Episcopal Church Hosp., Williamsburg, Virginia, w/ pneumonia, 5/12–14/62; on *Virginia* CRR, Drewry's Bluff, Virginia, 7/20/62; paid, Drewry's Bluff, 5/13/62–3/31/63; listed as having deserted on 3/31/63 payroll; apparently RTD since he was paid, Drewry's Bluff, 10/1/63–3/31/64; NFR.[237]

LITTLEPAGE, HARDIN BEVERLY: b. 3/8/41, "Piping Tree", King William Co., Virginia; parents, Lewis Littlepage and Caroline B. Ellett; educated at Rumford Military Academy; appointed by Virginia to USNA, 9/23/57;

resigned as an acting midshipman, USN, 4/25/61; appointed as an acting midshipman, CSN, 6/11/61; assigned to receiving ship, CSS *Confederate States*, '61; assigned to CSS *Virginia*, 1/1/62; served as captain of Gun #4 during Battle of Hampton Roads, Virginia, 3/8–9/62; took part in Battle of Drewry's Bluff, Virginia, 5/15/62; selected by Catesby ap Roger Jones to serve on CSS *Chattahoochee*, 5/16–11/23/62; master in line of promotion, 10/4/62; served on CSS *Atlanta*, Savannah River Squadron, '62–'63; passenger on steamer *Margaret and Jesse* when chased ashore by USS *Rhode Island*, 5/30/63; 2nd Lt., 8/5/63; served in France in unsuccessful attempt to secure ironclad contracts, 8/63–9/64; roomed w/former *Virginia* shipmate, Henry H. Marmaduke in Paris; served on CSS *Virginia II*, James River Squadron, '64–'65; 1st Lt., Provisional Navy, 6/2/64; assigned to Semmes naval brigade, '65; postwar: married Emily Sinclair Castleman; resided Washington, D.C.; worked in Naval War Records Office collecting and publishing Confederate naval primary resources, '89; commander, Confederate Naval Veterans Association; alive in 1907.[238]

LOGAN, PATRICK H.: b. ca. 1838, Ohio; occ. carpenter; enlisted at New Orleans, Louisiana, for the war as a pvt. in Co. C., 1st (Nelligan's) Louisiana Infantry, 5/27/61; described as 5'8", dark comp., black hair, hazel eyes; present until "discharged to join the *Merrimac*" per SO 18, Dept. of Norfolk, 1/14/62; served as a landsman, CSS *Virginia*; appears to have left crew of *Virginia* before 4/1/62; NFR.[239]

LONG, JAMES CROSBY: b. ca. 1840, Chattanooga, Tennessee; appointed by Tennessee to USNA; tendered resignation as an acting midshipman, USN, 5/9/61; resignation accepted 5/15/61; appointed as an acting midshipman, CSN, 7/3/61; assigned to receiving ship, CSS *Confederate States*, '61; assigned to CSS *Virginia*, 3/1/62; served in her powder division during Battle of Hampton Roads, Virginia, 3/8–9/62; took part in Battle of Drewry's Bluff, Virginia, 5/15/62; served on CSS *Richmond* and *Patrick Henry*, James River Squadron, '62–'63; assigned CSS *Savannah*, Savannah River Squadron, '63–'64; passed midshipman, 1/8/64; served on CSS *Albemarle*, North Carolina Squadron, '64; master in line of promotion, Provisional Navy, 6/2/64; served on the blockade runner, CSS *Owl*, '65; postwar; resided Tiskilva, Illinois, 1907; d. Chicago, Illinois, May 1910.[240]

LONG, WILLIAM: served as a 1st class fireman, CSS *Virginia*; reenlisted for the war, received $50 bounty as member of crew, *Virginia*, 3/25/62; in Bruton Parish Episcopal Church Hosp., Williamsburg, Virginia, w/ lumbago, 5/10–27/62; NFR; possibly "William A. Long" who was b. ca. 1840, Alabama; occ. mechanic, single; enlisted at Carrolton, Mississippi, for 1 year as a pvt.

in Co. K, 11th Mississippi Infantry, 4/29/61; mustered in at Lynchburg, Virginia, 5/13/61; in Gen. Hosp., Orange CH, Virginia, w/ fever, 8/19/61; discharged for disability, 9/14/61; described then as 5'6½", sallow comp., light hair, blue eyes; NFR.[241]

Loyd, William: served as a coal heaver, CSS *Virginia*; paid for service on the *Virginia* and Drewry's Bluff, Virginia, 4/1–8/12/62; on *Virginia* CRR, Drewry's Bluff, 7/20/62; listed as a POW on *Virginia* muster roll, 7/31/62; listed as discharged on final *Virginia* payroll, dated 9/30/62; NFR.[242]

Lyon, William W.: b. ca. 1834, Caswell Co., North Carolina; occ. carpenter; resided Rockingham Co., North Carolina; enlisted at Lawsonville, North Carolina, for 1 year as a pvt. in Co. K, 13th North Carolina Infantry, 5/30/61; mustered in at Suffolk, Virginia, 6/3/61; described as 5'8", light comp., light hair, blue eyes; present until "discharged by order of Gen. Huger having reenlisted in CSN for duty on the *Merrimack*," 2/14/62; served as a landsman, CSS *Virginia*; reenlisted for the war, received $50 bounty as member of crew, *Virginia*, 3/25/62; paid for service on the *Virginia*, 4/1–5/12/62; superannuated for rations, Drewry's Bluff, Virginia, 5/12–24/62; on *Virginia* CRR, Drewry's Bluff, 7/20/62; paid, Drewry's Bluff, 5/13/62–6/30/63; listed as having deserted on 6/30/63 payroll; NFR.[243]

Manly, Richard: served as a 1st class fireman, CSS *Virginia*; NFR; possibly "Richard C. Manley," who was b. ca. 1839, Ogelthorpe Co., Georgia; occ. farmer; enlisted at Macon, Georgia, for 1 year as a pvt. in Co. B, 5th Georgia Infantry, 5/10/61; described as 5'8", dark comp., black hair, black eyes; discharged for disability, 3/30/62; NFR.[244]

Marmaduke, Henry Hungerford: b. ca. 1842, Saline, Missouri; son of Meredith Miles Marmaduke, former Gov. of Missouri; appointed by Missouri to USNA, 9/20/58; tendered resignation as an acting midshipman, USN, 3/12/61; resignation accepted, 3/18/61; appointed as a midshipman, CSN, 5/8/61; served on CSS *McRae*, New Orleans Naval Station, '61; took part in attack on Federal blockading fleet at Head of the Passes, Mississippi River, 10/21/61; served on CSS *Calhoun* in action near Cairo/Belmont; assigned to CSS *Virginia*, 12/3/61; ordered to Elizabeth City, North Carolina, in charge of ordnance stores, 2/10/62; captain of *Virginia*'s Gun #2 and WIA during Battle of Hampton Roads, Virginia, 3/8–9/62; master in line of promotion, 10/4/62; served on CSS *Chattahoochee*, 5/13/62–5/7/63; 2nd Lt., 8/23/63; served in France in unsuccessful attempt to secure ironclad contracts, 8/63–9/64; roomed w/ former *Virginia* shipmate, Hardin B. Littlepage in Paris; 1st Lt., 1/6/64; 1st Lt., Provisional Navy, 6/2/64; served on CSS *Sampson* and *Chicora*, Charleston Squadron, '65; assigned Semmes naval brigade, 3/65;

captured, Saylor's Creek, Virginia, 4/6/9/65; sent to Old Capital Prison, Washington, D.C., then Johnson's Island, Ohio; released on oath, 6/20/65; described then as age 23, 5'9", fair comp., light hair, blue eyes; postwar; resided Washington, D.C.; superintendent of Consular Bureaus of South American Republics; captain, Colombian cruiser *Bogota*; later worked in U.S. Naval War Records Office for the collection and publication of Confederate naval primary resources; d. Washington, D.C., 11/14/1924; bur. Arlington National Cem.[245]

Marry, Peter: b. ca. 1839; enlisted at New Orleans, Louisiana, as a pvt. in Co. C, CSMC, 6/4/61; issued clothing in New Orleans, 1/62; assigned to Marine Guard, CSS *Virginia* before 5/12/62; advertised as a deserter, 6/12/62; described then as 5'6"; dark comp., dark hair, hazel eyes; apparently RTD by 3rd qtr. '62; assigned Marine Guard, CSS *Richmond*, James River Squadron, 3/3/63; slightly WIA in cheek during cutting out of USS *Underwriter*, 2/2/64; presumably RTD until new guard was sent on board the *Richmond*, 2/8/65, NFR; on *Virginia* muster roll as "Peter Many."[246]

Martin, Patrick: b. ca. 1838, Monaghan, Ireland; occ. seaman; enlisted at Mobile, Alabama, as a pvt. in Co. E, 8th Alabama Infantry, 5/6/61; described as 5'8", black hair, gray eyes; "transferred to CSN by order of Secretary of War," 2/12/62; shipped for 3 years as a seaman, Naval Rendezvous, Richmond, Virginia, 2/14/62; served as a seaman, CSS *Virginia*, '62; reenlisted for the war, received $50 bounty as member of crew, *Virginia*, 3/25/62; paid for service on the *Virginia* and Drewry's Bluff, Virginia, 4/1–6/30/62; on *Virginia* CRR, Drewry's Bluff, 7/20/62; volunteered for duty on CSS *Chattahoochee*, 8/1/62; served as captain of top, CSS *Chattahoochee* thru 6/12/63; transferred to CSS *Savannah*, Savannah River Squadron, 6/13/63; may have later served as a 1st class fireman, CSS *Resolute*, Savannah River Squadron, '63–'64; deserted near Savannah, Georgia, 5/1/64; captured and confined; NFR.[247]

Martin, Thomas: enlisted at Mobile, Alabama, for the war as a pvt. in Co. A, 12th Alabama Infantry, 6/4/61; present on 7–8/61 rolls; transferred to CSN per SO 89, A&IGO 62, para 10, 3/27/62; shipped for 1 year as a seaman, Naval Rendezvous, Richmond, Virginia, 3/31/62; reenlisted for 3 years, received $50 bounty, 4/21/62; served as a seaman, CSS *Virginia*; paid for service on the *Virginia* and Drewry's Bluff, Virginia, 4/1–5/15/62; listed as having deserted on final *Virginia* payroll, dated 9/30/62; may have later shipped for the war, Naval Rendezvous, Richmond, 3/12/63; may have later served as a landsman, CSS *Isondiga*, Savannah River Squadron, '63–'64; NFR.[248]

Mary, Michael: enlisted at New Orleans, Louisiana, as a pvt. in Co. B, CSMC, 5/3/61; transferred to Co. C, ca. 7/61; assigned to Marine Guard, CSS *Virginia*, ca. 11/61–5/12/62; on CSS *Drewry* CRR, 1st qtr. '63; assigned to Marine Guard, Rocketts Navy Yard, Richmond, Virginia, 2nd qtr. '63; assigned to Marine Guard, CSS *Charleston*, Charleston Squadron, 8/2/63; corp., 12/9/63; on same duty on 12/31/64 roll; NFR.[249]

Mason, George J.: served as a landsman, CSS *Virginia*; NFR; possibly "George Mason" who was b. ca. 1841; occ. student; enlisted at Hicksford, Virginia, for 1 year as a 1st corp. in Co. F, 5th Btn. Virginia Infantry, 5/4/61; mustered in at Jamestown, Virginia, 5/4/61; present until "discharged by order of the Secretary of War," 1/1/62; NFR.[250]

Mathews, Sterling N.: enlisted at New Orleans, Louisiana, for 1 year as a corp. in Co. K, 2nd Louisiana Infantry, 5/11/61; present until "transferred to Naval Dept.," 3/22/62; served as a landsman, CSS *Virginia*; superannuated for rations, Drewry's Bluff, Virginia, 5/12–24/62; on *Virginia* CRR, Drewry's Bluff, 7/20/62; paid, Drewry's Bluff, 5/13/62–6/30/63 and 10/1/63–3/31/64; served as a carpenter's mate, CSS *Virginia II*, James River Squadron, '64–'65; served as a seaman/pvt., Co. E, 1st Regt., Naval Brigade; paroled, Greensboro, North Carolina, 4/26/65; on *Virginia* muster roll in *ORN* as "Sterling V. Mathew."[251]

Maxwell, John: b. ca. 1823, Dundalk, Ireland; occ. mechanic; married; resided and enlisted at New Orleans, Louisiana, as a pvt. in Co. F, 1st (Nelligan's) Louisiana Infantry, 5/2/61; described as 5'8", fair comp., brown hair, hazel eyes; present until "discharged having volunteered to go aboard the *Merrimac*" per SO 18, Dept. of Norfolk, 2/3/62; served as a 1st class fireman, CSS *Virginia*; paid on the *Virginia* and Drewry's Bluff, Virginia, 2/18–7/25/62; on *Virginia* CRR, Drewry's Bluff, 7/20/62; paid as a 2nd class fireman, Drewry's Bluff, 7/26–8/12/62; listed as a POW on *Virginia* muster roll, 7/31/62; listed as discharged on final *Virginia* payroll, dated 9/30/62; may have later enlisted at Richmond, Virginia, as a pvt. in Co. C, 4th Btn. Virginia Infantry, Local Defense (Naval Btn.), 6/20/63; NFR; not on *Virginia* muster roll in *ORN*.[252]

May, George: b. ca. 1838, Missouri; occ. sailor; single; resided New Orleans, Louisiana; enlisted at Camp Moore, Louisiana, for the war as a pvt. in Co. G, 10th Louisiana Infantry, 7/22/61; present until court-martialed for failure to obey lawful order and showing contempt for a superior officer, 10/15/61; sentenced to confinement and 5 days' hard labor; present until "transferred to CSS *Merrimac*," 1/2/62; served as captain of top, CSS *Virginia*; paid as a gun captain on the *Virginia*, 4/1–

5/12/62; superannuated for rations, Drewry's Bluff, Virginia, 5/12–24/62; on *Virginia* CRR, Drewry's Bluff, 7/20/62; paid, Drewry's Bluff, 5/13–6/30/62; volunteered to serve on CSS *Chattahoochee*, 8/1/62; paid for service as a gunner's mate, CSS *Chattahoochee*, 8/1–12/31/62 and 4/1–6/12/63; transferred to CSS *Savannah*, Savannah River Squadron, 6/13/63; surrendered and paroled, Lynchburg, Virginia, 4/9/65.[253]

McAdams, Francis: served as a coal heaver, CSS *Virginia;* reenlisted for 3 years, received $50 bounty, 4/21/62; on *Virginia* CRR, Drewry's Bluff, Virginia, 7/20/62; paid for service, Drewry's Bluff, 1/1/63–2/29/64; paid for service as a landsman, CSS *Albemarle*, North Carolina Squadron, 7/1–9/30/64; NFR; possibly "Frank McAdams" who enlisted at New Orleans, Louisiana, for the war as a 3rd sgt. in Co. E, 15th Louisiana Infantry, 5/27/61; present on all rolls to 7/61.[254]

McBride, William E.: b. ca. 1843, Kentucky; occ. boatman; single; resided Louisville, Kentucky; enlisted at New Orleans, Louisiana, for 1 year as a pvt. in Co. F, 1st (Nelligan's) Louisiana Infantry, 5/2/61; present thru 1/62; discharged "having volunteered to go aboard the *Merrimack*," 2/3/62; served as a landsman, CSS *Virginia*; superannuated for rations, Drewry's Bluff, Virginia, 5/12–24/62; NFR.[255]

McCarty, Neil: b. ca. 1835, Ireland; occ. ship's captain; married; resided Maryland; enlisted at Norfolk, Virginia, for the war as a pvt. in the United Artillery (1st Co. E, 41st Virginia Infantry) and received $50 bounty, 3/4/62; "volunteered to go onboard the Ironclad Steamer *Virginia*," 3/6/62; manned gun on CSS *Virginia* during Battle of Hampton Roads, Virginia, 3/8–9/62; on CRRs, 12/22/62, 6/3/63, 9/14/63, 11/2/63 and 3/18/64; absent, on detached duty w/ CSN, per SO 63, A&IGO 64, dated 3/16/64; requested transferred to CSN, 11/30/64; discharged to receive an appointment in CSN per SO 303, A&IGO 64, para 11, 12/22/64; listed as having deserted, 12/22/64; entered Federal lines, 12/26/64; sent to Washington, D.C., took oath, furnished transport to Norfolk, Virginia, 1/3/65; postwar: occ. ship's captain; married; resided Norfolk Co., Virginia, '70.[256]

McClure, Michael: served as a coal heaver, CSS *Virginia*; reenlisted for the war, received $50 bounty as member of crew, *Virginia*, 3/25/62; appears to have left crew of *Virginia* before 4/1/62; NFR; possibly "M. McClure," who enlisted at New Orleans, Louisiana, for the war as a pvt. in Co. A, 15th Louisiana Infantry, 6/30/61; present on 9–10/61 rolls; NFR.[257]

McCoy, Thomas: b. ca. 1826, Ireland; occ. laborer; single; resided New Orleans, Louisiana; enlisted at Camp Moore, Louisiana, for the war as a pvt. in Co. G, 10th Louisiana Infantry, 7/22/61; present until "transferred to CSS *Merrimac*," 1/2/62; served as a coal heaver, CSS *Virginia*; paid for service on the Virginia, 4/1–5/12/62; superannuated for rations, Drewry's Bluff, Virginia, 5/12–24/62; on *Virginia* CRR, Drewry's Bluff, 7/20/62; paid, Drewry's Bluff, 5/13–7/23/62; discharged, 7/23/62; NFR.[258]

McCRADY, JOHN: served as a coal heaver, CSS *Virginia;* reenlisted for 3 years, received $50 bounty, 4/21/62; on *Virginia* CRR, Drewry's Bluff, Virginia, 7/20/62; paid for service on the *Virginia* and Drewry's Bluff, 4/1/62–6/30/63; served as a boatswain's mate, CSS *Baltic*, Mobile Squadron, '63; appointed by Alabama as boatswain, 1/13/64; boatswain, Provisional Navy, 6/2/64; served on CSS *Tennessee*, Mobile Squadron, '64; took part in Battle of Mobile Bay, 8/15/64; captured; escaped from hosp. barracks, 11/15/64; NFR; also found as "McCredie."[259]

McCUBBINS, JOHN: shipped for the war as a coal heaver, Naval Rendezvous, Richmond, Virginia, 4/7/62; served as a coal heaver, CSS *Virginia*; reenlisted for 3 years, received $50 bounty, 4/21/62; may have later enlisted at Richmond, Virginia, for the war as pvt. in 2nd Co. H, 41st Btn. Virginia Cavalry, 6/13/63; absent w/ permission, 8/23/63; NFR.[260]

McDEVETT, CHARLES: b. ca. 1836, Londonderry, Ireland; occ. painter; enlisted at Union Springs, Alabama, as a pvt. in Co. D, 3rd Alabama Infantry, 4/26/61; described as 5'6"; discharged, 2/31/62; served as a landsman, CSS *Virginia*; reenlisted for the war, received $50 bounty as member of crew, *Virginia*, 3/25/62; on *Virginia* CRR, Drewry's Bluff, Virginia, 7/20/62; paid, Drewry's Bluff, 10/1/62–9/30/63; paid for service, naval det. on special expedition from Drewry's Bluff to Charleston Naval Station, 10/1–31/63; in Gen. Hosp., Danville, Virginia, w/ colitis, 4/8/65; NFR; on *Virginia* muster roll in *ORN* as "Charles McDevitt."[261]

McGINNIS, PATRICK: occ. laborer; resided New Orleans, Louisiana; enlisted at New Orleans as a pvt. in Co. C, CSMC, 6/4/61; probably served at Pensacola Navy Yard until Co. C transferred to Gosport Navy Yard, Portsmouth, Virginia, 11/29/61; assigned to Marine Guard, CSS *Virginia*, ca. 11/61–5/12/62; served Co. C, Drewry's Bluff, Virginia. until transferred to Wilmington, North Carolina, 2/17/64; paid for service on CSS *Arctic*, North Carolina Squadron, 7/1–9/30/64; POW, Ft. Fisher, North Carolina, 1/15/65; arrived Point Lookout, Maryland, 1/22/65; requested release as a conscripted foreigner with no interest in the South, 4/10/65; took oath and released, 5/14/65; also found as "Pat. Megginnis."[262]

McGOWAN, JOHN: b. ca. 1837, Dundalk, Ireland; occ. seaman; single; resided and enlisted at New Orleans, Louisiana, for 1 year as a pvt. in Co. F, 1st (Nelligan's) Louisiana Infantry, 5/2/61; described as 5'10", fair comp., brown hair, gray eyes; present until discharged "having volunteered to go aboard the *Merrimack*" per SO 20, Dept. of Norfolk, 1/30/62; served as quarter gunner, CSS *Virginia*; reenlisted for the war, received $50 bounty as member of crew, *Virginia*, 3/25/62; paid for service on the *Virginia*, 4/1–5/12/62; superannuated for rations, Drewry's Bluff, Virginia, 5/12–24/62; on *Virginia* CRR, Drewry's Bluff, 7/20/62; paid, Drewry's Bluff, 5/13–9/30/62; listed as discharged on 12/31/62 payroll; NFR; may later enlisted for the war as a pvt. in Co. C, 10th Btn. Virginia Heavy Artillery, 10/16/62; NFR.[263]

McGraw, Patrick: enlisted at West Point, Virginia, as a pvt. in Co. H, 53rd Virginia Infantry, 7/26/61; "transferred to Steamer *Merrimac*" per SO 62, Dept. of Norfolk, 3/1/62; served as a landsman, CSS *Virginia*; superannuated for rations, Drewry's Bluff, Virginia, 5/12–24/62; on *Virginia* CRR, Drewry's Bluff, 7/20/62; paid, Drewry's Bluff, 5/12/62–6/20/63; listed as having deserted on payroll, dated 6/30/63; NFR.[264]

McHenry, William: served as an ordinary seaman, CSS *Virginia*; appears to have left crew of *Virginia* before 4/1/62; NFR; possibly, "William S. McHenry" who was b. ca. 1844, Madison, Georgia; occ. student; enlisted at Madison for the war as a pvt. in Co. G, Cobb's Legion Georgia Infantry, 7/29/61; present, 8/61; present, 1/1/62; in Winder Gen. Hosp., Richmond, Virginia, 10/23/62; discharged to accept a commission as an officer, per SO 262, A&IGO 62, para 10, 11/8/62; described then as 5'4½", light comp., light hair, blue eyes; NFR.[265]

McKue, John: b. ca. 1833; occ. miner; enlisted at Chester, Virginia, for 1 year as a pvt. in Co. I, 14th Virginia Infantry, 5/11/61; mustered in at Richmond, Virginia, 5/11/61; present until "transferred to naval service of Confederate States of America by order of Gen. Magruder and placed on board steamer *Merrimack*," 1/2/62; served as a seaman, CSS *Virginia* before 4/1/62; listed as having deserted on final *Virginia* payroll, dated 9/30/62; may have later enlisted at Naval Ordnance Works, Richmond, Virginia, as a pvt. in Co. A, 4th Btn. Virginia Infantry, Local Defense (Naval Btn.), 8/15/63; on CRR, 4th qtr. '64; NFR; also found as "McCue."[266]

McLaughlin, Ephraim Kirby: b. 10/30/35, Baltimore, Maryland; parents, Andrew McLaughlin and Frances Barnum; enlisted at Richmond, Virginia, for the war as a pvt. in Capt. Snowden's Co., 1st Btry., Maryland Artillery, 8/14/61; present, 9–11/61; discharged having provided S.R. Berry as a substitute, 12/7/61; reenlisted for the war, received $50 bounty as member of crew, *Virginia*, 3/25/62; assigned as acting master's mate, CSS *Virginia*, 4/4/62; paid thru 5/12/62; POW, USS *Susquehanna*, Norfolk, Virginia, '62; paroled, 5/11/62; NFR.[267]

McLaughlin, Peter: enlisted as a pvt. in Co. C, CSMC, 5/23/61; served at New Orleans Naval Station, Louisiana, 1/62; assigned to Marine Guard, CSS *Virginia* before 5/12/62; on Co. C CRR, 3rd qtr. '62; listed as having deserted on final *Virginia* payroll, dated 9/30/62; may have later served as a fireman, CSS *Webb*, '65; POW, Parish St. Bernard, Louisiana, 4/25/65; POW, New Orleans; paroled, 5/23/65.[268]
McMoore, Michael: shipped for 3 years as a landsman, Naval Rendezvous, Richmond, Virginia, 2/15/62; billeted on receiving ship, CSS *Confederate States;* served as a landsman, CSS *Virginia*; KIA during Battle of Drewry's Bluff, Virginia, 5/15/62; on *Virginia* muster roll in *ORN* as "Michael M. Moore."[269]
McNamee, Christopher: served as a 1st class fireman, CSS *Virginia*; reenlisted for the war, received $50 bounty as member of crew, *Virginia*, 3/25/62; paid for service on the *Virginia* and Drewry's Bluff, Virginia, 4/1/62–7/30/63; NFR; possibly "Christopher McNamee," who enlisted at New Orleans, Louisiana, for the war as a pvt. in Co. A, 15th Louisiana Infantry, 6/30/61; present on 9–10/61 rolls; NFR.[270]
McQuinn, John: served as a 1st class fireman, CSS *Virginia*; reenlisted for the war, received $50 bounty as member of crew, *Virginia*, 3/25/62; appears to have left crew of *Virginia* before 4/1/62; NFR; possibly "John E. McQuin," who was b. ca. 1840, New Orleans, Louisiana; occ. clerk; enlisted at New Orleans for the war as a pvt. in Co. A, 15th Louisiana Infantry, 6/30/61; described as 5'6", fair comp., black hair, brown eyes; present on 9–10/61 rolls; discharged on surgeon's certificate of disability, Portsmouth, Virginia, 12/11/61; NFR; also found as "McQueen."[271]
Meads, Charles C.: b. ca. 1838; shipped for the war as a yeoman, Naval Rendezvous, Richmond, Virginia; served as a yeoman, CSS *Virginia*; paid for service on the *Virginia*, 4/1–5/12/62; superannuated for rations, Drewry's Bluff, Virginia, 5/12–24/62; on *Virginia* CRR, Drewry's Bluff, 7/20/62; paid, Drewry's Bluff, 5/13–5/19/62; listed as discharged on final *Virginia* payroll, dated 9/30/62; NFR; may have been detailed from CSN at Rocketts Navy Yard, Richmond, Virginia, as a pvt. in Co. E, 4th Btn. Virginia Infantry, Local Defense (Naval Btn.), 6/23/63; present, 8/4/64; detailed to work as a carpenter at Rocketts Navy Yard, 1/27/65; NFR.[272]
Meiere, Julius Ernest: b. 11/25/33, New Haven, Connecticut; father, Julius Meiere, French & German instructor at Yale University; appointed by D.C. to USMC as a Lt., 4/16/55; took part in capture of Admiral Marin's Mexican Squadron; served as acting U.S. consul to Vera Cruz, Mexico; married Nannie Buchanan, daughter of Franklin Buchanan, at

Washington Navy Yard, 4/3/61; wedding attended by President Lincoln and Senator Stephen A. Douglas; assigned to Marine Barracks, Washington, D.C., at start of Civil War; tendered resignation, 4/23/61; dismissed from USMC, 5/6/61; appointed as a 1st Lt., CSMC, 5/8/61; assigned to Savannah Naval Station, '61–'62; capt., 12/5/61; served on CSS *Virginia* after Battle of Hampton Roads, Virginia; took part in Battle of Drewry's Bluff, Virginia, 5/15/62; commanded marine detachment, Mobile Naval Station, 10/8/62–8/64; captured, Mobile Bay, Alabama, 8/8/64; escaped prison, New Orleans, Louisiana, 10/13/64; surrendered, Mobile, Alabama, 5/4/65; released on oath, Key West, Florida, 5/20/65; postwar: graduated Medical College of University of New York, '69; practiced medicine in east then settled in Leadville, Colorado, '78; appointed city physician, '79; president, Lake Co. & Cripple Creek Medical Societies; d. of pneumonia, Sister's Hospital, Cripple Creek, Colorado, 12/3/1905.[273]

Mercer, James E.: paid for service as a seaman, CSS *Ellis*, North Carolina Squadron, 8/2–10/3/61; served as a seaman, CSS *Virginia*; paid for service on the *Virginia*, 4/1–5/12/62; superannuated for rations, Drewry's Bluff, Virginia, 5/12–24/62; on *Virginia* CRR, Drewry's Bluff, 7/20/62; paid, Drewry's Bluff, 5/13–9/13/62; discharged, 9/13/62; NFR.[274]

Mercer, Thomas P.: b. ca. 1826, North Carolina; occ. mariner; married; resided Elizabeth City, North Carolina; paid for service as a seaman, CSS *Ellis*, North Carolina Squadron, 8/2–10/3/61; served as a seaman, CSS *Virginia*; appears to have left crew of *Virginia* before 4/1/62; POW, Goldsboro, North Carolina, date unknown; NFR.[275]

Merriam, James: b. ca. 1836, England; occ. seaman; single; resided and enlisted at Mobile, Alabama, for 1 year as a pvt. in Co. E, 8th Alabama Infantry, 5/8/61; described as 5'6", fair comp., dark hair, hazel eyes; present thru 8/61; transferred to CSN by order of secretary of war; shipped for 3 years as a seaman, Naval Rendezvous, Richmond, Virginia, 2/14/62; served as a seaman, CSS *Virginia*; reenlisted for 3 years, received $50 bounty, 4/21/62; paid for service on the *Virginia* and Drewry's Bluff, Virginia, 4/1–5/23/62; listed as having deserted, 5/23/62; NFR; also found as "Merrian" and "Marriam," on *Virginia* muster roll in *ORN* as "James Marrian."[276]

Messick, Zadock Wesley: b. ca. 1837, York Co., Virginia; parents, John Messick and Ann Mariah Linton of Messick Point, York Co.; occ. sailor; single; resided and enlisted at Williamsburg, Virginia, for 1 year as a pvt. in Co. I, 32nd Virginia Infantry, 5/27/61; on convalescence leave at home, 6–8/61; "transferred to CSN to join the *Virginia*," 1/1/62; served as an ordinary seaman, CSS *Virginia*; paid for service on the *Virginia* and

Drewry's Bluff, Virginia, 4/1–8/28/62; superannuated for rations, Drewry's Bluff, 5/12–24/62; paid as a landsman, 8/29–9/30/62; discharged due to rheumatism; postwar: occ. waterman; resided Back River, Elizabeth City Co., Virginia; married Indiana Rowe, '86; resided Poquoson, Virginia, 1903; received pension, 1904; entered R.E. Lee Camp Soldiers' Home, Richmond, Virginia; d. 9/2/1916; bur., Hollywood Cem., Richmond, Virginia; on *Virginia* muster roll in *ORN* as "Wesley Messex."[277]

MILLER, LAFAYETTE: b. ca. 1838; occ. laborer; resided, Baltimore, Maryland: enlisted at Craney Island, Virginia, for 1 year as a pvt. in Co. B, 9th Virginia Infantry, 6/30/61; present thru 10/61; absent, in hosp., 11–12/61; discharged per SO 18, Dept. of Norfolk, 1/29/62; served as a landsman, CSS *Virginia*; appears to have left crew of *Virginia* before 4/1/62; NFR.[278]

MILLS, BENJAMIN B.: b. ca. 1843; resided Washington Co., North Carolina; enlisted at Plymouth, North Carolina, for the war as a pvt. in Co. G, 1st North Carolina Infantry, 6/24/61; present until transferred to CSN per SO 17, A&IGO 62, para 23, 2/3/62; served as a landsman, CSS *Virginia*; reenlisted for the war, received $50 bounty as member of crew, *Virginia*, 3/25/62; superannuated for rations, Drewry's Bluff, Virginia, 5/12–24/62; NFR.[279]

MINOR, ROBERT DABNEY: b. 9/21/27, "Sunning Hill", Louisa Co., Virginia; parents, Garrett Minor and Eliza McWilliams; resided Missouri; appointed by Missouri to USN as a midshipman, 2/25/41; passed midshipman, 8/10/47; married Ladonia Randolph, 12/17/50; assigned to Commodore Perry's flagship, USS *Susquehanna*, for visit to Japan, '53; master, 9/14/55; Lt., 9/15/55; assisted Matthew F. Maury at U.S. Naval Observatory; assigned to USS *Cumberland* at start of Civil War; tendered resignation and dismissed from USN, 4/22/61; joined Virginia State Navy, 4/61; appointed as a 1st Lt., CSN, 6/10/61; helped seize USS *St. Nicholas*, the brig *Monticello*, and schooner *Mary Pierce*, 6/29/61; assigned to CSN Bureau of Ordnance & Hydrography under older brother, George Minor, '61–'62; assisted Maury with electric torpedo experiments; worked with Gen. Robert E. Lee on construction of coastal defenses; led failed torpedo attack on Union warships at Newport News, 10/61; assigned Richmond Naval Station, 10–12/61; assigned as Flag Lt., CSS *Virginia*, 2/26/62; WIA leading boarding party to USS *Congress*; left crew, *Virginia*, 3/23/62; assisted John Mercer Brooke's work on a RR gun for Richmond & York River RR; commander, Richmond Naval Ordnance Works, 9/1/62; took part in Johnson's Island expedition, '63; commander, 4th Btn. Virginia Infantry, Local Defense (Naval Btn.) with rank of major,

CSA, 6/63; detailed, as Flag Lt., CSS *Virginia II*, James River Squadron, 5/64; took part in artillery duels with Union shore batteries, 8/17/64; assigned Richmond Naval Ordnance Works, 8/64; paroled, Richmond, Virginia, 5/3/65; postwar: worked for life insurance co., formed arms co. with Catesby Jones and John Mercer Brooke; later worked as an engineer, James River Improvement Co.; resided 6th Street, Richmond; d. of stroke, in home, 11/25/71; bur. prewar home, Fauquier Co., Virginia; reinterred beside wife, Hollywood Cem., Richmond, 1913.[280]

Mitchell, Richard Alex: b. ca. 1835, Caswell Co., North Carolina; occ. laborer; resided Alamance Co., North Carolina; enlisted at Graham, North Carolina, as a pvt. in Co. E, 13th North Carolina Infantry, 5/8/61; mustered in at Garysburg, North Carolina, 5/15/61; described as 6', fair comp., sandy hair, blue eyes; present until "discharged, reenlisted in naval service on the *Merrimac*," 2/19/62; served as a landsman, CSS *Virginia*; appears to have left crew of *Virginia* before 4/1/62; enlisted at Salisbury, North Carolina, for the war as a 3rd sgt. in Co. I, 57th North Carolina Infantry, 7/8/62; reduced in ranks to pvt., 3/63; KIA, Chancellorsville, Virginia, 5/4/63.[281]

Moore, George T.: b. ca. 1840, Twiggs Co., Georgia; occ. farmer; enlisted at Jeffersonville, Georgia, for 1 year as a pvt. in Co. C, 4th Georgia Infantry, 6/20/61; described as 5', light comp., light hair, gray eyes; present thru 11/61; in regt. hosp., Camp Jackson, Virginia, w/ int. fever, 11/23–24/61; present thru 1/62; granted furlough, 1/16–2/5/62; "transferred to *Merrimack*" per SO 12, Dept. of Norfolk, 2/10/62; served as a landsman, CSS *Virginia*; appears to have left crew of *Virginia* before 4/1/62; may have later served as an ordinary seaman, CSS *Atlanta*, Savannah River Squadron, '63; captured Warsaw Sound, 6/7/63; paroled, Ft. Norfolk, Virginia, 6/29/63; NFR; Georgia commissioner of pensions requested verification of service, 1916.[282]

Moore, John T.: b. ca. 1834, Twiggs Co., Georgia; occ. farmer; enlisted at Jeffersonville, Georgia, for 1 year as a pvt. in Co. C, 4th Georgia Infantry, 6/20/61; described as 6'1", dark comp., dark hair, gray eyes; present thru 9/61; in regt. hosp., Camp Jackson, Virginia w/ colica, 10/6–7/61; present until granted furlough in Georgia, 1/16–2/5/62; "transferred to *Merrimack*" per SO 12, Dept. Norfolk, 2/10/62; served as a landsman, CSS *Virginia*; superannuated for rations, Drewry's Bluff, Virginia, 5/12–24/62; reenlisted for the war as a pvt. in Co. C, 4th Georgia Infantry, 6/2/62; present thru 12/62; on CRR, 1st qtr. '63; KIA, Chancellorsville, Virginia, 5/5/63; Georgia commissioner of pensions requested verification of service, 1916.[283]

Moore, Robert W.: shipped for war, Naval Rendezvous, New Orleans, Louisiana, 2/24/62; served as a landsman, CSS *Virginia*; appears to have left crew of *Virginia* before 4/1/62; assigned to Jackson Naval Station, prior to 3rd quarter '62; NFR.[284]

Moriarty, Daniel: b. ca. 1839; enlisted at New Orleans, Louisiana, as a pvt. in Co. C, CSMC, 5/28/61; issued clothing in New Orleans, 1/62; assigned Marine Guard, CSS *Virginia* before 5/12/62; advertised as a deserter, 6/12/62; described then as 5'9", light comp., light hair, blue eyes; apparently RTD w/ Co. C; on Co. C CRRs, 3rd qtr. '62–8/2/63; on provision roll, CSS *Charleston*, Charleston Squadron, 8/4–15/63; NFR.[285]

Morris, William: b. ca. 1835, Tyrrell Co., North Carolina; occ. shoemaker; resided Tyrrell Co.; enlisted at Plymouth, North Carolina, for the war as a pvt. in Co. G, 1st North Carolina Infantry, 6/24/61; present until transferred to CSN per SO 17, A&IGO 62, para 23, 2/3/62; served as a landsman, CSS *Virginia*; reenlisted for the war, received $50 bounty as member of crew, *Virginia*, 3/25/62; paid for service on the *Virginia*, 4/1–5/12/62; superannuated for rations, Drewry's Bluff, Virginia, 5/12–24/62; on *Virginia* CRR, Drewry's Bluff, 7/20/62; paid, Drewry's Bluff, 5/13/62–3/31/64; served as an ordinary seaman, CSS *Fredericksburg*, James River Squadron, '64–'65; ordered to Naval Ordnance Works, Richmond, Virginia, 1/13/65; NFR.[286]

Morton, Edwin: b. ca. 1830, Napoleonville, Louisiana; occ. barkeeper; resided and enlisted at New Orleans, Louisiana, for 1 year as a corp. in Co. C, 1st (Nelligan's) Louisiana Infantry, 5/27/61; described as 5'4½", dark comp., brown hair, blue eyes; present thru 10/61; listed as "present, reduced from sgt., 11/15/61, now in ranks," on 11–12/61 rolls; "discharged by S.O. of Gen. Huger to join the *Merrimac*," 1/2/62; served as a landsman, CSS *Virginia*; reenlisted for the war, received $50 bounty as member of crew, *Virginia*, 3/25/62; paid for service on the *Virginia*, 4/1–5/12/62; superannuated for rations, Drewry's Bluff, Virginia, 5/12–24/62; on *Virginia* CRR, Drewry's Bluff, 7/20/62; paid, Drewry's Bluff, 5/13/62–9/20/63; paid for service, naval det. on special expedition from Drewry's Bluff to Charleston Naval Station, 10/1–31/63; paid for service as an ordinary seaman, CSS *Albemarle*, North Carolina Squadron, 7/1–9/30/64; captured, Plymouth, North Carolina, 10/31/64; POW, Camp Hamilton, Virginia, 11/16/64; sent to Point Lookout, Maryland, 11/25/64; took oath and paroled, 5/14/65; on *Virginia* muster roll in *ORN* as "Edward Morten."[287]

Mowle, Jacob R.: b. ca. 1836, New York; enlisted at Norfolk, Virginia, for 1 year as a pvt. in the United Artillery (1st Co. E, 41st Virginia Infantry),

4/19/61; present thru 3/6/62; "volunteered to go onboard the Ironclad Steamer *Virginia*," 3/6/62; manned gun on CSS *Virginia* during Battle of Hampton Roads, Virginia, 3/8–9/62; reenlisted for war, received $50 bounty, 3/10/62; on CRRs, 12/22/62, 6/3/63, 9/14/63, 11/2/63 and 3/18/64; in Chimborazo Hosp. #2, Richmond, Virginia, 7/1/64; in Chimborazo Hosp. #2 w/ int. fever, 8/10–10/17/64; in Chimborazo Hosp. #1, 12/6/64–2/7/65; deserted, entered Federal lines, Bermuda Hundred, Virginia, 4/5/65; sent to City Point, Virginia, and then Washington, D.C.; took oath, furnished transport to Norfolk; postwar: member, Pickett-Buchanan Camp, UCV, Norfolk; entered R.E. Lee Camp Soldiers' Home, Richmond, 5/13/1902; d. 5/5/1913; buried, Hollywood Cem., Richmond.[288]

Muirhead, Philip T.: b. ca. 1834, Mississippi; occ. laborer; enlisted at New Orleans, Louisiana, for 1 year as a pvt. in Co. A., 2nd Louisiana Infantry, 5/11/61; described as 6', dark comp., black hair, black eyes; present until "transferred to Naval Dept. by order of Secretary of War," 3/18/62; served as a landsman, CSS *Virginia*; appears to have left crew of *Virginia* before 4/1/62; enlisted at Richmond, Virginia, for the war as a pvt. in Co. F, 2nd Btn. Maryland Infantry, 5/23/62; present, 3/31/63; received $50 bounty; received 30-day furlough, 8/27/63; WIA, Cold Harbor, Virginia, 6/3/64; in Chimborazo Hosp. #5, Richmond, 6/5/64; listed as absent, from wounds, on 9–10/64 and 1–2/65 rolls; NFR.[289]

Mulroy, John: shipped for the war as a fireman, CSS *Forrest*, North Carolina Squadron, 7/25/61; served as a 1st class fireman, CSS *Virginia*; reenlisted for the war, received $50 bounty; paid for service on the *Virginia*, 4/1–5/12/62; superannuated for rations, Drewry's Bluff, Virginia, 5/12–24/62; on *Virginia* CRR, Drewry's Bluff, 7/20/62; paid, Drewry's Bluff, 5/13–11/30/62; NFR.[290]

Murphy, John: b. ca. 1832, Ireland; occ. laborer; resided New Orleans, Louisiana; single; enlisted at Camp Moore, Louisiana, for the war as a pvt. in Co. G, 10th Louisiana Infantry, 7/22/61; present until "transferred to *Merrimac*," 1/2/62; served as a seaman, CSS *Virginia*; paid for service as an ordinary seaman on the *Virginia*, 4/1–5/12/62; superannuated for rations, Drewry's Bluff, Virginia, 5/12–24/62; paid, Drewry's Bluff, 5/13/62–1/15/63; in Bruton Parish Episcopal Church Hosp., Williamsburg, Virginia, w/ int. fever, 2/6–26/63; may have later served as a seaman, CSS *Arctic*, North Carolina Squadron, '63; paid, Drewry's Bluff, 4/1–9/30/63 and 1/1–3/31/64; served on CSS *Virginia II*, James River Squadron, '65; deserted, Bermuda Hundred, Virginia, 3/13/65; sent to City Point, Virginia; took oath and sent to Washington, D.C.; furnished transport to New York City.[291]

Murphy, Patrick: shipped for 1 year as a landsman, Naval Rendezvous, Richmond, Virginia, 7/1/61; served as a landsman/2nd class fireman, CSS *Patrick Henry*, James River Squadron, '61; assigned to CSS *Jamestown*, James River Squadron, 1/10/62; paid, North Carolina Squadron, 1/15–2/28/62; served as an ordinary seaman, CSS *Virginia*; reenlisted for the war, received $50 bounty as member of crew, *Virginia*, 3/25/62; paid, Drewry's Bluff, Virginia, 6/1–11/30/62; may have later served as a coal heaver, CSS *Missouri*, *Webb*, and *Cotton*, Red River Squadron, '63; served as a 2nd class fireman, CSS *Virginia II*, James River Squadron, '64–'65; NFR.[292]

Myers, Joseph H.: b. ca. 1842, resided Washington Co., North Carolina; enlisted at Plymouth, North Carolina, for the war as a pvt. in Co. G, 1st North Carolina Infantry, 6/24/61; present until transferred to CSN per SO 17, A&IGO 62, para 23, 2/23/62; served as a landsman, CSS *Virginia*; reenlisted for the war, received $50 bounty as member of crew, *Virginia*, 3/25/62; superannuated for rations, Drewry's Bluff, Virginia, 5/12–24/62; NFR; may have later enlisted at Richmond, Virginia, for the war as a pvt. in Co. B, 24th Btn. Virginia Partisan Rangers, 7/20/62; discharged when unit disbanded by order of the secretary of war, 1/3/63; described then as age 17, occ. farmer; 5'7½", dark comp., light hair, black eyes; NFR.[293]

Nance, Richard: b. ca. 1822, Virginia; occ. sailor; married; resided Warwick Co., Virginia; enlisted at Norfolk, Virginia, for 1 year as a pvt. in the United Artillery (1st Co. E, 41st Virginia Infantry), 4/19/61; described as 5'8", blond hair, blue eyes; present until "detailed to go on *Merrimac*" per SO 9, Dept. of Norfolk, 1/16/62; enlisted in CSN; billeted on receiving ship, CSS *Confederate States*; manned gun on CSS *Virginia* during Battle of Hampton Roads, Virginia, 3/8–9/62; paid for service on the *Virginia* and Drewry's Bluff, Virginia, 4/1–8/31/62; reenlisted for the war, received $50 bounty; on *Virginia* CRR, Drewry's Bluff, 7/20/62; shipped for the war, Naval Rendezvous, Richmond, Virginia, 7/17/62; listed as having transferred as a 1st class fireman, CSS *Beaufort*, on *Virginia* muster roll, 7/31/62; in Howard's Grove Hosp., Richmond, Virginia, w/ varioloid, 12/12/62; d. 12/27/62; wife, Ann, received pension in '88; also found as "Nantz."[294]

Naughton, John: b. ca. 1832; occ. miner; enlisted at Chester, Virginia, as a pvt. in Co. I 14th Virginia Infantry, 5/11/61; present 7–8/61; arrested for sleeping on post, 11/19/61; transferred to CSN by order of Gen. Magruder to serve on ironclad CSS *Virginia*, 3/25/62; superannuated for rations, Drewry's Bluff, Virginia, 5/12–24/62; listed as having deserted on final *Virginia* payroll, dated 9/30/62; NFR; also found as "Norton;" on *Virginia* muster roll in *ORN* as "John Naughton."[295]

Nelson, John: shipped for the war as a seaman, CSS *Forrest*, North Carolina Squadron, 7/25/61; served as a seaman, CSS *Virginia*; reenlisted for the war, received $50 bounty as member of crew, *Virginia*, 3/25/62; paid for service on the *Virginia*, 4/1–5/12/62; superannuated for rations, Drewry's Bluff, Virginia, 5/12–24/62; on *Virginia* CRR, Drewry's Bluff, 7/20/62; paid, Drewry's Bluff, 5/13/62–6/30/63 and 10/1/63–3/31/64; paid for service as a seaman, CSS *Patrick Henry*, James River Squadron, 10/1–12/31/64; deserted, Washington, D.C., 4/17/65; furnished transport to New York City.[296]

Nelson, Rambling W.: b. ca. 1834, Florida; occ. merchant; enlisted at Jeffersonville, Georgia, for 1 year as a pvt. in Co. C, 4th Georgia Infantry, 6/20/61; described as 6'2", florid comp., red hair, blue eyes; present thru 8/61; in regt. hosp., Camp Jackson, Virginia, w/ diarrhea, 9/25/61; in regt. hosp. w/ diarrhea and hepatitis, 10/4/61; sent to Gen. Hosp., 10/5–11/12/61; present until transferred to CSS *Virginia* per SO 12, Dept. Norfolk, 2/10/62; served as a landsman, CSS *Virginia*; appears to have left crew of *Virginia* before 4/1/62; enlisted in 1st Maryland Cavalry, 5/11/62; POW, Bermuda Hundred, Virginia, 10/9/64; arrived Washington, D.C., 10/12/64; took oath and released; furnished transport to Baltimore, Maryland; postwar: resided Gainesville, Florida, 1905; Georgia commissioner of pensions requested verification of service, 1916; also found as "Robert Nelson;" on *Virginia* muster roll in *ORN* as "Reinbeim W. Nelson."[297]

Noon, Patrick: b. ca. 1833, Ireland; occ. laborer; married; resided New Orleans, Louisiana; enlisted at Camp Moore, Louisiana, for the war as a pvt. in Co. C., 10th Louisiana Infantry, 7/22/61; present until "transferred to Steamer *Virginia*," 1/3/62, formally transferred to CSN, 2/62; served as a coal heaver, CSS *Virginia*; listed as having deserted on final *Virginia* payroll, dated 9/30/62; NFR.[298]

Norris, William H.: served as master at arms, CSS *Virginia*; appears to have left crew of *Virginia* before 4/1/62. NFA.[299]

O'Haloran, Dixon Richardson: b. ca. 1840, Ireland; occ. laborer; enlisted at Camp McDonald, Georgia, for the war as a pvt. in Co. A, 22nd Georgia Infantry, 8/31/61; described as 5'8", fair comp., dark hair, blue eyes; deserted while on furlough in Charleston, South Carolina, 9/61; arrested and returned to unit in chains; transferred to CSN, 1/15/62; paid as an ordinary seaman, CSS *Virginia*, 1/16–2/18/62; NFR; Georgia commissioner of pensions requested verification of service, 1916; on *Virginia* muster roll in *ORN* as "Dixon R. O'Halloran."[300]

Oliver, Charles B.: b. ca. 1821, Massachusetts; appointed by Virginia as a master's mate, USN, 5/3/43; gunner, 6/1/46; married; resided Elizabeth River Parish, Norfolk Co., Virginia, '60; tendered resignation and dismissed from USN, 4/21/61; appointed gunner, CSN, 6/11/61; assigned to *Confederate States* and Gosport Navy Yard, '61; served on CSS *Virginia* during Battle of Hampton Roads, Virginia, 3/8–9/62; uncapped powder in *Virginia's* magazine when it was destroyed, 5/12/62; assigned to Drewry's Bluff, Virginia until 8/11/62; served on CSS *Richmond*, James River Squadron, '62–'63; Lt. for the war, 5/5/63; ordnance officer, Savannah River Squadron, '63–'64; 1st Lt., 1/6/64, 1st Lt., Provisional Navy, 6/2/64; postwar: occ. seaman; resided Richmond Co., Virginia, '70; member, Pickett-Buchanan Camp, UCV, Norfolk, Virginia; alive '86.[301]

Oliver, John: served as a landsman, CSS *Virginia*; appears to have left crew of *Virginia* before 4/1/62; NFR; possibly "John E. Oliver' who was b. ca. 1841, Caroline Co., Virginia; father, Reuben Oliver; occ. mason; single; enlisted at Jamestown, Virginia, for 1 year as a pvt. in Co. B, Virginia Hanover Light Artillery (Nelson's Co., Virginia Light Artillery), 7/22/61; described as 5'7¾", light comp., light hair, blue eyes; granted furlough, 2/28/62; d. Richmond, Virginia, 5/10/62; claim filed, certificate #12303, 7/17/63; $52 found due, 2/10/64.[302]

Oliver, Joseph L.: b. ca. 1842, North Carolina; resided Washington Co., North Carolina; occ. day laborer; married; enlisted at Plymouth, North Carolina, for the war as a pvt. in Co. G, 1st North Carolina Infantry, 6/24/61; present until transferred to CSN per SO 17, A&IGO 62, para 23, 2/3/62; served as a landsman, CSS *Virginia*; reenlisted for the war, received $50 bounty as member of crew, *Virginia*, 3/25/62; appears to have left crew of *Virginia* before 4/1/62; NFR.[303]

Ollsen, Jacob: served as a seaman, CSS *Virginia;* paid, Drewry's Bluff, Virginia, 5/12/62–6/30/63; on *Virginia* CRR, Drewry's Bluff, 7/20/62; paid for service, naval det. on special expedition from Drewry's Bluff to Charleston Naval Station, 10/1–31/63; NFR; may have later served as a quarter master, CSS *Virginia II*, James River Squadron, '64–'65, as "H. J. Olson."[304]

Palmer, Charles K.: enlisted as a pvt. in Co. K, 32nd Virginia Infantry, 5/14/61; joined Co. F. after Co. K disbanded, 6/15/61; transferred to CSS *Virginia*, 3/62; rejoined Co. K, 32nd Virginia Infantry, 8/31/62; NFR.[305]

Parish, Hillsman: b. ca. 1842, Raleigh, North Carolina; occ. seaman; resided and enlisted at Beaufort Co., North Carolina, for the war as a pvt. in Co. I, 3rd North Carolina Infantry, 5/10/61; described as 5'4", dark comp., dark hair, dark eyes; present until transferred to CSN per SO 17, A&IGO 62, para 23, 1/29/62; served as an ordinary seaman, CSS *Virginia*; reenlisted for the war, received $50 bounty as member of crew, *Virginia*, 3/25/62; superannuated for rations, Drewry's Bluff, Virginia, 5/12–24/62; enlisted at Charlottesville, Virginia, for the war as a pvt. in Co. B, 39th Btn. Virginia Cavalry, 8/25/62; present, 11–12/62; lost Sharps carbine, fined $45; present, 1–10/63; transferred to Maryland Line, 4/1/64; transfer revoked 4/6/64; POW, Petersburg, West Virginia, 7/28/64; took oath, Berlin, Maryland, 11/14/64; listed residence as New Bern, North Carolina, and destination as Norfolk, Virginia; also found as "Hilsman Parrish"; on *Virginia* muster roll in *ORN* as "Hinsman Parrish."[306]

Parrish, William: b. ca. 1816, Virginia; occ. pilot; married; served as acting master, CSN, 10/14/61; served at Gosport Navy Yard, '61; assigned to CSS *Virginia*, 1/1/62; served during Battle of Hampton Roads, Virginia, 3/8–9/62; discharged 5/14/62; NFR.[307]

Patrick, James A.: served as a landsman, CSS *Virginia*; reenlisted for the war, received $50 bounty as member of crew, *Virginia*, 3/25/62; superannuated for rations, Drewry's Bluff, Virginia, 5/12–24/62; NFR.[308]

Perkinson, Edward: b. ca. 1835, Chesterfield Co., Virginia; occ. painter; resided and enlisted at Petersburg, Virginia, for 1 year as a pvt. in Co. C, 41st Virginia Infantry, 5/9/61; described as 5'7½", light comp., red hair; mustered in, 5/11/61; absent, sick in quarters, 11–12/61 rolls; listed as AWOL, 2/24/62; discharged by SO 69, Dept. of Norfolk, 3/29/62; served as a landsman, CSS *Virginia*; appears to have left crew of *Virginia* before 4/1/62; may have later enlisted at Petersburg for the war as a pvt. in 2nd Co. E, 10th Virginia Cavalry, 11/1/63; on CRR, 6/1/64; listed as AWOL, 1/27/65; NFR; postwar: resided Little Church Road, Petersburg; alive, 7/1915.[309]

Perry, John: b. ca. 1825, Pennsylvania; resided and enlisted at Mobile, Alabama, for the war as a pvt. in Co. A, 12th Alabama Infantry, 6/4/61; discharged and transferred to CSN per SO 89, A&IGO 62, para 10, 3/27/62; shipped for 1 year, Naval Rendezvous, Richmond, Virginia, 3/31/62; served as a seaman, CSS *Virginia*; reenlisted for 3 years, received $50 bounty, 4/21/62; paid for service on the *Virginia* and Drewry's Bluff, Virginia, 4/1–6/30/62; volunteered for duty, CSS *Chattahoochee*, 8/1/62; served as master of arms, CSS *Chattahoochee* thru 6/12/63; transferred to CSS *Savannah*, Savannah River Squadron, 6/13/63; served as a boatswain's mate, CSS *Chicora*, Charleston Squadron, '63; captured, Morris Island, South Carolina, 9/7/63; POW, Hilton Head, South Carolina; sent to Ft. Columbus, New

York, 10/9/63; sent to Point Lookout, Maryland, 10/25/63; took oath and paroled, 2/12/64; acting boatswain's mate, '64; took part in expedition to capture USS *Water Witch*, 6/3/64; boatswain, date unknown; commanded Shell Bluff Battery, Savannah Naval Station, '65; NFR; postwar, reportedly resided Philadelphia, Pennsylvania, 1910.[310]

Peterson, Andrew G.: b. ca. 1835, England; shipped for 1 year as a seaman, Naval Rendezvous, New Orleans, Louisiana, 9/27/61; paid for service, New Orleans Naval Station, 9/27/61–3/31/62; served as a boatswain's mate, CSS *Virginia*; reenlisted for the war, received $50 bounty as member of crew, *Virginia*, 3/25/62; paid for service on the *Virginia*, 4/1–5/12/62; superannuated for rations, Drewry's Bluff, Virginia, 5/12–24/62; on *Virginia* CRR, Drewry's Bluff, 7/20/62; paid, Drewry's Bluff, 5/13/62–3/31/63; listed as having deserted, 1/1/63; NFR.[311]

Pettit, Edmond: enlisted at Mobile, Alabama, for the war as a pvt. in Co. A, 12th Alabama Infantry, 6/4/61; present on 7–8/61 rolls; transferred to CSN per SO 89, A&IGO 62, para 10, 3/27/62; shipped for 1 year as an ordinary seaman, Naval Rendezvous, Richmond, Virginia, 3/31/62; served as an ordinary seaman, CSS *Virginia*; reenlisted for 3 years, received $50 bounty, 4/21/62; NFR; on *Virginia* muster roll in *ORN* as "Edmond Pettet."[312]

Phillips, Dinwiddie Brazier: b. 6/7/26, Fauquier Co., Virginia; parents, Col. William Fowke Phillips and Edith Harrison Ashmore Cannon; graduated, Jefferson Medical College, Philadelphia, Pennsylvania; appointed by Virginia to USN as an assistant surgeon, 11/8/47; married; resided in Portsmouth, Virginia; passed assistant surgeon, USN; assigned to Portsmouth Navy Hospital at start of Civil War; tendered resignation 4/18/61; resignation accepted, 5/6/61; appointed as a surgeon, CSN, 6/10/61; medical director, Wise Legion, White Sulphur Springs, West Virginia; served on CSS *Nashville*, '61; assigned to Richmond Naval Station, 10/1/61–1/26/62; assigned to CSS *Virginia*, 1/27/62; took part in Battle of Hampton Roads, Virginia, 3/8–9/62; served on CSS *Richmond*, James River Squadron, '62–'63; served on CSS *Tennessee*, Mobile Squadron, '63–'64; NFR; postwar: married Minnie F. Walden; resided near Madison Run station, Orange Co., Virginia.[313]

Phillips, Jerome B.: enlisted at Memphis, Tennessee, as a pvt. in CSMC, 9/2/61; probably transferred from Capt. Hays's det., Co. D, to Co. C, Pensacola Navy Yard, ca. 11/61; transferred with Co. C to Gosport Navy Yard, Portsmouth, Virginia, ca. 11/29/61; assigned to Marine Guard, CSS *Virginia*, ca. 11/61–ca. 5/12/62; on CRR, 3rd qtr. '62; NFR.[314]

Pitt, Lorenzo D.: b. ca. 1843, Portsmouth, Virginia; occ. apprentice painter; single; resided Portsmouth; enlisted at Naval Hosp., Portsmouth for 1 year as a pvt. in Co. D, 9th Virginia Infantry, 4/27/61; mustered in at Naval Hosp., 5/5/61; described as 5'9", light comp., light hair, blue eyes; present until discharged per SO 18, Dept. of Norfolk, 1/29/62; served as an ordinary seaman, CSS *Virginia*; paid for service on the *Virginia*, 4/1–5/12/62; reenlisted for the war, received $50 bounty; on *Virginia* CRR, Drewry's Bluff, Virginia, 7/20/62; shipped for the war as an ordinary seaman, Naval Rendezvous, Richmond, Virginia, 7/21/62; paid, Drewry's Bluff, 5/13–11/30/62; served on CSS *Patrick Henry*, James River Squadron, '64; acting master's mate, date unknown; served on CSS *Albemarle*, North Carolina Squadron, '64; acting master's mate, Provisional Navy, 6/2/64; name found on list of deserters at Suffolk, Virginia, 4/5/65; NFR; postwar: occ. painter; married; resided Portsmouth; d. 1883; bur. Oak Grove Cem., Portsmouth.[315]

Porter, Christopher: b. ca. 1833, Ireland; occ. mechanic; resided Norfolk, Virginia; enlisted at Craney Island, Norfolk Co., Virginia, for 1 year as a pvt. in Co. H, 6th Virginia Infantry, 5/8/61; mustered in at Craney Island, 6/30/61; described as 5'4", ruddy comp., red hair, gray eyes; present until discharged per SO 69, Dept. of Norfolk, 3/28/62; served as a landsman, CSS *Virginia*; appears to have left crew of *Virginia* before 4/1/62; NFR; may have later enlisted at Richmond, Virginia, for the war as a pvt. in Co. C, 1st Btn. Virginia Infantry, 11/9/63; present, 11–12/63, on CRR, 7/1/64; present, 9/64–2/28/65; POW, Petersburg, Virginia, 4/3/65; sent to City Point, Virginia; arrived Hart's Island, New York, 4/7/65; took oath and released, 6/20/65; described then as 5'5", light comp., dark hair, hazel eyes, residence Norfolk.[316]

Powers, James: served as an ordinary seaman, CSS *Virginia;* reenlisted for the war, received $50 bounty as member of crew, *Virginia*, 3/25/62; appears to have left crew of *Virginia* before 4/1/62; shipped for the war, Naval Rendezvous, Richmond, Virginia, 8/3/62; paid for service as a landsman, CSS *Tuscaloosa*, Mobile Squadron, 4/1–5/20/63; listed as having deserted, Savannah, Georgia, '65; NFR; possibly "James Powers," who enlisted at New Orleans, Louisiana, for the war as a pvt. in Co. A, 15th Louisiana Infantry, 6/30/61; present, 9–10/61; granted 40-day furlough from Jackson Hosp., Richmond, Virginia, 8/5/63; NFR.[317]

Powers, William Riley: b. ca. 1838, Buncombe Co., North Carolina; occ. farmer; resided Buncombe Co.; enlisted at Asheville, North Carolina, as a pvt. in Co. F, 14th North Carolina Infantry, 5/3/61; mustered in at Raleigh, North Carolina, 5/23/61; described as 6', light comp., sandy hair, gray eyes; sick, in camp, 7–8/61 rolls; present until "discharged by order of Gen. Huger to go onboard the *Merrimac*," 2/18/62; served as a landsman, CSS *Virginia*; reenlisted for the war, received $50 bounty; paid for service on the *Virginia*, 4/1–5/12/62; on *Virginia* CRR, Drewry's Bluff, Virginia, 7/20/62; paid, Drewry's Bluff, 5/13–12/31/62; listed as having deserted, 1/5/63; enlisted at Asheville for the war as a pvt. in Co. D, 7th Btn. North Carolina Cavalry, 5/1/63; present until transferred to Co. C, 6th North Carolina Cavalry, 8/3/63; present thru 12/64; NFR.[318]

Price, William M.: b. ca. 1837, Edgecombe Co., North Carolina; occ. carpenter and cabinetmaker; resided Edgecombe Co.; enrolled at Tarboro, North Carolina, for 1 year as a pvt. in Co. G, 13th North Carolina Infantry, 5/8/61; mustered in Garysburg, North Carolina, 5/10/61; described as 5'6", dark comp., black hair, black eyes; present until "transferred to Confederate Navy," 2/20/62; served as a landsman, CSS *Virginia*; paid, Drewry's Bluff, Virginia, 5/12/62–1/5/63; on *Virginia* CRR, Drewry's Bluff, 7/20/62; paid, Drewry's Bluff, 4/1–6/30/63; listed as having deserted on 6/30/63 payroll; NFR.[319]

Pritchett, Adam: b. ca. 1835, Washington, North Carolina; resided Washington Co.; enlisted at Plymouth, North Carolina, for the war as a pvt. in Co. G, 1st North Carolina Infantry, 6/24/61; present until transferred to CSN, 2/3/62; served as an ordinary seaman, CSS *Virginia*; reenlisted for the war, received $50 bounty as member of crew, *Virginia*, 3/25/62; appears to have left crew of *Virginia* before 4/1/62; NFR; on *Virginia* muster roll in *ORN* as "Adam Pritchard."[320]

Pryde, Nicholas B.: b. ca. 1838; New York; occ. sailor; married; resided and enlisted at New Orleans, Louisiana, for the war as a pvt. in Co. D, 15th Louisiana Infantry, 7/18/61; present, 7–10/61; "transferred to the Navy," 2/62 roll; served as a sail-master's mate, CSS *Virginia*; reenlisted for the war, received $50 bounty as member of crew, *Virginia*, 3/25/62; paid for service on the *Virginia*, 4/1–5/12/62; listed as transferred on final *Virginia* payroll, dated 9/30/62; served as a quarter gunner on the cruiser, CSS *Florida*, '62–'63; took part in *Clarence-Tacony-Archer* expedition off Brazilian coast, '63; promoted acting master's mate by Lt. C.W. Read, CSN, ca. 5/6/63; later served on *Clarence*; captured off Portland, Maine, 6/27/63; confined at Ft. Preble, Maine, escaped from Ft. Warren, Massachusetts, 8/10/63; NFR; on *Virginia* muster roll in *ORN* as "Nicholas P. Pryde."[321]

Purcell, Thomas: enlisted at Montgomery, Alabama, as a pvt. in CSMC, 4/12/61; probably served at Pensacola Navy Yard until Co. C transferred to Gosport Navy Yard, Portsmouth, Virginia, 11/29/61; assigned to Marine

Guard, CSS *Virginia*, ca. 11/61–5/12/62; paid for service, CSS *Virginia*, 1/1–5/12/62; on CRR, 3rd qtr. '62; NFR.[322]

Rainey, Theophilus: b. ca. 1841, Prince George Co., Virginia; occ. laborer; resided Petersburg, Virginia, Plum St. between Cross and Canal Streets; enlisted at Petersburg for 1 year as a pvt. in Co. K, 12th Virginia Infantry, 5/4/61; described as 6', light comp., light hair, gray eyes; present until discharged for naval service per SO 69, Dept. of Norfolk, 3/28/62; served as a landsman, CSS *Virginia*; appears to have left crew of *Virginia* before 4/1/62; enlisted at Petersburg for the war as a pvt. in the Petersburg Horse Artillery (Graham's Co., Virginia Horse Artillery), 6/25/62; served as a teamster, 2/1/63–3/23/65; NFR; postwar: resided Petersburg; received pension, 4/30/1902.[323]

Ramsay, Henry Ashton: b. Washington, D.C.; parents, Charles Rufus Ramsay and Caroline Henry Ashton; appointed by D.C. to USN as 3rd assistant engineer, 5/21/53; 2nd assistant engineer, 6/26/56; served on USS *Merrimack* under chief engineer Alban C. Stimers (chief engineer, USS *Monitor* during Battle of Hampton Roads, Virginia), '59; 1st assistant engineer, 8/2/59; refused to take oath to U.S. and dismissed from USN, 5/6/61; appointed as engineer, CSN, 6/19/61; served at Gosport Navy Yard, '61; married Julia Cooke of Norfolk, daughter of Dr. Armistead T. Mason Cooke, '61; acting chief engineer, 12/2/61; assigned to CSS *Virginia*, 1/1/62; took part in Battle of Hampton Roads, Virginia, 3/8–9/62; assigned to Charlotte Naval Works, '62–'65; transferred to CSA as Lt. Col., '65; surrendered with Gen. Johnston's army; paroled, Charlotte, North Carolina, 5/3/65; postwar: occ. consulting engineer; resided Baltimore, Maryland; member, Buchanan Camp, UCV, Baltimore; d. at home, Baltimore, Maryland, 3/25/1916.[324]

Reardon, John W.: b. ca. 1840; occ. farmer; enlisted at Jacksonport, Arkansas, for 1 year as a pvt. in 1st (Colquitt's) Arkansas Infantry, 5/15/61; mustered in at Lynchburg, Virginia, 5/14/61; reenlisted for 2 years, received $50 bounty, 1/23/62; transferred to CSN per SO 17, A&IGO 62, para 23, 2/3/62; served as an ordinary seaman, CSS *Virginia*; reenlisted for the war, received $50 bounty as member of crew, *Virginia*, 3/25/62; superannuated for rations, Drewry's Bluff, Virginia, 5/12–24/62; NFR.[325]

Rice, Robert J.: served as a seaman/signal quarter master on CSS *Virginia*; slightly WIA during battle w/ USS *Monitor*, 3/9/62; reenlisted for the war, received $50 bounty as member of crew, *Virginia*, 3/25/62; listed as POW on *Virginia* muster roll, 7/31/62; however, paid for service on the *Virginia* and Drewry's Bluff, Virginia, 4/1/62–3/31/64; paid as a landsman starting 7/25/62; NFR; probably "Robert J. Rice," who enlisted at Lynchburg,

Virginia, as a corp. in the Lynchburg Artillery (Co. D, 38th Btn. Virginia Light Artillery), 4/23/61; absent, sick in hosp., Richmond, Virginia, 7–10/61 roll; discharged, 10/9/61.[326]

RICHARDSON, BENJAMIN ADWORTH: enlisted at Norfolk, Virginia, as a pvt. in the United Artillery (1st Co. C, 19th Btn. Virginia Heavy Artillery); 4/19/62; manned a gun on CSS *Virginia*; co. became an ind. btry., 10/1/62; on CRRs, 12/22/62, 6/3/63, 9/14/63, 11/2/63, and 3/18/64; in Chimborazo Hosp. #5, Richmond, Virginia, w/ int. fever, 8/9–10/8/64; absent sick, 10/10/64; present, 12/31/64; NFR.[327]

RIDDOCK, JOSEPH: b. ca. 1839, Yorkville, New York; occ. gas fitter; enlisted at Charleston, South Carolina, for the war as a corp. in Co. M, 1st South Carolina Infantry, 4/22/61; described as 5'7", light comp., auburn hair, gray eyes; mustered out at Richmond, Virginia, 7/9/61; enlisted at Richmond as a pvt. in Co. I, 1st (McCreary's) South Carolina Infantry, 7/20/61; 5th sgt., 9/7–12/31/61; 4th sgt., 12/31/61–1/17/62; "discharged by reason of order of Gen. Huger," 1/17/62; served as a landsman, CSS *Virginia*; paid for service on the *Virginia* and Drewry's Bluff, Virginia, 4/1–8/15/62; shipped for the war as a landsman, Naval Rendezvous, Richmond, 7/21/62; listed as discharged on final *Virginia* payroll, dated 9/30/62; may have later been captured on steamer *Mary Ann* by USS *Grand Gulf*, 3/18/64; POW, Camp Hamilton, Virginia; released, 4/24/64; NFR.[328]

RILEY, JOHN G.: b. ca. 1842, Ireland; occ. laborer; resided Cumberland Co., North Carolina; enlisted at Fayetteville, North Carolina, for the war as a pvt. in Co. C, 3rd North Carolina Infantry, 5/29/61; described as 5'9", fair comp., dark hair, hazel eyes; present until transferred to CSN per SO 17, A&IGO 62, para 23, 1/7/62; served as an ordinary seaman, CSS *Virginia* before 4/1/62; listed as having deserted on final *Virginia* payroll, dated 9/30/62; may have later served as a landsman, CSS *Tennessee*, Mobile Squadron, '64; POW, Mobile Bay, Alabama, 8/5/64; in St. Louis U.S. Army Gen. Hosp., New Orleans, Louisiana, w/ fistula, 9/27–10/22/64; sent to Ship's Island, Mississippi; exchanged 3/4/65; NFR.[329]

RILEY, OWEN: b. ca. 1836, Ireland; occ. iron molder; resided Baltimore, Maryland; served as a coal heaver, CSS *Virginia*; paid for service on the *Virginia*, 4/1–5/12/62; superannuated for rations, Drewry's Bluff, Virginia, 5/12–24/62; on *Virginia* CRR, Drewry's Bluff, 7/20/62; paid, Drewry's Bluff, 5/13–9/6/62 and 10/1/62–7/31/63; served as a fireman, CSS *Charleston*, Charleston Squadron, '65; deserted, Charleston, South Carolina, 2/18/65; sent to Hilton Head, South Carolina; described then as 5'11", dark comp., dark hair, hazel eyes; furnished transport to New York City.[330]

Roach, James I.: enlisted as a corp. in Co. G, 3rd Alabama Infantry, 4/26/61; detached duty w/ light artillery unit, 2/17/62; shipped for 3 years as a seaman, Naval Rendezvous, Richmond, Virginia, 2/19/62; served as a seaman, CSS *Virginia*; reenlisted for the war, received $50 bounty as member of crew, *Virginia*, 3/25/62; appears to have left crew of *Virginia* before 4/1/62; NFR.[331]

Rootes, Lawrence M.: b. 4/30/45, Gloucester, Virginia; parents, Thomas Reade Rootes, commander, USN/CSN, and Mary Overton Minor; appointed by Virginia as an acting midshipman, CSN, 7/31/61; assigned to Aquia Creek naval batteries, '61; paid for service, Richmond Naval Station, 10/1/61–1/31/62; assigned to CSS *Virginia*, 2/1/62; served as aide to Flag Officer Buchanan during Battle of Hampton Roads, Virginia, 3/8–9/62; paid for service, Richmond Naval Station, 5/11–6/30/62; assigned to Atlanta Naval Ordnance Works, '62–'63; served on CSS *Tuscaloosa*, Mobile Squadron, '63; served on CSS *Patrick Henry* and *Fredericksburg*, James River Squadron, '64–'65; assigned to Semmes naval brigade, '65; paroled, Greensboro, North Carolina, 4/28/65; NFR; postwar: married cousin, Mary Rootes.[332]

Rosler, John A.: b. ca. 1840; enlisted at Pasquotank Co., North Carolina, for 1 year as a pvt. in Co. K, 1st North Carolina Artillery, 4/22/61; "joined CSN on the *Forrest*," 7/61; shipped for 1 year as an ordinary seaman, CSS *Forrest*, North Carolina Squadron, 8/3/61; served as a landsman, CSS *Arctic*, North Carolina Squadron, '61; served as a coxswain, CSS *Virginia*; reenlisted for the war, received $50 bounty as member of crew, *Virginia*, 3/25/62; paid for service on the *Virginia* and Drewry's Bluff, Virginia, 4/1–6/30/62; on *Virginia* CRR, Drewry's Bluff, 7/20/62; volunteered for duty on CSS *Chattahoochee*, 8/1/62; coxswain, CSS *Chattahoochee* thru 6/12/63; transferred to CSS *Savannah*, Savannah River Squadron, 6/13/63; acting master's mate, 7/17/63; served on CSS *Sampson*, Savannah River Squadron, '63–'64; acting master's mate, Provisional Navy, 6/2/64; took part in expedition to capture USS *Water Witch*, 6/3/64; NFR; also found as "Rosser" and "Rossler."[333]

Ross, James: b. ca. 1829, Russia; occ. sailor; single; resided and enlisted at New Orleans, Louisiana, for the war as a pvt. in Co. D, 15th Louisiana Infantry, 7/18/61; present, 7–10/61; "transferred to the Navy," 2/62 roll; paid for service on receiving ship, CSS *Confederate States*, 1/1–9/62; served as an ordinary seaman, CSS *Virginia*; reenlisted for the war, received $50 bounty as member of crew, *Virginia*, 3/25/62; on *Virginia* CRR, Drewry's Bluff, Virginia, 7/20/62; paid, Drewry's Bluff, 12/29/62–9/30/63; shipped for the war, Naval Rendezvous, Richmond, 2/11/63; paid for service, naval det. on special expedition from Drewry's Bluff to Charleston Naval Station, 10/1–31/63; NFR; may have later enlisted at Rocketts Navy Yard, Richmond, Virginia, as a pvt. in Co. B, 4th Btn. Virginia Infantry, Local Defense (Naval Btn.), 6/21/63; present, 8/8/64; NFR.[334]

Rudd, James: b. ca. 1826, England; occ. brick mason; enlisted at Arkansas Co., Arkansas, for 1 year as a pvt. in Co. H, 1st (Colquitt's) Arkansas Infantry, 5/8/61; mustered in at Lynchburg, Virginia, 5/21/61; present, 6/61; transferred to CSN per SO 17, A&IGO 62, para 23, 2/3/62; served as a 2nd class fireman, CSS *Virginia*; reenlisted for the war, received $50 bounty as member of crew, *Virginia*, 3/25/62; paid for service on the *Virginia*, 4/1–5/12/62; superannuated for rations, Drewry's Bluff, Virginia, 5/12–24/62; on *Virginia* CRR, Drewry's Bluff, 7/20/62; paid, Drewry's Bluff, 5/13/62–3/31/64; reportedly served on CSS *Fredericksburg*, James River Squadron; paroled, Appomattox CH, 4/9/65; postwar: occ. brick layer; married; resided Chesterfield Co., Virginia; entered R.E. Lee Camp Soldiers' Home, Richmond, Virginia, 4/16/1903; discharged on own request, 5/4/1903; reentered Soldiers' Home, 10/12/1903; d. 9/30/1908; buried, Hollywood Cem., Richmond.[335]

Runnels, John: b. ca. 1835; resided Washington Co., North Carolina; enlisted at Plymouth, North Carolina, for the war as a pvt. in Co. G, 1st North Carolina Infantry, 6/25/61; present until transferred to CSN per SO 17, A&IGO 62, para 23, 2/3/62; served as a seaman, CSS *Virginia*; reenlisted for the war, received $50 bounty as member of crew, *Virginia*, 3/25/62; paid for service on the *Virginia* and Drewry's Bluff, Virginia, 4/1/62–9/30/63; on *Virginia* CRR, Drewry's Bluff, 7/20/62; paid for service, naval det. on special expedition from Drewry's Bluff to Charleston Naval Station, 10/1–31/63; NFR.[336]

Russell, George: b. ca. 1842, Georgia; occ. farm laborer; single; resided Pike Co., Georgia; enlisted at Covington, Georgia, for 1 year as a pvt. in Co. H, 3rd Georgia Infantry, 6/11/61; "transferred to CSS *Merrimac*," 1/13/62; served as a 2nd class fireman, CSS *Virginia*; appears to have left crew before 4/1/62; listed as having deserted on final *Virginia* payroll, dated 9/30/62; RTD w/ Co. H, 3rd Georgia Infantry; POW, Gettysburg, Pennsylvania, 7/2/63; sent to Ft. McHenry, Maryland, 7/6/63; sent to Ft. Delaware, Delaware; sent to Point Lookout, Maryland, 10/22/63; took oath and released, 6/5/65; Georgia commissioner of pensions requested verification of service, 1916.[337]

Ryal, Peter: b. Ireland; resided and enlisted at New Orleans, Louisiana, as a pvt. in Co. A, 10th Louisiana Infantry, 6/30/61; present thru 10/61;

served as a 2nd class fireman, CSS *Virginia*; reenlisted for 3 years, received $50 bounty, 4/21/62; paid for service on the *Virginia*, 4/1–5/12/62; superannuated for rations, Drewry's Bluff, Virginia, 5/12–24/62; on *Virginia* CRR, Drewry's Bluff, 7/20/62; paid, Drewry's Bluff, 5/13/62–7/30/63; deserted, Charleston, South Carolina, 2/18/65; sent to Hilton Head, South Carolina; furnished transport to New York City; on *Virginia* muster roll in *ORN* as "Peter Ryall."[338]

Ryan, James: served as a ship's carpenter, CSS *Virginia;* paid for service on the *Virginia* and Drewry's Bluff, Virginia, 4/1–6/15/62; listed as having deserted on final *Virginia* payroll, dated 9/30/62; may have later served as an ordinary seaman, CSS *Arctic*, North Carolina Squadron, '62–'63; shipped for the war as an ordinary seaman, Wilmington, North Carolina, 9/22/62; served as an ordinary seaman and seaman, CSS *North Carolina*, North Carolina Squadron, '64; paroled, Farmville, Virginia, 4/65.[339]

Ryan, John T.: b. ca. 1838, occ. machinist; enlisted at Manchester, Virginia, for 1 year as a pvt. in Co. I, 6th Virginia Infantry, 5/9/61; mustered in at Richmond, Virginia, 5/9/61; present until "discharged to labor on government works" per SO 259, A&IGO 61, para 5, 8/28/61; joined CSN; served as a seaman, CSS *Virginia*; reenlisted for the war, received $50 bounty as member of crew, *Virginia*, 3/25/62; paid for service on the *Virginia* and Drewry's Bluff, Virginia, 4/1/62–12/15/63; on *Virginia* CRR, Drewry's Bluff, 7/20/62; enlisted at Richmond as a sgt. in Co. A, 1st Btn. Virginia Infantry, 12/30/63; detailed to report to president of Richmond & Danville RR, 8/27/64–2/65; reduced to pvt., 12/1/64; NFR.[340]

Sailor, John: enlisted at New Orleans, Louisiana, as a pvt. in Co. B, CSMC, 4/15/61; served with Co. B at Pensacola Navy Yard, '61; transferred to Co. C; on Co. C CRR, 4th qtr. '61; assigned to Marine Guard, CSS *Virginia*, ca. 11/61–ca. 5/62; on CRR, 3rd qtr. '62; listed as having deserted before 9/30/62; NFR; also found as "Sealar."[341]

Salyer, Samuel: enlisted at Mayoch, North Carolina, as a pvt. in Co. L, 17th North Carolina Infantry (1st Org.), 5/13/61; deserted or transferred to CSN before 7/28/61; served as a landsman, CSS *Fanny*, North Carolina Squadron, '61; paid as a landsman, Gosport Navy Yard, Portsmouth, Virginia, 12/13/61; served as a landsman, CSS *Virginia*; d. at or near Petersburg, Virginia, 5/23/62; also found as "Salyear."[342]

Satchfield, Francis: b. ca. 1842; occ. boatman, resided Plum St., Petersburg, Virginia; enlisted at Petersburg for 1 year in Co. K, 12th Virginia Infantry, 5/4/61; described as 5'8", sandy hair, blue eyes; present until transferred to Co. C, 41st Virginia Infantry per SO 373, Dept. of Norfolk,

11/1/61; present until discharged having "enlisted on C.S. Steamer *Virginia*," 3/29/62; served as a landsman, CSS *Virginia*; appears to have left crew of *Virginia* before 4/1/62; enlisted at Camp Davis, North Carolina, for the war as a pvt. in Pegrams's Co., Virginia Light Artillery, 2/27/63; married, 8/64; present thru 10/64; on CRR, 11/30/64; NFR; postwar: widow, Mary, received pension, 1905.[343]

SAUNDERS, GEORGE: enlisted at Richmond, Virginia, for the war as a pvt. in Co. I, 1st (McCreary's) South Carolina Infantry, 7/20/61; transferred to CSN, 1/17/62; served as a landsman, CSS *Virginia*; superannuated for rations, Drewry's Bluff, Virginia, 5/12–24/62; NFR; may have later enlisted at Richmond as a pvt. in Co. A, 4th Btn. Virginia Infantry, Local Defense (Naval Btn.), 3/15/64; present, 8/2/64; NFR.[344]

SAUNDERS, THOMAS: served as a seaman on CSS *Virginia*; reenlisted for the war, received $50 bounty as member of crew, *Virginia*, 3/25/62; in Bruton Parish Episcopal Church Hosp., Williamsburg, Virginia, w/ venereal, 5/10–14/62; paid for service on the *Virginia* and Drewry's Bluff, Virginia, 4/1–6/30/62; on *Virginia* CRR, Drewry's Bluff, 7/20/62; volunteered for duty on CSS *Chattahoochee*, 8/1/62; served as a quarter gunner, CSS *Chattahoochee* thru 6/12/63; transferred to CSS *Savannah*, Savannah River Squadron, 6/13/63; NFR.[345]

SCHOLLS, JACOB S.: b. ca. 1820, Easton, Pennsylvania; enlisted at Pensacola Navy Yard for four years as a pvt., USMC, 8/27/44; married Bridget Thompson in Pensacola Navy Yard, 1/31/47; wife received Mexican War Pension, Certificate #3424; reenlisted as sgt. for four years at Pensacola, 7/14/48; 1st sgt., Co. C, USMC, during Mexican War; reenlisted at Pensacola for four years as an orderly sgt., 5/14/52; described then as 5'11½", dark comp., brown hair, hazel eyes, occ. soldier; discharged, Warrington, Florida, 11/20/56; enlisted at Montgomery, Alabama, in Co. C, CSMC, 3/25/61; sgt. on CRR, 3rd qtr. '61; 1st sgt., assigned to Marine Guard, CSS *Virginia*, ca. 11/61–5/12/62; paid for service, CSS *Virginia*, 4/1–5/12/62; d. 8/31/62, Camp Beall, Drewry's Bluff, Virginia, bur. Hollywood Cem., Richmond, Virginia; father of James Lawrence Scholls, musician Co. A, CSMC, and William Henry Scholls, musician Cos. B &C, CSMC; also found as "Shoals" and "Schools;" on *Virginia* muster roll in *ORN* as "Jacob S. Sholls."[346]

SCHROEDER, CHARLES: b. 1/22/36, Portsmouth, Virginia; parents, Antonius & Theresa Schroeder; appointed by Virginia to USN as a 3rd assistant engineer, '53; 2nd assistant engineer, 5/9/57; 1st assistant engineer, 8/2/59; assigned to USS *Saranac* at start of Civil War; tendered resignation from USN, 5/2/61; dismissed 5/18/61; married Mary E. City, '61; appointed as 1st assistant engineer, CSN, 6/19/61; served on steamer *St. Nicholas*, Chesapeake Bay, '61; served on CSS *Rappahannock*, '61; paid for service, Richmond Naval Station, 1/1–3/10/62; served on CSS *Virginia*, 3/11–4/2/62; paid for service, Richmond Naval Station, 5/1–6/30/62; engineer, 7/15/62; assigned to Mobile Naval Station, '62; chief engineer, 10/2/62; served on CSS *Richmond*, James River Squadron, '63; took part in Johnson's Island expedition, '63; served on the cruiser, CSS *Chickamauga*, '64; assigned to James River Squadron, '64; chief engineer, Provisional Navy, 6/2/64; service abroad, '64–'65; postwar: occ. mercantile business for 2 years in Halifax, Nova Scotia, with John Taylor Woods; engineer for 5 years for Pacific Steamship Mail Co. in China; returned to U.S.; joined former shipmate E.V. White's hardware co. in Portsmouth, Virginia, '73; member, Portsmouth City Council; d. 5/12/1910; bur. Schroeder Plot, Cedar Grove Cem., Portsmouth, Virginia.[347]

Schwartz, Henry: enlisted at New Orleans, Louisiana, as a pvt. in Co. B, CSMC, 4/18/61; ordered transferred to Co. C by Col. Tyler, 7/31/61, but order canceled; transferred to Co. C, 4th qtr. '61; with Co. C at Pensacola Navy Yard until it transferred to Gosport Navy Yard, Portsmouth, Virginia, 11/29/61; assigned to Marine Guard, CSS *Virginia*, ca. 11/61–5/12/62; paid for service, CSS *Virginia*, 1/1–5/12/62; paid commuted transportation, Richmond to Mobile ($20.54) on evidence of honorable discharge, 6/7/62; also found as "Swartz;" on *Virginia* muster roll in *ORN* as "Henry Swaroz."[348]

Scott, James R.: paid as a 1st class fireman, Richmond Naval Station, 1/1–2/9/62; served as a 1st class fireman, CSS *Virginia*; on *Virginia* CRR, Drewry's Bluff, Virginia, 7/20/62; paid for service on the *Virginia* and Drewry's Bluff, 3/10–8/12/62; superannuated for rations, Drewry's Bluff, 5/12–24/62; listed as discharged on final *Virginia* payroll, dated 9/30/62; paid, Richmond Naval Station, 10/1–12/31/62; NFR.[349]

Scott, John: b. ca. 1834, Scotland; occ. seaman; enlisted at Camp Walker, Louisiana, for 1 year as a pvt. in Co. E, 2nd Louisiana Infantry, 5/9/61; described as 5'5", fair comp., light hair, light eyes; absent, on detached service under Gen. Magruder, 6/61; present, 7–12/61; "transferred to service aboard the *Merrimac*," 1/31/62; served as a quarter gunner, CSS *Virginia*; paid for service on the *Virginia* and Drewry's Bluff, Virginia, 4/1–8/31/62; superannuated for rations, Drewry's Bluff, 5/12–24/62; NFR.[350]

Scott, William: enlisted as a pvt. in Co. G, 3rd Alabama Infantry, 4/26/61; discharged by SO 18, A&IGO 62, para 14, 1/23/62; served as a landsman, CSS *Virginia*; paid for service on the *Virginia* and Drewry's Bluff, Virginia, 4/1–5/15/62; superannuated for rations, Drewry's Bluff, 5/12–24/62;

listed as having deserted on final *Virginia* payroll, dated 9/30/62; NFR; may have later served as a landsman, CSS *Savannah*, Savannah River Squadron, '62; served as an ordinary seaman, CSS *Raleigh*, James River Squadron, '63; served as an ordinary seaman and seaman, CSS *Arctic*, North Carolina Squadron, '63–'64; NFR.[351]

Scultatus, George: b. ca. 1836, Germany; enlisted at Norfolk, Virginia, for 1 year as a pvt. in the United Artillery (1st Co. E, 41st Virginia Infantry), 4/19/61; present thru 3/6/62; "volunteered to go onboard the Ironclad Steamer *Virginia*," 3/6/62; manned gun on CSS *Virginia* during Battle of Hampton Roads, Virginia, 3/8–9/62; reenlisted for war, received $50 bounty, 3/10/62; on CRRs, 12/22/62, 6/3/63, 9/14/63, 11/2/63 and 3/18/64; absent, on detached duty w/ CSN, per SO 63, A&IGO 64, dated 3/16/64; present, 12/31/64; in Gen. Hosp. #9, Richmond, Virginia, w/ scabies, 3/21/65; in Chimborazo Hosp. #3, 3/22/65; paroled, Appomattox CH, 4/9/65; postwar: occ. policeman; married; resided Norfolk next to William Crosby, formerly of the United Artillery and CSS *Virginia*; member, Pickett-Buchanan Camp, UCV, Norfolk.[352]

Semple, James Allen: b. 2/24/19, Cedar Hill, New Kent Co., Virginia; parents, Col. George Semple and Elizabeth Russell Holt; attended College of William and Mary; married Letitia Tyler, daughter of President John & Letitia Christian Tyler; appointed by Virginia to USN as a purser, 10/12/44; assigned to USRS *New York* at start of Civil War; tendered resignation from USN, 5/19/61; appointed paymaster, CSN 6/61; finally dismissed from USN, 7/15/61; asst. chief, CSN Bureau of Provisions and Clothing, Richmond, '61; served on CSS *McRae*, New Orleans Naval Station, '61; assigned as paymaster, CSS *Virginia*, 1/1/62; commanded *Virginia*'s powder division during Battle of Hampton Roads, Virginia, 3/8–9/62; attached to Drewry's Bluff, Virginia and CSS *Roanoke*, '62–'64; chief, CSN Bureau of Provisions and Clothing, '64–'65; traveled w/ Jefferson Davis's party after fall of Richmond; was to take $86,000 in gold, concealed in false carriage bottom, to an eastern port to be shipped to a Confederate agent abroad; disposition of money unclear; d. 1886; bur. Bruton Parish Cem., Williamsburg, Virginia.[353]

Sharp, Charles: b. 1829, Virginia; father, W.W. Sharp; graduated University of Virginia, '50; occ. lawyer; married; resided Norfolk, Virginia; friend of Capt. Thomas Kevill, United Artillery; enlisted in an artillery co., 4/61; discharged due to chronic kidney inflammation and varicose veins, 6/61; civilian adjutant for artillery co. until 2/62; did not apply for exemption after Conscription Act passed 2/62; enlisted at Norfolk for the war as a pvt. in the United Artillery (1st Co. E, 41st Virginia Infantry), 3/4/62; received $50 bounty; on CSS *Virginia*, 3/8/62; discharged due to severe kidney trouble, 3/62; clerk, QM Dept., Richmond, Virginia, 10/14/62–early '64; postwar: occ. lawyer; resided Norfolk.[354]

Sharpe, Andrew Jackson: b. 3/13/38, Grundy, Tennessee; parents, Daniel Hiram Sharpe and Charlotte Tucker; married Nancy C. Blakely, Lafayette, Alabama, 12/10/57; occ. farmer; enlisted at New Orleans, Louisiana, for 1 year as a pvt. in Co. I, 2nd Louisiana Infantry, 5/9/61; described as 6'1", fair comp., dark hair, gray eyes; reenlisted at Camp Magruder, Virginia, for 9 months as a pvt. in Co. I, 2nd Louisiana Infantry, 8/1/61; present until "transferred to the Steamer *Virginia*," 3/15/62; served as a landsman, CSS *Virginia*; reenlisted for the war, received $50 bounty; paid for service as a seaman on the *Virginia*, 3/10–5/12/62; superannuated for rations, Drewry's Bluff, Virginia, 5/12–24/62; on *Virginia* CRR, Drewry's Bluff, 7/20/62; paid, Drewry's Bluff, 5/13–9/30/62 and 1/1–6/30/63; listed as having deserted on 6/30/63 payroll; may have later enlisted as a pvt. in Co. B, 2nd Arkansas Battery, Louisiana Artillery; paroled, Shreveport, Louisiana, 6/9/65; NFR; postwar: married Mary E. Brown, Texas, '67; married Sarah Jane Taylor, Texas, 11/20/77; d. Franklin, Robertson Co., Texas, 2/5/1928; on *Virginia* muster roll in *ORN* as "Andrew J. Spark."[355]

Shavor, Jacob: b. ca. 1841, Germany; occ. printer; single; resided and enlisted at Montgomery, Alabama, as a pvt. in CSMC, 4/4/61; probably served at Pensacola Navy Yard until Co. C transferred to Gosport Navy Yard, Portsmouth, Virginia, 11/29/61; assigned to Marine Guard, CSS *Virginia*, ca. 11/61–5/12/62; paid for service, CSS *Virginia*, 1/1–5/12/62; on CRR, 3rd qtr. '62; NFR; may have later served as a landsman, CSS *Arctic*, North Carolina Squadron, '63, and CSS *Raleigh*, James River Squadron, '64; also found as "Shaver."[356]

Sheffield, James M.: b. ca. 1827, Henry Co., North Carolina; occ. farmer; resided Mecklenburg Co., North Carolina; enlisted at Ranalesburg, North Carolina, as a pvt. in Co. B, 13th North Carolina Infantry, 6/23/61; mustered in 8/31; described as 6'3"; present until "discharged for naval service on CSS *Merrimack*," 2/15/62; served as a landsman, CSS *Virginia*; paid for service on the *Virginia* and Drewry's Bluff, Virginia, 1/15–6/15/62; superannuated for rations, Drewry's Bluff, 5/12–24/62; listed as discharged on final *Virginia* payroll, dated 9/30/62; may have later enlisted at Richmond, Virginia, for the war as a pvt. in Hardwicke's Co., Virginia Light Artillery, 12/13/63; on CRRs, 1–2/64; WIA and captured, Winchester, Virginia, 9/29/64; in U.S. Army Field Hosp., Winchester w/ gunshot, right lung until 11/8/64; in

Gen. Hosp., Frederick, Maryland, 11/9/64; in hosp., Baltimore, Maryland, 11/23/64; in Hammond Gen. Hosp., Point Lookout, Maryland, 1/28/65; released from hosp., 4/7/65; took oath and released, 6/21/65; d. '95; bur. Oakwood Cem., Raleigh, North Carolina.[357]

Sheriff, Benjamin R.: enlisted as an ordinary seaman, USN, 4/15/61; assigned to USS *Allegheny*; deserted, 4/30/61; served as a quarter gunner, CSS *Virginia;* reenlisted for the war, received $50 bounty as member of crew, *Virginia*, 3/25/62; listed as having deserted on final *Virginia* payroll, dated 9/30/62; served at Charleston Naval Station; paid for service as a ship's steward and seaman, CSS *Palmetto State*, Charleston Squadron, 1/1–4/25/64; listed as having deserted, 4/25/64; apparently RTD since he was paid for service as a ship's steward, CSS *Palmetto State*, 10/1–12/31/64; NFR; postwar: resided Baltimore, Maryland, 1907.[358]

Shever, James: served as officer's cook, CSS *Virginia*; reenlisted for the war, received $50 bounty as member of crew, *Virginia*, 3/25/62; on *Virginia* CRR, Drewry's Bluff, Virginia, 7/20/62; listed as having transferred as an officer's cook w/ Capt. James H. Rochelle on *Virginia* muster roll dated 7/31/62, but paid for service on the *Virginia* and Drewry's Bluff, 4/1–10/10/62; paid as a landsman, Drewry's Bluff, 10/11–11/11/62, NFR.[359]

Simms, Charles Carroll: b. 1811, Alexandria, Virginia; parents, John Douglas Simms and Eleanor Carroll Brent; appointed by Virginia to USN as a midshipman, 10/9/39, passed midshipman, 7/2/45; master, 1/15/54; Lt., 8/12/54; married Elizabeth Nourse of Georgetown, Washington, D.C., 11/13/59; assigned to Washington Navy Yard at start of Civil War; tendered resignation and dismissed from USN, 4/22/61; served as Lt., Virginia State Navy, '61; appointed as a 1st Lt., CSN, 6/10/61; assigned to Gosport Navy Yard, Portsmouth, Virginia, '61; commanded CSS *Richmond*, Potomac River defenses, '61; helped seized USS *St. Nicholas*, brig *Monticello* and schooner *Mary Pierce*, 6/29/61; served on *St. Nicholas*, '61; assigned to Richmond Naval Station, 10/1–11/29/61; assigned to CSS *Virginia*, 11/30/61; detached to command CSS *Appomattox* during Battle of Roanoke Island, 2/6/62; served on CSS *Virginia,* commanded bow gun and first division, 3/8/62; served as executive officer, 3/9/62; left *Virginia*, 4/12/62; served on CSS *Nansemond*, James River Squadron, '62; ordered to CSS *Atlanta*, 10/6/62; served on the cruiser, CSS *Florida* reporting for duty 10/25/62; assigned to Selma Naval Station, '62–'64; assigned to Mobile Squadron, '64–'65; 1st Lt., Provisional Navy, 6/2/64; commanded CSS *Baltic* and *Nashville*, '64; surrendered, 5/4/65; paroled, 5/10/65; NFR.[360]

Sinclair, Arthur, IV: b. ca. 1837, Norfolk, Virginia; parents, Arthur Sinclair, commander, USN/CSN, and Lelia Imogene Dawley; grandfather, Commodore Arthur Sinclair, War of 1812 veteran; entered USN at 13; served four years under father on USS *Supply* and on Commodore Perry's expedition to Japan, '52–'55; occ. merchant; resided Norfolk, Virginia; enlisted at Norfolk Co., Virginia, for 1 year as a pvt. in Co. G, 6th Virginia Infantry, 4/19/61; detailed to special duty w/ QM Dept., 5/61; served as an acting master's mate, CSS *Winslow*, North Carolina Squadron, 5–10/61; formally discharged from CSA, 8/18/61; took part in defense of Ft. Hatteras, North Carolina; assigned to receiving ship, CSS *Confederate States*, '61; assigned as a captain's clerk, CSS *Virginia*, 2/1/62; served during Battle of Hampton Roads, Virginia, 3/8–9/62; left crew, *Virginia*, 4/1/62; appointed by Virginia as an acting master, 8/6/62; served under his father on CSS *Mississippi*, New Orleans Naval Station, '62; served on the cruiser, CSS *Alabama*, '62–'64; Lt. for war, 8/25/63; 1st Lt., 1/6/64; 1st Lt., Provisional Navy, 6/2/64; took part in engagement with USS *Kearsarge* off Cherbourg, France, 6/10/64; served on the cruiser, CSS *Rappahannock*, Calais, France, '64; granted leave of absence for health while abroad, '64–'65; detailed to CSS *Texas*, under construction in Scotland, when war ended; postwar: resided Norfolk; married Drusilla Willet; published an account of his time on *Alabama* in '95; entered R.E. Lee Camp Soldiers' Home, Richmond, Virginia, 1/4/1904; released on own request, 1/9/1907; d. Baltimore, Maryland, 11/15/1925.[361]

Skerrit, James: enlisted at Richmond, Virginia, for the war as a pvt. in Co. I, 1st (McCreary's) South Carolina Infantry, 7/29/61; present until "transferred to Navy Dept.," 1/17/62; served as a landsman, CSS *Virginia*; superannuated for rations, Drewry's Bluff, Virginia, 5/12–24/62; NFR.[362]

Slade, Benjamin: b. ca. 1843; occ. laborer; resided and enlisted at Petersburg, Virginia, for 1 year as a pvt. in Co. K, 12th Virginia Infantry, 5/4/61; described as 5'10", light comp., light hair, blue eyes; reenlisted for 2 years, $50 bounty due, 2/1/62; discharged for naval service per SO 69, Dept. of Norfolk, 3/29/62; served as a landsman, CSS *Virginia*; appears to have left crew of *Virginia* before 4/1/62; drafted at Huger Barracks, Norfolk, as a pvt. in Co. F, 16th Virginia Infantry, 4/11/62; AWOL, 15 days; found guilty and fined one month's pay, 2/23/63; WIA, Chancellorsville, Virginia, 5/3/63; present, 7/63–2/28/64; POW, Spottsylvania, Virginia, 5/6/63; arrived Point Lookout, Maryland, from Belle Plain, 5/17/63; sent to Elmira, New York, 7/27/64; released on oath, 6/14/65.[363]

Smith, George N.: b. ca. 1832; occ. seaman; enlisted at Tunstalls, Virginia, as a pvt. in the Pamunkey Artillery (Jones's Co., Virginia Heavy Artillery), 5/21/61; paid thru 12/31/61; enlisted at Yorktown, Virginia, for the war as a pvt. in the Magruder Light Artillery (Page's Co., Virginia Light Artillery), 1/27/62; "transferred to Steamer *Virginia*," 2/9/62; served as a landsman, CSS *Virginia*; reenlisted for the war, received $50 bounty as member of crew, *Virginia*, 3/25/62; appears to have left crew of *Virginia* before 4/1/62; paid, Drewry's Bluff, Virginia, 10/1–12/31/63; captured on blockade runner, CSS *Greyhound* by USS *Connecticut*, 5/12/64; POW, Camp Hamilton, Virginia, 5/16/64; released and sent to Baltimore, Maryland, 6/14/64; may have later served as a carpenter on the blockade runner, CSS *Stag*, '64–'65; captured off Smithville, North Carolina, 1/19/65; sent to Ft. Monroe, Virginia; sent to Point Lookout, Maryland, 2/1/65; took oath and paroled, 5/7/65.[364]

Smith, James F.: enlisted at Memphis, Tennessee, as a pvt. in CSMC, 9/17/61; on Co. C CRR, 4th qtr. '61; assigned to Marine Guard, CSS *Virginia*, ca. 11/61–ca. 5/12/62; corp., 4th qtr. '62; sgt., 1/23/63; 1st sgt., 9/18/63; pvt. on Co. A CRR, 1st qtr. '64; assigned Wilmington Naval Station, 2/16/64; paid for service, CSS *Arctic*, North Carolina Squadron, 7/1–9/30/64; 1st sgt., Battery Buchanan, ca. 11/4–12/31/64; in hosp., Fayetteville, North Carolina, 1/23–2/28/65; NFR; may have later reported to Federal authorities, Plymouth, North Carolina, 5/25/65; took oath, 6/10/65; gave his former residence and destination as New York; on *Virginia* muster roll in *ORN* as "J.T. Smith."[365]

Smith, John David, Jr.: b. ca. 1840, Virginia; occ. painter; single; resided and enlisted at Norfolk, Virginia, for 1 year as a pvt. in the United Artillery (1st Co. E, 41st Virginia Infantry), 4/19/61; present thru 3/6/62; "volunteered to go onboard the Ironclad Steamer *Virginia*," 3/6/62; manned gun on CSS *Virginia* during Battle of Hampton Roads, Virginia, 3/8–9/62; reenlisted for war, received $50 bounty, 3/10/62; on CRRs, 12/22/62, 6/3/63, 9/14/63, 11/2/63 and 3/18/64; in Chimborazo Hosp. #2, Richmond, Virginia, w/ int. fever, 9/9–16/64; granted 35-day furlough to Plymouth, North Carolina, 9/17/64; in Gen. Hosp. #9, Richmond, 9/28–29/64; absent sick, in Chimborazo Hosp. on 10/10/64 roll; in Chimborazo Hosp. #1 w/ ascites, 11/29–12/20/64; absent sick, on furlough to Goldsboro, North Carolina, on 12/31/64 roll; deserted, Bermuda Hundred, Virginia, 3/22/65; sent to City Point, Virginia, 3/22/65; sent to Washington, D.C., 3/27/65; took oath, furnished transport to Norfolk, 3/29/65; listed residence as Goldsboro or Plymouth, North Carolina.[366]

Smyth, Samuel Bell: enlisted at Montgomery, Alabama, as a pvt. in CSMC, 4/13/61; probably served at Pensacola Navy Yard until Co. C transferred to Gosport Navy Yard, 11/29/61; assigned to Marine Guard, CSS *Virginia*, ca. 11/61–ca. 5/12/62; listed as sgt. on Co. C CRR, 1/23/63; paid for extra duty, QM Dept., 1/13/64; on daily duty at commissaries, 7–12/64; paroled, Farmville, Virginia, between 4/11–21/65; also found as "Smith."[367]

Southall, Benjamin F.: b. ca. 1840, Twiggs Co., Georgia; occ. farmer; enlisted at Marion, Georgia, for 1 year as a pvt. in Co. C, 4th Georgia Infantry, 4/25/61; described as 5'6", dark comp., dark hair, gray eyes; present thru 8/61; in regt. hosp., Camp Jackson, Virginia, w/ typhoid fever, 9/9–15/61 and 9/24–26/61; in regt. hosp. w/int. fever, 11/1–9/61; detailed to work at artillery btry., Pig Point, Virginia, 12/61; transferred to CSS *Virginia* per SO 28, Dept. of Norfolk, 2/10/62; served as landsman on CSS *Virginia*; superannuated for rations, Drewry's Bluff, Virginia, 5/12–24/62; enlisted near Richmond, Virginia, for the war as a pvt. in Co. G, 48th Georgia Infantry, 6/7/62; listed as having deserted, Richmond, 6/24/62; enlisted at Culpepper, Virginia, for the war as a pvt. in Co. G, 10th Georgia Infantry, 11/17/62; attached to African Church Hosp., Winchester, Virginia, as a nurse, 9/20/62; in Gen. Hosp. #9, Richmond, w/ GSW (ball lodged in right thigh), 7/19/63; in Chimborazo Hosp. #4, Richmond, 7/20/63; granted 35-day furlough, 7/25/63; paid, 9/63–2/64; listed as AWOL in Georgia, 9/64–2/65 rolls; NFR; widow, Marriah, received Florida pension, 1903.[368]

Sparks, Charles M.: b. ca. 1835, Kentucky; parents, W. and Nancy Sparks; occ. farmer; married; resided Greenville, Kentucky; enlisted at Memphis, Tennessee, as a pvt. in CSMC, 8/8/61; probably transferred from Capt. Hays's det., Co. D, to Co. C, Pensacola Navy Yard, ca. 11/29/61; assigned to Marine Guard, CSS *Virginia*, ca. 11/61–ca. 5/12/62; on Co. C CRR, 3rd qtr. '62; listed as having deserted on final *Virginia* payroll, dated 9/30/62; NFR.[369]

Spence, Alexander: b. ca. 1834, Virginia; occ. cooper, single; resided and enlisted at Norfolk, Virginia, for the war as a pvt. in the United Artillery (1st Co. E, 41st Virginia Infantry) and received $50 bounty, 3/4/62; "volunteered to go onboard the Ironclad Steamer *Virginia*," 3/6/62; manned gun on CSS *Virginia* during Battle of Hampton Roads, Virginia, 3/8–9/62; on CRR, Drewry's Bluff, Virginia, 12/22/62, 6/2/63, 9/14/63 and 10/30/63; POW, Howlett's Farm, Virginia, 7/11/64; sent to Point Lookout, Maryland; transferred to Elmira, New York, 8/3/64; paroled, Elmira, 5/13/65; described then as 5'9½", fair comp., brown hair, blue eyes.[370]

Spence, Charles H.: b. Norway; enlisted at Norfolk, Virginia, for the war as a pvt. in the United Artillery (1st Co. E, 41st Virginia Infantry) and received $50 bounty, 3/4/62; "volunteered to go onboard the Ironclad Steamer *Virginia*," 3/6/62; manned gun on CSS *Virginia* during Battle of Hampton Roads, Virginia, 3/8–9/62; on CRR, Drewry's Bluff, Virginia, 12/22/62, 6/2/63, 9/14/63 and 10/30/63; deserted and POW, City Point, Virginia, 9/28/64; described then as 5'7", dark comp., brown hair, blue eyes, took oath and sent to Washington, D.C., 9/29/64; furnished transport to Baltimore, Maryland.[371]

Spence, Robert: shipped for the war as an ordinary seaman, Naval Rendezvous, New Orleans, Louisiana, 9/5/61; served as a seaman, CSS *Virginia*; reenlisted for the war, received $50 bounty as member of crew, *Virginia*, 3/25/62; paid for service on the *Virginia* and Drewry's Bluff, Virginia, 4/1/62–3/31/64; superannuated for rations, Drewry's Bluff, 5/12–24/62; NFR.[372]

Sprague, Chauncey Alexander: b. 1/15/37, Concord, New York; parents, Parris A. Sprague and Elizabeth Rector; occ. teacher, single; resided Newton Co., Georgia; enlisted at Atlanta, Georgia, for the war as a pvt. in Co. B, Cobb's Legion, Georgia Cavalry, 8/14/61; value of horse, $250, saddle provided by State of Georgia; present thru 12/61; "transferred to *Virginia*," 3/20/62; reenlisted for the war, received $50 bounty; served as a landsman, CSS *Virginia*; on *Virginia* CRR, Drewry's Bluff, Virginia, 7/20/62; paid, Drewry's Bluff, 5/12–12/14/62; listed as having deserted on 12/14/62; may have later served as a 1st Lt., C.S. Nitre and Mining Corps; paroled, Talladega, Alabama, 5/22/65; NFR.[373]

Stack, Garrett N.: enlisted at Mobile, Alabama, as a pvt. in CSMC, 8/5/61; probably served at Pensacola Navy Yard until Co. C transferred to Gosport Navy Yard, Portsmouth, Virginia, 11/29/61; assigned to Marine Guard, CSS *Virginia*, ca. 11/61–ca. 5/12/62; paid for service, CSS *Virginia*, 1/1–5/12/62; discharged by 8/17/62; NFR; postwar; married, Sarah J. Boseley, Mobile, 9/2/69; found on *Virginia* muster roll in *ORN* as "Garnett N. Stack."[374]

Steen, John: b. ca. 1827, Philadelphia, Pennsylvania; resided and enlisted at Norfolk, Virginia, for 1 year as a pvt. in the United Artillery (1st Co. E, 41st Virginia Infantry), 4/19/61; present thru 3/6/62; "volunteered to go onboard the Ironclad Steamer *Virginia*," 3/6/62; manned gun on CSS *Virginia* during Battle of Hampton Roads, Virginia, 3/8–9/62; reenlisted for war, received $50 bounty, 3/10/62; on CRRs, 12/22/62, 6/3/63, 9/14/63, 11/2/63 and 3/18/64; present, 12/31/64; paroled, Appomattox

CH, 4/9/65; postwar: occ. laborer/waterman; resided Norfolk; received pension, witnessed by *Virginia* shipmate, Andrew J. Dalton, Norfolk justice of peace, 2/8/1910; also found as "Stien."[375]

Stevens, James H.: served as a landsman, CSS *Virginia*; on *Virginia* CRR, Drewry's Bluff, Virginia, 7/20/62; listed as a POW on *Virginia* muster roll, dated 7/31/62 but paid, Drewry's Bluff, 5/12–9/15/62; listed as discharged on final *Virginia* payroll, dated 9/30/62; possibly "James H. Stevens" who b. ca. 1842, Portsmouth, Virginia; occ. boatman; resided, Portsmouth; enlisted at Glebe Schoolhouse, Norfolk Co., Virginia, for 1 year as a pvt. in 1st Co. G; 41st Virginia Infantry, 6/16/61; described as 5'9", ruddy comp., red hair, blue eyes; present thru 2/62; reenlisted for war, received $50 bounty, 3/62; transferred out of regt., 4/19/62, listed as having deserted at Btry. No. 3 near Richmond, Virginia, but actually captured, Fair Oaks, Virginia, 6/1/62; sent from Ft. Monroe, Virginia, to Ft. Delaware, Delaware, 6/5/62; exchanged Aiken's Landing, Virginia, 8/5/62; unit transferred as Co. I, 61st Virginia Infantry, 8/8/62; present, 8/62–7/14/63; POW, Falling Waters, Maryland, 7/14/63; arrived Old Capital Prison, Washington, D.C., 8/1/63; took oath, 12/13/63; released and sent north, 12/20/63; NFR; d. 1928; bur. Beechwood Cem., Boykins, Virginia.[376]

Stewart, David: served as captain of hold, CSS *Virginia*, '61–'62, reenlisted for the war, received $50 bounty as member of crew, *Virginia*, 3/25/62; paid for service on the *Virginia*, 4/1–5/12/62; superannuated for rations, Drewry's Bluff, Virginia, 5/12–24/62; on *Virginia* CRR, Drewry's Bluff, 7/20/62; captured by Potomac River flotilla trying to deliver mail between Virginia and Maryland, 10/8/62; paroled, Aiken's Landing, Virginia, 11/2/62; paid for service at Drewry's Bluff, 5/13/62–3/26/64; served on CSS *Virginia II*, James River Squadron, '64; served on the cruiser, CSS *Tallahassee* and *Arctic*, North Carolina Squadron, '64; acting gunner, Provisional Navy, date unknown; gunner, date unknown; served on CSS *Columbia*, Charleston Squadron, '64–'65; assigned Semmes naval brigade, '65; paroled, Greensboro, North Carolina, 4/28/65.[377]

Stillman, Eleazor: b. ca. 1846, Norfolk, Virginia; parents, Eleazor Stillman and Adelia Manning; single; enlisted at Norfolk for 1 year as a pvt. in the United Artillery (1st Co. E, 41st Virginia Infantry), 5/2/61; present thru 3/6/62; "volunteered to go onboard the Ironclad Steamer *Virginia*," 3/6/62; manned gun on CSS *Virginia* during Battle of Hampton Roads, Virginia, 3/8–9/62; reenlisted for war, received $50 bounty, 3/10/62; on CRR, Drewry's Bluff, Virginia, 12/22/62, 6/2/63, 9/14/63 and 10/30/63; transferred to Co. D, 1st Confederate Engineer Troops,

1/2/64; listed as having deserted, 1/5/64; NFR; postwar: occ. waterman; married, ca. '68; resided Norfolk.[378]

STROUP, ALFORD A.: b. ca. 1840, Gaston Co., North Carolina; occ. farmer; resided Cleveland Co., North Carolina; enlisted at White Plains, North Carolina, as a pvt. in Co. D, 14th North Carolina Infantry, 4/26/61; mustered in at Garysburg, North Carolina, 6/10/61; described as 5'8", dark comp., dark hair, gray eyes; present until "transferred to the *Merrimac*," 2/15/62; served as a landsman, CSS *Virginia*; reenlisted for the war, received $50 bounty as member of crew, *Virginia*, 3/25/62; superannuated for rations, Drewry's Bluff, Virginia, 5/12–24/62; NFR; may have later enlisted at Chattanooga, Tennessee, as a pvt. in Co. D, 3rd Confederate Engineer Troops, 5/13/62; present thru 6/62; present, 1/63–2/64; NFR; found on *Virginia* muster roll in *ORN* as "Alfred A. Stroup."[379]

STURGES, JOHN J.: b. ca. 1839, New Jersey; occ. sailor; enlisted at Chuckatuck, Virginia, for 1 year as a pvt. in Co. F, 9th Virginia Infantry, 5/18/61; mustered in, 5/22/61; described as 5'10", light comp., light hair, blue eyes; present until discharged per SO 30, Dept of Norfolk, for "service on *Merrimac*," 2/12/62; served as an ordinary seaman, CSS *Virginia*; reenlisted for the war, received $50 bounty as member of crew, *Virginia*, 3/25/62; in Bruton Parish Episcopal Church Hosp., Williamsburg, Virginia, w/ syphilis, 5/10/62; transferred to hosp., Richmond, Virginia; still listed as in hosp. on *Virginia* muster roll, 7/31/62; paid for service on the *Virginia* and Drewry's Bluff, Virginia, 4/1/62–6/30/63; paid for service, naval det. on special expedition from Drewry's Bluff to Charleston Naval Station, 10/1–31/63; transferred to CSS *Chicora*; NFR; on *Virginia* muster roll in *ORN* as "John P. Sturgess."[380]

TABB, GEORGE E., JR.: b. ca. 1841, Virginia; parents, John H. and Margret Tabb; father, farmer, worth $60,000 in '60; occ. engineer; single; resided Gloucester, Virginia; enlisted at Gloucester CH, as a 3rd sgt. in Co. A, 34th Virginia Infantry, 5/8/61; described as 5'9", light comp., light hair, black eyes; present, 5/61–2/62; reenlisted at Gloucester for war, received $50 bounty, 2/15/62; detailed to Signal Light Dept. by order of Gen. Magruder, 2/62; on detached duty, Gosport Navy Yard, Portsmouth, Virginia, 3/3/62; served on CSS *Virginia* during Battle of Hampton Roads, Virginia, 3/8–9/62; ordered to City Point, Richmond, Virginia, 4/15/62; RTD w/ 34th Virginia Infantry at Chaffin's Bluff, Virginia, 7/25/62; transferred as a pvt. to Co. A, 5th Cavalry, 8/62; recommended by Lt. Robert D. Minor, CSN, for a commission based on his service aboard *Virginia*, 12/5/62; appointed 2nd Lt., CSA Signal Corps, 12/16/62; assigned to duty w/ Lt. Gen. Polk, 7/9/63; traveled to Mobile, Alabama, 10/30/63; POW, Raccoon

Mountain, Georgia, 11/24/63; sent to Nashville, Tennessee, 11/27/63; sent to Louisville, Kentucky, 11/29/63; sent to Johnson's Island, Ohio, 12/1/63; released on oath, 6/13/65; postwar: d. Colorado.[381]

Tattnall, Josiah: b. 11/9/1795, family estate Bonaventura near Savannah, Georgia; son of Josiah and Harriet Fenwick Tattnall; father, gov. of Georgia; orphaned at age 9; sent to England to be educated under maternal grandfather's supervision, ca. 1805; appointed by Georgia as a midshipman, USN, 1/1/12; assigned to USS *Constellation*; participated in repulse of a British boat expedition by the battery on Craney Island, Norfolk, Virginia, 6/23/13; served on the USS *Epervier*, *Constellation* and *Ontario*, Mediterranean Squadron during war with Algiers; Lt., 4/1/18; assigned to USS *Macedonian*, Pacific Squadron; married his cousin, Henriette Fenwick Jackson, 9/6/21; first Lt. of USS *Jackal*, Commodore Porter's squadron operating against pirates in West Indies, '23; served on USS *Constitution* and *Brandywine* in the Mediterranean, '24–'26; assigned to USS *Erie*, West Indies squadron, where he made surveys of the fortifications on the Tortuous Reef, '28; commanding USS *Grampus*, '31; escorted Gen. Santa Anna back to Mexico after he had been captured in Texas while commanding USS *Pioneer*, '35; commander, 2/25/38; commanded Boston Navy Yard; commanded USS *Fairfield* and *Saratoga*; commanded USS *Spitfire* during Mexican War; commanded Boston Navy Yard, '48–'50; captain, 2/5/50; commanded USS *Saranac*, *Independence* and naval station on the Great Lakes '50–'57; flag officer, '57; assumed command of naval forces in Indian Ocean and China Sea; assisted English/French fleets in their attack on Chinese forts on the Pei-ho River in violation of U.S. neutrality, 6/25/59; explained his actions by saying, "Blood is thicker than water!"; brought first Japanese ambassadors to U.S., '60; President James Buchanan's guest of honor at White House dinner; commanded Sackett's Harbor Navy Yard, New York, when Georgia seceded; tendered resignation from USN, 2/20/61; resignation accepted 2/21/61; appointed capt., CSN, 3/26/61; given command of Georgia State Navy; responsible for defense of South Carolina and Georgia coasts, '61–'62; replaced Capt. Buchanan and commanded CSS *Virginia* until ordering her destruction, 5/11/62; commanded naval defenses of Georgia, '62–'63; relieved of command afloat and limited to shore duty, 3/63; commanded Savannah Station, '63–'65; paroled 5/9/65; postwar, res. Nova Scotia '66–'70; returned to U.S. and became inspector of ports, Savannah, Georgia; d. 6/14/71; bur. 6/16/71, family estate Bonaventura near Savannah, Georgia; USS *Tattnall* (DD-125), commissioned, 6/26/1919, and USS *Tattnall* (DDG-19), commissioned, 4/13/1963, were named in his honor.[382]

Tattnall, Paulding: served on Savannah Naval Station; paid for service as secretary to Flag Officer Josiah Tattnall on CSS *Virginia*, 4/1–5/12/62; served on Savannah Naval Station, '62–'64; NFR.[383]

Tetterton, William Ropheus: enlisted at Plymouth, North Carolina, for the war as a pvt. in Co. H, 1st North Carolina Infantry, 7/29/61; present until transferred to CSN per SO 17, A&IGO 62, para 23, 2/3/62; served as a landsman, CSS *Virginia*; reenlisted for the war, received $50 bounty as member of crew, *Virginia*, 3/25/62; paid for service on the *Virginia*, 4/1–5/12/62; on *Virginia* CRR, Drewry's Bluff, Virginia, 7/20/62; paid, Drewry's Bluff, 5/13/62–2/29/64; assigned to CSS *Neuse*, 3–10/64; NFR; also found as "Tittenton;" on *Virginia* muster roll in *ORN* as "William R. Titleton."[384]

Tharpe, Marcellas Augustus: b. 9/15/40, Doles, Georgia; parents, Benjamin A. Tharpe and Susan Chappel; occ. schoolteacher; enlisted at Marion, Georgia, for 1 year as a pvt. in Co. C, 4th Georgia Infantry, 4/25/61; described as 5', light comp., dark hair, gray eyes; 3rd corp., 6/30–8/31/61; present thru 10/61; granted furlough 11/5–25/61; present until transferred to CSS *Virginia* per SO 28, Dept. Norfolk, 2/10/62; served as landsman, CSS *Virginia*; reenlisted for the war, received $50 bounty; paid for service on the *Virginia*, 4/1–5/12/62; on *Virginia* CRR, Drewry's Bluff, Virginia, 7/20/62; paid, Drewry's Bluff, 5/13/62–9/30/63; paid for service, naval det. on special expedition from Drewry's Bluff to Charleston Naval Station, 10/1–31/63; transferred to Co. A, 1st Regt., Naval Brigade as a coxswain; served on CSS *Virginia II*, James River Squadron, '64–'65; ordered to Naval Ordnance Works, Richmond, Virginia, 1/13/65; surrendered, Greensboro, North Carolina, 4/26/65; postwar; married Maria Elizabeth McClendon, of Sumpter Co., Georgia, 10/15/65; d. Worth Co., Georgia, 1/9/1901; Georgia commissioner of pensions requested verification of service, 1916; "on *Virginia* muster roll in *ORN* as "M.A. Tharp."[385]

Thayer, Martin G.: b. ca. 1839; occ. apprentice blacksmith, Southside RR shops; resided and enlisted at Petersburg, Virginia, for 1 year as a 1st sgt. in Co. K, 12th Virginia Infantry, 5/4/61; described as 5'8", dark comp., black hair, black eyes; reduced to pvt., 8/21/61; court-martialed, pay stopped for 1 month, per GO 56, Dept. of Norfolk, 11/10/61; extra duty in Engineer Dept., 1–2/62; discharged per SO 69, Dept. of Norfolk, 3/29/62; served as a landsman, CSS *Virginia*; appears to have left crew of *Virginia* before 4/1/62; enlisted at Richmond, Virginia, as 4th sgt., Co. E, 1st Btn. Virginia Infantry, Local Defense, 6/17/63; present, 12/10/64; conscripted as a pvt. in Co. H, 15th Virginia Infantry, 1/8/65; assigned to regt., 1/9/65; detailed to repair engines on Southside RR, 2/9/65; listed as having deserted near New Bern, North Carolina, 2/3/65; NFR.[386]

THOM, REUBEN TRIPLETT: b. ca. 1820, Fredericksburg, Virginia; parents, Reuben Triplett Thom and Eleanor Reade; veteran of Mexican War, served as capt. of Alabama volunteers and 2nd Lt., 13th U.S. Infantry; married Basilisa Valdes, Orizaba, Mexico, 6/10/48; quartermaster general, Alabama, '61, traveled to North to purchase arms; commissioned as an artillery capt., CSA; first officer appointed to CSMC, 3/25/61; on recruiting duty in several cities, spring, '61; assigned to Pensacola Navy Yard, 6/61; led 55 Marines in engagement at Ship's Island, Mississippi, 7/9/61; assigned to New Orleans Naval Station, '61; commanded Co. C, CSMC when it arrived Norfolk, Virginia, 12/7/61; ordered to receiving ship, CSS *Confederate States* pending assignment to combatant; assigned as commander, Marine Guard, CSS *Virginia*, 2/62; commanded guns #8 & #9 during Battle of Hampton Roads, Virginia, 3/8–9/62; took part in Battle of Drewry's Bluff, Virginia, 5/15/62; on recruiting duty in several cities, 2–6/63; in Gen. Hosp. #4, Richmond, Virginia, w/ int. fever, 10/8/63; RTD, 11/1/63; left CSMC due to poor health, fall '63; 2nd Lt., 11th Alabama Infantry on detached duty as assistant inspector general on Brig. Gen. Richard L. Page's staff, Mobile, Alabama; resigned on surgeon's certificate of disability, 5/24/64; POW, Ft. Morgan, Alabama, 8/23/64; sent to Ft. Lafayette, New York; transferred to Ft. Delaware, Delaware, 3/13/65; paroled, 6/10/65; d. 12/25/73; bur. City Cem., Montgomery, Alabama.[387]

THOMAS, GEORGE: b. ca. 1839, North Carolina; occ. tinner; single; resided Raleigh, North Carolina; enlisted at Goldsboro, North Carolina, for the war as a pvt. Co. D, 2nd North Carolina Infantry, 5/29/61; transferred to CSN per SO 17, A&IGO 62, para 23, 1/28/62; served as a seaman, CSS *Virginia*; reenlisted for the war, received $50 bounty as member of crew, *Virginia*, 3/25/62; paid for service as an ordinary seaman on the *Virginia*, 4/1–5/12/62; superannuated for rations, Drewry's Bluff, Virginia, 5/12–24/62; on *Virginia* CRR, Drewry's Bluff, 7/20/62; in Howard's Grove Hosp., Richmond, Virginia, w/ varioloid, 12/28/62–1/20/63; paid, Drewry's Bluff, 5/13/62–3/31/64; took part in naval det. on special expedition from Drewry's Bluff to Charleston Naval Station, 10/1–31/63; served on CSS *Savannah*; served in Co. F, 2nd Naval Brigade; paroled, Greensboro, North Carolina, 4/26/65.[388]

TINSLEY, JEFFERSON M.: b. ca. 1843, Georgia; parents, Alanson Jefferson Tinsley and Nancy Black; occ. silversmith; resided Newton Co., Georgia; enlisted at Covington, Georgia, for 1 year as a pvt. in Co. H, 3rd Georgia

Infantry, 4/25/61; "transferred to steamer *Merrimac*," 1/13/62; served as a 1st class fireman, CSS *Virginia*; superannuated for rations, Drewry's Bluff, Virginia, 5/12–24/62; listed as having deserted on final *Virginia* payroll, dated 9/30/62; enlisted Newton Co., Georgia, for the war as a pvt. In Co. B, 13th Georgia Cavalry, 5/12/62; captured near Knoxville, Tennessee, 9/22/63; arrived, Military Prison, Louisville, Kentucky, 10/11/63; discharged and forwarded to Camp Morton, Indiana, having "enlisted in Federal Army," 10/13/63; NFR; Georgia commissioner of pensions requested verification of service, 1916; on *Virginia* muster roll in *ORN* as "Jefferson W. Tinsley."[389]

Tolson, Valentine: b. 6/6/41, Carteret Co., North Carolina; parents, Gideon Tolson and Jane Weeks; shipped for the war as an ordinary seaman, CSS *Forrest*, North Carolina Squadron, 8/3/61; served as a seaman, CSS *Virginia;* reenlisted for the war, received $50 bounty as member of crew, *Virginia*, 3/25/62; paid for service on the *Virginia*, 4/1–5/12/62; in Bruton Parish Episcopal Church Hosp., Williamsburg, Virginia, w/ venereal, 5/10–14/62; listed as in hosp. on *Virginia* muster roll, 7/31/62; paid, Drewry's Bluff, Virginia, 5/13/62–6/30/63; paid for service, naval det. on special expedition from Drewry's Bluff to Charleston Naval Station, 10/1–31/63; served in Co. G, Naval Brigade; POW, Burkeville Junction, Virginia; paroled, 4/21/65.[390]

Traylor, Thomas A.: b. ca. 1843, Dinwiddie Co., Virginia; occ. apprentice shoemaker; single; resided and enlisted at Petersburg, Virginia, for 1 year as a pvt. in Co. K, 12th Virginia Infantry, 5/4/61; described as 5'7", dark comp., dark hair, blue eyes; present until discharged per SO 18, Dept of Norfolk, 1/29/62; served as a landsman, CSS *Virginia*; appears to have left crew of *Virginia* before 4/1/62; enlisted at Richmond, Virginia, as a pvt. in Co. C, 18th Btn. Virginia Heavy Artillery, 7/28/63; detailed mending shoes for btn. at Btry. #8, Richmond, 8/31/64; present thru 12/31/64; listed as a deserter in Suffolk, Virginia, 2/11/65; NFR.[391]

Truesdale, Stephen Pickett: b. ca. 1839, South Carolina; enlisted at Charleston, South Carolina, for the war as a corp. in Co. M, 1st South Carolina Infantry, 4/22/61; mustered out at Richmond, Virginia, 7/9/61; enlisted at Richmond as a pvt. in Co. I, 1st (McCreary's) South Carolina Infantry, 7/20/61; transferred to CSN, 1/17/62; served as a landsman, CSS *Virginia*; reenlisted for the war, received $50 bounty; paid for service on the *Virginia* and Drewry's Bluff, Virginia, 4/1–11/30/62; shipped for the war, Naval Rendezvous, Richmond, 7/24/62; paid for service, Charleston Naval Station, 2/20–3/23/63; served on CSS *Charleston*, Charleston Squadron, '65; POW, Charleston, South Carolina; took oath and released, 3/18/65;

described then as 5'9", dark comp., light hair, blue eyes; postwar resided, Duval Co., Florida; received Florida pension, 1915; also found as "Truesdill;" on *Virginia* muster roll in *ORN* as "Stephen P. Truesdell."[392]

Turner, Robert G.: b. ca. 1841, England; occ. blacksmith; single; resided Yorktown, Virginia; enlisted at Norfolk, Virginia, for 1 year as a pvt. in Co. H, 6th Virginia Infantry, 4/19/61; mustered in at Craney Island, Virginia, 6/30/61; described as 5'7", dark comp., brown hair, brown eyes; present until discharged per SO 18, Dept. of Norfolk, 1/29/62, served as a landsman, CSS *Virginia*; reenlisted for the war, received $50 bounty as member of crew, *Virginia*, 3/25/62; paid for service on the *Virginia*, 4/1–5/12/62; superannuated for rations, Drewry's Bluff, Virginia, 5/12–24/62; on *Virginia* CRR, Drewry's Bluff, 7/20/62; paid, Drewry's Bluff, 5/13/62–9/30/63; paid for service, naval det. on special expedition from Drewry's Bluff to Charleston Naval Station, 10/1–31/63; POW, 4/3/65; in Jackson Gen. Hosp., Richmond, Virginia, w/chronic diarrhea, 4/8/65; turned over to Provost Marshall, 4/14/65; listed as having deserted from hosp., 4/28/65; NFR; post-war: occ. blacksmith; single; resided Norfolk, '70; d. 11/16/93.[393]

Turner, Robinson: enlisted at Memphis, Tennessee, as a pvt. in Co. C, CSMC, 9/13/61; probably served with Co. C at Pensacola Navy Yard, 7–11/61; assigned to Marine Guard, CSS *Virginia*, ca.11/61–ca. 5/62; listed as having deserted on final *Virginia* payroll, dated 9/30/62; NFR.[394]

Tynan, John W.: b. Virginia; appointed by Virginia to USN as a 3rd assistant engineer, 11/21/57, 2nd assistant engineer, 8/2/60; tendered resignation from USN, 4/20/61; dismissed from USN, 5/6/61; appointed as a 2nd assistant engineer, CSN, 6/15/61; served on steamer *St. Nicholas*, '61; acting 1st assistant engineer, 12/21/61; assigned to CSS *Virginia*, 12/9/61; assigned to engine room during Battle of Hampton Roads, Virginia, 3/8–9/62; served on CSS *Chattahoochee*, 5/13/62–'63; acting chief engineer, 5/22/63; assigned to Savannah Naval Station, '63–'64; served on the cruiser, CSS *Tallahassee*, '64; married Margaret A. Martin, Chatham, Georgia, 10/5/64; NFR.[395]

Volentine, John C.: b. ca. 1840, Ireland; occ. clerk; enlisted at New Orleans, Louisiana, for 1 year as a 3rd corp. in Co. C., 2nd Louisiana Infantry, 5/11/61; described as 5'5", fair comp., brown hair, gray eyes; absent, on special service under Gen. John Bankhead Magruder, 6/30/61; returned to unit, 7/61; 2nd corp., 11/1/61; absent, sick in Williamsburg, Virginia, on 1/62 roll; present until "transferred to C.S. Steamer *Merrimac*," 3/21/62; billeted on receiving ship, CSS *Confederate States*; served as a landsman, CSS *Virginia*; appears to have left crew of *Virginia* before 4/1/62; NFR.[396]

Vosmus, Owen D.: served as a 1st class fireman, CSS *Virginia*; reenlisted for the war, received $50 bounty as member of crew, *Virginia*, 3/25/62; appears to have left crew of *Virginia* before 4/1/62; probably enlisted at Hartsville, Tennessee, for the war as a pvt. in Co. G, 2nd (Duke's) Kentucky Cavalry, 8/1/62; listed as AWOL on 8–9/62 rolls; dropped from rolls as a deserter, 12/62; NFR; also found as "Vosemus."[397]

Wainwright, John William: b. ca. 1836, Virginia; occ. sailor; single; resided York Co., Virginia; enlisted at Yorktown, Virginia, as a pvt. in Co. C, 115th Virginia Militia, 6/24/61; acting orderly sgt., 11/61; transferred to CSN, 1/1/62; served as a seaman, CSS *Virginia*; paid for service on the *Virginia* and Drewry's Bluff, Virginia, 4/1–5/12/62; superannuated for rations, Drewry's Bluff, 5/12–24/62; listed as having deserted, 6/1/62; served as a seaman, CSS *Arctic*, North Carolina Squadron, '62–'63; NFR; postwar: occ. oysterman; married; resided York Co., '70.[398]

Waldeck, Louis: b. ca. 1838; enlisted at Charleston, South Carolina, for the war as a pvt. in Co. L, 1st (McCreary's) South Carolina Infantry, 8/27/61; present, 11–12/61; transferred to CSS *Virginia* per SO 12, Dept. of Norfolk, dated 1/18/62; KIA, 3/8/62; also found as "Waltick."[399]

Walker, George W.: b. ca. 1843, Tyrrell Co., North Carolina; occ. mariner; resided Pasquotank Co., North Carolina; enrolled at Elizabeth City, North Carolina, for the war as a 4th corp. in Co. A, 8th North Carolina Infantry, 7/22/61; mustered in 9/14/61; "transferred to C.S. Navy"; served as a seaman, CSS *Sea Bird*, North Carolina Squadron, '61; served as a seaman, CSS *Virginia;* reenlisted for the war, received $50 bounty as member of crew, *Virginia*, 3/25/62; paid for service on the *Virginia*, 4/1–5/12/61; superannuated for rations, Drewry's Bluff, Virginia, 5/12–24/62; on *Virginia* CRR, Drewry's Bluff, 7/20/62; paid, Drewry's Bluff, 5/13/62–9/30/63; paid for service, naval det. on special expedition from Drewry's Bluff to Charleston Naval Station, 9/8–10/31/63; assigned to CSS *Neuse*, North Carolina Squadron, 3-10/64; served as a quarter master, CSS *Virginia II*, James River Squadron, '64; POW, Bermuda Hundred, 10/17/64; served as a quarter master, Co. E, 1st Regt., Naval Brigade; paroled, Greensboro, North Carolina, 4/26/65.[400]

Walker, William G.: b. ca. 1842, Northumberland Co., Virginia; occ. mechanic/carpenter; enlisted at Norfolk, Virginia, for 1 year as a pvt. in Co. H, 6th Virginia Infantry, 4/19/61; mustered in at Craney Island, Virginia, 6/30/61; described as 5'4", light comp., light hair, gray eyes; present thru 12/61; AWOL, 12/26/61; court-martialed, forfeited 1 month pay; reenlisted for war, $50 bounty due, 2/24/62; discharged per SO 69,

Dept. of Norfolk, 3/30/62, served as a landsman, CSS *Virginia*; appears to have left crew of *Virginia* before 4/1/62; served as an ordinary seaman, CSS *Savannah*, Savannah River Squadron, '62–'63; paid for service as an ordinary seaman, CSS *Albemarle*, North Carolina Squadron, 7/1–9/30/64; served in Co. F, Naval Btn.; POW, Harper's Farm, Virginia, 4/6/65; sent to City Point, Virginia, and then to Point Lookout, Maryland, 4/14/65.[401]

Walling, Isaac Huff: b. 7/21/36, Keyport, New Jersey; parents, Isaac W. Walling and Mary E. Huff; occ. professional diver; resided Keyport; "when war commenced was mate of wrecking schooner *Henry W. Johnson*, belonging to Johnson & Higgins, of New York. Came to Norfolk from Baltimore on business for his captain and was unable to return"; enlisted at Norfolk, Virginia, for 1 year as a pvt. in the United Artillery (1st Co. E, 41st Virginia Infantry), 4/19/61; on special duty, raising hull of *Merrimack* from Elizabeth River, summer '61; reenlisted for war, received $50 bounty, 3/62; unit became 1st Co. C, 19th Btn. Virginia Heavy Artillery, 4/19/62; became an independent btry., 10/1/62; on CRRs, 12/22/62, 6/3/63, 9/14/63, 11/2/63 and 3/18/64; absent sick, 10/10/64; deserted, 12/22/64; obtained refuge on USS *Onongaga*; entered Federal lines at Bermuda Hundred, Virginia; sent to Washington, D.C.; took oath, furnished transport to Norfolk.[402]

Walton, John W.: b. ca. 1841, Portsmouth, Virginia; parents, John and Julia Walton; occ. apprentice painter; single; resided Portsmouth, Virginia; enlisted at Naval Hosp., Portsmouth, for 1 year as a pvt. in Co. D, 9th Virginia Infantry, 4/27/61; described as 5'4", dark comp., dark hair, dark eyes; present until discharged per SO 72, Dept. of Norfolk, 4/12/62; served as a coal heaver, CSS *Virginia*; reenlisted for the war, received $50 bounty as member of crew, *Virginia*, 3/25/62; paid for service on the *Virginia*, 4/1–5/12/62; superannuated for rations, Drewry's Bluff, Virginia, 5/12–24/62; on *Virginia* CRR, Drewry's Bluff, 7/20/62; paid, Drewry's Bluff, 5/13/62–6/30/63; NFR; postwar: occ. sign painter/painter; married; business address, 216 Queen St., Portsmouth; resided Glasgow St. near Virginia Ave.; alive, 1909.[403]

Walton, Robert H.: served as a 1st class fireman, CSS *Virginia*; paid for service on the *Virginia*, 4/1–5/12/62; superannuated for rations, Drewry's Bluff, Virginia, 5/12–24/62; on *Virginia* CRR, Drewry's Bluff, 7/20/62; shipped for the war, Wilmington, North Carolina, 11/18/62; paid, Drewry's Bluff, 5/13/62–3/31/63; listed as discharged, 2/23/63; NFR; not on *Virginia* muster roll in *ORN*; possibly "Robert H. Walton" who was b. ca. 1834, Portsmouth, Virginia; occ. mechanic/painter; enlisted at Portsmouth for 1 year as a pvt. in Co. B, 3rd Virginia Infantry, 4/20/61; described as

5'8", light comp., dark hair, blue eyes; present until discharged for disability, 1/5/62; paroled, Farmville, Virginia, 4/11–21/65.[404]

Ward, William H.: b. ca. 1841, Alamance Co., North Carolina; occ. laborer; resided Alamance Co.; enlisted at Graham, North Carolina, as a pvt. Co. E, 13th North Carolina Infantry, 5/1/61; mustered in at Garysburg, North Carolina, 5/15/61; described as 6'5", fair comp., dark hair, dark eyes; present thru 10/61; absent, on detached guard duty, Smithfield, Virginia, 11–12/61 rolls; discharged "having reenlisted in naval service on the *Merrimac*," 2/19/62; served as landsman, CSS *Virginia*; discharged, 5/8/62; NFR.[405]

Waters, John: paid as a seaman, Richmond Naval Station, 1/1–2/15/62; served as captain of top, CSS *Virginia;* reenlisted for the war, received $50 bounty as member of crew, *Virginia*, 3/25/62; on *Virginia* CRR, Drewry's Bluff, Virginia, Virginia, 7/20/62; paid for service on the *Virginia* and Drewry's Bluff, 4/1–9/30/62; paid as a seaman, Richmond Naval Station, 10/1–12/31/62; paid, Drewry's Bluff, 1/1–6/30/63; assigned to receiving ship CSS *Confederate States*, '63; acting gunner, 8/19/63; paid for service, Richmond Naval Station, 10/1–12/31/63; served on CSS *Nansemond*, James River Squadron, '63; took part in Johnson's Island expedition, '63; paid for service as a gunner, CSS *Raleigh*, James River Squadron, 3/1–5/31/64; served on CSS *Arctic*, North Carolina Squadron, '64; gunner, Provisional Navy, 6/2/64; NFR.[406]

Waters, Robert: served as a coal heaver, CSS *Virginia*; reenlisted for the war, received $50 bounty as member of crew, *Virginia*, 3/25/62; paid for service on the *Virginia*, 4/1–5/12/62; on *Virginia* CRR, Drewry's Bluff, Virginia, Virginia, 7/20/62; paid, Drewry's Bluff, 5/13/62–7/30/63; shipped for the war at Orange CH, Virginia, 3/31/64; deserted, City Point, Virginia; sent to Washington, D.C., 4/12/65; NFR.[407]

Watson, Oliver: enlisted at Memphis, Tennessee, as a pvt. in CSMC, 9/28/61; probably transferred from Capt. Hays's det., Co. D, to Co. C, Pensacola Navy Yard, ca. 11/27/61; assigned to Marine Guard, CSS *Virginia*, ca. 11/61–ca. 5/62; sgt., 5/9/63; 1st sgt., 10/1/63; reduced to pvt., 12/4/64; on duty with Co. C on 12/31/64 roll; NFR.[408]

Webb, James: served as officer's steward, CSS *Virginia*; NFR.[409]

Wenzel, Joseph: enlisted at Memphis, Tennessee, as a pvt. in CSMC, 8/22/61; probably transferred from Capt. Hays's det., Co. D, to Co. C, Pensacola Navy Yard, ca. 11/27/61; assigned to Marine Guard, CSS *Virginia*, ca. 11/61–ca. 5/62; listed as having deserted on final *Virginia* payroll, dated 9/30/62; on Co. C CRRs thru 6/30/63; NFR; also found as "Wentzel"; on *Virginia* muster roll in *ORN* as "John Wenzel."[410]

Whalen, Edward: b. ca. 1837, occ. cooper; enlisted at Conwaysboro, South Carolina, for the war as a pvt. in Co. F, 1st (McCreary's) South Carolina Infantry, 8/12/61; present, 11–12/61l; transferred to CSS *Virginia* per SO 12, Dept. of Norfolk, 1/17/62; served as a cooper, CSS *Virginia*; paid for service on the *Virginia*, 4/1–5/12/62; reenlisted for the war, received $50 bounty; superannuated for rations, Drewry's Bluff, Virginia, 5/12–24/62; on *Virginia* CRR, Drewry's Bluff, 7/20/62; paid, Drewry's Bluff, 5/13–9/30/62 and 1/1–3/31/64; served on CSS *Patrick Henry*, James River Squadron, '65; deserted, Richmond, Virginia, 4/14/65; sent to City Point, Virginia, and then to Washington, D.C., 4/17/65; described then as 5'7", light comp., dark hair, blue eyes; furnished transport to New York City.[411]

Whelin, Edward: enlisted at Richmond, Virginia, for the war as a pvt. in Co. I, 1st (McCreary's) South Carolina Infantry, 8/23/61; "transferred to Navy Dept," 1/17/62; served as a 2nd class fireman, CSS *Virginia*; paid for service on the *Virginia*, 4/1–5/12/62; superannuated for rations, Drewry's Bluff, Virginia, 5/12–24/62; on *Virginia* CRR, Drewry's Bluff, 7/20/62; paid as a 2nd class fireman, Drewry's Bluff, 5/13–7/25/62 and as a coal heaver, 7/26–9/30/62; NFR; may have later enlisted at Charleston, South Carolina, for the war as a pvt. in Co. C, 1st (Charleston) Btn. South Carolina infantry, 1/21/63; present thru 8/63; NFR; not on *Virginia* muster roll in *ORN*; also found as "Wheelan."[412]

Whitaker, William C.M.: b. ca. 1831; resided Chowan Co.; enlisted at Edenton, North Carolina, for the war as a pvt. in Co. A, 1st North Carolina Infantry, 4/22/61; transferred to CSN per SO 17, A&IGO 62, para 23, 2/1/62; served as an ordinary seaman, CSS *Virginia*; reenlisted for the war, received $50 bounty as member of crew, *Virginia*, 3/25/62; on *Virginia* CRR, Drewry's Bluff, Virginia, 7/20/62; paid for service on the *Virginia* and Drewry's Bluff, 4/1–12/31/62; discharged, 12/14/62; NFR; on *Virginia* muster roll in *ORN* as "W.C.M. Whitacker."[413]

White, Elsberry Valentine: b. 1839, Wilkinson Co., Georgia; served in City Light Guard, Columbus, Georgia; transferred to CSN; appointed by Virginia as an acting 3rd assistant engineer, CSN, 1/16/62; assigned to CSS *Virginia*, 1/18/62; served on gun deck in charge of engine room gong and speaking tube during Battle of Hampton Roads, Virginia, 3/8–9/62; one of a party of ten selected to destroy *Virginia*; took part in Battle of Drewry's Bluff, Virginia, 5/15/62; served on CSS *Baltic*, Mobile Squadron, '62; resigned CSN, 8/29/62; returned to Columbus, Georgia; invented machinery used to manufacture nearly all buttons and buckles subsequently used by CSA; enrolled at Columbus as a 1st Lt. in Co. F, 19th Btn. Georgia

Infantry (State Guards), 8/3/63; served at Atlanta, Georgia, under Gen. Hood, '64; captured, Columbus, postwar: founded E.V. White & Co., mill/ general supply business; president, Tidewater Insurance Co., Norfolk; founded, Norfolk National Bank; resided Portsmouth, Virginia; member, Pickett-Buchanan Camp, UCV, Norfolk; commander, Norfolk militia; d. 1919; bur. Cedar Grove Cem., Portsmouth, Virginia.[414]

Whitten, James: enlisted at Mobile, Alabama, as a pvt. in CSMC, 8/13/61; probably served at Pensacola Navy Yard until Co. C transferred to Gosport Navy Yard, Portsmouth, Virginia, 11/29/61; assigned to Marine Guard, CSS *Virginia*, ca. 11/61–ca. 5/12/62; listed as having deserted on final *Virginia* payroll, dated 9/30/62; on Co. C CRR, 3rd qtr. '62; listed as having deserted before 9/30/62; NFR.[415]

Wilkins, Willis A.: served as a carpenter's mate, CSS *Virginia*; superannuated for rations, Drewry's Bluff, Virginia, 5/12–24/62; NFR; probably "Willis A. Wilkins" who was b. ca. 1836; occ. carpenter; married; resided Portsmouth, Virginia; postwar: retail grocer; resided Portsmouth, '70.[416]

Williams, George: b. ca. 1833, New York; occ., sailor; resided New Orleans, Louisiana; enlisted at Camp Moore, Louisiana, for the war as a pvt. in Co. G., 10th Louisiana Infantry, 7/22/61; present until "discharged to CSS *Merrimac*," 1/2/62; served as a 2nd class fireman on CSS *Virginia*; paid for service on the *Virginia* and Drewry's Bluff, Virginia, 4/1–7/20/62; superannuated for rations, Drewry's Bluff, 5/12–24/62; listed as having deserted on 7/20/62; may have later served on CSS *Virginia II*, James River Squadron, '64; deserted, Bermuda Hundred, 10/8/64 sent to City Point, Virginia; took oath and furnished transport to New York City.[417]

Williams, Hezekiah (or Henry): b. ca. 1829, Virginia; occ. pilot; married; resided Norfolk, Virginia; served as a civilian pilot, CSS *Virginia;* took part Battle of Hampton Roads, Virginia, 3/8–9/62; served on CSS *Richmond*, James River Squadron, '64–'65; NFR; postwar: occ. pilot; resided Norfolk, Virginia.[418]

Williams, John Q.A.: b. 1/29/37, Portsmouth, Virginia; occ. upholsterer; single; resided Portsmouth; enlisted at Pig Point, Virginia, as a pvt. in Co. G, 9th Virginia Infantry, 6/13/61; mustered in, 6/30/61; described as 6', dark comp., dark hair, blue eyes; present until discharged per SO 20, Dept. of Norfolk, 2/1/62; served as a surgeon's steward, CSS *Virginia*; paid for service on the *Virginia*, 4/1–5/12/62; superannuated for rations, Drewry's Bluff, Virginia, 5/12–24/62; on *Virginia* CRR, Drewry's Bluff, 7/20/62; in Howards Grove Hosp., Richmond, Virginia, w/ variola, 12/28/62–2/24/63; paid, Drewry's Bluff, 5/13/62–3/31/64; in Jackson Gen. Hosp.,

Richmond, 4/7/65; captured and sent to U.S. Army Prison (Liddy), Richmond, Virginia; postwar: occ. coach trimmer; resided 4th near Lincoln St., Portsmouth; member, Stonewall Camp, UCV, Portsmouth; d. 11/7/92; bur. Oak Grove Cem., Portsmouth.[419]

WILLIAMS, JOHN: enlisted at New Orleans, Louisiana, for the war as a pvt. in Co. F., 15th Louisiana Infantry, 6/21/61; present until transferred to CSN, 2/62; served as a seaman, CSS *Virginia*; reenlisted for the war, received $50 bounty as member of crew, *Virginia*, 3/25/62; appears to have left crew of *Virginia* before 4/1/62; shipped for the war as a seaman, Naval Rendezvous, Richmond, Virginia, 7/24/62; served as a quarter gunner, CSS *Jamestown*, James River Squadron, '62; paid, Drewry's Bluff, Virginia, 1/1–3/31/63; may have later served as a seaman, CSS *Tuscaloosa*, Mobile Squadron, '63; may have later served as a seaman, CSS *Missouri*, *Webb* and *Cotton*, Red River Squadron, '63; NFR.[420]

WILLIAMS, PETER: b. ca. 1827, England; occ. seaman; resided and enlisted at Mobile, Alabama, as a pvt. in Co. E, 8th Alabama Infantry, 5/6/61; described as 5'9", red hair, gray eyes; transferred to CSN by order of secretary of war, 2/12/62; shipped for 3 years as a seaman, Naval Rendezvous, Richmond, Virginia, 2/16/62; served as a seaman, CSS *Virginia*, '62; reenlisted for the war, received $50 bounty as member of crew, *Virginia*, 3/25/62; appears to have left crew of *Virginia* before 4/1/62; may have later served as a coal heaver, CSS *Missouri*, *Webb* and *Cotton*, Red River Squadron, '63; paid for service as a seaman and cook, CSS *Chattahoochee*, 4/8–6/30/64; captured on blockade runner, CSS *Greyhound* by USS *Connecticut*, 5/12/64; POW, Camp Hamilton, Virginia, 5/16/64; released and sent to Baltimore, Maryland, 6/14/64; deserted, Savannah, Georgia, 4/27/65; described then as 5'9", dark comp., dark hair, gray eyes; NFR.[421]

WILLIAMS, PLEASANT H.: b. ca. 1836, Roanoke Co., Virginia; occ. farmer; enlisted at Salem, Virginia, for 1 years as a pvt. in 1st Co. A, 9th Virginia Infantry, 5/14/61; mustered in at Lynchburg, Virginia, 5/16/61; described as 5'7", light comp., brown hair, blue eyes; transferred to CSN, 3/28/62; reenlisted for 2 years, $50 bounty due, 2/13/62; present until discharged per SO 69, Dept. of Norfolk, 3/29/62; served as a landsman, CSS *Virginia*; paid for service on the *Virginia* and Drewry's Bluff, Virginia, 4/1–5/15/62; superannuated for rations, Drewry's Bluff, 5/12–24/62; listed as having deserted on final *Virginia* payroll, dated 9/30/62; enlisted at Salem, Virginia, for the war as a pvt. in the Salem Artillery (Griffin's Co., Virginia Light Artillery), 6/21/62; present, 6–9/62; furloughed from Gen. Hosp., Staunton, Virginia, 10/62–3/63; corp., 9/10/63; present until transferred to 2nd Virginia Cavalry, enlisting at Montpelier, Virginia, for the war as a pvt. in Co. D, 2/64; absent, on horse detail, 2/29/64; listed as AWOL, 3/9/64; listed as having deserted, 6/64; NFR.[422]

Wilson, Charles H.: b. ca. 1838; enlisted at Washington, North Carolina, for 1 year as a pvt. in Co. K, 1st North Carolina Artillery, 5/8/61; present until "joined CSN on *Forrest*"; shipped for 1 year as a seaman, CSS *Forrest*, North Carolina Squadron, 8/3/61; served as a seaman, CSS *Virginia*; reenlisted for the war, received $50 bounty as member of crew, *Virginia*, 3/25/62; paid, Drewry's Bluff, Virginia, 6/1–8/15/62; may have later served as a seaman, CSS *Morgan*, Mobile Squadron, '62–'63; paid for service as a quarter master, CSS *Chicora*, Charleston Squadron, 7/1–9/30/63; paid for service, Charleston Naval Station, 10/1–31/63; paid for service as a quarter master, CSS *Chicora*, 1/1–3/31/64; NFR.[423]

Wilson, John A.: b. ca. 1824, Danzig, Germany; occ. bridge carpenter; enlisted at Onslow Co., North Carolina, for the war as a pvt. in Co. E, 3rd North Carolina Infantry, 5/13/61; described as 5'7", light comp., light hair, gray eyes; discharged by order of War Dept. to enable him to join the navy," per SO 92 A&IGO 61, para 9, 7/9/61; paid as a seaman, CSS *Sea Bird*, North Carolina Squadron, 7/14–11/30/61; paid, Gosport Navy Yard, Portsmouth, Virginia, 2/18/62; served as captain of hold/seaman, CSS *Virginia*; reenlisted for the war, received $50 bounty as member of crew, *Virginia*, 3/25/62; superannuated for rations, Drewry's Bluff, Virginia, 5/12–24/62; may have later served as a seaman, CSS *Morgan*, Mobile Squadron, '62–'63; served as a quarter master, CSS *Chicora*, Charleston Squadron, '63–'64; NFR.[424]

Witz, William: b. ca. 1826, Germany; occ. sailor; single; resided New Orleans, Louisiana; enlisted at Camp Moore, Louisiana, for the war as a pvt. in Co. G., 10th Louisiana Infantry, 7/22/61; present until "transferred to CSS *Merrimac*," 1/2/62; served as an ordinary seaman, CSS *Virginia* before 4/1/62; listed as having deserted on final *Virginia* payroll, dated 9/30/62; on *Virginia* muster roll in *ORN* as "Witts."[425]

Wood, John Taylor: b. 8/13/30, Ft. Snelling, Iowa Territory (present-day St. Paul, Minnesota); parents, Robert Crook Wood, assistant surgeon general, U.S. Army, and Anne Mackall Taylor, eldest daughter of President Zachary Taylor; aunt married Jefferson Davis; appointed by Kentucky to USN as a midshipman, 4/7/47; passed midshipman, 6/10/53; master, 9/15/55; Lt., 9/16/55; married Lola MacKubin, 11/26/56; served on USS *Cumberland*; taught gunnery tactics at USNA at start of Civil War; tendered resignation and dismissed from USN, 4/2/61; appointed 1st Lt., CSN, 10/4/61; served at naval batteries, Evansport, Virginia, '61; assigned to CSS

Virginia, 11/25/65; commanded stern gun during Battle of Hampton Roads, Virginia, 3/8–9/62; took part in Battle of Drewry's Bluff, Virginia, 5/15/62; earned reputation as commerce raider, received thanks from Confederate Congress; terrorized Northeast coast, destroying or capturing 31 vessels; burned ship *Alleghania*, 10/29/62; appointed colonel, CSA cavalry, 1/26/63; captured USS *Satellite* and *Reliance*, 8/23/63; captured schooners *Golden Rod*, *Coquette* and *Two Brothers*, 8/24/63; cmdr., 9/21/63; captured/destroyed USS *Underwriter*, 2/2/64; took part in capture of Plymouth, North Carolina, 4/20/64; commanded the cruiser, CSS *Tallahassee*, '64; cmdr., Provisional Navy, 5/13/64; capt., Provisional Navy, 2/10/65; aide to Jefferson Davis, '63–'65; accompanied Jefferson Davis after fall of Richmond, Virginia; escaped to Cuba; postwar: formed mercantile company and shipping/marine insurance firm; resided Halifax, Nova Scotia; d. of muscular rheumatism, 7/19/04; bur. Camp Hill Cem., Halifax.[426]

Wood, Levin H.: b. ca. 1834, Alamance Co., North Carolina; occ. grocer; resided Caswell Co., North Carolina; enlisted at Yanceyville, North Carolina, as a pvt. in Co. A, 13th North Carolina Infantry, 4/29/61; mustered in at Raleigh, North Carolina, 5/18/61; described as 6', dark comp., light hair, gray eyes; "transferred to CSN for duty on the CSS *Merrimac*," 2/19/62; billeted on receiving ship, CSS *Confederate States*; served as a landsman, CSS *Virginia*; superannuated for rations, Drewry's Bluff, Virginia, 5/12–24/62; enlisted at Trenton, North Carolina, for the war as a pvt. in Co. A, 8th Btn. North Carolina Partisan Rangers, 11/1/62; present thru 8/63; transferred to Co. F, 66th North Carolina Infantry, 10/2/63; present thru 10/64; paroled Greensboro, North Carolina, 3/9/65.[427]

Wright, George: b. ca. 1820, Virginia; occ. sailor; married; resided Richmond, Virginia; served as a civilian pilot, CSS *Virginia*; took part in Battle of Hampton Roads, Virginia, 3/8–9/62; on list of officers of CSS *Richmond*, James River Squadron, '64.[428]

Wright, Sidney R.: b. ca. 1837, Caswell Co., North Carolina; occ. carpenter; resided Caswell Co.; enlisted at Leasburg, North Carolina, as a sgt. in Co. D, 13th North Carolina Infantry, 5/1/61; mustered in at Raleigh, North Carolina, 5/5/61; described as 6'3", light comp., dark hair, dark eyes; absent, sick in hosp., 5/61; reduced to pvt., 10/61; present until "transferred to CSN for duty on the CSS *Merrimac*," 2/15/62; served as a landsman, CSS *Virginia*; superannuated for rations, Drewry's Bluff, Virginia, 5/12–24/62; apparently RTD w/ regt.; in Chimborazo Hosp. #1, Richmond, Virginia, w/ diarrhea, 6/6/62; transferred to Gen. Hosp., Petersburg, Virginia, 6/26/62; enlisted at Yanceyville, North Carolina, for the war as a pvt. in Co. B, 4th North Carolina

Cavalry, 3/6/63; present, 5–6/63; absent, home on detail, 7–8/63; present, 9–12/63; POW, Alexandria, Virginia, 1/12/64; sent to Old Capital Prison, Washington, D.C., 1/13/64; sent to Point Lookout, Maryland, 2/3/64; exchanged, 2/10/64; on rolls at Camp Lee, Virginia, 2/17/65; NFR.[429]

Young, Ephraim: b. ca. 1839, Matthews Co., Virginia; occ. farmer; enlisted at Craney Island, Virginia, for 1 year as a pvt. in Co. H, 6th Virginia Infantry, 8/1/61; described as 6', light comp., light hair, gray eyes; present until discharged per SO 18, Dept. of Norfolk, 1/29/62; served as a landsman, CSS *Virginia*; paid for service on the *Virginia*, 4/1–5/12/62; reenlisted for the war, received $50 bounty; superannuated for rations, Drewry's Bluff, Virginia, 5/12–24/62; on *Virginia* CRR, Drewry's Bluff, 7/20/62; paid, Drewry's Bluff, 5/13/62–9/30/63; paid for service, naval det. on special expedition from Drewry's Bluff to Charleston Naval Station, 10/1–31/63; served as a landsman, CSS *Fredericksburg*, James River Squadron, '64–'65; ordered to Naval Ordnance Works, Richmond, Virginia, 1/13/65; paroled, 4/25/65; postwar: member, Healy-Claybrook Camp, UCV, Lot, Virginia; d. c.1916.[430]

Young, William H.: b. ca. 1837; occ. lawyer; resided Granville Co., North Carolina; enlisted at Oxford, North Carolina, as a pvt. in 2nd Co. D, 12th North Carolina Infantry, 4/22/61; mustered in at Garysburg, North Carolina, 5/16/61; promoted corp.; transferred to CSN, 6/7/61; paid for service as a seaman on receiving ship, CSS *Confederate States*, 1/1–9/62; paid, North Carolina Squadron, 1/10–3/6/62; served as a boatswain's mate, CSS *Virginia*; paid for service on the *Virginia* and Drewry's Bluff, Virginia, 4/1–6/30/62; superannuated for rations, Drewry's Bluff, 5/12–24/62; on *Virginia* CRR, Drewry's Bluff, 7/20/62; volunteered for duty on CSS *Chattahoochee*, 8/1/62; paid for service, CSS *Chattahoochee*, 8/1–12/31/62 and 4/1–5/15/63; listed as having deserted, 5/13/63; NFR.[431]

Others Reportedly Aboard the CSS *Virginia*

The following individuals *may* have served on the *Virginia*. However, none of their names appear in the muster roll in the *ORN* or in any of the pay records of the *Virginia*.

Brady, J.T.: claimed he served on the *Virginia*; raised in Charleston, South Carolina; reportedly served on CSS *Lady Davis*, Charleston Squadron, then shipped on CSS *Patrick Henry*, James River Squadron, and later assigned Drewry's Bluff, Virginia; reportedly helped capture two Union gunboats with John Taylor Wood on the Rappahannock River below Fredericksburg, Virginia; postwar: resided Kosse, Texas, 1908.[432]

Cline, William R.: wrote a *Southern Historical Society Proceedings* article in which he claimed to have been aboard (apparently as a marine) when the *Virginia* was launched; name does not appear in CSMC records (although many have been lost), and he was not a member of the United Artillery.[433]

Donegan, Samuel W.: reportedly served on the *Virginia* and took part in the Battle of Hampton Roads, 3/8–9/62; b. 7/3/34, Huntsville, Alabama; enlisted at New Orleans, Louisiana, for 1 year as a pvt. in Green's Co., Louisiana Artillery, 4/26/61; present, 7/61–4/62; reenlisted for 2 years, 4/30/62; granted leave per SO 161, HQ, Army of Northern Virginia, para 2, 7/24/62; in Gen. Hosp. #7, Richmond, Virginia; discharged due to kidney/bladder problems, 10/4/62; enlisted as a pvt. in Co. K, 2nd Louisiana Cavalry, date unknown; surrendered, New Orleans, 5/26/65; paroled, Alexandria, Louisiana, 6/3/65; d. Confederate Soldiers' Home, Mountain Creek, Alabama, 9/3/1916.[434]

Fentress, William Henry: NFR

Holsenback, Alex: shipped for the war at Orange CH, Virginia, 4/3/64; NFR.[435]

Holzman, Benjamin: reportedly served as a gunner, CSS *Virginia*; probably Benjamin Holzman, b. 4/3/44; enlisted at Norfolk, Virginia, for 1 year as a pvt. in the United Artillery (1st Co. E, 41st Virginia Infantry), 5/12/61; present thru 3/10/62; reenlisted for war, received $50 bounty, 3/10/62; NFR; d. 1/4/1922; bur. Hebrew Rest Cemetery, Shreveport, Louisiana.[436]

Langhorne, James K.: NFR

Mabey, James: a British citizen who claimed to be on *Virginia*.

Mackenzie, John Malcolm: b. ca. 1847; resided Cameron, Louisiana; reportedly shipped as a cabin boy on CSS *Sumter*, New Orleans Naval Station, 5/61–1/62; ordinary seaman, 1/62; took passage on English ship from Gibraltar to New Orleans; reportedly went to Norfolk, Virginia, joined crew of the *Virginia* and took part in the Battle of Hampton Roads, 3/8–9/62; supposedly took part in the Battle of Drewry's Bluff, Virginia, 5/15/62; reportedly served on CSS *Alabama* and took part in engagement with USS *Kearsarge* off Cherbourg, France, 6/10/64; postwar: served as a seaman on several English merchant vessels and then joined the Chilean navy; finally resided in Tacoma, Washington; occ. fireman; member, Pickett Camp, UCV, Tacoma; d. 6/27/1916.[437]

McClellan, John: attended veterans' gathering at Gettysburg, Pennsylvania, 1913; claimed he served on the *Virginia*.[438]

Penn, Thomas R.: enlisted at Camp Wight for the war as a pvt. in Co. G, 4th Georgia Infantry, 5/16/62; present thru 6/62; present 11–12/62; paid, on CRRs, 2/4/63 and 3/24/63; paid, 1–3/64; transferred to CSN per SO 94, HQ, Army of Northern Virginia, para 18; shipped for the war at Orange CH, Virginia, 3/31/64; NFR; Georgia commissioner of pensions requested verification of service, 1916.[439]

Reed, Charles C.: enlisted at Fort Drewry, Virginia, for the war as a pvt. in the United Artillery, 3/28/63; on CRRs, 6/3/63, 9/14/63 and 11/2/63; absent, detached duty as corp. w/ CSN, per SO 63, A&IGO, dated 3/16/64; present on 12/31/64; paroled, Appomattox CH, 4/9/65.[440]

Ring, John L.: enlisted at Charleston, South Carolina, for six months as pvt. in Co., M, 1st South Carolina Infantry, 4/22/61; mustered out, Richmond, Virginia, 7/9/61; enlisted at Richmond for the war as a pvt. in Co. I, 1st (McCreary's) South Carolina Infantry, 7/20/61; transferred to CSN, 1/17/62; unofficial compilation says to CSS *Virginia*; may have later enlisted as a pvt. in Co. F, 4th Virginia Cavalry, 3/1/62; deserted 3/15/63; took Oath March, '64; sent to Philadelphia, Pennsylvania; NFR.[441]

Roberts, Henry: served as a deckhand, CSS *Elton McDonald*.[442]

Savage, John H.: b. ca. 1835; enlisted at Charleston, South Carolina, for the war as a corp. in Co. M, 1st South Carolina Infantry, 4/22/61; mustered out at Richmond, Virginia, 7/9/61; enlisted at Richmond as a pvt. in Co. I, 1st (McCreary's) South Carolina Infantry, 7/20/61; transferred to CSN, 1/17/62; unofficial compilation says to CSS *Virginia*; in Gen. Hosp. #4, Wilmington, North Carolina, 5/19–20/62; NFR.[443]

Appendix XIII

Chronology of the CSS *Virginia*

1795

November 9

Josiah Tattnall was born on his family estate, Bonaventura, near Savannah, Georgia.

1800

September 17

Franklin Buchanan was born in Baltimore, Maryland.

1801

April 2

The U.S. Navy purchased the Gosport Navy Yard from the Commonwealth of Virginia.

1812

January 1

Josiah Tattnall was appointed by Georgia as a midshipman, USN, and assigned to the USF *Constellation*.

1813

June 22

Josiah Tattnall fought during the battle of Craney Island.

1818

April 1
Josiah Tattnall was promoted lieutenant, USN.

1819

February 24
James Allen Semple was born at Cedar Hill, New Kent County, Virginia.

1821

April 15
Catesby ap Roger Jones was born in Fairfield, Virginia.

1826

April 5
Thomas Kevill was born in Sligo, Ireland.

1826

June 7
Dinwiddie Brazier Phillips was born in Fauquier County, Virginia.

September 20
Hunter Davidson was born in Georgetown, Washington, D.C.

December 18
John Mercer Brooke was born at Fort Brooke, Florida.

1827

September 21
Robert D. Minor was born at Sunning Hill, Louisa County, Virginia.

1830

John Taylor Wood was born at Fort Snelling, Iowa Territory.

1833

November 25
Julius Ernest Meiere was born at New Haven, Connecticut.

1836

Charles Schroeder was born at Portsmouth, Virginia.

June 18
Catesby Jones was appointed by Virginia to the USN as a midshipman.

1837

January 15
Chauncey Alexander Sprague was born in Concord, New York.

January 29
John Q.A. Williams was born in Portsmouth, Virginia.

March 3
Benjamin Simms Herring was born in Duplin County, North Carolina.

March 8
John Dunlop was born in Portland, Maine.

August 17
Douglas French Forrest was born in Fairfax, Virginia.

1838

February 25
Josiah Tattnall was promoted commander, USN.

March 13
Andrew Jackson Sharpe was born in Grundy, Tennessee.

November 7
Julius Durand was born in Elizabethtown, New York.

1839

October 9
Charles Carroll Simms was appointed by Virginia as a midshipman, USN.

December 10
Walter Raleigh Butt was born in Portsmouth, Virginia.

1840

February 6
Marcellus Augustus Tharpe was born in Doles, Georgia.

September 15
Cornelius Sidney Hampton was born in Franklin County, Kentucky.

July 17
Eugenius Alexander Jack was born in Portsmouth, Virginia.

November 4
Hugh Lindsay enlisted as carpenter, USN.

1841

Janurary 5
Elijah Wilson Flake was born in Anson County, North Carolina.

February 25
Robert D. Minor was appointed by Missouri to the USN as a midshipman.

March 8
Hardin Beverly Littlepage was born at Piping Tree, King William County, Virginia.

April 14
Robert Chester Foute was born in Greenville County, Tennessee.

October 19
John Pembroke Jones was appointed by Virginia as a midshipman.

1842

March 16
James M. Harvey was born in Selma, Alabama.

April 14
Carey J. Hall was born in Norfolk, Virginia.

1843

May 3
Charles Oliver was appointed by Virginia as a master's mate, USN.

1844

Augst 27
Jacob Scholls enlisted at Pensacola Navy Yard for four years as a private, USMC.

October 5
John Patrick Kevill was born at Charlestown, Massachusetts.

October 10
Franklin Buchanan was appointed first superintendent, U.S. Naval Academy, Annapolis, Maryland.

October 12
James Semple was appointed by Virginia to the USN as purser.

1845

April 30
Lawrence M. Rootes was born in Gloucester County, Virginia.

July 2
Charles Carroll Simms was promoted to passed midshipman.

1846

June 1
Charles B. Oliver was named gunner, USN.

1847

March 22

Commander Josiah Tattnall led the Mosquito Flotilla of small steamer- and schooner-rigged gunboats to attack Fort San Juan d'Ulua guarding the sea approaches to Vera Cruz.

April 7

John Taylor Wood was appointed by Kentucky to the USN as midshipman.

November 8

Dinwiddie Phillips was appointed by Virginia to the USN as an assistant surgeon.

1848

September 14

Jacob Scholls reenlisted as sergeant, USMC, for four years at Pensacola, Florida.

1850

Catesby Jones was appointed master, USN.

February 5

Josiah Tattnall was promoted captain.

1852

April 17

Gosport Navy Yard was the scene of experiments that tested the capacity of iron vessels in resisting the force of shells and shot. The tests "proved that iron is not so invulnerable as many as heretofore supposed, and unsuited to such purposes."

1853

May 21

Henry Ashton Ramsay was appointed to the USN as third assistant engineer.

December 9

Hugh Lindsay was dismissed from the USN.

December 24

Marshall P. Jordan was appointed by Virginia as a third assistant engineer.

Chronology of the CSS *Virginia*

1854

June 15
The USS *Merrimack* was launched at Charleston Navy Yard, Boston Harbor.

August 12
Charles Carroll Simms was promoted lieutenant, USN.

1855

March 1
John Pembroke Jones was promoted lieutenant, USN.

April 16
Julius Meiere was appointed by Washington, D.C., to the USMC as lieutenant.
John Taylor Wood was promoted to lieutenant, USN.

June 14
The USS *Merrimack* was launched at Charlestown Navy Yard.

June 21
The USS *Merrimack*'s machinery arrived at Charlestown Navy Yard for installation.

September 14
Robert D. Minor was appointed lieutenant, USN.

1856

February 20
The USS *Merrimack* was commissioned.

February 25
The USS *Merrimack* left Boston Harbor for her first cruise.

March 2
The USS *Merrimack*'s log noted: "Ship rolling very deeply. Engines racing badly."

June 26
Ashton Ramsay was promoted to second assistant engineer, USN.

September 25
Charles Kirby King Jr. was appointed by Washington, D.C., to the USNA.

1857

May 7
The USS *Roanoke* (one of five *Merrimack*-class frigates) was commissioned at Gosport Navy Yard.

May 9
Charles Schroeder was promoted to second assistant engineer, USN.

September 23
Hardin Beverly Littlepage was appointed by Virginia to the USNA.

March 28
Charles H. Hasker was appointed by Virginia as a boatswain, USN.

May 16
Algernon S. Garnett was appointed by Virginia as an assistant surgeon.

May 9
Marshall Jones was appointed second assistant engineer, USN.

August 2
Ashton Ramsay and Charles Schroeder were promoted first assistant engineers, USN.

November 21
John Tynan was appointed by Virginia to USN as a third assistant engineer.

1858

September 30
Henry H. Marmaduke was appointed by Missouri to the USNA.

1859

June 25
Captain Josiah Tattnall assisted the Anglo-French fleet in its attack on Chinese forts on the Pei-ho River in violation of U.S. neutrality. Tattnall explained his actions by saying, "Blood is thicker than water."

1860

February 16

The USS *Merrimack* arrived at Gosport Navy Yard and was immediately placed in ordinary for an overhaul and repair of her engines.

August 11

Benjamin S. Simms was appointed from North Carolina to the USN as third assistant engineer.

1861

January 13

John Randolph Eggleston, while at sea aboard the USS *Wyandotte*, tendered his resignation from the USN.

January 22

John Randolph Eggleston's resignation from USN was accepted.

February 2

Congressman William P. Chilton of Alabama introduced a resolution "that the Committee on Naval Affairs be instructed into the propriety of constructing by this government of two iron-plated frigates and such iron-plated gunboats as may be necessary to protect the Commerce and provide the safety of the Confederacy."

February 10

Andrew Joseph Dalton enlisted at Charleston, South Carolina, as a private in Company C, 1st South Carolina Artillery.

February 14

Confederate Congress authorized "the Committee on Naval Affairs to procure...all such persons versed in naval affairs as they may deem it advisable to consult with."

February 18

In his inaugural address, President Jefferson Davis suggested "that for the protection of our harbors and commerce on the high seas a Navy adapted to those objects will be required."

February 20

Josiah Tattnall tendered his resignation from the USN.

The Confederate Department of the Navy was established.

February 21

Stephen Russell Mallory of Florida was appointed secretary of the navy for the Confederacy.

Northampton County, Virginia native William Bramwell Colonna enlisted in the 1st South Carolina Infantry.

Josiah Tattnall's resignation was accepted.

March 3

Josiah Tattnall was appointed captain, CSN.

March 23

The USS *Cumberland* arrived in Hampton Roads.

March 25

Reuben Thom was the first officer appointed to the CSMC.

Jacob Scholls enlisted at Montgomery, Alabama, in Company C, CSMC.

March 29

Samuel N. Aenchbacker became the seventh man to enlist in the CSMC.

March 31

U.S. secretary of the navy Gideon Welles ordered 250 men from the Brooklyn Navy Yard to reinforce the Gosport Navy Yard.

April 2

John Taylor Wood tendered his resignation and was dismissed from the USN.

April 4

Jacob Shavor enlisted as a private in the CSMC.

Benjamin Slade enlisted for one year as a private in Company K, 12th Virginia Infantry.

April 3

Lieutenant Julius Meiere, USMC, married Nannie Buchanan at Washington Navy Yard. President Abraham Lincoln and Senator Stephen Douglas were among the many notable guests who attended the wedding.

April 4

Thomas A. Traylor enlisted at Petersburg as a private in Company K, 12th Virginia Infantry.

Martin G. Thayer enlisted at Petersburg for one year as first sergeant in Company K, 12th Virginia Infantry.

April 5

John Randolph Eggleston was appointed first lieutenant, CSN.

April 10

The USS *Pawnee*, captained by Commander Stephen Rowan, departed Hampton Roads for relief of Fort Sumter. Secretary of the Navy Gideon Welles ordered Flag Officer Charles Stewart McCauley, commandant of the Gosport Navy Yard, to place the USS *Merrimack* in condition to steam to a Northern yard, "or, in case of danger from unlawful attempts to take possession of her, that she may be placed beyond their reach." Welles instructed McCauley to guard public property within Gosport Navy Yard and to "exercise your judgement in discharging the responsibilities that resolves on you." The secretary also cautioned the commandant that "there should be no steps taken to give needless alarm, but it may be best to order most of the shipping to sea or other stations."

April 11

Commander James Alden was ordered to report to Flag Officer McCauley to take command of the USS *Merrimack.*

McCauley telegraphed Welles that it would take a month to revitalize the *Merrimack*'s dismantled engines.

John C. Volentine enlisted at New Orleans, Louisiana, as a third corporal in Company C, 2nd Louisiana Infantry.

April 12

Welles considered McCauley "feeble and incompetent for the crisis" and detailed the U.S. Navy's chief engineer, Benjamin Franklin Isherwood, to Gosport to prepare the *Merrimack* for sea.

Andrew J. Dalton participated in the bombardment of Fort Sumter.

Thomas Purcell enlisted at Montgomery, Alabama, as a private, CSMC.

April 13

Samuel Bell Smyth enlisted in Company C, CSMC.

April 14

Commander James Alden and Chief Engineer B.F. Isherwood arrived at Gosport Navy Yard. Isherwood immediately began work on the *Merrimack*'s engine system.

Andrew J. Dalton witnessed the evacuation of Fort Sumter's Union garrison.

April 15

Benjamin R. Sheriff enlisted as an ordinary seaman, USN.

APRIL 16

Flag Officer Garrett J. Pendergrast, commander of the USS *Cumberland*, was ordered to keep his ship in Gosport Navy Yard "and, in case of invasion, insurrection, or violence of any kind, to suppress it, repelling assault by force."

Secretary Welles ordered McCauley to remove all public property from Gosport. Besides the *Merrimack*, the *Dolphin*, *Germantown* and *Plymouth* were the only ships in relatively good condition to warrant removal.

APRIL 17

Virginia seceded from the Union.

Benjamin F. Isherwood completed emergency repairs to the *Merrimack* and reported to McCauley that the frigate would be ready to leave port the next day.

The citizens of Norfolk and Portsmouth organized a "Vigilant Committee."

Robert D. Minor resigned his commission aboard the USS *Cumberland*.

APRIL 18

Chief Engineer B.F. Isherwood reported the *Merrimack* ready for sea; however, McCauley, as commandant of Gosport Navy Yard, refused to release the frigate from the yard.

John Letcher, governor of the commonwealth of Virginia, ordered Major General William Booth Taliaferro of the Virginia Militia to assume command of troops assembling in the Norfolk area and to occupy Gosport Navy Yard.

Governor Letcher named Catesby Jones and Robert Pegram as captains in the Virginia State Navy. Pegram was instructed by the governor to "assume command of the naval station, with authority to organize naval defenses, enroll and enlist seaman and marines, and temporarily appoint warrant officers, and to do and perform whatever may be necessary to preserve and protect the property of the commonwealth and of the citizens of Virginia."

Flag Officer Hiram Paulding, USN, was detailed to take command of Gosport Navy Yard. Welles ordered that on "no account should the arms and munitions be permitted to fall into the hands of insurrectionists… should it finally become necessary, you will, in order to prevent that result, destroy the property." Paulding left the Washington Navy Yard with one hundred marines onboard the eight-gun steamer *Pawnee*.

Dinwiddie Phillips resigned from the USN.

April 19

General Taliaferro advised yard commandant C.S. McCauley that he planned to assume possession of Gosport Navy Yard on behalf of the "sovereign state of the Commonwealth of Virginia."

Thomas Kevill and the United Artillery occupied Fort Norfolk. The fort's magazine contained over 250,000 pounds of gunpowder.

Flag Officer McCauley refused to allow Lieutenant Thomas O. Selfridge of the USS *Cumberland* to take the brig *Dolphin* to Craney Island and stop the sinking of any more ships in the channel by the Virginians.

Franklin Buchanan resigned from the USN.

William G. Walker enlisted at Norfolk as a private in Company H, 6th Virginia Infantry.

Arthur Sinclair IV enlisted as a private in Company G, 6th Virginia Infantry.

April 20

John Tynan tendered his resignation from the USN.

12:00 p.m.: Flag Officer C.S. McCauley dismissed workmen from Gosport Navy Yard.

1:00 p.m.: Gosport's loyal workmen, marines and sailors began to scuttle ships and destroy property that could not be removed.

8:00 p.m.: Flag Officer Paulding arrived at Gosport with 100 marines and 350 men of the 3rd Massachusetts Volunteers. The *Pawnee* was stocked with combustibles, including forty barrels of gunpowder, eleven tanks of turpentine, twelve barrels of cotton waste and 181 flares.

The yard was partially destroyed by the work of Paulding's command. The USS *Pennsylvania*, *Germantown*, *Raritan*, *Columbia*, *Dolphin*, *Delaware*, *Columbus*, *Plymouth* and *Merrimack* were either burned or sunk. The USF *United States* was abandoned as a "venerable relic."

As the ships burned in the harbor, buildings throughout the yard were set on fire. The two huge ship houses were quickly engulfed by flame; Ship House A contained the partially completed seventy-four-gun *New York*, which was consumed by fire as she sat on the stocks. Since everything would not burn, sailors and marines rushed through the yard laying powder trails to destroy the valuable machinery and facilities. When efforts to break off the trunnions of the over one thousand cannons in the yard with sledgehammers proved futile, the guns were spiked with wrought-iron nails. Two officers, Commander John Rodgers and Captain Horatio Gouverneur Wright, were assigned the task of mining the granite dry docks. Their work was purportedly foiled by a petty officer who did not wish the explosion to damage nearby homes of his prewar friends.

April 21

4:30 a.m.: Paulding's command left the burning Gosport Navy Yard on board the USS *Yankee*, *Pawnee* and *Cumberland*.

Confederates occupied Gosport Navy Yard. The Richmond press gloated over the yard's capture with all of the supplies and facilities left behind by the Union, stating, "We have enough to build a navy of iron-plated ships."

Charles Oliver tendered his resignation and was dismissed from the USN.

April 22

Charles Carroll Simms tendered his resignation and was dismissed from the USN.

William C.M. Whitaker enlisted at Edenton, North Carolina, for the war as a private in Company A, 1st North Carolina Infantry.

Stephen Pickett Truesdale enlisted at Charleston, South Carolina, as a corporal in Company M, 1st South Carolina Infantry.

Flag Officer French Forrest, CSN, assumed command of Gosport Navy Yard.

Franklin Buchanan was dismissed from the USN.

John Rosler enlisted in Company K, 1st North Carolina Artillery.

William Young enlisted at Oxford, North Carolina, in the 2nd Company D, 12th North Carolina Infantry.

William Whitaker enlisted at Edenton, North Carolina, as a private in Company A, 1st North Carolina Infantry.

April 23

Flag Officer Hiram Paulding, aboard the USS *Pawnee*, arrived in Washington, D.C., and reported the loss of Gosport Navy Yard.

John H. Ingraham was appointed midshipman, CSN.

April 25

Jefferson M. Tinsley enlisted at Covington, Georgia, for one year as a private in Company H, 3rd Georgia.

Marcellus Augustus Tharpe enlisted at Marion, Georgia, in Company C, 4th Georgia.

Benjamin F. Southall enlisted as a private in Company C, 4th Virginia Infantry.

April 26

Secretary Stephen Mallory advised President Jefferson Davis to "adopt a class of vessels hitherto unknown to naval services. The perfection of a warship would doubtless be a combination of the greatest known ocean speed with the greatest known floating battery and power of resistance."

Alfred A. Stroup enlisted in Company D, 14th North Carolina Infantry.

April 29

John Pembroke Jones resigned and was dismissed from the USN while serving aboard the USS *Richmond*.

Major General Robert E. Lee commissioned the CS Receiving Ship (CSRS) *Confederate States* (formerly the USF *United States*) as the first ship in the Virginia State Navy.

Lieutenant Catesby Jones was assigned to command water batteries on Jamestown Island.

Levin Wood enlisted at Yanceyville, North Carolina, as a private in Company A, 13th North Carolina Infantry.

April 30

Benjamin R. Sheriff deserted from the USS *Alleghany*.

May 2

John Pembroke Jones was appointed first lieutenant, CSN.

Charles Schroeder tendered his resignation from the USN.

Eleazor Stillman enlisted in the United Artillery.

May 3

Michael May enlisted at New Orleans in the CSMC.

May 4

Theophilius Rainey enlisted at Petersburg in Company K, 12th Virginia Infantry.

May 6

Dinwiddie Phillips's resignation was accepted by the USN.

John Tynan was dismissed from the USN.

Ashton Ramsay refused to take the oath to the United States and was dismissed from the USN.

May 8

George E. Tabb Jr. enlisted at Gloucester Court House, Virginia, in Company A, 34th Virginia Infantry.

May 9

John Ryan enlisted at Manchester, Virginia, for one year as a private in Company I, 6th Virginia Infantry.

John Scott enlisted at Camp Walker, Louisiana, for one year as a private in Company E, 2nd Louisiana Infantry.

May 10

Naval Secretary Stephen Mallory advised the Confederate Congress's Committee on Naval Affairs, "I regard the possession of an iron-armored ship as a matter of the first necessity. Such a vessel at this time could traverse the entire coast of the United States, prevent all blockades, and encounter, with a fair prospect of success, their entire Navy…But inequality of numbers may be compensated by invulnerability; and thus not only does economy but naval success dictate the wisdom and expediency of fighting with iron against wood."

Confederate Congress appropriated $2 million for the acquisition of armored vessels built in Great Britain or France.

Robert D. Minor was appointed first lieutenant, CSN.

Dinwiddie Phillips was appointed as an assistant surgeon, CSN.

May 13

Catesby Jones's resignation from the USN was accepted.

John A. Wilson enlisted in Company E, 13th North Carolina.

John Q.A. Williams enlisted at Pig Point, Virginia, as a private in Company G, 9th Virginia Infantry.

May 14

Mallory instructed Captain Lawrence Rousseau and Captain Duncan Ingraham to identify sources of iron plate in the South.

May 15

James Crosby Long's resignation as an acting midshipman was accepted by the USN.

John Reardon enlisted at Jacksonport, Arkansas, as a private in the 1st (Colquitt's) Arkansas Infantry.

John Tynan was appointed as second assistant engineer.

May 18

Charles Schroeder was dismissed from the USN.

A contract was issued to raise the scuttled USS *Merrimack* from the Elizabeth River.

John J. Sturges enlisted at Chuckatuck, Virginia, for one year as a private in Company F, 9th Virginia.

May 19

Ashton Ramsay was appointed engineer, CSN.

The gunboats USS *Monticello* and *Thomas Freeborn* shelled Confederate batteries at Sewell's Point and were forced to retire by Confederate counter-battery fire. Flag Officer French Forrest and Major General Walter Gywnn used 196 of the over 1,000 cannons found at Gosport Navy Yard to defend the entrance to the Elizabeth River.

Charles Schroeder was appointed first assistant engineer, CSN.

James Semple tendered his resignation from the USN.

May 21

George N. Smith enlisted in the Pamunkey Light Artillery.

May 27

Zadock Wesley Messick enlisted in Company I, 32nd Virginia Infantry at Williamsburg, Virginia.

Union troops occupied Newport News Point and constructed the fortified position called Camp Butler. Federal control of this point closed the riverine link between Norfolk and Richmond.

May 28

Daniel Moriarty enlisted at New Orleans, Louisiana, as a private in Company C, CSMC.

May 29

John Riley enlisted at Fayetteville, North Carolina, for the war as a private in Company C, 3rd North Carolina.

George Thomas enlisted at Goldsboro, North Carolina, for the war as a private in the 2nd North Carolina Infantry.

May 30

The *Merrimack* was raised and moved into Gosport's Dry Dock # 1. Contracts were issued to raise the *Germantown* and *Plymouth* from the Elizabeth River.

June 1

John Volentine was assigned to special service under Brigadier General John Bankhead Magruder.

June 3

Mallory instructed Lieutenant John Mercer Brooke to develop an ironclad design for construction in the South.

June 4

Patrick McGinnis enlisted as a private in Company C, CSMC.

June 7
William Young was transferred to the CSN.

June 10
Richard Curtis, Wythe Rifles, fought at the Battle of Big Bethel.

June 11
Charles Oliver was named gunner, CSN.

June 12
Thomas Kevill was promoted to captain of the United Artillery.

June 15
The CSRS *Confederate States* was organized as a school ship and armed with nineteen guns to also serve as a block ship.

June 16
James Stevens enlisted at Glebe School House, Norfolk County, Virginia, for one year as a private in Company G, 41st Virginia Infantry.

June 18
The CSS *Teaser*, commanded by Lieutenant Robert Randolph Carter, was assigned to help defend the James River and observe Union naval operations at the river's mouth.

June 23
James M. Sheffield enlisted as a private in Company B, 13th North Carolina Infantry.

June 24
A report on Confederate homefront ironclad design was completed by Lieutenant John Mercer Brooke, Naval Constructor John Luke Porter and Chief Engineer William Price Williamson. The panel recommended that the *Merrimack* be transformed into an ironclad.
John William Wainwright enlisted at Yorktown in the 115th Virginia Militia.

June 25
John Runnels enlisted at Plymouth, North Carolina, for the war as a private in Company G, 1st North Carolina.

June 28

Gabriel Allen enlisted in Company G, 1st North Carolina at Plymouth, North Carolina.

Lieutenant John Mercer Brooke began to study and experiment with projectiles and rifled guns.

The side-wheeler steamer *St. Nicholas* was captured by a force commanded by Captain George Hollins, CSN, on the Potomac River. Hollins's command included Algernon S. Garnett, Charles Carroll Simms and Robert D. Minor.

June 29

The steamer *St. Nicholas* captured schooners *Margaret* and *Mary Pierce* and the brig *Monticello*.

June 30

Robert Turner enlisted for one year as a private in Company K, 6th Virginia Infantry.

July 1

Patrick Murphy shipped for one year as landsman, Naval Rendezvous, Richmond, Virginia.

July 3

James Crosby Long was appointed acting midshipman, CSN.

Charles H. Wilson shipped for one year as seaman, CSS *Forrest*, North Carolina Squadron.

July 5

Benjamin S. Herring resigned from the USN aboard the USS *Richmond*.

July 6

Marshall P. Jones was appointed first assistant engineer, CSN.

July 8

Barron Carter was appointed acting midshipman, CSN.

July 9

Reuben Thom commanded a detachment of fifty-five marines during the capture of Ship's Island, Mississippi.

John Wilson was discharged by order of the War Department to enable him to join the CSN.

July 11

Flag Officer French Forrest was ordered to begin work on the *Merrimack* conversion project.

July 15

James Semple was dismissed from the USN.

July 18

James Ross enlisted for the war as a private in Company D, 15th Louisiana Infantry.

Naval Secretary Mallory reported to the Confederate Congress that the *Merrimack* should be converted into an ironclad at the cost of $172,523.

July 22

George Walker enrolled at Elizabeth City, North Carolina, for the war as a fourth corporal in Company A, 8th North Carolina Infantry.

July 24

Tredegar Iron Works received the contract to produce iron plate for the *Merrimack* conversion project.

July 25

Joseph Fisher and John Nelson shipped for the war as seamen, John Mulroy as a fireman and John Jones as a landsman aboard the CSS *Forrest*, North Carolina Squadron.

July 28

John Cunningham was captured on board the privateer CSS *Petrel.*

July 30

Lawrence Rootes was appointed by Virginia as an acting midshipman, CSN.

August 1

Ephiram Young enlisted at Craney Island for one year as a private in Company H, 6th Virginia Infantry.

August 2

Seaman Thomas P. Mercer and Seaman James E. Mercer were assigned to the CSS *Ellis.*

John Rosler shipped for one year as an ordinary seaman, CSS *Forrest.*

August 3

Volentine Tolson shipped for the war as an ordinary seaman aboard the CSS *Forrest.*

August 5

Garrett N. Stack enlisted at Mobile, Alabama, as a private, CSMC.
Charles M. Sparks enlisted at Memphis, Tennessee, as private, CSMC.

August 10

Silas Gaskill shipped for the war as an ordinary seaman and was detailed to the CSS *Forrest.*

August 12

Acting midshipman William James Craig's resignation from the USN was accepted.

August 13

James Whitten enlisted at Mobile, Alabama, as a private, CSMC.

August 14

Chauncey Alexander Sprague enlisted in Company B, Cobb's Legion, Georgia Cavalry. The value of his horse was noted at $250.

August 17

James F. Smith enlisted in Company C, CSMC.

August 27

Andrew G. Peterson shipped for one year as a seaman, Naval Rendezvous, New Orleans, Louisiana.
Arthur Sinclair IV was discharged from the CSA.

August 28

Thomas Jones enlisted in Company C, CSMC, at Mobile, Alabama.

August 29

Flag Officer Samuel Barron, CSN, surrendered Hatteras Inlet to Flag Officer Silas Horton Stringham, commander of the North Atlantic Blockading Squadron.

September 2

Jerome B. Phillips enlisted as a private in CSMC.

September 5

Franklin Buchanan was named chief, Bureau of Orders and Details, CSN.

Robert Spence shipped for the war as an ordinary seaman, Naval Rendezvous, New Orleans, Louisiana.

September 7

Alexander Halstead enlisted as a substitute for John Anderson at Craney Island in Company D, 9th Virginia Infantry.

September 13

The CSS *Patrick Henry* (previously the bay steamer *Yorktown*) shelled Newport News Point. This fast side-wheeler was appropriated by the Confederate navy and armed with ten heavy guns, including two 32-pounder rifles and a 10-inch shellgun. The *Patrick Henry* featured a one-inch iron shield to protect her engines and was the most powerful Confederate warship in Virginia waters until the emergence of the CSS *Virginia*.

Robinson Turner enlisted at Memphis, Tennessee, in Company C, CSMC.

September 17

Private Allen Gillmore was discharged from the 12th Virginia Infantry, having supplied a substitute.

September 24

John H. Ingraham was appointed acting master.

September 28

Oliver Watson and Joseph Wenzel enlisted at Memphis, Tennessee, as privates, CSMC.

October 1

Dinwiddie Phillips was assigned to Richmond Naval Station, Richmond, Virginia.

October 4

John Taylor Wood was appointed lieutenant, CSN.

October 10

James E. Lindsay was held as a prisoner of war at Fort Warren, Massachusetts.

October 12

Lieutenant Catesby Jones reported to Secretary of the Navy Stephen Mallory about iron-plating tests on Jamestown Island. The report noted that the sloped sides of the *Merrimack*'s casemate would greatly enhance the shot-proof qualities of the shield; however, it must be clad with four inches of iron, preferably with two layers of two-inch plate. These findings forced Tredegar Iron Works to rework its machinery to produce two-inch iron plate.

October 13

Flag Officer L.M. Goldsborough, commander of the North Atlantic Blockading Squadron, advised Secretary of the Navy Gideon Welles, "Nothing, I think, but very close work can possibly be of service in accomplishing the destruction of the *Merrimack*."

Goldsborough ordered the steam tugs *Dragon* and *Zouave* to remain in constant company with the sail-powered USS *Congress* and *Cumberland* "so as to tow them into an advantageous position in case of an attack from the *Merrimack* or any other quarter."

October 14

William Parrish was named acting master, CSN.

October 15

George May was court-martialed for failure to obey lawful order and showing contempt for a superior officer; he was sentenced to confinement and five days' hard labor.

October 21

Midshipman Henry Marmaduke served aboard the CSS *McRae* during its attack on the Federal blockading fleet at Head of the Passes, Mississippi.

October 23

Cornelius Sidney Hampton enlisted at Mobile, Alabama, as a private in Company C, CSMC.

John Delley was discharged from Company B, 4th Texas Infantry, "to work in another government branch."

October 30

Alfred William King shipped for the war as a seaman, Naval Rendezvous, Mobile, Alabama.

November 1
John Volentine was promoted to second corporal.

November 7
John Pembroke Jones served in the Battle of Port Royal Sound, South Carolina. Acting midshipman Barron Carter served as an aide to Flag Officer Josiah Tattnall, commander of the Confederate squadron, during this engagement. Tattnall's squadron was dispersed by the USS *Ottawa*, *Penbina*, *Seneca* and *Pawnee.*

November 8
Franklin B. Dornin was appointed as captain's clerk.

November 19
Private John Naughton, Company I, 14th Virginia, was arrested for sleeping on post.

November 29
Company C, CSMC, was transferred to Gosport Navy Yard, Portsmouth, Virginia.

November 18
Chief Surgeon Algernon Garnett was detailed to serve as surgeon of the *Merrimack.*

November 25
Mallory approved the first armor-plate shipment to Gosport Navy Yard for use in cladding the *Merrimack.*
Lieutenant John Taylor Wood was detailed to the *Merrimack* and assigned the task of recruiting crew members.

November 26
Flag Officer Josiah Tattnall engaged the Union fleet with four gunboats in Cockspur Roads below Savannah, Georgia, in an effort to lure the Federal fleet under the guns of Fort Pulaski.

November 30
Lieutenant Charles Carroll Simms was assigned to the *Merrimack.*
John Wilson was paid for service as seaman aboard the CSS *Sea Bird.*

December 2
Ashton Ramsay was appointed acting chief engineer, CSN.

December 3
Officers John Randolph Eggleston, Henry H. Marmaduke, Marshall P. Jordan and E.A. Jack were detailed to the *Merrimack*.

December 7
Captain Reuben T. Thom's Company C, USMC, arrived at Gosport Navy Yard.

December 9
John W. Tynan was appointed first assistant engineer, CSN.

December 20
Thomas Dumphrey was sentenced to twenty days' hard labor by order of court-martial for insubordination while on guard duty.

December 21
John Tynan was assigned to the CSS *Virginia*.

December 26
Flag Officer Franklin Buchanan visited Gosport Navy Yard to review the *Merrimack*'s conversion.

1862

John H. Leonard was confined and sentenced to thirty days' labor and loss of ten dollars pay.

January 1
Henry Ashton Ramsay and John William Wainwright were assigned to the CSS *Virginia*.

January 2
Coal heaver Thomas McCoy was detailed to the CSS *Virginia*.
William Witz and George Williams were "transferred to CSS *Merrimac*."

January 10
Commander William Smith, captain of the USS *Congress*, reported, "I have not yet devised any plan to defend us against the *Merrimack* unless it be with hard knocks."
E.V. White was appointed by Virginia as an acting third assistant engineer.

January 13
Jefferson M. Tinsley transferred to the CSS *Virginia* as a first class fireman.

January 17
Joseph Riddock was "discharged by order of Gen. Huger."
George Saunders, Edward Whelen and James Skerrit were transferred to the CSN.

January 18
E.V. White and Louis Waldeck were assigned to the CSS *Virginia*.

January 25
Flag Officer French Forrest wrote to Major General Benjamin Huger, "I have just learned that one of the enemy's vessels has been driven ashore with several hundred gallons of oil on board…We are without oil for the *Merrimack* and the importance of supplying this deficiency is too obvious for me to urge anything more in its support."

January 27
Dinwiddie Phillips was assigned to the CSS *Virginia*.
George N. Smith enlisted at Yorktown, Virginia, as a private in the Magruder Light Artillery.

January 28
Captain John Marston, commander of the USS *Roanoke* and senior officer in Hampton Roads while Flag Officer L.M. Goldsborough was in North Carolina waters, responded to Secretary Welles's suggestion that the USS *Congress* be transferred to Boston, stating, "as long as the *Merrimack* is held as a rod over us, I would by no means recommend that she should leave this place."
George Thomas was transferred to the CSS *Virginia*.

January 29
Robert Turner was transferred to the CSS *Virginia*.

January 31
John Scott was "transferred to service aboard the *Merrimac*."

February 1
William C.M. Whitaker was transferred to the CSN.
Arthur Sinclair IV was assigned as captain's clerk, CSS *Virginia*.

John Q.A. Williams was discharged from the CSA to serve as surgeon's steward on the CSS *Virginia.*

February 3

Adam Pritchett was transferred from the 1st North Carolina to the CSN.

February 6

Charles Carroll Simms was detailed to command the CSS *Appomattox*, North Carolina Squadron.

February 7

Lubricating oil for the *Merrimack* arrived at Gosport Navy Yard from Richmond.

February 8

The CSS *Curlew* was disabled by Union gunfire, ran aground and was destroyed during the Battle of Roanoke Island. Flag Officer William Lynch's North Carolina Squadron retreated up the Pasquotank River.

Roanoke Island surrendered to Union forces.

Flag Officer Franklin Buchanan ordered Commander John Randolph Tucker to keep the CSS *Patrick Henry* and *Jamestown* positioned off Mulberry Island Point, in constant readiness to cooperate with the *Merrimack* when that ship was ready for service.

George N. Smith was "transferred to Steamer *Virginia.*"

February 10

Henry H. Marmaduke was named master in line of promotion, CSN.

Marcellus Augustus Tharpe was transferred to the CSS *Virginia.*

Flag Officer Lynch's squadron was destroyed by Union forces near Elizabeth City, North Carolina. Several Confederate vessels attempted an escape to Norfolk, Virginia, via the Dismal Swamp Canal at the South Mills Lock. The CSS *Raleigh* and *Beaufort* entered the canal; however, the CSS *Appomattox* was too large and was scuttled as an obstruction.

Flag Officer Franklin Buchanan advised Secretary Mallory that the *Merrimack* had not yet received her crew "not withstanding all my efforts to procure them from the Navy."

Andrew Dalton discharged from 1st South Carolina Artillery.

February 11

Food and other supplies were stored in the *Merrimack.*

Peter Williams was transferred from the 8th Alabama Infantry to the CSN.

FEBRUARY 13

The *Merrimack* was floated in dry dock for the first time.

FEBRUARY 14

Patrick Hughes shipped for three years, Naval Rendezvous, Richmond, Virginia.

William W. Lyon, 13th North Carolina, was "discharged by order of Gen. Huger having re-enlisted in CSN for duty on the *Merrimack*."

FEBRUARY 15

Michael Moore and Patrick Martin shipped for three years as landsmen, Naval Rendezvous, Richmond, Virginia.

James M. Sheffield was discharged for naval service on the CSS *Merrimac*.

William C. Little was "transferred to Steamer *Merrimack*."

Flag Officer Josiah Tattnall's gunboat squadron attacked Union batteries on Venus Point on the Savannah River.

FEBRUARY 16

August Lembler and Peter Williams shipped for war as seamen, Naval Rendezvous, Richmond, Virginia.

FEBRUARY 17

James Merrian shipped for three years as seaman, Naval Rendezvous, Richmond.

The *Merrimack* was launched, commissioned and rechristened as the CSS *Virginia*.

Gosport Navy Yard commandant French Forrest ordered Executive Officer Catesby Jones "to receive on board the *Virginia* immediately after dinner today the officers and men attached to the vessel with baggage, hammocks, etc., and have the ship put into order throughout. She will remain where she is to coal and receive her powder."

FEBRUARY 19

Levin Wood was "transferred to CSN for duty on the CSS *Merrimac*."

William Ward was discharged from the 13th North Carolina, "having re-enlisted in naval service on the *Merrimac*."

FEBRUARY 20

William W. Price was transferred to the CSN.

February 21

Captain John Marston of the USS *Roanoke* reported to Gideon Welles "by a dispatch from General Wool, I learn that the *Merrimack* will positively attack Newport News within five days, acting in conjunction with the *Jamestown* and *Yorktown* from the James River, and the attack will be at night."

February 23

When he heard that the *Merrimack* had finally been launched, Captain Gershom Jacques Van Brunt, commander of the USS *Minnesota*, advised the Navy Department that "the sooner she gives us the opportunity to test her strength the better."

February 24

Robert W. Moore shipped for war, Naval Rendezvous, New Orleans, Louisiana.

Flag Officer Franklin Buchanan was named commander of the Confederate James River Defenses with the CSS *Virginia* as his flagship. This command also included the CSS *Patrick Henry*, *Jamestown*, *Teaser*, *Raleigh* and *Beaufort*.

William G. Walker was court-martialed and forfeited one month's pay.

February 25

The *Virginia* started loading coal and ammunition.

Flag Officer French Forrest reported that the lack of gunpowder would delay the *Virginia*'s sortie. Catesby Jones notified John Mercer Brooke that the *Virginia* was not sufficiently protected below the water line. "We are least protected where we need it. The constructor should have put on six inches of [iron] where we have one."

February 26

Robert D. Minor was assigned as flag lieutenant aboard the CSS *Virginia*.

John H. Ingraham was selected as boarding officer when the CSS *Nashville* captured the merchant ship *Robert Gilifilan*.

CSS *Germantown* was moved to the entrance of the Elizabeth River to serve as a blockship. This powerless floating battery was armed with seven cannons and fitted with a sand-filled bulkhead seven feet thick.

February 27

Flag Officer French Forrest reported that the lack of gunpowder delayed any attack of the *Virginia* against the Union blockaders.

February 28

Flag Officer Franklin Buchanan arrived at Gosport Navy Yard to assume his command. He found that the *Virginia* was still not ready for combat.

March 2

Flag Officer Franklin Buchanan requested that Major General John Bankhead Magruder support his planned attack on Newport News Point.

March 3

Major General John Bankhead Magruder advised General Robert E. Lee that the Army of the Peninsula would not be able to cooperate with Buchanan's attack on Newport News Point. Magruder believed that "no one ship can produce such an impression upon the troops at Newport News as to cause them to evacuate." The Army of the Peninsula commander advocated that the *Virginia* be deployed as a floating battery guarding the James River.

March 4

Charles Sharp and Andrew Dalton enlisted in the United Artillery.

March 6

Captain Thomas Kevill and the Untied Artillery (1st Co. E, 41st Virginia Infantry) "volunteer[ed] to go onboard the Ironclad Steamer *Virginia*." Kevill assumed command of one of the ironclad's IX-inch Dahlgrens.

The last gunpowder shipment arrived for the CSS *Virginia*.

March 7

The CSS *Virginia* was ready for sea trials. A heavy gale kept the unseaworthy ironclad at Gosport Navy Yard.

Major General John Bankhead Magruder advised Buchanan that the Army of the Peninsula would not cooperate with the *Virginia*'s planned attack on Newport News.

Franklin Buchanan met with the commanders of his escort-armed tugs, Lieutenant Joseph W. Alexander of the CSS *Raleigh* and Lieutenant William H. Parker of the CSS *Beaufort*. Buchanan detailed his plans to attack Newport News Point and reminded them that if the battle turned against them, he would hoist a new signal: "Sink Before Surrender."

Lieutenant John D. Minor observed Union fleet dispositions late in the afternoon: "I reconnoitered the enemy off Newport News and Old Point and was glad to report that they were not in such force as I had been led to suppose."

March 8

10:00 a.m.: The *Virginia*'s casemate was coated with a thick layer of "ship's grease" to help deflect shot.

10:45 a.m.: Franklin Buchanan inspected his ironclad's engines and her ability to withstand his proposed ramming tactics with chief engineer Ashton Ramsay. When Ramsay advised Buchanan that all of the machinery was securely braced and the ten-mile trip downriver would be sufficient to test the engines' reliability, Buchanan declared, "I am going to ram the *Cumberland*. I am told she has the new rifled guns, the only ones in their whole fleet we have cause to fear. The moment we are out in the Roads, I'm going to make right for her and ram her."

11:00 a.m.: Buchanan hoisted his flag officer's red pennant over the *Virginia* and ordered the crew to cast off. Workmen dashed off the ship without completing many minor details.

11:30 a.m.: The banks of the Elizabeth River thronged with thousands of cheering citizens watching the ironclad move through the river.

12:30 p.m.: The ironclad's trial run proved that the *Virginia* was as unmanageable as a "water-logged" log. The warship was slow (five knots) and ran so close to the river bottom (twenty-two-foot draft) that a towline from the CSS *Beaufort* was needed to help the huge warship round a bend in the river.

1:00 p.m.: As the *Virginia* passed Craney Island, Buchanan informed the crew, "Sailors in a few minutes you will have the long-awaited opportunity to show your devotion to your country and our cause...The Confederacy expects every man to do his duty, beat to quarters."

1:30 p.m.: The *Virginia* dropped her towline from the *Beaufort* and entered Hampton Roads at high tide.

2:20 p.m.: The *Virginia* and her consorts, *Raleigh* and *Beaufort*, exchanged fire with Union forces at Newport News Point. The *Beaufort* fired the first Confederate shot of the day. The first shot from the *Virginia*'s forward 7-inch Brooke rifle struck the *Cumberland*'s starboard rail, injuring several marines.

2:55 p.m.: The *Virginia* and the fifty-two-gun sailing frigate traded salvoes. Shot from the *Congress* bounced off the Confederate ironclad like "pebble stones." The *Virginia* incurred no significant damage. The Confederate ironclad replied and unleashed her broadside of four guns against the *Congress*. Hot shot and shell ignited two fires on the hapless frigate, and the *Congress* appeared critically damaged.

3:00 p.m.: The *Virginia* steamed on toward the *Congress*.

3:05 p.m.: The *Virginia* crossed the anti-torpedo obstructions surrounding the *Cumberland* and rammed the sloop of war. The *Cumberland* was mortally wounded; the ramming was made worse by a simultaneous shot from the *Virginia*'s bow rifle, which killed ten men. The Union warship immediately began to sink with the *Virginia*'s ram trapped within the *Cumberland*'s hull.

Charles Dunbar became the first man killed aboard the *Virginia* during the Battle of Hampton Roads when a shell from the *Cumberland* struck the sill of the forward port while Dunbar was reloading the 7-inch Brooke rifle.

3:06 p.m.: As the weight of the *Cumberland* rested on the ram, the *Virginia*'s engines refused to reverse, and the Confederate ironclad began to settle. The poorly mounted ram broke off and freed the *Virginia*.

3:08 p.m.: The *Virginia* floated fifty yards apart from the *Cumberland* as the two ships continued to bombard each other.

3:10 p.m.: The USS *Minnesota* ran aground off Salter's Creek.

3:35 p.m.: The *Cumberland* sank.

3:40 p.m.: The CSS *Virginia*, due to her deep draft and poor steering, was forced to steam up the James River to turn around. While this maneuver was executed, Lieutenant John Taylor Wood struck the *Congress* with several shells from the 7-inch stern Brooke rifle. The *Virginia* also destroyed two Union transports and captured one other, which was anchored along a wharf.

4:00 p.m.: The USS *Monitor* neared Cape Henry Lighthouse. Louis Stodder recorded in the *Monitor*'s logbook that he "heard heavy firing in the distance."

The USS *Roanoke* ran aground on the Middle Ground Shoal.

4:05 p.m.: The CSS *Virginia* steamed within two hundred yards of the stranded *Congress*. The *Congress*'s stern was quickly demolished, and the main deck was "literally reeking with slaughter." The USS *Zouave* was struck by several shells from the ironclad, which destroyed the tug's figurehead and pilothouse. When a shell hit the *Zouave*'s rudderpost, the disabled vessel broke off action and fled the scene.

4:20 p.m.: Lieutenant Joseph B. Smith, acting commander of the *Congress*, was struck by a shell fragment, which tore off his head and a portion of his shoulder. The ship's command devolved onto the shoulders of Lieutenant Austin Pendergrast.

4:40 p.m.: The USS *Congress* surrendered.

5:00 p.m.: The CSS *Raleigh* and *Beaufort* steamed alongside the *Congress*. The Union frigate was boarded to remove the wounded and to complete the ship's destruction when rifle and cannon fire from Camp Butler, commanded by Brigadier General Joseph King Fenno Mansfield, forced the Confederate gunboats to back away from the *Congress*.

The USS *Roanoke* floated off Middle Ground Shoal and moved away from the action to the protection of Fort Monroe.

5:05 p.m.: Flag Officer Franklin Buchanan ordered Lieutenant Robert Dabney Minor to take the *Virginia*'s remaining cutter over to the *Congress* to complete the frigate's destruction. A volley of musketry hit the boat, and Robert Minor was seriously wounded. The CSS *Teaser*, commanded by Lieutenant William Webb, picked up the survivors in a bold dash.

5:20 p.m.: Buchanan, standing atop the *Virginia* and engaged by the Union actions under a flag of truce, shot at the Federal soldiers on the shore. He was shot with return fire and severely wounded in the thigh. The flag officer was carried below and ordered Catesby Jones to "plug hot shot into her and don't leave her until she is afire." Jones assumed command of the *Virginia*.

5:45 p.m.: The *Congress* was destroyed by hot shot and shell. The *Virginia* then left the Federal frigate burning "stem to stern."

6:00 p.m.: Jones steered the Confederate ironclad back into Hampton Roads to destroy the grounded Union frigates. Shells from the *Virginia* damaged the USS *Minnesota* and *St. Lawrence*. The *St. Lawrence* had a shell lodged in her mainmast, and over a dozen shells struck the *Minnesota*.

7:00 p.m.: General Wool reported the destruction of the *Congress* and *Cumberland* to Washington, D.C., via telegraph. Wool's chief of staff, Colonel LeGrand Cannon, noted that the land-based armament at Fort Monroe was "as useless as musket-balls against the ironclad." Major General George B. McClellan immediately replied and authorized Wool to abandon Camp Butler on Newport News Point. McClellan telegraphed, "The performances of the *Merrimac* places a new aspect upon everything, and may very probably change my whole plan of campaign, just on the eve of execution."

8:00 p.m.: Darkness and receding tide compelled Jones to steam the *Virginia* to her mooring at Sewell's Point. As the *Congress* spars and ropes "glittered against the dark sky in dazzling lines of fire," Jones vowed to destroy the Federal fleet the next day.

The *St. Lawrence* floated and was towed toward Fort Monroe.

9:00 p.m.: The USS *Monitor* entered Hampton Roads. Worden met Captain John Marston of the USS *Roanoke* and acting commander of Union naval forces in Hampton Roads. Marston rescinded the orders he had received from Gideon Welles to immediately send the *Monitor* to Washington, D.C. He recognized that the best way to stop any Confederate ironclad assault against Washington was to defend the wooden frigates in Hampton Roads. Marston ordered Worden to station the *Monitor* near the *Minnesota* and to protect that warship from the *Virginia*.

10:00 p.m.: Lieutenant John Worden wrote to his wife, "The *Merrimac* has caused sad work amongst our vessels. She can't hurt us."

The crew of the *Virginia* received their supper.

Catesby Jones inspected the *Virginia* for battle damage. The smokestack was riddled, a leak in the bow was discovered, two IX-inch Dahlgren barrels were partially shot off, several iron plates were cracked and much of the superstructure equipment—flag staffs, cutters, railings, etc.—were lost. Jones did not know that the ram was missing. The *Virginia*'s acting commander believed that his ironclad could fight again the next day.

11:00 p.m.: The USS *Monitor* anchored next to the *Minnesota*. As the *Congress* continued to send an eerie glow across Hampton Roads, one of the *Virginia*'s crew "chanced to be looking in the direction of the *Congress* when there passed a strange-looking craft, brought out in bold relief by the light of the burning ship, which at once he proclaimed to be the *Ericsson*."

11:05 p.m.: Worden and Greene went on board the *Minnesota* and met with Captain G.J.H. Van Brunt. Van Brunt doubted that the ironclad could aid the *Minnesota*. Paymaster Keeler recalled, "The idea of assistance or protection being offered to the huge thing by the little pygmy at her side seemed absolutely ridiculous."

Franklin B. Dorin was appointed by Maryland as an acting midshipman, CSN.

MARCH 9

12:00 a.m.: The USS *Congress* exploded.

2:00 a.m.: Captain Van Brunt attempted to float the USS *Minnesota* at high tide, but the frigate remained struck in the mud.

5:30 a.m.: William Cline reported that the crew of the CSS *Virginia* "began the day with two jiggers of whiskey and a hearty breakfast."

6:00 a.m.: The *Virginia* slipped her mooring at Sewell's Point but could not steam into Hampton Roads due to heavy fog.

8:00 a.m.: The CSS *Virginia* entered Hampton Roads and moved toward the *Minnesota*.

8:30 a.m.: Lieutenant Hunter Davidson fired the first shot of the day from the forward 7-inch Brooke gun at the *Minnesota*. The range was one thousand yards. Another shot quickly followed, "exploding on the inside of the ship, causing considerable destruction and setting the ship on fire." It appeared that the *Virginia* would make short work of the *Minnesota*. The USS *Monitor* then moved from alongside the *Minnesota* and blocked the *Virginia*'s approach to the stranded frigate.

8:35 a.m.: The *Monitor* and *Virginia* began circling each other in concentric circles, testing their opponent's armor. The battle was primarily fought at a range of less than one hundred yards. Often, the ironclads almost touched each other as each endeavored to gain an advantage.

10:05 a.m.: The *Monitor* broke off action and steamed onto a shoal to reload ammunition in the turret.
10:10 a.m.: The *Virginia* steamed toward the *Minnesota*.
10:15 a.m.: The *Virginia* began shelling the *Minnesota* but ran aground.
10:30 a.m.: The *Monitor* began bombarding the *Virginia*.
11:15 a.m.: The *Virginia* pulled itself off the shoal. Lieutenant Jones decided to ram the *Monitor* and maneuvered the *Virginia* into position.
11:45 a.m.: The *Monitor* eluded ramming and was only hit with a glancing blow. This action caused no damage to the *Monitor*; however, the *Virginia* developed a new leak at her bow. The *Monitor* also fired both of her XI-inch Dahlgrens at the *Virginia* when she was rammed. The shot struck just above the stern pivot gun port, which forced the *Virginia*'s iron shield in three inches.
11:50 a.m.: When the *Monitor* avoided the *Virginia*'s attempt to ram her, the Union ironclad moved away from the action. The *Virginia* steamed toward the *Minnesota*. The *Minnesota* and the armed tug *Dragon* were shelled. Several shells struck the *Minnesota* and ignited a fire. One shell struck the *Dragon*. The *Dragon*'s boiler burst, and the tug, which had been alongside the *Minnesota* to tow that vessel to safety, sunk.
12:10 p.m.: The *Monitor* attempted to ram the *Virginia*. A steering malfunction caused the *Monitor* to miss the fantail of the *Virginia*; as the *Monitor* passed the stern of the *Virginia*, a shell from the 7-inch Brooke gun commanded by Lieutenant John Taylor Wood struck the *Monitor*'s pilothouse. Worden was wounded, and the *Monitor* broke off action.
12:25 p.m.: Lieutenant Catesby Jones believed that the *Monitor* had retreated and given up the fight, and it steamed toward the *Minnesota*.
12:30 p.m.: The *Virginia* could not get closer than one mile to the *Minnesota*. The pilots warned Lieutenant Jones that the tide was falling fast. Jones concurred with the ironclad's officers and decided to return to the Elizabeth River.
12:40 p.m.: Lieutenant Samuel Dana Greene assumed command of the *Monitor*. The ironclad was brought back into action. Greene mistook the *Virginia*'s course toward Sewell's Point as a sign of defeat. The *Monitor* did not pursue the *Virginia* and steamed to a defensive position near the *Minnesota*.

March 10

Daniel Knowles and George Scutatus reenlisted for the war and received a $50 bounty.
John Mercer Brooke began producing wrought-iron, steel-tipped, armored piercing "bolts" for 6.4- and 7-inch Brooke rifles.
Confederate Secretary of State Judah Benjamin sent propaganda messages

to European nations stating that "success of our iron-clad steamer *Virginia* (late the *Merrimac*) in destroying three first class frigates in her first battle, evinces our ability to break for ourselves the much-vaunted blockade, and ere the lapse of ninety days we hope to drive from our waters the whole blockading fleet."

The CSS *Virginia* was placed in dry dock.

John Steen reenlisted for the war and received a $50 bounty.

March 11

Lieutenant Robert J. Minor advised John Mercer Brooke that he "deserved the gratitude and thanks of the Confederacy for the plan of the now celebrated *Virginia*."

Assistant Engineer Charles Schroeder was detailed to the CSS *Virginia*.

March 12

The chief engineer of the Army of the Potomac, Brigadier General John G. Barnard, informed the assistant secretary of the navy, G.V. Fox, that "the possibility of the *Merrimac* appearing paralyzes the movement of this army." Fox replied that the *Monitor* should be able to defeat the Confederate ironclad; however, "great dependence on her" was not wise.

Flag Officer L.M. Goldsborough advised Major General G.B. McClellan that the James River was closed to Union operations and that the U.S. Navy could not attack the Confederate water batteries at Yorktown and Gloucester Point guarding the entrance to the York River.

Captain Augustus Hermann Drewry's Southside Artillery was assigned to Drewry's Bluff.

March 21

Flag Officer Josiah Tattnall was named commander of the CSS *Virginia*.

John Volentine was transferred to the CSS *Virginia*.

March 22

Charles W. Harrison shipped for one year as a seaman aboard the CSS *Jamestown*.

March 24

Captain Sidney Smith Lee, formerly executive officer of Gosport Navy Yard, was ordered to replace Flag Officer French Forrest as commandant of Gosport. Forrest was reassigned as head of the Office of Orders and Details.

March 25

The CSS *Virginia* crew members receive a $50 bounty for their service.

Edmond Pettit shipped for the war as an ordinary seaman.

March 29

Flag Officer Josiah Tattnall arrived at Gosport Navy Yard to assume command of the *Virginia* while she was still in dry dock. Secretary Mallory instructed Tattnall, "Do not hesitate or wait for orders, but strike when, how, and where your judgment may dictate."

Benjamin Slade was discharged from naval service.

March 30

William G. Walker was transferred to the CSS *Virginia*.

April 1

Frank Anderson was reassigned from the *Virginia* to the CSS *Beaufort*.

April 4

The CSS *Virginia* left dry dock after the completion of significant repairs and improvements, including additional iron plating below the waterline and the installation of a twelve-foot-long, steel-tipped ram. The ram was designed by John Mercer Brooke to strike beneath the *Monitor*'s armor belt and penetrate the Union ironclad's half-inch hull.

Ephraim Kirby McLaughlin was assigned as acting master's mate, CSN.

April 7

John McCubbins shipped for war as a coal heaver.

April 8

General R.E. Lee, military advisor to President Jefferson Davis, asked Secretary Mallory to order the *Virginia* to strike at the Union transports in the York River.

April 11

The CSS *Virginia* was finally ready for her third foray into Hampton Roads. The Confederate ironclad was supported by the James River Squadron: CSS *Patrick Henry*, *Jamestown*, *Teaser*, *Beaufort* and *Raleigh*.

6:00 a.m.: The *Virginia* left Gosport Navy Yard and steamed down the Elizabeth River to Sewell's Point.

7:10 p.m.: The *Virginia* entered Hampton Roads. The Federal transports fled the harbor to the protection of Fort Monroe. The *Monitor*, which had just been reinforced by the iron-hulled USRMS *Naugatuck* (*Stevens Battery*), stayed in the channel between Fort Monroe and the Rip Raps. The Union ironclad had strict orders not to engage the *Virginia* unless the Confederate ironclad moved into the open waters of the Chesapeake Bay. Tattnall refused to take his ironclad out of Hampton Roads, and the *Monitor* would not accept the *Virginia*'s challenge.

3:00 p.m.: With all of the Union fleet's attention placed on the moves of the CSS *Virginia*, the CSS *Jamestown*, commanded by Lieutenant Joseph Barney, captured two brigs and an Accomac schooner and towed them to Norfolk.

4:00 p.m.: The CSS *Virginia*, flying the captured transport's flag upside down under her own colors as an act of disdain, fired several shells at the *Naugatuck* and returned to Gosport Navy Yard.

Landsman Jonathan W. Agnew was discharged from the crew of the CSS *Virginia*.

APRIL 12

Charles Carroll Simms was assigned to the CSS *Nansamond*.

Confederate gunboats CSS *Teaser*, *Raleigh* and *Jamestown* were transferred from Norfolk up the James River to support the Confederate Warwick–Yorktown right riverine flank at Mulberry Island.

APRIL 19

Benjamin Richardson enlisted in the United Artillery.

Captain Thomas Kevill and the United Artillery were reassigned to the Sewell's Point Battery.

MAY 1

The CSS *Richmond* was launched at Gosport Navy Yard.

Secretary Stephen R. Mallory ordered all moveable equipment to be transferred from Gosport Navy Yard to yards being established at Richmond, Virginia, and Charlotte, North Carolina.

MAY 2

Elisha R. Johnson was appointed as acting gunner, CSN.

MAY 5

The CSS *Jamestown* towed a brig containing heavy guns and ordnance intended for the CSS *Richmond* to Richmond. The CSS *Patrick Henry* towed the *Richmond* to the Rocketts Navy Yard at Richmond.

May 6

President Abraham Lincoln, accompanied by Brigadier General Egbert L. Viele, Secretary of the Treasury Salmon P. Chase and Secretary of War Edwin M. Stanton, arrived during the evening at Fort Monroe. The president had been invited to Fort Monroe by Major General John E. Wool in conjunction with Wool's desire to strike against Norfolk.

May 3

General Joseph E. Johnston ordered the abandonment of the Confederate Warwick–Yorktown Line. During the evening, the Confederates covered their retreat with a massive bombardment of Union siege lines.

May 7

The CSS *Virginia* emerged into Hampton Roads but could not induce the USS *Monitor* into combat.

President Lincoln met with Major General John Ellis Wool and Flag Officer Louis M. Goldsborough at Fort Monroe, Virginia, to plan how to capture Norfolk and how to enable the U.S. Navy to open the James River in support of General G.B. McClellan's drive toward Richmond.

May 8

The USS *Galena*, accompanied by USS *Aroostook* and *Port Royal*, steamed into the James River and began bombardment of Fort Boykin on Burwell's Bay and Fort Huger on Hardin's Bluff.

The USS *Monitor* and the *Naugatuck*, supported by the USS *Susquehanna*, *San Jacinto*, *Dacotah* and *Seminole*, began shelling the batteries on Sewell's Point. The CSS *Virginia* steamed down the Elizabeth River from Gosport Navy Yard to contest the Union advance. While it appeared a second conflict between the two ironclads might occur, Goldsborough ordered the Federal squadron to withdraw to its anchorage. The *Virginia* stayed out in Hampton Roads for several hours hoping to engage the *Monitor*. When no action ensued, Flag Officer Josiah Tattnall ordered the *Virginia* back into the Elizabeth River.

President Lincoln observed the entire event from the ramparts of Fort Wool and was disappointed with the naval action. Lincoln conducted a personal reconnaissance of the Ocean View area and selected a site for a Union landing.

William Ward was discharged from the crew of the *Virginia*.

May 9

2:00 p.m.: Wool organized an embarkation of six thousand Union troops. The soldiers were loaded on canal boats to be later ferried across the

Chesapeake Bay to Ocean View, Virginia. President Lincoln had selected this landing site the night before. Brigadier General J.F.K. Mansfield commanded the first wave. General Wool was in overall command of the expeditionary force.

4:00 p.m.: The Confederates held a council of war, chaired by Flag Officer George N. Hollins, concerning the fate of Norfolk and Gosport Navy Yard. The Confederate army asked the CSS *Virginia* to stay at Craney Island for ten days to cover the Confederate retreat from Norfolk.

5:00 p.m.: President Abraham Lincoln ordered the USS *Monitor* to reconnoiter at Sewell's Point. The *Monitor* discovered the batteries abandoned. Lincoln ordered Wool to initiate the landing at Ocean View.

6:00 p.m.: The USS *Galena* reached Jamestown Island.

MAY 10

7:00 a.m.: The Union army completed landing at Ocean View, Virginia.

8:00 a.m.: John Pembroke Jones, flag lieutenant of the CSS *Virginia*, reported to Flag Officer Josiah Tattnall that the Confederate flag no longer flew over the Sewell's Point batteries.

5:00 p.m.: General Wool's troops occupied Norfolk.

7:00 p.m.: Lieutenant J.P. Jones reported to Flag Officer Josiah Tattnall that Portsmouth was abandoned and Gosport Navy Yard was in flames.

Flag Officer Josiah Tattnall realized that an effort must be made to get the *Virginia* up the James River toward Richmond. The pilots advised that this could only be achieved if the huge ironclad could reduce her draft from twenty-three feet to eighteen feet so she could cross Harrison's Bar. The crew immediately went to work throwing coal, ballast and everything else overboard except the ironclad's guns and ammunition.

MAY 11

Even though the *Virginia*'s draft had been lightened to twenty feet, the pilots informed Catesby Jones that she could not get across the bar. The pilots noted that the wind was from the west rather than the east, blowing the water away from the bar and making it even shallower. Since the lightening had made the *Virginia* "no longer an ironclad [it] was therefore unable to engage the Federal fleet it had to be scuttled."

2:00 a.m.: The *Virginia* steamed across the mouth of the Elizabeth River from Sewell's Point to Craney Island and was grounded.

4:58 a.m.: The CSS *Virginia* was scuttled.

11:00 a.m.: President Abraham Lincoln steamed up the Elizabeth River past the wreck of the *Virginia* aboard the USS *Baltimore* to view the smoldering ruins of Gosport Navy Yard.

May 12

The CSS *Virginia*'s crew arrived in Richmond.
Peter Marry was advertised as a deserter.
Commander John Rodgers's squadron shelled Fort Huger.
The Confederates abandoned Fort Boykin and Fort Huger.

May 13

Henry H. Marmaduke and John W. Tynan were detailed to the CSS *Chattahoochee*.

Lieutenant Catesby Jones was ordered to report with the *Virginia*'s crew to Commander Ebenezer Farrand at Drewry's Bluff. The crew began the construction of two additional gun emplacements.

May 14

Rodgers's James River flotilla, which consisted of the USS *Galena*, *Monitor*, *Naugatuck*, *Aroostook* and *Port Royal*, arrived at Harrison's Landing, Virginia.
Confederates continued work constructing fortifications atop Drewry's Bluff.
Pilot William Parrish was discharged.

May 15

6:00 a.m.: Rodgers's flotilla left its anchorage near the mouth of Kingsland Creek and steamed toward Drewry's Bluff.

7:45 a.m.: With the *Galena* in the lead, Rodgers placed his flotilla within six hundred yards of the bluff. The river was very narrow at this point and also blocked by obstructions placed by the Confederates only a few days before. Rodgers swung the *Galena*'s broadside toward the Confederate batteries. The *Galena* received two hits while completing this maneuver and quickly became the primary target of the Confederate batteries. The Confederates encountered significant problems. The 10-inch Columbiad, loaded with a double charge of powder, recoiled off its platform when the first shot was fired. The mud and log casemate protecting the 7-inch Brooke gun collapsed from the first shot's vibrations.

9:00 a.m.: Confederate plunging shot had begun to take effect on the *Galena*. Lieutenant William Jeffers moved the *Monitor* virtually abreast of the *Galena* in an effort to draw some of the Confederate shot away from the larger ironclad. The *Monitor*'s turret, however, did not permit the ironclad to elevate her two XI-inch Dahlgrens sufficiently to hit the Confederate batteries. The *Monitor* was struck three times by Confederate shot before she backed downstream.

10:00 a.m.: The *Naugatuck*'s Parrott rifle burst and forced the vessel out of action.

11:15 a.m.: An eight-inch shell crashed through the *Galena*'s bow gun port and exploded. The shell ignited a cartridge, then being handled by a powder monkey, killing three men and wounding several others. The explosion sent smoke billowing out of the ironclad's gun ports.

11:30 a.m.: The *Galena* slipped her cables and retreated downriver. Confederate gunners atop the bluff gave three hearty cheers when the *Galena* broke off action.

11:45 a.m.: The Battle of Drewry's Bluff was over.

Landsman Michael Moore was killed in action during the Battle of Drewry's Bluff.

JUNE 1

James Stevens was captured at Fair Oaks, Virginia.

JUNE 5

James Stevens was sent to Fort Delaware, Delaware, as a POW.

JUNE 11

George Russell enlisted at Covington, Georgia, for one year as a private, Company H, 3rd Georgia Infantry.

JUNE 29

During the Battle of Savage Station, Virginia, Lieutenants John Mercer Brooke and Robert Dabney Minor commanded an armored railway gun (7-inch Brooke rifle). The gun was protected by an iron-plated shield and operated on the Richmond & York River Railroad.

JULY 17

Richard Nance shipped for war, Naval Rendezvous, Richmond, Virginia.

JULY 20

Joseph H. Myers enlisted as a private in Company B, 24th Battalion, Virginia Partisan Rangers.

JULY 21

Lorenzo D. Pitt shipped for war as an ordinary seaman.

JULY 24

John Williams and Stephen Pickett Truesdale shipped for war at Naval Rendezvous, Richmond, Virginia.

July 31

James Shiver was transferred as an officer's cook with Captain James H. Rochelle.

Richard Nance was transferred as a first class fireman to the CSS *Beaufort*.

August 1

Seaman Patrick Martin volunteered for service aboard the CSS *Chattahoochee* and served as captain of the top.

Coxswain John Rosler volunteered for duty on the CSS *Chattahoochee*.

William Young volunteered for duty on the CSS *Chattahoochee*.

George May volunteered for service as gunner's mate aboard the CSS *Chattahoochee*.

Seaman August Lembler volunteered for duty on the CSS *Chattahoochee* and served as captain of forecastle.

Seaman John Perry volunteered for duty aboard the CSS *Chattahoochee* as master of arms.

Thomas Saunders volunteered for duty aboard the CSS *Chattahoochee* as quarter gunner.

August 5

James Stevens was exchanged at Aikens Landing, Virginia.

August 6

Arthur Sinclair IV was appointed acting master, CSN.

August 12

Edward Harrington was discharged from the Confederate navy.

August 16

E.A. Jack was promoted first assistant engineer.

August 20

Benjamin F. Southall was attached to African Church Hospital, Winchester, Virginia, as a nurse.

August 29

E.V. White resigned his CSN commission.

August 31

Sergeant Jacob Scolls died at Camp Beall, Drewry's Bluff, Virginia.

September 30

Zadock Wesley Messick was discharged from the CSN due to rheumatism. Benjamin R. Sheriff was listed as deserted on the final *Virginia* payroll.

October 2

Charles Schroeder was appointed chief engineer, CSN.

October 6

Charles Carroll Simms was assigned to the CSS *Atlanta*.

October 8

David Stewart was captured by the Potomac River Flotilla trying to deliver mail between Virginia and Maryland.

October 25

Charles Carroll Simms reported for duty, CSS *Florida*.

October 29

John Taylor Wood commanded a cutting-out expedition that captured and burned the *Alleganian*.

November 1

Levin Wood enlisted at Trenton, North Carolina, as a private in Company A, 8th Battalion, North Carolina Partisan Rangers.

November 2

David Stewart was paroled at Aikens Landing, Virginia.

December 5

George E. Tabb Jr. was recommended by Lieutenant Robert D. Minor for a commission based on his service aboard the CSS *Virginia*.

December 12

First Class Fireman Richard Nance entered Howard's Grove Hospital, Richmond, Virginia, with varioloid and died two weeks later.

December 14

William C.M. Whitaker was discharged from the CSN.

December 16

George M. Tabb Jr. was appointed second lieutenant, CSA, Signal Corps.

December 22

Private Thomas Jones, CSMC, was detailed to the CSS *Richmond*, James River Squadron.

1863

January 1

Seaman Andrew G. Peterson was listed as a deserter.

January 26

John Taylor Wood was appointed colonel, CSA cavalry.

February 1

Theophilius Rainey was assigned as teamster, Petersburg Horse Artillery.

February 11

James Ross shipped for war as an ordinary seaman, Naval Rendezvous, Richmond, Virginia.

February 14

William O. Hall was found guilty of desertion and sentenced to six months' labor with a twelve-pound ball and three-foot chain attached to his left ankle, forfeited six months' pay and received twenty-nine lashes on his back. The "court was lenient because accused may have intended to return to the army at some time."

February 23

Benjamin Slade was found guilty of being AWOL and fined one month's pay.

February 27

Francis Satchfield enlisted at Camp Davis, North Carolina, for the war as a private in Pegram's Company, Virginia Light Artillery.

March 22

Hugh Lindsay resigned from the CSN while assigned to the North Carolina Squadron.

March 27

John R. Joliff and Eugenius Turner Henderson were killed in a boiler explosion on board the CSS *Chattahoochee*. William J. Craig was injured during the explosion.

April 5

Charles Peter Eanes was detailed for special duty with the Petersburg & Weldon Railroad.

May 3

Benjamin Slade was wounded in action during the Battle of Chancellorsville.

May 4

Richard A. Mitchell was killed in action at Chancellorsville, Virginia.

May 5

John T. Moore was killed in action at Chancellorsville, Virginia.

May 6

Quarter gunner Nicholas B. Pryde was promoted to acting master mate by Lieutenant C.W. Read, CSN.

May 9

Oliver Watson was promoted to sergeant, CSMC.

May 13

William Young was listed as having deserted.

May 20

James Powers was paid for service as a landsman, CSS *Tuscaloosa*, Mobile Bay Squadron, Mobile, Alabama.

May 29

Catesby Jones was appointed commander for gallant and meritorious conduct aboard the CSS *Virginia*.

June 10

James Edwards was appointed by Alabama as third assistant engineer and assigned to the CSS *Morgan*, Mobile Bay Squadron.

June 12

Thomas Saunders was transferred to the CSS *Savannah*, Savannah River Squadron.

June 13

George May, John A. Rosler and Patrick Martin were transferred to the CSS *Savannah*.

June 1

John A. Rosler was promoted to acting master's mate.

Lieutenant Joseph W. Alexander and George T. Moore were captured at Wassau Sound while serving aboard the CSS *Atlanta.*

June 27

Acting master's mate Nicholas Pryde, CSS *Clarence*, was captured off Portland, Maine.

July 2

George Russell was captured during the Battle of Gettysburg.

July 3

Elijah Wilson Flake was wounded in action during the Battle of Gettysburg.

July 11

James C. Cronin was appointed boatswain while serving on the CSS *Savannah.*

July 14

James Stevens was captured at Falling Waters, Maryland.

July 19

Benjamin F. Southall was in General Hospital #9, Richmond, Virginia, with a gunshot wound (ball lodged in his right thigh).

August 1

James Stevens arrived at Old Capitol Prison, Washington, D.C.

August 2

Private Michael Mary was assigned to the Marine Guard, CSS *Charleston.*

August 3

William Riley Powers was transferred to Company C, 6th North Carolina Cavalry.

E.V. White enrolled at Columbus as a first lieutenant in Company F, 19th Battalion, Georgia Infantry.

August 7

Boatswain's mate John Perry was captured at Morris Island, Charleston, South Carolina.

August 23

John Taylor Wood captured the gunboats USS *Satellite* and *Reliance.*

August 24

John Taylor Wood aboard the *Reliance* captured the schooners *Golden Rod, Coquette* and *Two Brothers.*

August 25

Arthur Sinclair IV was promoted to lieutenant, CSN.

August 29

Charles Hasker, along with Lieutenant John Payne and two other unknown volunteers from the CSS *Chicora*, escaped from the CSS *Hunley* when she sank during testing.

September 4

Frederick Archer enlisted in Company D, 3rd Maryland Cavalry, USA.

September 8

George Walker was assigned to a naval detachment on a special expedition from Drewry's Bluff to Charleston Naval Station, Charleston, South Carolina.

September 22

Jefferson M. Tinsley was captured near Knoxville, Tennessee.

September 30

Charles H. Wilson was paid for service as quartermaster, CSS *Chicora*, Charleston Squadron, Charleston, South Carolina.

October 1

Jacob Ollsen, John Runnels, Robert Turner, Ephiram Young, Marcellus Augustus Tharpe, John J. Sturges, George Thomas and James Ross joined the naval detachment on a special expedition from Drewry's Bluff to Charleston Naval Station, Charleston, South Carolina.

Sterling Mathews was assigned as a carpenter's mate, CSS *Virginia II.*

Frank Anderson began a thirty-day assignment as a painter on the CSS *Indian Chief.*

Oliver Watson was promoted to first sergeant, CSMC.

October 13

Jefferson M. Tinsley was discharged from Military Prison, Louisville, Kentucky, and forwarded to Camp Morton, Indiana, having "enlisted in the Federal Army."

November 24

Lieutenant George E. Tabb Jr. was captured at Raccoon Mountain, Georgia.

December 9

Samuel Aenchbacker was given extra duty as a carpenter, Camp Beall, Drewry's Bluff, Virginia.

December 13

James Stevens took the loyalty oath and was sent north.

December 17

Paymaster Steward Charles J. Creekmur's last recorded service was on this day aboard the CSS *Roanoke.*

1864

January 2

Andrew Dalton was transferred to Company D, 1st Confederate Engineer Troops.

January 13

John McGrady was appointed by Alabama as boatswain, CSN.

February 2

John Taylor Wood commanded a cutting-out expedition at New Bern, North Carolina, that resulted in the capture and destruction of the USS *Underwriter.*

Private William Bell, CSMC, known as "an excellent man, tried and faithful," was killed during the cutting-out of the USS *Underwriter*. Private Perry Marry, CSMC, was also wounded in the cheek during the capture of this Union gunboat.

Thomas Duncan was captured at New Bern, North Carolina.

February 6

Henry H. Marmaduke was promoted to first lieutenant.

February 27

Thomas Duncan arrived at Point Lookout POW Camp, Maryland.

March 3

Atley A. Cooper died in South Carolina Hospital, Petersburg, Virginia, of chronic diarrhea.

March 10

William R. Tetterton and George Walker were assigned to the CSS *Neuse.*

March 18

Joseph Riddock was captured on the steamer *Mary Ann* by the USS *Grand Gulf.*

March 31

Robert Waters was shipped for war at Orange Court House, Virginia.

Charles H. Wilson was paid for service as quartermaster, CSS *Chicora*, Charleston Squadron.

April 12

Andrew H. Forrest was captured onboard the blockade runner CSS *Greyhound* by the USS *Connecticut.*

April 20

John Taylor Wood participated in the capture of Plymouth, North Carolina.

April 24

Joseph Riddock was released as a POW from Camp Hamilton, Virginia.

May 1

Edward Applewhite entered Chimorazo Hospital with secondary syphilis.

May 6

Benjamin Slade was captured at Spotsylvania Court House, Virginia.

May 12

George N. Smith and Peter Williams were captured aboard the blockade runner CSS *Greyhound* by the USS *Connecticut.*

May 11

Jonathan Agnew was wounded in action during the Battle of Yellow Tavern.

May 13

John Hickey was released on oath from Point Lookout, Maryland.
John Taylor Wood was promoted to commander, Confederate Provisional Navy.

May 14

John Driscoll, Samuel N. Aenchbacker and Hugh Aird were assigned to marine guard, CSS *Fredericksburg*.

May 16

George N. Smith and Peter Williams were held as POWs at Camp Hamilton, Virginia.

May 24

Captain Reuben Thom resigned on a surgeon's certificate of disability as assistant inspector general, Brigadier General Richard L. Page's staff, Mobile, Alabama.

May 31

George Washington City was named acting chief engineer of the CSS *Chattahoochee*.

June 2

Franklin Buchanan was named admiral, Confederate Provisional Navy.
Charles Oliver was appointed first lieutenant, CSN.
Elisha R. Johnson was promoted to gunner, Confederate Provisional Navy.
John Waters was promoted to gunner, Confederate Provisional Navy.

June 3

John Perry and John A. Rosler took part in an expedition to capture the USS *Water Witch*.
Philip Muirhead was wounded in action at Cold Harbor, Virginia.

June 10

Arthur Sinclair IV participated in the engagement between the CSS *Alabama* and USS *Kearsage* off Cherbourg, France.

June 14

George N. Smith was released and sent to Baltimore, Maryland.

June 30
Peter Williams was paid for service as a seaman and cook, CSS *Chattahoochee*.

July 1
Francis McAdams was assigned to the CSS *Albemarle*.

July 11
Alexander Spence was captured at Howlett's Farm.

July 26
Edward Larkin was appointed first lieutenant, Confederate Provisional Navy.

July 27
Benjamin Slade was sent to Elmira, New York.

August 3
Alexander Spence was transferred to the Elmira POW Camp, New York.

August 5
Franklin Buchanan was wounded and captured during the Battle of Mobile Bay, Alabama.
John G. Riley, CSS *Tennessee*, was captured during the Battle of Mobile Bay, Alabama.

August 8
Julius Meiere was captured at Fort Gaines, Mobile Bay, Alabama.

August 17
Robert D. Minor, while serving as flag lieutenant of the CSS *Virginia II*, engaged in an artillery duel with Union shore batteries on the James River.

August 23
Reuben Thom was captured at Fort Morgan, Mobile Bay, Alabama.

August 27
John T. Ryan was detailed to report to the president of the Richmond & Danville Railroad.

August 31
Thomas Traylor was detailed to mending shoes for the 18th Battalion Virginia Heavy Artillery at Battery #8, Richmond, Virginia.

September 12

William P. Craig's court-martial sentence for being AWOL was suspended by the secretary of war.

September 28

Charles Spence deserted at City Point, Virginia.

September 29

Charles Spence took the loyalty oath at Washington, D.C., and was transported to Baltimore, Maryland.

September 30

Alfred W. King, William G. Walker and James Cullington were paid for service aboard the CSS *Albemarle*.

October 7

John Barclay was aboard the CSS *Florida* when the commerce raider was captured by the USS *Wachusett* off Bahia, Brazil.

October 13

Julius Meiere escaped from prison at New Orleans, Louisiana.

October 18

Lieutenant Joseph W. Alexander was exchanged.

October 22

Laurence Hinds was wounded in action—left hand and forearm—while aboard the CSS *Fredericksburg*.

October 31

Edwin Morton was captured at Plymouth, North Carolina, following the sinking of the CSS *Albemarle*.

November 8

James M. Sheffield was wounded in the right lung by a gunshot; he was captured at Winchester, Virginia.

November 15

Boatswain John McCrady, after his capture during the Battle of Mobile Bay, escaped from the hospital barracks.

December 4
Oliver Watson was reduced to private, CSMC.

December 22
George A. Knight deserted and entered the Union lines.

December 31
John Nelson was paid for service aboard the CSS *Patrick Henry*, James River Squadron, Richmond, Virginia.

1865

January 13
William Morris, Ephiram Young, Thomas Hansell and Marcellus Augustus Tharpe were ordered to Naval Ordnance Works, Richmond, Virginia.

January 15
John Hickey, Brice Harralson and Patrick McGinnis were captured at Fort Fisher, Nort Carolina.

January 19
George N. Smith was captured off Smithfield, North Carolina.

January 24
Walter Raleigh Butt was named commander of the CSS *Nansemond.*
Edward Larkin was wounded during the explosion of the CSS *Drewry*.

January 28
John F. Higgins's right leg was amputated above the knee.

February 1
George N. Smith was sent to Point Lookout, Maryland, as a POW.

February 9
Martin G. Tharpe was detailed to repair engines on the Southside Railroad.

February 10
John Taylor Wood was promoted to captain, Confederate Provisional Navy.

February 18
Owen Riley deserted from the CSS *Charleston*, Charleston Squadron.

February 28
Albert A. Jones deserted.

March 4
John G. Riley was exchanged at Ship's Island, Mississippi.

March 9
Levin Wood was paroled at Greensboro, North Carolina.

March 13
Reuben Thom was transferred to Fort Delaware, Delaware.
John Murphy deserted at Bermuda Hundred, Virginia.

March 18
Stephen Pickett Truesdale took the loyalty oath and was released from a POW camp.

March 23
Daniel Knowles resigned his commission due to poor health and a desire to "get off the James River."

March 28
Wilson Harrell was exchanged at Aiken's Landing, Virginia.

April 1
James Hannon took the loyalty oath and was paroled at Fort Monroe, Virginia.

April 3
Christopher Porter was captured at Petersburg, Virginia.
Alex Armstrong was captured near Richmond, Virginia.

April 5
James Leahy surrendered to the officer of the USS *Monadnock* in the James River.
William Johnson deserted at Bermuda Hundred, Virginia.

April 6
August W. Lembler was captured near Burkeville, Virginia.
William G. Walker was captured at Harper's Farm.

April 8

Charles McDevett was in the General Hospital, Danville, Virginia, with colitis.

Robert Turner was in the Jackson General Hospital, Richmond, Virginia, with chronic diarrhea.

Alex Armstrong was in Jackson General Hospital, Richmond, Virginia, with chronic diarrhea.

John Delley was in Jackson General Hospital, Richmond, Virginia.

April 9

Catesby Jones was paroled aboard the USS *Stockdale*, Mobile Bay, Alabama.

April 10

Patrick McGinnis requested his release as a conscripted foreigner with no interest in the South.

Charles Carroll Simms was paroled.

April 14

Robert Chester Foute was arrested in Washington, D.C., following the assassination of President Abraham Lincoln and held in Washington Street Federal Prison, Alexandria, Virginia, for the following two months.

William G. Walker was imprisoned at Point Lookout, Maryland.

April 21

Valentine Tolson was paroled after his capture at Burkeville Junction, Virginia.

April 25

Ephiram Young was paroled.

May 3

Ashton Ramsay was paroled at Charlotte, North Carolina.

May 9

Josiah Tattnall was paroled.

May 11

Patrick Hughes was paroled at Mobile, Alabama.

June 10

Reuben Thom was paroled.

December 5
William Gray was paroled at St. Marks, Florida.

1871

June 14
Josiah Tattnall died in Savannah, Georgia.

June 16
Josiah Tattnall was buried at Bonaventura near Savannah, Georgia.

1873

September 1
Douglas French Forrest was ordained an Episcopal minister.

December 25
Reuben Thom died in Montgomery, Alabama.

1874

April 11
Franklin Buchanan died at his residence, the Rest, in Talbot County, Maryland.

1877

June 17
Catesby Jones was killed in an argument with his neighbor.

1879

April 19
Charles Hasker entered R.E. Lee Camp Soldiers' Home, Richmond, Virginia.

1882

November 7
John Q.A. Williams died in Portsmouth, Virginia.

1893

June 9
Dr. Algernon Garnett was appointed medical director of the Department of Arkansas, Medical Corps, SCV.

November 16
Robert Turner died in Norfolk, Virginia.

1894

April 20
James Madison Harvey died in Mineola, Texas.

1896

September 17
John Huddleston entered the R.E. Lee Camp Soldiers' Home, Richmond, Virginia.

1898

July 8
Charles Hasker died in Richmond, Virginia.

1900

April 19
William O. Hall received his pension for the wound he received at the Crater.

1901

January 9
Marcellus Augustus Tharpe died in Worth County, Georgia.

September 10
Samuel Aenchbacker died from a stroke in Columbus, Georgia.

December 6
Charles J. Creekmur died in Portsmouth, Virginia.

1902

May 3
Douglas French Forrest died of a heart attack in Ashland, Virginia.

1903

April 16
James Rudd entered the R.E. Lee Camp Soldiers' Home, Richmond, Virginia.

April 28
William Colonna died of paralysis in the R.E. Lee Camp Soldiers' Home, Richmond, Virginia.

1904

January 4
Arthur Sinclair IV entered the R.E. Lee Camp Soldiers' Home, Richmond, Virginia.

July 19
John Taylor Wood died of muscular rheumatism in Halifax, Nova Scotia.

July 28
Robert Chester Foute died in San Francisco, California.

1905

Julius Meiere died of pneumonia at Sister's Hospital, Cripple Creek, Colorado.

1907

January 9
Arthur Sinclair IV was released from the R.E. Lee Camp Soldiers' Home at his own request.

1908

September 30
James Rudd died in Richmond, Virginia.

1910

February 8
John Steen received his pension.

May 12
Charles Schroeder died in Portsmouth, Virginia.

May 25
John Pembroke Jones died in Pasadena, California.

1911

June 30
John Huddleston committed suicide by tying iron quoits from the Soldiers' Home around his waist and jumping into Byrd Park Pond.

December 18
E.A. Jack died in Alton, Illinois.

1913

February 16
Hunter Davidson died in Pitrayu, Paraguay.

1915

Benjamin Herring died at home in Tallahassee, Florida.

1916

February 12
Carey Hall died in Portsmouth, Virginia.

March 25
Ashton Ramsay died at his home in Baltimore, Maryland.

September 2
Zadock Wesley Messick died.

1919

June 26
The USS *Tattnall* (DD-125) was commissioned.

1920

February 4
Robert Hite died of chronic bronchitis in Petersburg, Virginia.

1924

September 12
Richard Curtis refused to sign his pension for religious reasons. His pension was approved anyway.

November 14
Henry Marmaduke died in Washington, D.C.

December 19
John Higgins died in Crittenden, Virginia.

1925

November 15
Arthur Sinclair IV died in Baltimore, Maryland.

1928

Andrew Jackson Sharpe died in Franklin, Texas.

1941

January 3
John Patrick Kevill died in Norfolk, Virginia.

1942

March 31
The USS *Buchanan* (DD-484), sponsored by Admiral Franklin Buchanan's great-granddaughter, Hildreth Meiere, was commissioned.

1963

April 13
The USS *Tattnall* (DDG-19) was commissioned.

Notes

Chapter 1

1. Barthell, *Mystery of the* Merrimack, 12.
2. Porter, *Norfolk County*, 8.
3. *Boston Daily Evening Transcript*, June 15, 1855.
4. Beese, Virginia *and* Monitor, 6.
5. Ibid., 26–27.
6. Ibid., 27.
7. Ibid., 28–30.
8. Griffiths, "New War Steamers," 302–03.
9. *New York Times*, April 22, 1856.
10. Griffiths, "New War Steamers," 302.
11. Ibid.
12. Jones Papers, Virginia Historical Society.
13. Griffiths, "New War Steamers," 302.

Chapter 2

1. *Norfolk* [Virginia] *Herald*, October 23, 1820.
2. Ibid., February 9, 1825.
3. Ibid., August 11, 1837.
4. Lull, *Navy Yard at Gosport*, 42–43.

5. *Norfolk* [Virginia]*Beacon*, July 20, 1824.
6. *Daily Southern Argus* [Norfolk, Virginia], April 17, 1852.
7. Lull, *Navy Yard at Gosport*, 43.
8. Porter, *Norfolk County*, 12.
9. Ibid., 12-–13.
10. U.S. Department of the Navy, *Official Records*, series 1, vol. 4:277–78, (hereafter cited as *ORN*).
11. Welles, *Diary*, 1:41–46; Niven, *Gideon Welles*, 340–45; Long, "Gosport Affair," 155–72.
12. *ORN*, 1, 4:277–78.
13. Ibid.
14. Ibid., 279.
15. Scharf, *Confederate States*, 130–32 (page citations are to the reprint edition); Porter, *Norfolk County*, 12–13.
16. Long, "Gosport Affair," 155–72; Niven, *Gideon Welles*, 340–45.
17. Long, "Gosport Affair," 166.
18. Selfridge, *Memoirs*, 26–27.
19. Scharf, *Confederate States*, 130.
20. Ibid., 131; Porter, *Norfolk County*, 13.
21. Selfridge, *Memoirs*, 27–28.
22. Ibid.
23. Flanders, *John L. Porter*, 44.
24. Peters, *Recollections*, 1.
25. Ibid., 3.
26. Connor, "North's Fiasco," 30–31.
27. Selfridge, *Memoirs*, 32–33.
28. Porter, *Norfolk County*, 13.
29. Connor, "North's Fiasco," 30.
30. Everett Collection, Massachusetts Historical Society.
31. Porter, *Norfolk County*, 13.
32. Selfridge, *Memoirs*, 34.

Chapter 3

1. Porter, *Norfolk County*, 15; Scharf, *Confederate States*, 133; Peters, *Recollections*, 3.
2. Peters, 4.
3. Ibid.

4. *ORN*, 2, 2: 69.
5. *Richmond Daily Enquirer*, April 22, 1861.
6. Still, *Iron Afloat*, 18 (page citations are to the reprint edition); *ORN*, 2, 1:765.
7. Porter, *Norfolk County*, 25.
8. *ORN*, 2, 2:78.
9. Luraghi, *Confederate Navy*, 10–13; Still, *Iron Afloat*, 5–7.
10. *ORN*, 2, 2:757.
11. Ibid., 67–69.
12. Hagerman, "Lord of the Turtle Boats," 66–75.
13. Hogg, *History of Artillery*, 72–73.
14. Ibid., 57–59.
15. Baxter, *Ironclad Warship*, 49; Luraghi, *Confederate Navy*, 90.
16. Baxter, *Ironclad Warship*, 73–86.
17. Baxter, *Ironclad Warship*, 49; Ballard, "British Battleships," 168–86.
18. Delafield, Mordecai and McClellan, *Report*, 1:168.
19. Baxter, *Ironclad Warship*, 48–52, 211–19.
20. Rodman, *Reports*, 17.
21. Ripley, *Artillery and Ammunition*, 91.
22. *Journal of the Congress of the Confederate States of America*, 5:760.
23. Brooke, *John M. Brooke*, 231.
24. *ORN*, 2, 2: 66–67.
25. Ibid., 70.
26. Still, *Iron Afloat*, 14.
27. *ORN*, 2, 2: 95.
28. Still, *Iron Afloat*, 12.
29. Ibid.
30. Brooke, "John Mercer Brooke," 2:763.
31. Brooke, "*Virginia* or *Merrimac*," 32–33.
32. *ORN*, 2, 1:784; Brooke, *Her Real Projector*, 32–33.
33. *ORN*, 2, 1: 784.
34. Ibid.
35. *ORN*, 2, 2: 175.
36. Brooke, *John M. Brooke*, 235.
37. Brooke, *Her Real Projector*, 716; Porter, *Norfolk County*, 331–32; *ORN*, 2, 1:784.
38. *ORN*, 2, 2:147–75.

Chapter 4

1. U.S. Department of the Navy, "Subject File of the Confederate States Navy, 1861-1865, File HA, Miscellaneous, Box 160, Narrative of H. Ashton Ramsay, Chief Engineer" (hereafter cited as "Ramsay Narrative").
2. *ORN*, 2, 2:175.
3. Beese, Virginia *and* Monitor, 10.
4. Beese, Virginia *and* Monitor, 13–14; Porter, *Norfolk County*, 335–36.
5. Baxter, *Ironclad Warship*, 79; Beese, Virginia *and* Monitor, 14; Flanders, *The Merrimac*, 35; Barnard, *Sea Coast Defense*, 34.
7. Porter, *Norfolk County*, 330; Beese, Virginia *and* Monitor, 14–15; Baxter, *Ironclad Warship*, 79; Brooke, *Her Real Projector*, 8; Porter, *Norfolk County*, 336.
8. Porter, *Norfolk County*, 337.
9. Brooke, "Plan and Construction," 1:715.
10. Brooke, *John M. Brooke*, 239.
11. Jones to Minor, September 1861, Robert Dabney Minor Papers, Minor Family Papers Collection, Virginia Historical Society (hereafter cited as Minor Papers).
12. Porter, *Incidents and Anecdotes*, 79.
13. Jones, "Report of Ordnance Experiments"; Brooke, *John M. Brooke*, 240–41; Jones, "Iron-Clad *Virginia*," 301–02.
14. Ibid.
15. Ibid.
16. *Charleston* [South Carolina] *Mercury*, October 30, 1861.
17. Dew, *Ironmaker to the Confederacy*, 47.
18. Dew, *Ironmaker to the Confederacy*, 117; Brooke, *John M. Brooke*, 241; *ORN*, 2, 2:152.
19. Brooke, *John M. Brooke*, 243; Luraghi, *Confederate Navy*, 97; Woodward, *Alabama Blast Furnaces*, 123, 126.
20. French Forrest Letterbook, Southern Historical Collection, Louis Round Wilson Library, University of North Carolina (hereafter cited as Forrest Letterbook); Letterbook, Tredegar Rolling Mill and Foundry Collection.
21. Ramsay, "Most Famous of Sea Duels," 11.
22. *ORN*, 1, 2:53.
23. Ibid., 67.
24. Ibid., 186.
25. Hogg, *History of Artillery*, 59; Ripley, *Artillery and Ammunition*, 128; *ORN*, 1, 2:186–87.
26. Ripley, *Artillery and Ammunition*, 128.

27. Luraghi, *Confederate Navy*, 242.
28. *ORN*, 2, 6:786.
29. Brooke, *John M. Brooke*, 241–42.
30. Porter, *Norfolk County*, 336; Wise, *End of an Era*, 193–94; *ORN*, 1, 6:776–77.
31. Porter, *Norfolk County*, 336.
32. Brooke, *Her Real Projector*, 3–5.
33. Porter, *Norfolk County*, 338.
34. Still, *Iron Afloat*, 21.
35. Ramsay Narrative.
36. White, *First Iron-Clad Naval Engagement*, 2.
37. Jack, *Memoirs*, 7.
38. Brooke, *John M. Brooke*, 247.
39. Jones, *Civil War at Sea*, 1:223.
40. Still, *Confederate States Navy*, 135.
41. Eggleston, "Captain Eggleston's Narrative," 168.
42. *ORN*, 1, 6:766.
43. Wood, "First Fight of the Ironclads," 1:695.
44. *ORN*, 2, 2:137.
45. Jones, *Captain Roger Jones*, 265.
46. Wood, "First Fight of the Ironclads," 1:695.
47. Porter, *Norfolk County*, 337.
48. Wood, "First Fight of the Ironclads," 1:694.
49. *ORN*, 1, 7:758–59.
50. Isherwood, *Experimental Researches*, 1:177, 213–16.
51. Ramsay Narrative.
52. Beese, Virginia *and* Monitor, 26.
53. Ramsay Narrative.
54. Jones, "Services of the *Virginia*," 66.
55. Porter, *Norfolk County*, 337.
56. Brooke, *John M. Brooke*, 257.
57. *Charleston* [South Carolina] *Mercury*, March 19, 1862.
58. Scharf, *Confederate States*, 152.
59. Porter, *Norfolk County*, 338.
60. Cline, "Ironclad Ram *Virginia*," 244.
61. Forrest Letterbook.

Chapter 5

1. Jack, *Memoirs*, 10.
2. Brooke, *Her Real Projector*, 30.
3. Ibid.
4. Porter, *Norfolk County*, 241; Beese, Virginia *and* Monitor, 10.
5. *Mobile* [Alabama] *Register*, February 14, 1862.
6. U.S. War Department, *War of the Rebellion*, series 1, vol. 9, 188 (hereafter referred to as *OR*).
7. McClellan, *McClellan's Own Story*, 202–03.
8. *ORN*, 1, 5:748.
9. *OR*, 1, 4:620–21.
10. Selfridge, *Memoirs*, 39.
11. *ORN*, 1, 6:333–34.
12. Goldsborough, *Narrative*, 1022–34.
13. *ORN*, 1, 6:363.
14. Ibid., 525.
15. Ibid., 526.
16. Ibid., 661.
17. McDonald, "How I Saw the *Monitor*," 548.
18. *ORN*, 1, 6:536.
19. *ORN*, 1, 9: 726–27.
20. Ibid., 363.
21. Selfridge, "*Merrimac* and *Cumberland*," 180.
22. *ORN*, 1, 6:375.
23. Selfridge, "*Merrimac* and *Cumberland*," 180.
24. Stuyvesant, "How the *Cumberland* Went Down," 205–06.
25. *ORN*, 1, 6:672.
26. *Mobile* [Alabama] *Register*, August 11, 1861.
27. Brooke, *John M. Brooke*, 247.
28. *Lynchburg Virginian*, September 12, 1861.
29. Forrest Letterbook.
30. Donnelly, *CS Marine Corps*, 15.
31. Still, *Confederate States Navy*, 157.
32. *OR*, 1, 51:345.
33. *New York Times*, February 14, 1862.
34. Fox, *Confidential Correspondence*, 1:285; *OR*, 1, 4:620; Cannon, *Personal Reminiscences*, 75.
35. Welles, "First Iron-Clad *Monitor*," 19–20.

36. *OR*, 1, 5:42.
37. *OR*, 1, 11, part 1:129.
38. Brooke, *John M. Brooke*, 240.
39. Baxter, *Ironclad Warship*, 129; Brooke, *John M. Brooke*, 240.
40. Ramsay Narrative.
41. Ibid.
42. Eggleston, "Captain Eggleston's Narrative," 170.
43. *ORN*, 1, 6:776–77.
44. Brooke, *John M. Brooke*, 246.
45. Jones, *Services of the* Virginia, 67; Brooke, *John M. Brooke*, 247.
46. Franklin Buchanan Letterbook, 1861–1863, Southern Historical Collection, Louis Round Wilson Library, University of North Carolina (hereafter cited as Buchanan Letterbook).
47. *OR*, 1, 9:44.
48. *OR*, 1, 51, part 11:480.
49. *OR*, 1, 9:50.
50. Ibid.
51. *ORN* 1, 6:778.
52. Buchanan Letterbook.
53. Jones, *Services of the* Virginia, 66.
54. Norris, *Story of the* "Virginia," 205 (page citations are to the reprint edition).
55. Henderson, *41st Virginia Infantry*, 13.
56. Lewis, *Admiral Franklin Buchanan*, 180–81.
57. Norris, *Story of the* "Virginia," 205.
58. Jones, *Services of the* Virginia, 67–68.
59. Scharf, *Confederate States*, 113.
60. Minor Papers.
61. *OR*, 1, 9:57.
62. *ORN*, 1, 6:780–81.
63. Jones, *Civil War at Sea*, 1:395–96.
64. *ORN*, 1, 6:540–41.

CHAPTER 6

1. Littlepage, "Career of *Merrimac-Virginia*," 44.
2. Jones, *Services of the Virginia*, 67–68.
3. Eggleston, "Captain Eggleston's Narrative," 170.
4. Norris, *Story of the* "Virginia," 206.

5. Parker, *Recollections*, 252.
6. Ibid., 252–53.
7. James Keenan Letter, March 10, 1862, Georgia Department of Archives and History, Atlanta, Georgia.
8. Phillips, "Career of the Iron-Clad *Virginia*," 201; Colston, "Watching the *Merrimac*," 1:712.
9. Phillips, "Career of the Iron-Clad *Virginia*," 201.
10. Littlepage, "Career of *Merrimac-Virginia*," 44.
11. *Norfolk Day Book*, March 10, 1862.
12. Wood, "First Fight of the Ironclads," 1:696.
13. Ramsay Narrative.
14. Ibid.
15. Ibid.
16. Ibid.
17. Ibid.
18. Norris, *Story of the* "Virginia," 206.
19. Ibid.
20. Littlepage, "Career of *Merrimac-Virginia*," 44.
21. Wood, "First Fight of the Ironclads," 1:696; Littlepage, "Career of *Merrimac-Virginia*," 44.
22. Eggleston, "Captain Eggleston's Narrative," 170; Phillips, "Career of the Iron-Clad *Virginia*," 201.
23. Reblen, "U.S.S. *Cumberland*," 45–46.
24. *Boston Journal*, March 13, 1862.
25. Gautier, "Combat Naval de Hampton-Roads," 807.
26. Still, *Iron Afloat*, 29.
27. Davis, *Duel*, 84.
28. Reaney, "*Zouave* sided the *Congress*," 1:714–15.
29. Ramsay Narrative.
30. McDonald, "How I Saw the *Monitor*," 548.
31. *ORN*, 1, 7:23.
32. Littlepage, "Career of *Merrimac-Virginia*," 44.
33. Selfridge, *Memoirs*, 46.
34. Ibid., 48.
35. Ibid.
36. *New York Times*, March 14, 1862.
37. Alger, "*Congress* and *Merrimac*," 688.
38. Eggleston, "Captain Eggleston's Narrative," 170–71.
39. Curtis, *History of Famous Battle*, 8 (page citations are to the reprint edition).

40. Alger, "*Congress* and *Merrimac*," 689.
41. Eggleston, "Captain Eggleston's Narrative," 171.
42. Norris, *Story of the* "Virginia," 205.
43. Ramsay, *Most Famous of Sea Duels*, 21.
44. Curtis, *History of Famous Battle*, 8.
45. Moore, *Rebellion Retold*, 4:272.
46. Ibid.
47. Jones, *Services of the* Virginia, 68.
48. Wood, "First Fight of the Ironclads," 1:698.
49. Kell, *Recollections of a Naval Life*, 282.
50. Jack, *Memoirs*, 14.
51. Ramsay, *Most Famous of Sea Duels*, 14.
52. Littlepage, "Career of *Merrimac-Virginia*," 46.
53. Selfridge, *Memoirs*, 46.
54. Ibid., 57.
55. Ramsay, *Most Famous of Sea Duels*, 14.
56. Littlepage, "Career of *Merrimac-Virginia*," 46.
57. Curtis, *History of Famous Battle*, 8.
58. Jones, *Services of the Virginia*, 68.
59. Ramsay, *Most Famous of Sea Duels*, 15; Littlepage, "Career of *Merrimac-Virginia*," 46.
60. Ramsay, *Most Famous of Sea Duels*, 15.
61. Eggleston, "Captain Eggleston's Narrative," 171.
62. Scharf, *Confederate States*, 137.
63. Jones, *Services of the Virginia*, 69.
64. Ramsay, *Most Famous of Sea Duels*, 15.
65. Jack, *Memoirs*, 16.
66. Littlepage, "Career of *Merrimac-Virginia*,"46.
67. Stuyvesant, "How the *Cumberland* Went Down," 210.
68. O'Neil, "Engagement Between the *Cumberland* and *Merrimack*," 893.
69. Quoted in *Sinking of the* Cumberland.
70. Jones, *Services of the* Virginia, 68.
71. White, *First Iron-Clad Naval Engagement*, 4.
72. Littlepage, *Career of the* Merrimac-Virginia, 47.
73. Colston, "Watching the *Merrimac*," 714.
74. James Kean Letter, March 10, 1862, Georgia Department of Archives and History, Atlanta, Georgia.
75. Colston, "Watching the *Merrimac*," 714.
76. Buchanan Letterbook.

77. Ibid.
78. McIntire, "U.S.S. *Congress*," 49–50.
79. Reaney, "*Zouave* sided the *Congress*," 168-170.
80. Parker, *Recollections*, 254.
81. Alger, "*Congress* and *Merrimac*," 690.
82. Parker, *Recollections*, 254.
83. Ibid.
84. *OR*, 1, 9:78.
85. *ORN*, 1, 7:35.
86. Parker, *Recollections*, 254.
87. Buchanan Letterbook.
88. Kell, *Recollections of a Naval Life*, 282–83.
89. Eggleston, "Captain Eggleston's Narrative," 173.
90. Ransom, "*Monitor* and the *Merrimac*," 111.
91. Norris, *Story of the "*Virginia,*"* 208.
92. Kell, *Recollections of a Naval Life*, 283.
93. Norris, *Story of the "*Virginia,*"* 217.
94. Eggleston, "Captain Eggleston's Narrative," 173.
95. Rae, "Little *Monitor* Saved Our Lives," 34.
96. Colston, "Watching the *Merrimac*," 714.
97. Jones, *Services of the* Virginia, 68.
98. Drake, "CSS *Virginia*," 52.

Chapter 7

1. Kell, *Recollections of a Naval Life*, 283.
2. *Daily Press* [Newport News, Virginia], March 9, 1962.
3. Rae, "Little *Monitor* Saved Our Lives," 34.
4. Gautier, "Combat Naval de Hampton-Roads," 810.
5. *ORN*, 1, 7:35.
6. Cannon, *Personal Reminiscences*, 85–86.
7. Sears, *McClellan Papers*, 198–99.
8. Ibid.
9. Dahlgren, *Memoirs*, 359–60.
10. Welles, *Diary*, 1:51–52.
11. Ibid., 1:61–64.
12. Sears, *McClellan Papers*, 199.
13. White, *First Iron-Clad Naval Engagement*, 4–5.

14. Norris, *Story of the "Virginia,"* 206.
15. Buchanan Letterbook.
16. White, *First Iron-Clad Naval Engagement*, 5.
17. Jones, *Services of the* Virginia, 70–71.
18. Wood, "First Fight of the Ironclads," 1:700.
19. Kell, *Recollections of a Naval Life*, 283.
20. Phillips, "Career of the Iron-Clad *Virginia*," 200.
21. Ramsay, *Most Famous of Sea Duels*, 11–12; Jones, *Services of the* Virginia, 71; Phillips, "Career of the Iron-Clad *Virginia*," 200.
22. Phillips, "Career of the Iron-Clad *Virginia*," 205.
23. Ibid., 210.
24. Ibid., 155.
25. Ibid., 211.
26. Ibid., 205.
27. Ramsay, *Most Famous of Sea Duels*, 12.
28. Osborne, *Twenty-ninth Regiment*, 60.
29. Colston, "Watching the *Merrimac*," 714.
30. *Sumter Republican* [Americus, Georgia], April 11, 1862. Acting midshipman William C. Hutter resigned from the U.S. Navy on April 20, 1861. He immediately joined the Virginia State Navy and was appointed an acting midshipman, Confederate navy, on June 21, 1861. Hutter served aboard the CSRS *Confederate States* until transferred to the North Carolina Squadron. Following the Battle of Roanoke Island, Hutter was detailed to the CSS *Raleigh*. William Hutter was killed on March 8, 1862, by Union musket fire from Newport News Point while the CSS *Raleigh* and CSS *Beaufort* were alongside the USS *Congress* accepting the Federal frigate's surrender.
31. Jones, *Services of the* Virginia, 70–71.

CHAPTER 8

1. Cline, "Ironclad Ram *Virginia*," 246.
2. Curtis, *History of Famous Battle*, 10.
3. Ibid.
4. Ramsay, *Most Famous of Sea Duels*, 12.
5. Jones, *Services of the* Virginia, 71.
6. Littlepage, "*Merrimac* vs. *Monitor*," 335; Eggleston, "Captain Eggleston's Narrative," 174; Curtis, *History of Famous Battle*, 11; Porter, *Norfolk County*, 361; Wood, "First Fight of the Ironclads," 1:701.

7. Beese, Virginia *and* Monitor, 37.
8. *ORN*, 1, 6:516–17.
9. Ericsson, "Building of the *Monitor*," 1:735.
10. Lewis, "Life on the *Monitor*," 258; Ellis, Monitor *of the Civil War*, 19.
11. *ORN*, 2, 1:148.
12. Greene, "In the *Monitor's* Turret," 1:720.
13. Daly, *Aboard the U.S.S.* Monitor, 8
14. Greene, *In the* Monitor*'s Turret*, 720.
15. Daly, *Aboard the U.S.S.* Monitor, 31.
16. Lewis, *Life on the* Monitor, 20.
17. Greene, *In the* Monitor*'s Turret*, 721.
18. White, *First Iron-Clad Naval Engagement*, 6.
19. Ramsay, *Most Famous of Sea Duels*, 12.
20. Jack, *Memoirs*, 16.
21. Littlepage, Merrimac *vs.* Monitor, 336.
22. Ibid.
23. White, *First Iron-Clad Naval Engagement*, 13.
24. Jack, *Memoirs*, 16.
25. White, *First Iron-Clad Naval Engagement*, 13.
26. Wood, "First Fight of the Ironclads,"1:703.
27. Norris, *Story of the* "Virginia," 219–20.
28. Eggleston, "Captain Eggleston's Narrative," 175-176.
29. Wood, "First Fight of the Ironclads,"1:702.
30. Eggleston, "Captain Eggleston's Narrative," 176.
31. Ramsay, *Most Famous of Sea Duels*, 12.
32. Eggleston, "Captain Eggleston's Narrative," 176.
33. *ORN*, 2, 1:157.
34. Ramsay, *Most Famous of Sea Duels* 12.
35. Phillips, "Career of the Iron-Clad *Virginia*," 209.
36. Ramsay, *Most Famous of Sea Duels*, 13.
37. White, *First Iron-Clad Naval Engagement*, 7.
38. Wood, "First Fight of the Ironclads,"1:702.
39. Daly, *Aboard the U.S.S. Monitor*, 36.
40. Ibid.
41. Ibid.
42. Phillips, "Career of the Iron-Clad *Virginia*," 209.
43. White, *First Iron-Clad Naval Engagement*, 7.
44. Eggleston, "Captain Eggleston's Narrative," 176.
45. Ramsay, *Most Famous of Sea Duels*, 14.

46. Wood, "First Fight of the Ironclads," 1:703.
47. Ibid.
48. Littlepage, *Merrimac vs. Monitor*, 336.
49. Curtis, *History of Famous Battle*, 12.
50. Norris, *Story of the "Virginia,"* 219.
51. Daly, *Aboard the U.S.S. Monitor*, 36.
52. Greene, "I Fired the First Gun," 103.
53. Ibid.
54. Jones, *Services of the* Virginia, 72.
55. Eggleston, "Captain Eggleston's Narrative," 176.
56. Ibid.
57. Ramsay, *Most Famous of Sea Duels*, 12.
58. Eggleston, "Captain Eggleston's Narrative," 177.
59. Jones, *Services of the Virginia*, 72.
60. Samuel Dana Greene Manuscript, *U.S. Naval Academy Trident*, Spring 1942, 44 (hereafter cited as Greene Manuscript).
61. Stimers, "Engineer Aboard the *Monitor*," 35.
62. *Great Rebellion Scrapbooks*, 1893.
63. Greene, "I Fired the First Gun," 109.
64. Greene Manuscript, 44.
65. Ramsay Narrative.
66. Norris, *Story of the "Virginia,"* 219; Parker, *Recollections*, 267; Foute, "Echoes from Hampton Roads," 246.
67. Wood, "First Fight of the Ironclads," 1:705.
68. *OR*, 4, 1:980.
69. John Taylor Wood Papers, Southern Historical Collection, Louis Round Wilson Library, University of North Carolina, Chapel Hill, North Carolina (hereafter cited as Wood Papers); Norris, *Story of the "Virginia,"* 220.
70. Wise, *End of an Era*, 202–04.
71. Ramsay Narrative.
72. Ibid.
73. Norris, *Story of the "Virginia,"* 221.
74. Colston, "Watching the *Merrimac*," 714.
75. *OR*, 1, 5:14.
76. *Norfolk Day Book*, March 10, 1862.
77. *OR*, 1, 51, 2:499.
78. Ramsay Narrative.
79. *OR*, 1, 11, 1:126.

Chapter 9

1. *ORN*, 1, 7:99–100.
2. Dahlgren, *Memoirs*, 307.
3. *ORN*, 1, 9:31.
4. *OR*, 1, 4:27.
5. *OR*, 1, 11, 1:8.
6. *ORN*, 1, 7:83; Welles, *Diary*, 3:473; Fox, *Confidential Correspondence*, 1:249.
7. Daly, *Aboard the U.S.S.* Monitor, 50.
8. *ORN*, 1, 9:25.
9. *ORN*, 1, 7:780–81.
10. Wood Papers.
11. Buchanan Letterbook.
12. Cline, "Ironclad Ram *Virginia*," 247.
13. "Facts About the *Merrimac*," *Scientific American*, June 6, 1862, 73.
14. Item 88, RG 45, National Archives, Washington, D.C.
15. Ibid.
16. Ibid.
17. Brooke, *John M. Brooke*, 256.
18. Brooke, *Her Real Projector*, 31.
19. Ibid.
20. Wood, "First Fight of the Ironclads," 1:705–06.
21. Parker, *Recollections*, 272.
22. *ORN*, 1, 7:757.
23. Ibid.
24. *OR*, 1, 9:14.
25. *ORN*, 1, 7:747.
26. Ibid., 748.
27. Ibid., 749.
28. Ibid., 748.
29. *OR*, 1, 9:31.
30. Daly, *Aboard the U.S.S.* Monitor, 64.
31. *OR*, 1, 40, part 2:387.
32. *OR*, 1, 5:50–51.
33. *OR*, 2, 3:53.
34. *OR*, 1, 40, 2:387.
35. Ibid., 390.
36. *ORN*, 1, 7:755.
37. Flanders, *The* Merrimac, 59.

38. Ramsay Narrative.
39. Wood, "First Fight of the Ironclads," 1:706.
40. Minor Papers; Wood, "First Fight of the Ironclads," 1:706.
41. *OR*, 2, 9:404.
42. Minnish, "Reminiscences," Virginia War Museum.
43. Woodward, *Mary Chesnut's Civil War*, 401.
44. Early, *Narrative*, 59.

Chapter 10

1. *OR*, 2, 11, 3:430.
2. *OR*, 1, 40, 2:391.
3. *ORN*, 1, 7:764–65.
4. Norris, *Story of the* "Virginia," 212.
5. *ORN*, 1, 7:764.
6. Foute, "Echoes from Hampton Roads," 247–48.
7. Parker, *Recollections*, 267.
8. Ibid.
9. Daly, *Aboard the U.S.S.* Monitor, 40.
10. Ibid., 46.
11. *ORN*, 1, 7:219.
12. Daly, *Aboard the U.S.S.* Monitor, 52.
13. *ORN*, 1, 7:224.
14. Ibid.
15. *New York Herald*, April 15, 1862.
16. Daly, *Aboard the U.S.S.* Monitor, 53.
17. *ORN*, 1, 7:225.
18. Ibid., 223.
19. Norris, *Story of the* "Virginia," 212.
20. Wood Papers.
21. Norris, *Story of the* "Virginia," 211.
22. *OR*, 1, 51, 2:539.
23. Ramsay, *Most Famous of Sea Duels*, 12.
24. Jones, *Services of the* Virginia, 73.
25. *OR*, 1, 11, 3:456.
26. Jones, *Services of the* Virginia, 73.
27. Ramsay, *Most Famous of Sea Duels*, 12.
28. *OR*, 1, 11, part 3:456.

29. Johnston, *Narrative*, 116.
30. *ORN*, 2, 1:633.
31. *OR*, 1, 11, 3:478.

Chapter 11

1. *ORN*, 1, 3:785.
2. Wood Papers.
3. Daly, *Aboard the U.S.S.* Monitor, 73.
4. Eggleston, "Captain Eggleston's Narrative," 177.
5. *OR*, 1, 11, 3:80.
6. *ORN*, 1, 7:326.
7. Johnson, *Rear Admiral John Rodgers*, 198.
8. Eggleston, "Captain Eggleston's Narrative," 178.
9. Curtis, *History of Famous Battle*, 16.
10. Ibid.
11. Eggleston, "Captain Eggleston's Narrative," 178.
12. Jones, *Services of the* Virginia, 73.
13. Wood, "First Fight of the Ironclads," 1:709.
14. Wood Papers.
15. *ORN*, 1, 7:332.
16. Daly, *Aboard the U.S.S.* Monitor, 73.
17. *OR*, 1, 11, 3:153.
18. *Baltimore American*, May 11, 1862.
19. Curtis, *History of Famous Battle*, 16.
20. Wood, "First Fight of the Ironclads," 1:710.
21. Ibid.
22. Scharf, *Confederate States*, 222.
23. Ibid.; *OR*, 1, 11, 3:160.
24. Viele, "Trip with Lincoln," 819.
25. Ibid.

Chapter 12

1. Norris, *Story of the* "Virginia," 211–12.
2. Wood, "First Fight of the Ironclads," 1:710.
3. Scharf, *Confederate States*, 223.

4. Ramsay, *Most Famous of Sea Duels*, 12.
5. Wood, "First Fight of the Ironclads," 1:710.
6. Scharf, *Confederate States*, 223.
7. Curtis, *History of Famous Battle*, 17.
8. White, *First Iron-Clad Naval Engagement*, 8.
9. Wood, "First Fight of the Ironclads," 1:710.
10. White, *First Iron-Clad Naval Engagement*, 8.
11. Ramsay, *Most Famous of Sea Duels*, 12.
12. Curtis, *History of Famous Battle*, 17.
13. Higgins, "Brilliant Career of the *Merrimac*," 357.
14. Ibid.
15. Ibid.
16. *ORN*, 1, 7:337.
17. Stephen R. Mallory Diary, Library of Congress, Washington, D.C.
18. Brooke, *John M. Brooke*, 258.
19. Franklin, *Memoirs of a Rear Admiral*, 182–83.
20. Daly, *Aboard the U.S.S.* Monitor, 119–20.
21. *OR*, 1, 11, 1:835.
22. Ibid.
23. *OR*, 1, 11, 3:165.
24. *ORN*, 1, 7:354–355.
25. Ibid., 357.
26. *OR*, 1, 51, 1:507.
27. Coski, *Capital Navy*, 41.
28. *ORN*, 2, 1:636.
29. *ORN*, 1, 7:799.
30. Mann, "New Light," 89.
31. Wood Papers.
32. Scharf, *Confederate States*, 713.
33. Soley, "Navy in the Peninsular Campaign," 2:269–70.
34. Ibid., 269.
35. *ORN*, 1, 7:369.
36. Mann, "New Light," 92–93.
37. Scharf, *Confederate States*, 764.
38. *ORN*, 1, 7:357.
39. Mann, "New Light," 92.
40. Scharf, *Confederate States*, 714.
41. *ORN*, 1, 7:370.
42. Scharf, *Confederate States*, 715.

43. *ORN*, 1, 7:357.
44. Daly, *Aboard the U.S.S.* Monitor, 192–93.
45. Mann, "New Light," 95.
46. Ibid.
47. Littlepage, "With the Crew of the *Virginia*," 42.
48. Coski, *Capital Navy*, 47.
49. *ORN*, 1, 7:370.
50. *Richmond Examiner*, May 21, 1862.
51. Coski, *Capital Navy*, 47.
52. *Richmond Whig*, May 22, 1862.

Chapter 13

1. Daly, *Aboard the U.S.S.* Monitor, 39.
2. White, *First Iron-Clad Naval Engagement*, 15.
3. Ibid.
4. Eggleston, "Captain Eggleston's Narrative," 178.
5. *ORN*, 1, 7:334.
6. *Richmond Enquirer*, July 1, 1862.
7. *ORN*, 1, 8:339.
8. Ibid.
9. Still, *Confederate States Navy*, 134.
10. Ibid., 203.
11. Ibid., 209.
12. *ORN*, 1, 9:806.
13. Daly, *Aboard the U.S.S.* Monitor, 184.
14. *ORN*, 1, 6:745.
15. Still, *Iron Afloat*, 167.
16. Still, *Confederate States Navy*, 167.
17. *ORN*, 2, 1:625.
18. Still, *Confederate States Navy*, 193.
19. Ibid.
20. Semmes, *Memoirs of Service*, 803–04 (page citations are to the reprint edition).
21. Scharf, *Confederate States*, 747.
22. *Norfolk Virginian*, October 7, 1874, 1.
23. Ibid., June 2, 1876, 1.
24. Norris, *Story of the* "Virginia," 233.

APPENDIX I

1. Norris, *Story of the* "Virginia," 234.

APPENDIX II

1. *ORN*, 2, 2:67.
2. Brooke, *John M. Brooke*, 286.
3. Flanders, *John L. Porter*, 7.
4. Still, *Iron Afloat*, 15.

APPENDIX III

1. Still, *Iron Afloat*, 24.
2. *ORN*, 1, 7:764.
3. Eggleston, "Captain Eggleston's Narrative," 168.
4. Ramsay, *Most Famous of Sea Duels*, 11.
5. Lewis, *Admiral Franklin Buchanan*, 166.
6. Ibid., 163; Dudley, *Going South*, 8.
7. Welles, *Diary*, 1:19; Swartz, "Franklin Buchanan," 62–66.

APPENDIX VI

1. Jack, *Memoirs*, 10.

APPENDIX XII

1. *Columbus* [Georgia] *Enquirer-Sun*, September 10–11, 1901; U.S. Bureau of the Census, "Population Schedules of the Seventh Census of the United States of America, 1850, Muscogee County, Georgia," Microfilm Series, M432, Roll 79, RG 29, National Archives (NA), (hereafter cited as *1850 "County (Co.) Name, State Name," Census*, followed by M432 and Roll Number); U.S. War Department, "Records Relating to Confederate Naval and Marine Personnel, Hospital and Prison Records of Naval Personnel," Microfilm Series, M260, Roll 1, Record Group (RG) 109, NA, (hereafter

cited as *Hospital & Prison Records*, followed by M260 and Roll Number 1–4); *ORN*, 2, 1:310, 315; Donnelly, *Service Record*, 2 (hereafter cited as *Confederate Enlisted Marines*).

2. U.S. War Department, "Compiled Service Records (CSRs) of Confederate Soldiers Who Served in Organizations from the State of Virginia, 9th Infantry," Microfilm Series, M324, Roll 475, RG 109, NA, (hereafter cited as *9th Virginia Infantry CSRs*, followed by M324 and Roll Number); Trask, *9th Virginia Infantry*, 49; Nicholas and Servis, *Powhatan, Salem and Courtney*, 214; *ORN*, 2, 1:310; U.S. Department of the Navy, "Richmond Station & James River Squadron, C.S. Navy," Microfilm Series, T829, Roll 172, RG 45, NA (hereafter cited as *Richmond & James River Payrolls*); U.S. War Department, "CSRs of Confederate Soldiers Who Served in Organizations from the State of Virginia, 5th Cavalry," Microfilm Series, M324, Roll 52, RG 109, NA, (hereafter cited as *5th Virginia Cavalry CSRs*); Driver, *5th Virginia Cavalry*, 178.
3. U.S. War Department, "CSRs of Confederate Soldiers Who Served in Organizations From the State of Louisiana, 8th Infantry," Microfilm Series, M320, Roll 187, RG 109, NA; Booth, *Louisiana Confederate Soldiers*, 1:33; *ORN*, 2, 1:310; U.S. Department of the Navy, "Papers Pertaining to Vessels of or Involved with the Confederate States of America, File V-5, CSS *Virginia*," Microfilm Series, M909, Roll 30, RG 45, NA (hereafter cited as *Vessel Papers*); *Richmond & James River Payrolls.*
4. Donnelly, *Confederate Enlisted Marines*, 2; *ORN*, 2, 1:310; Scharf, *Confederate States*, 688; *ORN*, 2, 1:315.
5. U.S. Department of the Navy, *CSN Register*, 2 (page citations are to the reprint edition); *ORN*, 2, 1:308; U.S. Department of the Navy, "New Orleans Station and Fleet, C.S. Navy," Microfilm Series, T829, Roll 169, RG 45, NA (hereafter cited as *New Orleans Payrolls*); *Richmond & James River Payrolls*.
6. Hamersly, *Complete General Navy Register*, 22; *CSN Register*, 3; Dudley, *Going South*, 40; *Richmond & James River Squadron*.
7. U.S. War Department, "CSRs of Confederate Soldiers Who Served in Organizations From the State of Alabama, 8th Infantry," Microfilm Series, M311, Roll 171, RG 109, NA, (hereafter cited as *8th Alabama Infantry CSRs*, followed by M311 and Roll Number); U.S. Department of the Navy, "1861–1865, Shipping Articles of C.S. Navy," Microfilm Series, T829, Roll 173, RG 45, NA (hereafter cited as *Shipping Articles*); U.S. Department of the Navy, "1861–1862, Officers & Men, North Carolina Station, C.S. Navy," Microfilm Series, T829, Roll 170, RG 45,

NA (hereafter cited as *North Carolina Payrolls*); *Vessel Papers*; U.S. Department of the Navy, "Subject File of the Confederate States Navy, 1861–1865, File NA, CSS *New Orleans*–CSS *Yorktown*," Microfilm Series, M1091, Roll 17, RG 45, NA (hereafter cited as *File NA-2*); *ORN*, 2, 1:309; *Richmond & James River Payrolls*; *CSN Register*, 3.

8. U.S. War Department, "CSRs of Confederate Soldiers Who Served in Organizations From the State of North Carolina, 1st Infantry," Microfilm Series, M270, Roll 90, RG 109, NA (hereafter cited as *1st North Carolina Infantry CSRs*, followed by M270 and Roll Number); Manarin and Jordan, *North Carolina Troops*, 3:211; *ORN*, 2, 1:310; *Richmond & James River Payrolls*.
9. U.S. War Department, "CSRs of Confederate Soldiers Who Served in Organizations From the State of Louisiana, 1st (Nelligan's) Infantry," Microfilm Series, M320, Roll 72, RG 109, NA, (hereafter cited as *1st (Nelligan's) Louisiana Infantry CSRs*, followed by M320 and Roll Number); Booth, *Louisiana Confederate Soldiers*, 1:51; *ORN*, 2, 1:309; *North Carolina Payrolls*; *Vessel Papers*; *Richmond & James River Payrolls*; U.S. War Department, "CSRs of Confederate Soldiers Who Served in Organizations From the State of Virginia, 39th Battalion (Btn.) Cavalry," Microfilm Series, M324, Roll 199, RG 109, NA, (hereafter cited as *39th Btn. Virginia Cavalry CSRs*).
10. U.S. War Department, "CSRs of Confederate Soldiers Who Served in Organizations From the State of South Carolina, 1st (McCreary's) Infantry," Microfilm Series, M267, Roll 126, RG 109, NA (hereafter cited as *1st (McCreary's) South Carolina Infantry CSRs*, followed by M267 and Roll Number); Salley, *South Carolina Troops*, 1:351–54, 356, 406; *ORN*, 2, 1:310; *Vessel Papers*; *ORN*, 2, 1:281; *Richmond & James River Payrolls*; *North Carolina Payrolls*; U.S. Department of the Navy, "Subject File of the Confederate States Navy, 1861–1865, File NA, CSS *Alabama*–CSS *Neuse*," Microfilm Series, M1091, Roll 17, RG 45, NA (hereafter cited as *File NA-1*); U.S. Department of the Navy, "Charleston Station and Fleet, C.S. Navy," Microfilm Series, T829, Roll 166, RG 45, NA (hereafter cited as *Charleston Payrolls*); *Hospital & Prison Records*, M260, Roll 1.
11. U.S. Bureau of the Census, "Population Schedules of the Eighth Census of the United States of America, 1860, Norfolk County, Virginia," Microfilm Series, M653, Roll 1366, RG 29, NA (hereafter cited as *1860 "County (Co.) Name* (or City Name), *State Name," Census*, followed by M653 and Roll Number); U.S. War Department, "CSRs of Confederate Soldiers Who Served in Organizations from the State of Virginia, 41st Infantry," Microfilm Series, M324, Roll 860, RG 109, NA (hereafter cited as *41st Virginia Infantry CSRs*, followed by M324 and Roll Number); U.S. War Department, "CSRs

of Confederate Soldiers Who Served in Organizations from the State of Virginia, 19th Btn. Heavy Artillery," Microfilm Series, M324, Roll 243, RG 109, NA, (hereafter cited as *19th Btn. Virginia Heavy Artillery CSRs*, followed by M324 and Roll Number); U.S. War Department, "CSRs of Confederate Soldiers Who Served in Organizations from the State of Virginia, Heavy Artillery, Kevill's Company," Microfilm Series, M324, Roll 315, RG 109, NA (hereafter cited as *Kevill's Company CSRs*); Henderson, *41st Virginia Infantry*, 86; Weaver, *Assorted Heavy Artillery*, 75; Fiveash, Virginia (Merrimac), 21–22; U.S. Bureau of the Census, "Population Schedules of the Ninth Census of the United States of America, 1870, Norfolk County, Virginia," Microfilm Series, M593, Roll 1666, RG 29, NA (hereafter cited as *1870 "County (Co.) Name* (or City Name), *State Name," Census*, followed by M593 and Roll Number).

12. U.S. War Department, "CSRs of Confederate Soldiers Who Served in Organizations from the State of Georgia, 4th Infantry," Microfilm Series, M266, Roll 180, RG 109, NA, (hereafter cited as *4th Georgia Infantry CSRs*, followed by M266 and Roll Number); U.S. War Department, "Records Relating to Confederate Naval and Marine Personnel, Reference Cards and Papers Relating to Naval and Marine Personnel," Microfilm Series, M260, Roll 5, Record Group (RG) 109, NA, (hereafter cited as *CSN & CSMC Personnel*, followed by M260 and Roll Number); *ORN*, 2, 1:309; *Vessel Papers*; *File NA-2*, M1091, Roll 18; *Richmond & James River Payrolls*.
13. *CSN & CSMC Personnel*, M260, Roll 5; *ORN*, 2, 1:309; *Vessel Papers*; *File NA-2*, M1091, Roll 18; *Richmond & James River Payrolls*; *Hospital and Prison Records*, M260, Roll 1.
14. *Shipping Articles*; *ORN*, 2, 1:310; *Richmond & James River Payrolls*.
15. *1860 Washington Co., North Carolina Census*, M653, Roll 917; *1st North Carolina Infantry CSRs*, M270, Roll 90; Manarin, *North Carolina Troops*, 3:211; *ORN*, 2, 1:310; *Vessel Papers*; *Richmond & James River Payrolls*.
16. Donnelly, *Confederate Enlisted Marines*, 4–5; *ORN*, 2, 1:310, 315; *Hospital & Prison Records*, M260, Roll 1.
17. *1860 Elizabeth City Co., Virginia Census*, M653, Roll 1343; U.S. War Department, "CSRs of Confederate Soldiers Who Served in Organizations from the State of Virginia, 32nd Infantry," Microfilm Series, M324, Roll 780, RG 109, NA (hereafter cited as *32nd Virginia Infantry CSRs*, followed by M324 and Roll Number); Jensen, *32nd Virginia Infantry*, 173; *ORN*, 2, 1:309; *Richmond & James River Payrolls*.
18. *ORN*, 2, 1:285; *North Carolina Payrolls*; U.S. Department of the Navy, "Wilmington Station, Marine Corps, and Miscellaneous," Microfilm

Series, T829, Roll 165, RG 45, NA (hereafter cited as *Wilmington Payrolls*); *CSN & CSMC Personnel*, M260, Roll 5; *ORN*, 1, 6:781; *ORN*, 2, 1:309; *Richmond & James River Payrolls*; *File NA-2*, M1091, Roll 18.

19. Donnelly, *Confederate Enlisted Marines*, 5; *ORN*, 2, 1:311; *Richmond & James River Payrolls*; *Hospital & Prison Records*, M260, Roll 1.

20. U.S. War Department, "CSRs of Confederate Soldiers Who Served in Organizations from the State of Virginia, 57th Infantry," Microfilm Series, M324, Roll 980, RG 109, NA, (hereafter cited as *57th Virginia Infantry CSRs*); Sublett, *57th Virginia Infantry*, 48; *CSN & CSMC Personnel*, M260, Roll 5; *ORN*, 2, 1:309; *File NA-2*, M1091, Roll 18; *Richmond & James River Payrolls*; *North Carolina Payrolls*; *Confederate Veteran*, April 1928, 159; *Confederate Veteran*, November 1931, 439.

21. *57th Virginia Infantry CSRs*, M324, Roll 980; Sublett, 48; *CSN & CSMC Personnel*, M260, Roll 5; *ORN*, 2, 1:309; *File NA-2*, M1091, Roll 18; *North Carolina Payrolls*; *Richmond & James River Payrolls*; *Hospital & Prison Records*, M260, Roll 1.

22. U.S. War Department, "CSRs of Confederate Soldiers Who Served in Organizations from the State of Louisiana, 2nd Infantry," Microfilm Series, M320, Roll 102, RG 109, NA (hereafter cited as *2nd Louisiana Infantry CSRs*, followed by M320 and Roll Number); Booth, *Louisiana Confederate Soldiers*, 1:123; *CSN & CSMC Personnel*, M260, Roll 5; *ORN*, 2, 1:310; *North Carolina Payrolls*; *Richmond & James River Payrolls*.

23. *1st North Carolina Infantry CSRs*, M270, Roll 90; *Vessel Papers*.

24. *41st Virginia Infantry CSRs*, M324, Roll 860; Henderson, *41st Virginia Infantry*, 88; *CSN & CSMC Personnel*, M260, Roll 5; ORN, 2, 1:310; *Richmond & James River Payrolls*.

25. *41st Virginia Infantry CSRs*, M324, Roll 860; Henderson, *41st Virginia Infantry*, 88; *19th Btn. Virginia Heavy Artillery CSRs*, M324, Roll 243; *Kevill's Company* C*SRs*, M324, Roll 315; Weaver, *Assorted Heavy Artillery*, 75; Stewart, *History of Norfolk County*, 138.

26. *CSN & CSMC Personnel*, M260, Roll 5; *ORN*, 2, 1:309; *Richmond & James River Payrolls*; *ORN*, 2, 1:311.

27. *1st North Carolina Infantry CSRs*, M270, Roll 90; Manarin, *North Carolina Troops*, 3:224; *File NA-2*, M1091, Roll 18; *ORN*, 2, 1:309; *Richmond & James River Payrolls*; *Charleston Payrolls*.

28. *41st Virginia Infantry CSRs*, M324, Roll 860; Henderson, *41st Virginia Infantry*, 89; *Kevill's Company* C*SRs*; Weaver, *Assorted Heavy Artillery*, 75; Fiveash, Virginia (Merrimac), 21–22; *1870 Norfolk Co., Virginia Census*, M593, Roll 1666.

29. Donnelly, *Confederate Enlisted Marines*, 7; *ORN*, 2, 1:310; Scharf, *Confederate States*, 688.

31. U.S. War Department, "CSRs of Confederate Soldiers Who Served in Organizations from the State of Alabama, 12th Infantry," Microfilm Series, M311, Roll 209, RG 109, NA (hereafter cited as *12th Alabama Infantry CSRs*, followed by M311 and Roll Number); *Shipping Articles*; *ORN*, 2, 1:309; *North Carolina Payrolls*; *Richmond & James River Payrolls*; *File NA-2*, M1091, Roll 18; *Hospital & Prison Records*, M260, Roll 1; *1850 Rutherford Co., Tennessee Census*, M432, Roll 894.

32. *1860 City of Baltimore, Maryland Census*, M653, Roll 459; Hartzler, *Marylanders in the Confederacy*, 86; *CSN Register*, 13; *ORN*, 2, 1:309; *Richmond & James River Payrolls*; *Hospital & Prison Records*, M260, Roll 1.

33. Donnelly, *Confederate Enlisted Marines*, 8; *ORN*, 2, 1:310, 315; Scharf, *Confederate States*, 688; *1850 Duval Co., Florida Census*, M432, Roll 58.

34. *9th Virginia Infantry CSRs*, M324, Roll 475; Trask, *9th Virginia Infantry*, 52; Hartzler, *Marylanders in the Confederacy*, 87; *CSN & CSMC Personnel*, M260, Roll 5; *ORN*, 2, 1:309; *Richmond & James River Payrolls*; *File NA-2*, M1091, Roll 18; *Wilmington Payrolls*; *Hospital & Prison Records*, M260, Roll 1.

35. U.S. War Department, "CSRs of Confederate Soldiers Who Served in Organizations from the State of Alabama, 3rd Infantry," Microfilm Series, M311, Roll 103, RG 109, NA (hereafter cited as *3rd Alabama Infantry CSRs*, followed by M311 and Roll Number); Roll 103; *CSN & CSMC Personnel*, M260, Roll 5; *ORN*, 2, 1:309; *North Carolina Payrolls*; *Richmond & James River Payrolls*.

36. U.S. War Department, "CSRs of Confederate Soldiers Who Served in Organizations from the State of Louisiana, 1st Heavy Artillery," Microfilm Series, M320, Roll 35, RG 109, NA; Booth, *Louisiana Confederate Soldiers*, 1:197; *CSN & CSMC Personnel*, M260, Roll 5; *ORN*, 2, 1:310; *Vessel Papers*; *Richmond & James River Payrolls*; *Shipping Articles*.

37. *1860 Roanoke Co., Virginia Census*, M653, Roll 1375; *9th Virginia Infantry CSRs*, M324, Roll 476; Trask, *9th Virginia Infantry*, 53; Nicholas and Servis, *Powhatan, Salem and Courtney*, 214; *41st Virginia Infantry CSRs*, M324, Roll 860; Henderson, *41st Virginia Infantry*, 90; *Kevill's Company CSRs*; Weaver, *Assorted Heavy Artillery*, 75; Fiveash, Virginia (Merrimac), 21–22.

38. Donnelly, *Confederate Enlisted Marines*, 11; *ORN*, 2, 1:310; *Richmond & James River Payrolls*.

39. Donnelly, *Confederate Enlisted Marines*, 11; *ORN*, 2, 1:311; *Richmond & James River Payrolls*.

40. *1860 City of Baltimore, Maryland Census*, M653, Roll 466; Hartzler, *Marylanders in the Confederacy*, 96; Donnelly, *Confederate Enlisted Marines*, 11; *ORN*, 2, 1:310; *Richmond & James River Payrolls*.
41. *Shipping Articles*; *ORN*, 2, 1:310; *Richmond & James River Payrolls*.
42. *1860 Bladen Co., North Carolina Census*, M653, Roll 888; *CSN & CSMC Personnel*, M260, Roll 5; *ORN*, 2, 1:310; *Vessel Papers*; *Richmond & James River Payrolls*.
43. *ORN*, 2, 1:310; *CSN & CSMC Personnel*, M260, Roll 5; *Vessel Papers*; *Richmond & James River Payrolls*; *Shipping Articles*; *ORN*, 2, 1:298; U.S. War Department, "CSRs of Confederate Soldiers Who Served in Organizations from the State of Georgia, Cobb's Legion" Microfilm Series, M266, Roll 581, RG 109, NA, (hereafter cited as *Cobb's Legion CSRs*, followed by M266 and Roll Number).
44. Donnelly, *Confederate Enlisted Marines*, 12; *ORN*, 2, 1:311, 316; *Richmond & James River Payrolls*.
45. *1st (Nelligan's) Louisiana Infantry CSRs*, M320, Roll 73; Booth, *Louisiana Confederate Soldiers*, 1:157; *ORN*, 2, 1:309; *North Carolina Payrolls*; *Richmond & James River Payrolls*; *ORN*, 2, 1:283; U.S. War Department, "CSRs of Confederate Soldiers Who Served in Organizations from the State of Virginia, 18th Infantry," Microfilm Series, M324, Roll 598, RG 109, NA; U.S. War Department, "CSRs of Confederate Soldiers Who Served in Organizations from the State of Virginia, Heavy Artillery, Epes' Company," Microfilm Series, M324, Roll 293, RG 109, NA; Robertson, *18th Virginia Infantry*, 43; Weaver, *Assorted Heavy Artillery*, 61.
46. Bergeron, "Buchanan, Franklin," in *Dictionary of American Military Biography*, 1:121–14; Delaney, "Buchanan, Franklin," in *Historical Times Illustrated Encyclopedia of the Civil War*, 86; Lewis, "Buchanan, Franklin," in *Dictionary of American Biography*, 3:206–07; Scharf, *Confederate States*, 155; Hamersly, *Complete General Navy Register*, 105; Dudley, *Going South*, 34; *CSN Register*, 25; U.S. Department of the Navy, *Dictionary Naval Fighting Ships*, 1:168–69.
47. *1860 Norfolk Co., Virginia Census*, M653, Roll 1366; *41st Virginia Infantry CSRs*, M324, Roll 860; Henderson, *41st Virginia Infantry*, 93; Fiveash, Virginia (Merrimac), 21–22; *Kevill's Company CSRs*; Weaver, *Assorted Heavy Artillery*, 76.
48. *9th Virginia Infantry CSRs*, M324, Roll 476; Trask, *9th Virginia Infantry*, 56; *CSN & CSMC Personnel*, M260, Roll 5; *ORN*, 2, 1:309; *Richmond & James River Payrolls*.
49. U.S. War Department, "CSRs of Confederate Soldiers Who Served in Organizations from the State of Virginia, 2nd Infantry," Microfilm Series,

M324, Roll 373, RG 109, NA; Frye, *2nd Virginia Infantry*, 87; *CSN & CSMC Personnel*, M260, Roll 5; *CSN Register*, 26; *ORN*, 1, 17:43; *ORN*, 2, 1:310; *Richmond & James River Payrolls*.

50. *1860 Norfolk Co., Virginia Census*, M653, Roll 1366; *41st Virginia Infantry CSRs*, M324, Roll 860; Henderson, *41st Virginia Infantry*, 93; Fiveash, Virginia (Merrimac), 21–22; *Kevill's Company CSRs*; Weaver, *Assorted Heavy Artillery*, 76; *1870 Norfolk Co., Virginia Census*, M593, 1666.

51. Delaney, "Butt, Walter R.," in *Historical Times Illustrated Encyclopedia*, 100; Hamersly, *Complete General Navy Register*, 120; *CSN Register*, 27; Dudley, *Going South*, 38; *ORN*, 2, 1:308; Jack, *Memoirs*, 12; *Richmond & James River Payrolls*; Scharf, *Confederate States*, 744; Coski, *Capital Navy*, 193, 221, 242–43; Stewart, *History of Norfolk County*, 147, 499–500; Burial marker.

52. Curtis, *History of Famous Battle*, 7; *ORN*, 2, 1:309; *Vessel Papers*; *Hospital & Prison Records*, M260, Roll 1; *File NA-2*, M1091, Roll 18; *Richmond & James River Payrolls*; *CSN Register*, 27.

53. *1st (McCreary's) South Carolina Infantry CSRs*, M267, Roll 127; Salley, *South Carolina Troops*, 357; *ORN*, 2, 1:309; *CSN & CSMC Personnel*, M260, Roll 5; *North Carolina Payrolls*; *Richmond & James River Payrolls*; *ORN*, 2, 1:298; *Charleston Payrolls*.

54. Donnelly, *Confederate Enlisted Marines*, 15; *ORN*, 2, 1:311, 316; *Richmond & James River Payrolls*.

55. Hamersly, *Complete General Navy Register*, 125; *CSN Register*, 29; Dudley, *Going South*, 52; Scharf, *Confederate States*, 619; Jack, *Memoirs*, 13; *ORN*, 2, 1:308; *North Carolina Payrolls*; *Richmond & James River Payrolls*.

56. *CSN & CSMC Personnel*, M260, Roll 5; *ORN*, 2, 1:310; *Vessel Papers*; *Richmond & James River Payrolls*.

57. *1860 Norfolk Co., Virginia Census*, M653, Roll 1366; *41st Virginia Infantry CSRs*, M324, Rolls 860, 861; Henderson, *41st Virginia Infantry*, 94; *19th Btn. Virginia Heavy Artillery CSRs*, M324, Roll 243; Fiveash, Virginia (Merrimac), 21–22; *Kevill's Company CSRs*; Weaver, *Assorted Heavy Artillery*, 76; *ORN*, 2, 1:311, 315; *Shipping Articles*; *1870 Norfolk Co., Virginia Census*, M593, Roll 1666.

58. *ORN*, 2, 1:310; *Vessel Papers*; *Richmond & James River Payrolls*.

59. *1860 Dinwiddie Co., Virginia Census*, M653, Roll 1342; *41st Virginia Infantry CSRs*, M324, Rolls 860, 861; *ORN*, 2, 1:310; *CSN & CSMC Personnel*, M260, Roll 5; Henderson, *41st Virginia Infantry*, 94.

60. *1st North Carolina Infantry CSRs*, M270, Roll 91; Manarin, *North Carolina Troops*, 3:146; *CSN & CSMC Personnel*, M260, Roll 5; *ORN*, 2, 1:309; *Richmond & James River Payrolls*; *Charleston Payrolls*.

61. U.S. War Department, "CSRs of Confederate Soldiers Who Served in Organizations from the State of Arkansas, 1st (Colquitt's) Infantry," Microfilm Series, M317, Roll 46, RG 109, NA (hereafter cited as *1st (Colquitt's) Arkansas Infantry CSRs*, followed by M317 and Roll Number); *ORN*, 2, 1:309; *Vessel Papers*; *File NA-2*, M1091, Roll 18; *Richmond & James River Payrolls*.
62. *41st Virginia Infantry CSRs*, M324, Rolls 860, 861; Henderson, *41st Virginia Infantry*, 94; *CSN & CSMC Personnel*, M260, Roll 5; *ORN*, 2, 1:309; *File NA-2*, M1091, Roll 18; *Richmond & James River Payrolls*; *ORN*, 2, 1:306; *Hospital & Prison Records*, M260, Roll 1.
63. *1850 Norfolk Co., Virginia Census*, M432, Roll 964; *41st Virginia Infantry CSRs*, M324, Rolls 860, 861; Henderson, *41st Virginia Infantry*, 95; Fiveash, Virginia (Merrimac), 21–22; *Kevill's Company CSRs*; Weaver, *Assorted Heavy Artillery*, 76.
64. Berent, "Georgians Crewed the *Virginia*," 123; Scharf, *Confederate States*, 665; Dudley, *Going South*, 46; *ORN*, 2, 1:308; *Richmond & James River Payrolls*; Jones, *Life and Services of Josiah Tattnall*, 138; *CSN Register*, 31.
65. *1860 Norfolk Co., Virginia Census*, M653, Roll 1366; Hartzler, *Marylanders in the Confederacy*, 110; Donnelly, *Confederate Enlisted Marines*, 17; *ORN*, 2, 1:310; *Richmond & James River Payrolls*.
66. *ORN*, 2, 1:310; *Vessel Papers*; *North Carolina Payrolls*; *Richmond & James River Payrolls*.
67. Hamersly, *Complete General Navy Register*, 145; *CSN Register*, 35; Dudley, *Going South*, 52; Scharf, *Confederate States*, 308, 334; *ORN*, 2, 1:308; *Richmond & James River Payrolls*; Hartzler, *Marylanders in the Confederacy*, 111.
68. *1860 Elizabeth City Co., Virginia Census*, M653, Roll 1343; *CSN Register*, 36; White, *First Iron-Clad Naval Engagement*, 16; Curtis, *History of Famous Battle*, 18.
69. *3rd Alabama Infantry CSRs*, M311, Roll 104; *CSN & CSMC Personnel*, M260, Roll 5; *ORN*, 2, 1:310; *Richmond & James River Payrolls*.
70. U.S. War Department, "CSRs of Confederate Soldiers Who Served in Organizations from the State of North Carolina, 13th Infantry," Microfilm Series, M270, Roll 212, RG 109, NA (hereafter cited as *13th North Carolina Infantry CSRs*, followed by M270 and Roll Number); Manarin, *North Carolina Troops*, 5:349; *CSN & CSMC Personnel*, M260, Roll 5; *ORN*, 2, 1:309; *File NA-2*, M1091, Roll 18; *North Carolina Payrolls*; *Richmond & James River Payrolls*; *Charleston Payrolls*.
71. Donnelly, *Confederate Enlisted Marines*, 19; *ORN*, 2, 1:311; *Richmond & James River Payrolls*.

72. *ORN*, 2, 1:309; *Vessel Papers*; *File NA-2,* M1091, Roll 18; *Richmond & James River Payrolls*; *CSN & CSMC Personnel*, M260, Roll 5; *Hospital & Prison Records*, M260, Roll 1; U.S. War Department, "CSRs of Confederate Soldiers Who Served in Organizations from the State of Louisiana, 15th Infantry," Microfilm Series, M320, Roll 267, RG 109, NA, (hereafter cited as *15th Louisiana Infantry CSRs*, followed by M320 and Roll Number); Booth, *Louisiana Confederate Soldiers*, 2:388.
73. *ORN*, 2, 1:310; *CSN & CSMC Personnel*, M260, Roll 5; *Richmond & James River Payrolls.*
74. *1860 Accomac Co., Virginia Census*, M653, Roll 1330; U.S. War Department, "CSRs of Confederate Soldiers Who Served in Organizations from the State of South Carolina, 1st (Butler's) Infantry," Microfilm Series, M267, Roll 111, RG 109, NA; *41st Virginia Infantry CSRs*, M324, Rolls 860, 862; Henderson, *41st Virginia Infantry*, 97; Fiveash, Virginia (Merrimac), 21–22; *Kevill's Company CSRs*; Weaver, *10th and 19th Btns*, 119; Weaver, *Assorted Heavy Artillery*, 76; Stewart, *History of Norfolk County*, 139; Robert E. Lee Camp Confederate Soldiers' Home (Richmond, Virginia), Applications for Admission, 1884–1941, State Government Records Collection, Library of Virginia, Richmond, Virginia (hereafter cited as *R.E. Lee Soldiers' Home Application*); Markham, *List of Confederate Veterans*, 4; *Pension Applications, Confederate Veterans and Widows*, Commonwealth of Virginia Acts of 1888, 1900, and 1902, State Government Records Collections, Library of Virginia, Richmond, Virginia (hereafter cited as *Virginia Pension Applications*).
75. U.S. War Department, "CSRs of Confederate Soldiers Who Served in Organizations from the State of Virginia, 12th Infantry," Microfilm Series, M324, Roll 518, RG 109, NA (hereafter cited as *12th Virginia Infantry CSRs*, followed by M324 and Roll Number); Henderson, *12th Virginia Infantry*, 118; *CSN & CSMC Personnel*, M260, Roll 5; *ORN*, 2, 1:309; *File NA-2,* M1091, Roll 18; *Richmond & James River Payrolls*; *Hospital & Prison Records*, M260, Roll 1.
76. *Vessel Papers*; *Shipping Articles.*
77. *1st (Colquitt's) Arkansas Infantry*, M317, Roll 47; *CSN & CSMC Personnel*, M260, Roll 5; *ORN*, 2, 1:309; *File NA-2*, M1091, Roll 18; *Richmond & James River Payrolls.*
78. *1860 City of New Orleans, Louisiana Census*, M653, Roll 415; *ORN*, 2, 1:311; Donnelly, *Confederate Enlisted Marines*, 21; *Richmond & James River Payrolls.*
79. *CSN Register*, 42; Dudley, *Going South*, 50; Jack, *Memoirs*, 12; *ORN*, 2, 1:308; *ORN*, 1, 17:868; *North Carolina Payrolls*; *Charleston Payrolls*; *Wilmington Payrolls.*

80. U.S. War Department, "CSRs of Confederate Soldiers Who Served in Organizations from the State of North Carolina, 14th Infantry," Microfilm Series, M270, Roll 223, RG 109, NA, (hereafter cited as *14th North Carolina Infantry CSRs*, followed by M270 and Roll Number); Manarin, *North Carolina Troops*, 5:446–447; *CSN & CSMC Personnel*, M260, Roll 5; *ORN*, 2, 1:309; *File NA-2*, M1091, Roll 18; *Richmond & James River Payrolls*; U.S. War Department, "CSRs of Confederate Soldiers Who Served in Organizations from the State of North Carolina, 6th Cavalry," Microfilm Series, M270, Roll 40, RG 109, NA (hereafter cited as *6th North Carolina Cavalry CSRs*); Manarin, *North Carolina Troops*, 2:471.
81. Trask, *9th Virginia Infantry*, 59, 108; *1860 Norfolk Co., Virginia Census*, M653, Roll 1366; *9th Virginia Infantry CSRs*, M324, Roll 478; Crew and Trask, *Grimes' Battery*, 88; *ORN*, 2, 1:308; *File NA-2*, M1091, Roll 18; *Richmond & James River Payrolls*.
82. Donnelly, *Confederate Enlisted Marines*, 22; *ORN*, 2, 1:310, 315; Scharf, *Confederate States*, 688; *Richmond & James River Payrolls*; *ORN*, 2, 1:312.
83. U.S. War Department, "CSRs of Confederate Soldiers Who Served in Organizations from the State of Texas, 4th Infantry," Microfilm Series, M323, Roll 285, RG 109, NA, (hereafter cited as *4th Texas Infantry CSRs*); *Wilmington Payrolls*; *North Carolina Payrolls*; *CSN & CSMC Personnel*, M260, Roll 5; *ORN*, 2, 1:309; Littlepage, *Midshipman Aboard the* Virginia, 46; *ORN*, 1, 17:864; *Vessel Papers*; *File NA-2*, M1091, Roll 18; *Richmond & James River Payrolls*; *CSN Register*, 43; *Hospital & Prison Records*, M260, Roll 1; *Charleston Payrolls*.
84. *1860 Norfolk Co., Virginia Census*, M653, Roll 1366; *41st Virginia Infantry CSRs*, M324, Rolls 860, 862; Henderson, *41st Virginia Infantry*, 98; Fiveash, Virginia (Merrimac), 21–22; *Kevill's Company CSRs*; Weaver, *Assorted Heavy Artillery*, 77; *1870 Norfolk Co., Virginia Census*, M593, Roll 1666; *Virginia Pension Applications*.
85. *41st Virginia Infantry CSRs*, M324, Roll 862; Henderson, *41st Virginia Infantry*, 98; *CSN & CSMC Personnel*, M260, Roll 5; *ORN*, 2, 1:310; *Richmond & James River Payrolls*; U.S. War Department, "CSRs of Confederate Soldiers Who Served in Organizations from the State of Virginia, 5th Btn. Infantry," Microfilm Series, M324, Roll 431, RG 109, NA, (hereafter cited as *5th Btn. Virginia Infantry CSRs*, followed by M324 and Roll Number); Gregory, *53rd and 5th Btn. Virginia Infantry*, 126.
86. U.S. War Department, "CSRs of Confederate Soldiers Who Served in Organizations from the State of Virginia, 38th Btn. Light Artillery," Microfilm Series, M324, Roll 257, RG 109, NA (hereafter cited as *38th Btn.*

Virginia Light Artillery CSRs, followed by M324 and Roll Number); Moore, *Fayette, Hampden, Thomas, and Blount's Artillery*, 150; *North Carolina Payrolls*; *Shipping Articles*; *ORN*, 2, 1:309; *Vessel Papers*; *CSN Register*, 43; *Richmond & James River Payrolls.*

87. *1st North Carolina Infantry CSRs*, M270, Roll 92; Manarin, *North Carolina Troops*, 3:214; *ORN*, 2, 1:309; *Vessel Papers*; *File NA-2*, M1091, Roll 18; *North Carolina Payrolls*; *Richmond & James River Payrolls.*

88. *1st North Carolina Infantry CSRs*, M270, Roll 92; Manarin, *North Carolina Troops*, 3:214; *ORN*, 2, 1:309; *Vessel Papers*; *File NA-2*, M1091, Roll 18; *North Carolina Payrolls*; *Richmond & James River Payrolls*; *Charleston Payrolls*; *ORN*, 2, 1:274; *Wilmington Payrolls.*

89. Donnelly, *Confederate Enlisted Marines*, 22–23; *ORN*, 2, 1:311; *Richmond & James River Payrolls.*

90. *1860 Chatham Co., Georgia Census*, M653, Roll 115; *ORN*, 2, 1:310; *Vessel Papers*; *Richmond & James River Payrolls*; *Shipping Articles*; *Charleston Payrolls*; *ORN*, 2, 1:313, 276; *Wilmington Payrolls*; *ORN*, 2, 1:311; U.S. War Department, "CSRs of Confederate Soldiers Who Served in Organizations from the State of Virginia, 4th Btn. Infantry, Local Defense," Microfilm Series, M324, Roll 475, RG 109, NA, (hereafter cited as *4th Btn. Virginia Infantry CSRs*); *Hospital & Prison Records*, M260, Roll 1.

91. *CSN Register*, 217; White, *First Iron-Clad Naval Engagement*, 16.

92. *1860 Macon Co., Alabama Census*, M653, Roll 14; Donnelly, *Confederate Enlisted Marines*, 23; *ORN*, 2, 1:310; Scharf, *Confederate States*, 688; *ORN*, 2, 1:315; *Richmond & James River Payrolls.*

93. Curtis, *History of Famous Battle*, 5, 6, 18; *32nd Virginia Infantry CSRs*, M324, Roll 781; Jensen, *32nd Virginia Infantry*, 180; *CSN & CSMC Personnel*, M260, Roll 5; *ORN*, 2, 1:309; *File NA-2*, M1091, Roll 18; *Richmond & James River Payrolls*; *Virginia Pension Applications.*

94. U.S. War Department, "CSRs of Confederate Soldiers Who Served in Organizations From the State of South Carolina, 1st Artillery," Microfilm Series, M267, Roll 58, RG 109, NA; Henderson, *41st Virginia Infantry*, 99; Fiveash, Virginia (Merrimac), 21–22; Weaver, *Assorted Heavy Artillery*, 77; U.S. War Department, "CSRs of Confederate Soldiers Who Served in Organizations Raised Directly by the Confederate Government, 1st Engineer Troops," Microfilm Series, M258, Roll 93, RG 109, NA (hereafter cited as *1st Engineer Troops CSRs*, followed by M258 and Roll Number); Stewart, *History of Norfolk County*, 139.

95. *ORN*, 2, 1:309; *File NA-2*, M1091, Roll 18; *North Carolina Payrolls*; *Richmond & James River Payrolls.*

96. Norman C. Delaney, "Davidson, Hunter," in *Historical Times Illustrated Encyclopedia of the Civil War*, 206-207; Hamersly, *Complete General Navy Register*, 192; Hartzler, *Marylanders in the Confederacy*, 125; Dudley, *Going South*, 40; *North Carolina Payrolls; ORN*, 2, 1:308; Curtis, *History of Famous Battle*, 11; Jack, *Memoirs*, 12; *CSN Register*, 46; *Richmond & James River Payrolls*; Coski, *Capital Navy*, 121-122, 124, 126, 242; Sumner A. Cunningham, ed. "Hunter Davidson," *Confederate Veteran*, June 1913, 307.
97. *1860 Prince William Co., Virginia Census*, M653, Roll 1373; *9th Virginia Infantry CSRs*, M324, Roll 476; Trask, *9th Virginia Infantry*, 61; *CSN & CSMC Personnel*, M260, Roll 5; *ORN*, 2, 1:309; *Richmond & James River Payrolls*; *4th Btn. Virginia Infantry CSRs*; *Hospital & Prison Records*, M260, Roll 1.
98. *32nd Virginia Infantry CSRs*, M324, Roll 782; Jensen, *32nd Virginia Infantry*, 180; *Shipping Articles*; *CSN & CSMC Personnel*, M260, Roll 5; *ORN*, 2, 1:309; *Vessel Papers*; *ORN*, 2, 1:290, 281; *Richmond & James River Payrolls*; Hartzler, *Marylanders in the Confederacy*, 126.
99. U.S. War Department, "CSRs of Confederate Soldiers Who Served in Organizations from the State of Georgia, 3rd Infantry," Microfilm Series, M266, Roll 167, RG 109, NA (hereafter cited as *3rd Georgia Infantry CSRs*, followed by M266 and Roll Number); *CSN & CSMC Personnel*, M260, Roll 5; *ORN*, 2, 1:310; *Vessel Papers*; *North Carolina Payrolls*; *Richmond & James River Payrolls*; Henderson, *Georgia Confederate Soldiers*, 1:498.
100. Donnelly, *Confederate Enlisted Marines*, 24; *ORN*, 2, 1:310; Scharf, *Confederate States*, 688; *Richmond & James River Payrolls.*
101. Donnelly, *Confederate Enlisted Marines*, 24; *ORN*, 2, 1:310; *Richmond & James River Payrolls.*
102. U.S. War Department, "CSRs of Confederate Soldiers Who Served in Organizations from the State of South Carolina, 1st Infantry," Microfilm Series, M267, Roll 145, RG 109, NA (hereafter cited as *1st South Carolina Infantry CSRs*, followed by M267 and Roll Number); *1st (McCreary's) South Carolina Infantry CSRs*, M267, Roll 128; Salley, *South Carolina Troops*, 359; *CSN & CSMC Personnel*, M260, Roll 5; *ORN*, 2, 1:309; *Vessel Papers*; *File NA-2*, M1091, Roll 18; *North Carolina Payrolls*; *Richmond & James River Payrolls*; U.S. War Department, "CSRs of Confederate Soldiers Who Served in Organizations from the State of South Carolina, 27th Infantry," Microfilm Series, M267, Roll 357, RG 109, NA.
103. *4th Texas Infantry CSRs*; *Wilmington Payrolls*; *CSN & CSMC Personnel*, M260, Roll 5; *ORN*, 2, 1:309; *ORN*, 2, 2:311; *Vessel Papers*; *File NA-2*, M1091, Roll 18; *North Carolina Payrolls*; *Richmond & James River Payrolls*; *Hospital & Prison Records*, M260, Roll 1.

104. Donnelly, *Confederate Enlisted Marines*, 25–26; *ORN*, 2, 1:311; *Richmond & James River Payrolls*.
105. *CSN & CSMC Personnel*, M260, Roll 5; *ORN*, 2, 1:309; *File NA-2*, M1091, Roll 18; *North Carolina Payrolls*; *Richmond & James River Payrolls*.
106. *ORN*, 2, 1:309; *Vessel Papers*; *Richmond & James River Payrolls*; *15th Louisiana Infantry CSRs*, M320, Roll 267.
107. *1860 Carteret Co., North Carolina Census*, M653, Roll 890; *North Carolina Payrolls*; *Shipping Articles*; *CSN & CSMC Personnel*, M260, Roll 5; *ORN*, 2, 1:309; *Richmond & James River Payrolls*; U.S. War Department, "CSRs of Confederate Soldiers Who Served in Organizations From the State of North Carolina, 3rd Cavalry," Microfilm Series, M270, Roll 20, RG 109, NA; U.S. War Department, "CSRs of Confederate Soldiers Who Served in Organizations from the State of North Carolina, 1st Artillery," Microfilm Series, M270, Roll 52, RG 109, NA, (hereafter cited as *1st North Carolina Artillery CSRs*, followed by M270 and Roll Number); Manarin, *North Carolina Troops*, 1:141, 2:256; 3:87.
108. *9th Virginia Infantry CSRs*, M324, Roll 478; Trask, *9th Virginia Infantry*, 62; *CSN & CSMC Personnel*, M260, Roll 5; *ORN*, 2, 1:309; *North Carolina Payrolls*; *Richmond & James River Payrolls*; U.S. War Department, "CSRs of Confederate Soldiers Who Served in Organizations From the State of Virginia, 24th Btn. Partisan Rangers," Microfilm Series, M324, Roll 180, RG 109, NA (hereafter cited as *24th Btn. Virginia Partisan Rangers CSRs*); Driver, *1st, 39th & 24th Virginia Btns*, 160; U.S. War Department, "CSRs of Confederate Soldiers Who Served in Organizations from the State of Virginia, 20th Cavalry," Microfilm Series, M324, Roll 163, RG 109, NA; Armstrong, *19th and 20th Virginia Cavalry*, 205; *1st Engineer Troops CSRs*, M258, Roll 93.
109. *1860 City of New Orleans, Louisiana Census*, M653, Roll 417; Donnelly, *Confederate Enlisted Marines*, 26; *ORN*, 2, 1:310; *Richmond & James River Payrolls*.
110. *ORN*, 2, 1:309; *Vessel Papers*; *File NA-2*, M1091, Roll 18; *North Carolina Payrolls*; *Richmond & James River Payrolls*; *ORN*, 1, 11:795; *Hospital & Prison Records*, M260, Roll 1.
111. *Shipping Articles*; *Wilmington Payrolls*; *CSN & CSMC Personnel*, M260, Roll 5; *ORN*, 2, 1:309; *File NA-2*, M1091, Roll 18; *North Carolina Payrolls*; *ORN*, 2, 1:300; *Richmond & James River Payrolls*.
112. *15th Louisiana Infantry CSRs*, M320, Roll 405; Booth, *Louisiana Confederate Soldiers*, 2:660; *ORN*, 2, 1:310; *Vessel Papers*; *Richmond & James River Payrolls*; *North Carolina Payrolls*.

113. *CSN Register*, 50; Scharf, *Confederate States*, 596; *ORN*, 2, 1:308; *Richmond & James River Payrolls*; *Hospital & Prison Records*, M260, Roll 1.
114. *1860 Washington Co., North Carolina Census*, M653, Roll 917; *1st North Carolina Infantry CSRs*, M270, Roll 92; Manarin, *North Carolina Troops*, 3:214; *ORN*, 2, 1:310; *Vessel Papers*; *Richmond & James River Payrolls*; *ORN*, 2, 1:277.
115. *41st Virginia Infantry CSRs*, M324, Rolls 860, 862; Henderson, *41st Virginia Infantry*, 101; Fiveash, Virginia (Merrimac), 21–22; *Kevill's Company* C*SRs*, M324, Roll 315; Weaver, *Assorted Heavy Artillery*, 77; Anderson, "Survivor of the First Battle Between Ironclads," 189.
116. Donnelly, *Confederate Enlisted Marines*, 28; *ORN*, 2, 1:310; *Richmond & James River Payrolls*; *ORN*, 2, 1:315.
117. *1860 Norfolk Co., Virginia Census*, M653, Roll 1366; *41st Virginia Infantry CSRs*, M324, Rolls 860, 862; Henderson, *41st Virginia Infantry*, 101; Fiveash, Virginia (Merrimac), 21–22; *Kevill's Company* C*SRs*, M324, Roll 315; Weaver, *Assorted Heavy Artillery*, 77.
118. Crews and. Parish, *14th Virginia Infantry*, 101; *ORN*, 2, 1:310; *Richmond & James River Payrolls*.
119. *1850 Baton Rouge Parish, Louisiana Census*, M432, Roll 229; *1860 City of New Orleans, Orleans Parish, Louisiana Census*, M653, Roll 419; *15th Louisiana Infantry CSRs*, M320, Roll 267; Booth, *Louisiana Confederate Soldiers*, 2:711; *CSN & CSMC Personnel*, M260, Roll 5; *ORN*, 1, 17:43; Curtis, *History of Famous Battle*, 8.
120. *Kevill's Company* C*SRs*, M324, Roll 315; Weaver, *Assorted Heavy Artillery*, 77; Fiveash, Virginia (Merrimac), 21–22.
121. *Wilmington Payrolls*; *ORN*, 2, 1:312, 309; *Vessel Papers*; M1091, Rolls 17, 18; *North Carolina Payrolls*; *Richmond & James River Payrolls*; *CSN & CSMC Personnel*, M260, Roll 5.
122. *41st Virginia Infantry CSRs*, M324, Rolls 860, 862; Henderson, *41st Virginia Infantry*, 101; *CSN & CSMC Personnel*, M260, Roll 5; *ORN*, 2, 1:309; *Vessel Papers*; *File NA-2,* M1091, Roll 18; Hartzler, *Marylanders in the Confederacy*, 136.
123. *1st (Nelligan's) Louisiana Infantry CSRs*, M320, Roll 75; Booth, *Louisiana Confederate Soldiers*, 2:715; *CSN & CSMC Personnel*, M260, Roll 5; *ORN*, 2, 1:309; *Vessel Papers*; *Hospital & Prison Records*, M260, Roll 1; *File NA-2,* M1091, Roll 18; *Richmond & James River Payrolls*; *ORN*, 1, 17:864; Burial marker.
124. *1st (Colquitt's) Arkansas Infantry CSRs*, M317, Roll 47; *ORN*, 2, 1:309; *Vessel Papers*; *File NA-2,* M1091, Roll 18; *North Carolina Payrolls*; *Richmond & James River Payrolls*; *CSN Register*, 53.

125. *12th Virginia Infantry CSRs*, M324, Roll 520; *CSN & CSMC Personnel*, M260, Roll 5; *ORN*, 2, 1:310; *Richmond & James River Payrolls*; Henderson, *12th Virginia Infantry*, 122.
126. *12th Alabama Infantry CSRs*, M311, Roll 210; *Shipping Articles*; *ORN*, 2, 1:310; *File NA-2*, M1091, Roll 18; *North Carolina Payrolls*; *Richmond & James River Payrolls*; *ORN*, 2, 1:276, 292; *CSN Register*, 54.
127. Donnelly, *Confederate Enlisted Marines*, 30; *ORN*, 2, 1:310, 315; *Richmond & James River Payrolls*; *1860 City of New Orleans, Orleans Parish, Louisiana Census*, M653, Roll 417.
128. *R. Bolling Batte Papers*, Personal Papers Collection, Library of Virginia, Richmond, Virginia (hereafter cited as *R. Bolling Batte Papers*); Hamersly, *Complete General Navy Register*, 232; Dudley, *Going South*, 37; *New Orleans Payrolls*; *ORN*, 2, 1:308; Jack, *Memoirs*, 12; *Richmond & James River Payrolls*; *CSN Register*, 54.
129. Donnelly, *Confederate Enlisted Marines*, 31; *ORN*, 2, 1:311; *Richmond & James River Payrolls; 1860 City of New Orleans, Orleans Parish, Louisiana Census*, M653, Roll 416.
130. *1860 Norfolk Co., Virginia Census*, M653, Roll 1366; U.S. War Department, "CSRs of Confederate Soldiers Who Served in Organizations from the State of Virginia, 6th Infantry," Microfilm Series, M324, Roll 438, RG 109, NA, (hereafter cited as *6th Virginia Infantry CSRs*, followed by M324 and Roll Number); *ORN*, 2, 1:310; *Richmond & James River Payrolls*; Cavanaugh, *6th Virginia Infantry*, 95.
131. *Shipping Articles*; *North Carolina Payrolls*; *ORN*, 2, 1:309; *File NA-2*, M1091, Roll 18; *Richmond & James River Payrolls*; *Charleston Payrolls*; *ORN*, 2, 1:311.
132. *Wilmington Payrolls*; *ORN*, 2, 1:310; *Richmond & James River Payrolls*.
133. *ORN*, 2, 1:309; *Vessel Papers*; *File NA-2*, M1091, Roll 18; *Richmond & James River Payrolls*; *North Carolina Payrolls*; *15th Louisiana Infantry CSRs*, M320, Roll 268.
134. *14th North Carolina Infantry CSRs*, M270, Roll 224; Manarin, *North Carolina Troops*, 5:417; *ORN*, 2, 1:309; *File NA-2*, M1091, Roll 18; U.S. War Department, "CSRs of Confederate Soldiers Who Served in Organizations from the State of North Carolina, 26th Infantry," Microfilm Series, M270, Roll 326, RG 109, NA; Manarin, *North Carolina Troops*, 7:593; *1870 Anson Co., North Carolina Census*, M593, Roll 1122.
135. *1860 Norfolk Co., Virginia Census*, M653, Roll 1366; *41st Virginia Infantry CSRs*, M324, Rolls 860, 863; Henderson, *41st Virginia Infantry*, 105; Fiveash, Virginia (Merrimac), 21–22; *File NA-2*, M1091, Roll 18; *Kevill's Company CSRs*, M324, Roll 315; Weaver, *Assorted Heavy Artillery*, 77.

136. *9th Virginia Infantry CSRs*, M324, Roll 479; Trask, *9th Virginia Infantry*, 65; U.S. War Department, "CSRs of Confederate Soldiers Who Served in Organizations from the State of Virginia, Light Artillery, Griffin's Company," Microfilm Series, M324, Roll 307, RG 109, NA, (hereafter cited as *Griffin's Company CSRs*, followed by M324 and Roll Number); *ORN*, 2, 1:309; *File NA-2*, M1091, Roll 18; *Richmond & James River Payrolls*; Nicholas and Servis, *Powhatan, Salem and Courtney*, 217.
137. *1860 Norfolk Co., Virginia Census*, M653, Roll 1366; *9th Virginia Infantry CSRs*, M324, Roll 479; Trask, *9th Virginia Infantry*, 65; *ORN*, 2, 1:309; *File NA-2*, M1091, Roll 18; *Richmond & James River Payrolls*; *Hospital & Prison Records*, M260, Roll 2; *Virginia Pension Applications*.
138. Forrest, *Odyssey in Gray*, 2; Hartzler, *Marylanders in the Confederacy*, 147; *CSN Register*, 63; U.S. War Department, "CSRs of Confederate Soldiers Who Served in Organizations from the State of Virginia, 17th Infantry," Microfilm Series, M324, Roll 586, RG 109, NA; *Wilmington Payrolls*; Forrest, "Odyssey in Gray," 129.
139. *CSN Register*, 64; Dudley, *Going South*, 45; *ORN*, 2, 1:308; Jack, *Memoirs*, 12; *Richmond & James River Payrolls*; *Richmond Times-Dispatch*, June 30, 1907; Foute, "Echoes from Hampton Roads," 246–48; Alexander and Alexander, "From Ironclads to Infantry," 9–15; Littlepage, "A Midshipman Abroad," 19.
140. *8th Alabama Infantry CSRs*, M311, Roll 174; *Shipping Articles*; *ORN*, 2, 1:309; *Vessel Papers*; *File NA-2*, M1091, Roll 18; *Richmond & James River Payrolls*.
141. U.S. War Department, "CSRs of Confederate Soldiers Who Served in Organizations from the State of Mississippi, 12th Infantry," Microfilm Series, M269, Roll 204, RG 109, NA; *Shipping Articles*; *ORN*, 2, 1:310; *Richmond & James River Payrolls; 1860 Bibb Co., Georgia Census*, M653, Roll 111.
142. *S. Bassett French Biographical Sketches*, Personal Papers Collection, Library of Virginia, Richmond, Virginia (hereafter cited as *S. Bassett French Sketches*); Hamersly, *Complete General Navy Register*, 274; *CSN Register*, 67; Dudley, *Going South*, 42; *ORN*, 2, 1:308; *North Carolina Payrolls*; *Richmond & James River Payrolls*; *Confederate Veteran*, June 1914, 16.
143. *1850 Carteret Co., North Carolina Census*, M432, Roll 623; *Shipping Articles*; *ORN*, 2, 1:309; *File NA-2*, M1091, Roll 18; *North Carolina Payrolls*; *Richmond & James River Payrolls*; *Charleston Payrolls*; *ORN*, 1, 11:794; *1870 Carteret Co., North Carolina Census*, M593, Roll 1127.
144. *15th Louisiana Infantry CSRs*, M320, Roll 268; *Shipping Articles*; *ORN*, 2, 1:310; *Vessel Papers*; *Richmond & James River Payrolls*.

145. *1860 City of New Orleans, Louisiana Census*, M653, Roll 417; Donnelly, *Confederate Enlisted Marines*, 36; *ORN*, 2, 1:310, 315; *Richmond & James River Payrolls*.
146. *Shipping Articles*; *Wilmington Payrolls*; *ORN*, 2, 1:309; *File NA-2*, M1091, Roll 18; *North Carolina Payrolls*; *Richmond & James River Payrolls*; *ORN*, 2, 1:284; *Charleston Payrolls*.
147. *41st Virginia Infantry CSRs*, M324, Rolls 860, 863; Henderson, *41st Virginia Infantry*, 107; Fiveash, Virginia (Merrimac), 21–22; Weaver, *10th & 19th Btns.*, 129; *Kevill's Company CSRs*, M324, Roll 315; Stewart, *History of Norfolk County*, 140; *1870 Norfolk Co., Virginia Census*, M593, 1666; Weaver, *Assorted Heavy Artillery*, 78.
148. *12th Virginia Infantry CSRs*, M324, Roll 521; *ORN*, 2, 1:310; *Richmond & James River Payrolls*; Henderson, *12th Virginia Infantry*, 126.
149. *ORN*, 2, 1:310; *Richmond & James River Payrolls*; *Hospital & Prison Records*, M260, Roll 2.
150. *Richmond & James River Payrolls*.
151. *ORN*, 2, 1:310; *Richmond & James River Payrolls*.
152. *9th Virginia Infantry CSRs*, M324, Roll 480; Trask, *9th Virginia Infantry*, 68; *ORN*, 2, 1:310; *Richmond & James River Payrolls*; *Hospital & Prison Records*, M260, Roll 2.
153. *41st Virginia Infantry CSRs*, M324, Rolls 860, 863; Fiveash, Virginia (Merrimac), 21–22; *Kevill's Company CSRs*, M324, Roll 315; Weaver, *Assorted Heavy Artillery*, 78; Stewart, *History of Norfolk County*, 140; Henderson, *41st Virginia Infantry*, 109.
154. *ORN*, 2, 1:310; *North Carolina Payrolls*; *Richmond & James River Payrolls*.
155. *1860 Norfolk Co., Virginia Census*, M653, Roll 1366; *9th Virginia Infantry CSRs*, M324, Roll 480; Trask, *9th Virginia Infantry*, 69, 110; *CSN & CSMC Personnel*, M260, Roll 6; *ORN*, 2, 1:309; *File NA-2*, M1091, Roll 18; *Richmond & James River Payrolls*; *Charleston Payrolls*; *Hospital & Prison Records*, M260, Roll 2; Stewart, *History of Norfolk County*, 148.
156. *Richmond & James River Payrolls*; *North Carolina Payrolls*; *File NA-2*, M1091, Roll 18.
157. Hartzler, *Marylanders in the Confederacy*, 163; *CSN Register*, 79; *ORN*, 2, 1:308; *Richmond & James River Payrolls*.
158. *41st Virginia Infantry CSRs*, M324, Rolls 860, 864; *CSN & CSMC Personnel*, M260, Roll 6; *ORN*, 2, 1:310; Henderson, *41st Virginia Infantry*, 110; *Richmond & James River Payrolls*; *Virginia Pension Applications*.
159. *6th Virginia Infantry CSRs*, M324, Roll 440; Cavanaugh, *6th Virginia Infantry*, 100; *9th Virginia Infantry CSRs*, M324, Roll 480; Trask, *9th Virginia*

Infantry, 69; *CSN & CSMC Personnel*, M260, Roll 6; *ORN*, 2, 1:309; *File NA-2,* M1091, Roll 18; *North Carolina Payrolls*; *Richmond & James River Payrolls*; *Charleston Payrolls*; *1870 Norfolk Co., Virginia Census*, M593, Roll 1667.

160. Donnelly, *Confederate Enlisted Marines*, 41; *ORN*, 2, 1:311; *Richmond & James River Payrolls*; *1870 Franklin Co., Kentucky Census*, M593, Roll 462.

161. *ORN*, 2, 1:309; *File NA-2,* M1091, Roll 18; *North Carolina Payrolls*; *Richmond & James River Payrolls*; *ORN*, 1, 11:794; *Hospital & Prison Records*, M260, Roll 2.

162. *12th Alabama Infantry CSRs*, M311, Roll 209; *Shipping Articles*; *ORN*, 2, 1:310; *File NA-2,* M1091, Roll 18; *Hospital & Prison Records*, M260, Roll 2; *Richmond & James River Payrolls*.

163. *13th North Carolina Infantry CSRs*, M270, Roll 214; Manarin, *North Carolina Troops,* 5:291; *CSN & CSMC Personnel*, M260, Roll 6; *ORN*, 2, 1:309; *File NA-2,* M1091, Roll 18; *North Carolina Payrolls*; *Richmond & James River Payrolls*; *1st North Carolina Artillery CSRs*, M270, Roll 53; Manarin, *North Carolina Troops,* 5:291; *1870 Caswell Co., North Carolina Census*, M595, Roll 1128.

164. *1860 Nansemond Co., Virginia Census*, M653, Roll 1365; *41st Virginia Infantry CSRs*, M324, Rolls 860, 864; *CSN & CSMC Personnel*, M260, Roll 6; *ORN*, 2, 1:309; *Richmond & James River Payrolls*; Henderson, *41st Virginia Infantry*, 111.

165. U.S. War Department, "CSRs of Confederate Soldiers Who Served in Organizations from the State of Louisiana, 10th Infantry," Microfilm Series, M320, Roll 219, RG 109, NA (hereafter cited as *10th Louisiana Infantry CSRs*, followed by M320 and Roll Number); Booth, *Louisiana Confederate Soldiers*, 3, bk. 1:199; *ORN*, 2, 1:309; *File NA-2,* M1091, Roll 18; *Richmond & James River Payrolls*.

166. *1860 Norfolk Co., Virginia Census*, M653, Roll 1366; *6th Virginia Infantry CSRs*, M324, Roll 440; Cavanaugh, *6th Virginia Infantry*, 101; *ORN*, 2, 1:310, 311; *Vessel Papers*; *Richmond & James River Payrolls*; *1870 Norfolk Co., Virginia Census*, M593, Roll 1667.

167. *Shipping Articles*; *ORN*, 2, 1:289, 310; *File NA-2,* M1091, Roll 18; *Richmond & James River Payrolls*.

168. *14th North Carolina Infantry CSRs*, M270, Roll 225; Manarin, *North Carolina Troops,* 5:438; *CSN & CSMC Personnel*, M260, Roll 6; *ORN*, 2, 1:309; *File NA-2,* M1091, Roll 18; *North Carolina Payrolls*; *Richmond & James River Payrolls*; U.S. War Department, "CSRs of Confederate Soldiers Who Served in Organizations from the State of North Carolina, 10th Btn. Heavy Artillery," Microfilm Series, M270, Roll 83, RG 109, NA; Manarin, *North Carolina Troops,* 5:438.

169. *1860 Mobile Co., Alabama Census*, M653, Roll 17; Donnelly, *Confederate Enlisted Marines*, 43; *ORN*, 2, 1:310, 315; *Richmond & James River Payrolls*.
170. *1860 Talladego Co., Alabama Census*, M653, Roll 24; *3rd Alabama Infantry CSRs*, M311, Roll 106; *CSN & CSMC Personnel*, M260, Roll 6; *ORN*, 2, 1:309; *File NA-2*, M1091, Roll 18; *North Carolina Payrolls*; *Richmond & James River Payrolls*; *ORN*, 2, 1:301.
171. *ORN*, 2, 1:309; *File NA-2*, M1091, Roll 18; *Richmond & James River Payrolls*.
172. Hamersly, *Complete General Navy Register*, 328; Dudley, *Going South*, 50; *New Orleans Payrolls*; *CSN Register*, 84; *ORN*, 2, 1:308; *Richmond & James River Payrolls*; *Charleston Payrolls*; W. B. Fort, "First Submarine in the Confederate Navy," *Confederate Veteran*, October 1918, 459–60; *R.E. Lee Camp Soldiers' Home Application*.
173. *CSN & CSMC Personnel*, M260, Roll 6; *ORN*, 2, 1:309; *Vessel Papers*; *Shipping Articles*; *Richmond & James River Payrolls*; U.S. War Department, "CSRs of Confederate Soldiers Who Served in Organizations From the State of Louisiana, 5th Infantry," Microfilm Series, M320, Roll 152, RG 109, NA; Booth, *Louisiana Confederate Soldiers*, 3, bk. 1:236.
174. *13th North Carolina Infantry CSRs*, M270, Roll 214; Manarin, *North Carolina Troops*, 5:350; *ORN*, 2, 1:309; *Vessel Papers*; *File NA-2*, M1091, Roll 18; *North Carolina Payrolls*; *Richmond & James River Payrolls*; *Hospital & Prison Records*, M260, Roll 2.
175. *2nd Louisiana Infantry CSRs*, M320, Roll 106; *CSN Register*, 86; *ORN*, 2, 1:309; *File NA-2*, M1091, Roll 18; Booth, *Louisiana Confederate Soldiers*, 3, bk. 1:263; *Richmond & James River Payrolls*; *Wilmington Payrolls*; *ORN*, 1, 17:868; Burial marker.
176. *9th Virginia Infantry CSRs*, M324, Roll 481; Trask, *9th Virginia Infantry*, 72; *ORN*, 2, 1:309; *Hospital & Prison Records*, M260, Roll 2; *Richmond & James River Payrolls*; *5th Virginia Cavalry CSRs*; Driver, *5th Virginia Cavalry*, 216.
177. *1st (Nelligan's) Louisiana Infantry CSRs*, M320, Roll 76; *CSN & CSMC Personnel*, M260, Roll 6; *ORN*, 2, 1:310; *North Carolina Payrolls*; *Richmond & James River Payrolls*; Booth, *Louisiana Confederate Soldiers*, 3, bk. 1:281.
178. Hamersly, *Complete General Navy Register*, 342; Dudley, *Going South*, 53; *CSN Register*, 87; Scharf, *Confederate States*, 597; *ORN*, 2, 1:308; Jack, *Memoirs*, 13; *Richmond & James River Payrolls*; *Richmond Times-Dispatch*, June 30, 1907; *Confederate Veteran*, November 1915, 513; *Florida Confederate Pension Application Files*, Record Group 137, Series 587, Bureau of Archives and Records Management [database on-line], Florida Department of State www.dos.state.fl.us/dlis/barm/Florida_CSA_Pension_Files.htm;

Internet (accessed February 1, 2000; hereafter cited as *Florida Confederate Pension Application*.)

179. *1st (McCreary's) South Carolina Infantry CSRs*, M267, Roll 129; Salley, *South Carolina Troops*, 387; *ORN*, 2, 1:310; *Richmond & James River Payrolls*.

180. Donnelly, *Confederate Enlisted Marines*, 45; *ORN*, 2, 1:310; *Richmond & James River Payrolls*; *ORN*, 2, 1:296, 297.

181. *9th Virginia Infantry CSRs*, M324, Roll 481; Hartzler, *Marylanders in the Confederacy*, 174; Trask, *9th Virginia Infantry*, 72; *CSN & CSMC Personnel*, M260, Roll 6; *ORN*, 2, 1:309; *Vessel Papers*; *File NA-2*, M1091, Roll 18; *North Carolina Payrolls*; *Richmond & James River Payrolls*; *Charleston Payrolls*; *Hospital & Prison Records*, M260, Roll 2; Application for Aid to Citizens of Virginia Wounded and Maimed during the Late War, While Serving as Soldiers or Marines, Commonwealth of Virginia Act of 1884, State Government Records Collection, Library of Virginia, Richmond, Virginia; *Virginia Pension Applications*.

182. *File NA-2*, M1091, Roll 18; *CSN & CSMC Personnel*, M260, Roll 6; *ORN*, 2, 1:309; *Richmond & James River Payrolls*; *Shipping Articles*; *ORN*, 1, 10:589; *Hospital & Prison Records*, M260, Roll 2.

183. *12th Virginia Infantry CSRs*, M324, Roll 522; *ORN*, 2, 1:310; *North Carolina Payrolls*; *Richmond & James River Payrolls*; Henderson, *12th Virginia Infantry*, 131; *Virginia Pension Applications*.

184. *ORN*, 2, 1:310; *Vessel Papers*; *Richmond & James River Payrolls*; U.S. War Department, "CSRs of Confederate Soldiers Who Served in Organizations From the State of Louisiana, 14th Infantry," Microfilm Series, M320, Roll 257, RG 109, NA; Booth, *Louisiana Confederate Soldiers*, 3, bk. 1:320.

185. *1st North Carolina Infantry CSRs*, M270, Roll 94; Manarin, *North Carolina Troops*, 3:216; *CSN & CSMC Personnel*, M260, Roll 6; *ORN*, 2, 1:309; *Vessel Papers*; *File NA-2*, M1091, Roll 18; *North Carolina Payrolls*; *Richmond & James River Payrolls*.

186. *1st (McCreary's) South Carolina Infantry CSRs*, M267, Roll 129; Salley, *South Carolina Troops*, 361; *CSN & CSMC Personnel*, M260, Roll 6; *ORN*, 2, 1:309; *Richmond & James River Payrolls*; *Shipping Articles*; *ORN*, 2, 1:277.

187. *41st Virginia Infantry CSRs*, M324, Roll 864; Henderson, *41st Virginia Infantry*, 115; *CSN & CSMC Personnel*, M260, Roll 6; *ORN*, 2, 1:310; *North Carolina Payrolls*; *Richmond & James River Payrolls*.

188. Donnelly, *Confederate Enlisted Marines*, 47; *ORN*, 2, 1:311; *Richmond & James River Payrolls*.

189. *ORN*, 2, 1:310; *Vessel Papers*; *Richmond & James River Payrolls*.

190. *41st Virginia Infantry CSRs*, M324, Roll 864; Henderson, *41st Virginia Infantry*, 115; *12th Virginia Infantry CSRs*, M324, Roll 523; *ORN*, 2, 1:310; *Richmond & James River Payrolls*; *R.E. Lee Soldiers' Home Application*; Markham, 15; Henderson, *12th Virginia Infantry*, 132.
191. *1860 Macon Co., Alabama Census*, M653, Roll 14; Donnelly, *Confederate Enlisted Marines*, 47; *ORN*, 2, 1:310; *Richmond & James River Payrolls*; Templeton, "The Last Roll," 468.
192. *8th Alabama Infantry CSRs*, M311, Roll 175; *Shipping Articles*; *ORN*, 2, 1:310; *Vessel Papers*; *Richmond & James River Payrolls*; *Hospital & Prison Records*, M260, Roll 2.
193. *ORN*, 2, 1:309; Littlepage, *Midshipman Abroad*, 46; *Vessel Papers*; *File NA-2*, M1091, Roll 18; *North Carolina Payrolls*; *Richmond & James River Payrolls*; *Hospital & Prison Records*, M260, Roll 2; *1860 Norfolk Co., Virginia Census*, M653, Roll 1366.
194. *CSN Register*, 97, Dudley, *Going South*, 46; *ORN*, 2, 1:308; R. Campbell, *Fire and Thunder*, 33; Coski, *Capital Navy*, 170; *Richmond & James River Payrolls*.
195. *North Carolina Payrolls*; *CSN & CSMC Personnel*, M260, Roll 6; *ORN*, 2, 1:309; *Wilmington Payrolls*; *File NA-2*, M1091, Roll 18; *ORN*, 1, 17:43; *Richmond & James River Payrolls*; *Hospital & Prison Records*, M260, Roll 2; *1860 Beaufort Co., North Carolina Census*, M653, Roll 887.
196. *9th Virginia Infantry CSRs*, M324, Roll 481; Trask, *9th Virginia Infantry*, 74; *ORN*, 2, 1:308; Jack, *Memoirs*, 13; *Richmond & James River Payrolls*; *Wilmington Payrolls*; *CSN Register*, 98; Cunningham, "Deaths in Stonewall Camp," 238; Burial marker.
197. *1860 Norfolk Co., Virginia Census*, M653, Roll 1366; *Shipping Articles*; *Virginia Pilot and Norfolk Landmark*, March 3, 1926; *ORN*, 2, 1:309; *File NA-2*, M1091, Roll 18; *North Carolina Payrolls*; *Richmond & James River Payrolls*; *CSN Register*, 99; Pollack, *Sketch Book of Portsmouth*, 193; Stewart, *History of Norfolk County*, 143.
198. *1860 Norfolk Co., Virginia Census*, M653, Roll 1366; *ORN*, 2, 1:309; *North Carolina Payrolls*; *Richmond & James River Payrolls*; Scharf, *Confederate States*, 774; *Charleston Payrolls*; *CSN Register*, 100.
199. *41st Virginia Infantry CSRs*, M324, Roll 864; Henderson, *41st Virginia Infantry*, 117; Weaver, *Assorted Heavy Artillery*, 79; *ORN*, 2, 1:309; *File NA-2*, M1091, Roll 18; Scharf, *Confederate States*, 715; *North Carolina Payrolls*; *Richmond & James River Payrolls*; *Kevill's Company* CSRs, M324, Roll 315; *ORN*, 2, 1:288, 298; *4th Btn. Virginia Infantry CSRs*, M324, Roll 416; *Charleston Payrolls*; *Hospital & Prison Records*, M260, Roll 2.

200. *ORN*, 2, 1:309; *Vessel Papers*; *North Carolina Payrolls*; *Richmond & James River Payrolls*; *ORN*, 2, 1:284; *Hospital & Prison Records*, M260, Roll 2.
201. *1860 Pasquotank Co., North Carolina Census*, M653, Roll 909; U.S. War Department, "CSRs of Confederate Soldiers Who Served in Organizations from the State of North Carolina, 17th Infantry," Microfilm Series, M270, Roll 250, RG 109, NA, (hereafter cited as *17th North Carolina Infantry CSRs*, followed by M270 and Roll Number); Manarin, *North Carolina Troops*, 6:184; *ORN*, 2, 1:309; *Vessel Papers*; *File NA-2,* M1091, Roll 18; *North Carolina Payrolls*; *Richmond & James River Payrolls*; *ORN*, 1, 17:864, 868; Burial marker.
202. *1860 Chesterfield Co., Virginia Census*, M653, Roll 1340; *41st Virginia Infantry CSRs*, M324, Roll 865; Henderson, *41st Virginia Infantry*, 117; *CSN & CSMC Personnel*, M260, Roll 6; *ORN*, 2, 1:310; *Richmond & James River Payrolls*; U.S. War Department, "CSRs of Confederate Soldiers Who Served in Organizations from the State of Virginia, Horse Artillery, Graham's Company," Microfilm Series, M324, Roll 304, RG 109, NA, (*Graham's Company CSRs*); Moore, *Graham's, Jackson's, and Lurty's Artillery*, 119.
203. Tyler, *Encyclopedia of Virginia Biography*, 3:60–62; Delaney, "Jones, Catesby ap Roger," in *Historical Times Illustrated Encyclopedia*, 402; Hamersly, *Complete General Navy Register*, 391; Dudley, *Going South*, 38; *ORN*, 2, 1:308; *Richmond & James River Payrolls*; Scharf, *Confederate States*, 130; Coski, *Capital Navy*, 49; *CSN Register*, 102–03.
204. *CSN & CSMC Personnel*, M260, Roll 6; *File NA-2,* M1091, Roll 18; *ORN*, 2, 1:309; Booth, *Louisiana Confederate Soldiers*, 3, bk. 1:467; *Richmond & James River Payrolls*; *North Carolina Payrolls*; *Hospital & Prison Records*, M260, Roll 2.
205. *North Carolina Payrolls*; *Shipping Articles*; *ORN*, 2, 1:309; *Vessel Papers*; *File NA-2,* M1091, Roll 18; *Richmond & James River Payrolls*; *ORN*, 2, 1: 276, 311.
206. *Confederate Veteran*, July 1910, 341; Hamersly, *Complete General Navy Register*, 393; *CSN Register*, 103; Dudley, *Going South*, 40; *ORN*, 2, 1:308; Curtis, *History of Famous Battle*, 16; *Richmond & James River Payrolls*; *Confederate Veteran*, November 1910, 533.
207. *ORN*, 2, 1:309; *Vessel Papers*; *File NA-2,* M1091, Roll 18; *Richmond & James River Payrolls*; *Hospital & Prison Records*, M260, Roll 2; *15th Louisiana Infantry CSRs*, M320, Roll 269; Booth, *Louisiana Confederate Soldiers*, 3, bk. 1:476.
208. Donnelly, *Confederate Enlisted Marines*, 52; *ORN*, 2, 1:310, 315; *Richmond & James River Payrolls*.
209. *1860 Newton Co., Georgia Census*, M653, Roll 133; *3rd Georgia Infantry CSRs*, M266, Roll 169; Berent, "Georgians Crewed the *Virginia*," 123; *CSN &*

CSMC Personnel, M260, Roll 6; *ORN*, 2, 1:309; *File NA-2*, M1091, Roll 18; *North Carolina Payrolls*; *Richmond & James River Payrolls*.

210. *3rd Georgia Infantry CSRs*, M266, Roll 169; Henderson, *Georgia Confederate Soldiers*, 1:471; *CSN & CSMC Personnel*, M260, Roll 6; *ORN*, 2, 1:309; *Richmond & James River Payrolls*; Berent, "Georgians Crewed the *Virginia*," 123.

211. *1860 Norfolk Co., Virginia Census*, M653, Roll 1366; *CSN Register*, 105; Dudley, *Going South*, 52; *ORN*, 2, 1:308; *North Carolina Payrolls*; *Richmond & James River Payrolls*; *Charleston Payrolls*.

212. Donnelly, *Confederate Enlisted Marines*, 54–55; *ORN*, 2, 1:311; *Richmond & James River Payrolls*.

213. *41st Virginia Infantry CSRs*, M324, Rolls 860, 865; *Kevill's Company* C*SRs*, M324, Roll 315; Weaver, *Assorted Heavy Artillery*, 79; Henderson, *41st Virginia Infantry*, 119; Stewart, *History of Norfolk County*, 143; *Virginia Pension Applications*; Burial marker.

214. *41st Virginia Infantry CSRs*, M324, Rolls 860, 865; Fiveash, Virginia (Merrimac), 21–22; Henderson, *41st Virginia Infantry*, 119; *19th Btn. Virginia Heavy Artillery CSRs*, M324, Roll 246; *Kevill's Company* C*SRs*, M324, Roll 315; Stewart, *History of Norfolk County*, 143, 817–18.

215. *1860 Mobile Co., Alabama Census*, M653, Roll 17; *Shipping Articles*; *CSN & CSMC Personnel*, M260, Roll 6; *ORN*, 2, 1:309; *File NA-2*, M1091, Roll 18; *North Carolina Payrolls*; *Richmond & James River Payrolls*; *Charleston Payrolls*; *ORN*, 2, 1:274; *Wilmington Payrolls*; *Hospital & Prison Records*, M260, Roll 2.

216. *CSN Register*, 108; Dudley, *Going South*, 49; *ORN*, 2, 1:308; *Richmond & James River Payrolls*.

217. *1860 Norfolk Co., Virginia Census*, M653, Roll 1366; *41st Virginia Infantry CSRs*, M324, Rolls 860, 865; Fiveash, Virginia (Merrimac), 21–22; Henderson, *41st Virginia Infantry*, 120; *Kevill's Company* C*SRs*, M324, Roll 315; Weaver, *Assorted Heavy Artillery*, 79.

218. *1860 Norfolk Co., Virginia Census*, M653, Roll 1366; *41st Virginia Infantry CSRs*, M324, Rolls 860, 865; Fiveash, Virginia (Merrimac), 21–22; Henderson, *41st Virginia Infantry*, 120; *19th Btn. Virginia Heavy Artillery CSRs*, M324, Roll 246; *Kevill's Company* C*SRs*, M324, Roll 315; Weaver, *Assorted Heavy Artillery*, 78; James, *Lower Norfolk Antiquary*, 5, part 1:126.

219. *CSN & CSMC Personnel*, M260, Roll 6; *ORN*, 2, 1:309; *North Carolina Payrolls*; *Richmond & James River Payrolls*.

220. *1860 Norfolk Co., Virginia Census*, M653, Roll 1366; *41st Virginia Infantry CSRs*, M324, Roll 865; *CSN Register*, 109; *Wilmington Payrolls*; *19th Btn. Virginia Heavy Artillery CSRs*, M324, Roll 246; Henderson, *41st Virginia Infantry*, 120.

221. *4th Georgia Infantry CSRs*, M266, Roll 184; *CSN & CSMC Personnel*, M260, Roll 6; Berent, "Georgians Crewed the *Virginia*," 123; *ORN*, 2, 1:309; *File NA-2,* M1091, Roll 18; *Richmond & James River Payrolls*; *Charleston Payrolls.*
222. Donnelly, *Confederate Enlisted Marines*, 58; *ORN*, 2, 1:310; *Richmond & James River Payrolls.*
223. *8th Alabama Infantry CSRs*, M311, Roll 176; *Shipping Articles*; *ORN*, 2, 1:310; *Vessel Papers*; *North Carolina Payrolls*; *Richmond & James River Payrolls*; U.S. War Department, "CSRs of Confederate Soldiers Who Served in Organizations from the State of Alabama, 35th Infantry," Microfilm Series, M311, Roll 362, RG 109, NA.
224. Donnelly, *Confederate Enlisted Marines*, 58; *ORN*, 2, 1:310, 315; *Richmond & James River Payrolls*; *ORN*, 2, 1:288; *Wilmington Payrolls.*
225. *ORN*, 2, 1:309; *File NA-2,* M1091, Roll 18; *CSN Register*, 104; *North Carolina Payrolls*; *Richmond & James River Payrolls*; *ORN*, 2, 1:284; *Charleston Payrolls.*
226. *1860 Currituck Co., North Carolina Census*, M653, Roll 895; *1st North Carolina Infantry CSRs*, M270, Roll 95; Manarin, *North Carolina Troops,* 3:217; *CSN & CSMC Personnel*, M260, Roll 6; *ORN*, 2, 1:310; *Richmond & James River Payrolls.*
227. *8th Alabama Infantry CSRs*, M311, Roll 176; *Shipping Articles*; *ORN*, 2, 1:309; *Vessel Papers*; *File NA-2,* M1091, Roll 18; *ORN*, 1, 17:864; *North Carolina Payrolls*; *Richmond & James River Payrolls*; *ORN*, 2, 1:304; *Wilmington Payrolls*; *Hospital & Prison Records*, M260, Roll 2.
228. *9th Virginia Infantry CSRs*, M324, Roll 482; Trask, *9th Virginia Infantry*, 78; *ORN*, 2, 1:310; *North Carolina Payrolls*; *Richmond & James River Payrolls.*
229. Hartzler, *Marylanders in the Confederacy*, 203; *1st North Carolina Infantry CSRs*, M270, Roll 95; Manarin, *North Carolina Troops,* 3:149; *ORN*, 1, 17:43; *ORN*, 2, 1:309; *File NA-2,* M1091, Roll 18.
230. *9th Virginia Infantry CSRs*, M324, Roll 482; Trask, *9th Virginia Infantry*, 78; *CSN & CSMC Personnel*, M260, Roll 6; *ORN*, 2, 1:309; *North Carolina Payrolls*; *Richmond & James River Payrolls.*
231. *ORN*, 2, 1:310; *Vessel Papers*; *North Carolina Payrolls*; *Richmond & James River Payrolls*; *15th Louisiana Infantry CSRs*, M320, Roll 269; Booth, *Louisiana Confederate Soldiers*, 3, bk. 1:743.
232. *ORN*, 2, 1:310; *North Carolina Payrolls*; *Richmond & James River Payrolls.*
233. *1860 Norfolk Co., Virginia Census*, M653, Roll 1366; Hamersly, *Complete General Navy Register*, 434; *CSN Register*, 114; *ORN*, 2, 1:308; *Richmond & James River Payrolls*; *Wilmington Payrolls.*
234. *CSN Register*, 114, Dudley, *Going South*, 42; *ORN*, 2, 1:308.

235. Donnelly, *Confederate Enlisted Marines*, 60; *ORN*, 2, 1:311; *Richmond & James River Payrolls*; Manarin, *North Carolina Troops*, 14:720.
236. *1860 Pasquotank Co., North Carolina Census*, M653, Roll 909; *1st North Carolina Infantry CSRs*, M270, Roll 95; Manarin, *North Carolina Troops*, 3:149; *ORN*, 2, 1:309; *Vessel Papers*; *File NA-2*, M1091, Roll 18; *Richmond & James River Payrolls*.
237. *14th North Carolina Infantry CSRs*, M270, Roll 226; Manarin, *North Carolina Troops*, 5:488; *ORN*, 2, 1:309; *Vessel Papers*; M1091, Roll 18; *Hospital & Prison Records*, M260, Roll 2; *Richmond & James River Payrolls*.
238. Littlepage, *With the Crew of the Virginia*; *CSN Register*, 115; Dudley, *Going South*, 49; *ORN*, 2, 1:308; Jack, *Memoirs*, 12; Coski, *Capital Navy*, 240; *Richmond Times-Dispatch*, June 30, 1907; *Richmond & James River Payrolls*; *Wilmington Payrolls*.
239. *1st (Nelligan's) Louisiana Infantry CSRs*, M320, Roll 77; Booth, *Louisiana Confederate Soldiers*, 3, bk. 1:781; *ORN*, 2, 1:309; *Richmond & James River Payrolls*.
240. *CSN Register*, 116; Dudley, *Going South*, 49; *ORN*, 2, 1:309; Jack, *Memoirs*, 12; *Richmond & James River Payrolls*; *Richmond Times-Dispatch*, June 30, 1907; *Confederate Veteran*, July 1910, 339.
241. *ORN*, 2, 1:310; *Vessel Papers*; *Richmond & James River Payrolls*; U.S. War Department, "CSRs of Confederate Soldiers Who Served in Organizations from the State of Mississippi, 11th Infantry," Microfilm Series, M269, Roll 195, RG 109, NA.
242. *ORN*, 2, 1:309; *CSN & CSMC Personnel*, M260, Roll 6; *File NA-2*, M1091, Roll 18; *Richmond & James River Payrolls*.
243. *13th North Carolina Infantry CSRs*, M270, Roll 216; Manarin, *North Carolina Troops*, 5:293; *ORN*, 2, 1:309; *Vessel Papers*; *File NA-2*, M1091, Roll 18; *North Carolina Payrolls*; *Richmond & James River Payrolls*.
244. *ORN*, 2, 1:309; U.S. War Department, "CSRs of Confederate Soldiers Who Served in Organizations from the State of Georgia, 5th Infantry," Microfilm Series, M266, Roll 197, RG 109, NA.
245. Tindall, "True Story of the *Virginia*," 2–3; Dudley, *Going South*, 47; *ORN*, 2, 1:308; Jack, *Memoirs*, 12; Scharf, *Confederate States*, 745; *Richmond & James River Payrolls*; Littlepage, *Midshipman Abroad*, 24; *Wilmington Payrolls*; *CSN Register*, 128; *Richmond Times-Dispatch*, June 30, 1907; *Confederate Veteran*, January 1925, 25.
246. Donnelly, *Confederate Enlisted Marines*, 68; *ORN*, 2, 1:310, 315; *Richmond & James River Payrolls*.
247. *8th Alabama Infantry CSRs*, M311, Roll 177; *Shipping Articles*; *ORN*, 2, 1:303, 309; *Vessel Papers*; *File NA-2*, M1091, Roll 18; *Richmond & James*

River Payrolls; *ORN*, 1, 15:734; *ORN*, 1, 17:864; *Wilmington Payrolls*; *ORN*, 2, 1:303.

248. *12th Alabama Infantry CSRs*, M311, Roll 209; *Shipping Articles*; *ORN*, 2, 1:309; *File NA-2*, M1091, Roll 18; *Richmond & James River Payrolls*; *ORN*, 2, 1:289.

249. Donnelly, *Confederate Enlisted Marines*, 69; *ORN*, 2, 1:310, 315; *Richmond & James River Payrolls*.

250. *ORN*, 2, 1:310; *5th Btn. Virginia Infantry CSRs*, M324, Roll 432; Gregory, *53rd and 5th Btn. Virginia Infantry*, 132.

251. *2nd Louisiana Infantry CSRs*, M320, Roll 108; Booth, *Louisiana Confederate Soldiers*, 3, bk. 1:913; *CSN & CSMC Personnel*, M260, Roll 6; *ORN*, 2, 1:309; *File NA-2*, M1091, Roll 18; *North Carolina Payrolls*; *Richmond & James River Payrolls*; *ORN*, 2, 1:312; *Hospital & Prison Records*, M260, Roll 3.

252. *1st (Nelligan's) Louisiana Infantry CSRs*, M320, Roll 78; Booth, 3, bk. 1:922; *Richmond & James River Payrolls*; *4th Btn. Virginia Infantry CSRs*.

253. *10th Louisiana Infantry CSRs*, M320, Roll 221; Booth, *Louisiana Confederate Soldiers*, 3, bk. 1:923; *CSN & CSMC Personnel*, M260, Roll 6; *ORN*, 2, 1:309; *File NA-2*, M1091, Roll 18; *North Carolina Payrolls*; *Richmond & James River Payrolls*; *ORN*, 1, 17:864; *ORN*, 2, 1:304; *Wilmington Payrolls*; *Hospital & Prison Records*, M260, Roll 3.

254. *ORN*, 2, 1:310; *File NA-2*, M1091, Roll 18; *CSN & CSMC Personnel*, M260, Roll 6; *Richmond & James River Payrolls*; *ORN*, 2, 274; *Wilmington Payrolls*; *15th Louisiana Infantry CSRs*, M320, Roll 270; Booth, *Louisiana Confederate Soldiers*, 3, bk. 1:1127.

255. *CSN & CSMC Personnel*, M260, Roll 6; *1st (Nelligan's) Louisiana Infantry CSRs*, M320, Roll 78; Booth, *Louisiana Confederate Soldiers*, 3, bk. 1:1127; *ORN*, 2, 1:310; *North Carolina Payrolls*; *Richmond & James River Payrolls*.

256. *41st Virginia Infantry CSRs*, M324, Rolls 860, 866; Hartzler, *Marylanders in the Confederacy*, 209; Henderson, *41st Virginia Infantry*, 123; *Kevill's Company* CSRs, M324, Roll 315; Weaver, *Assorted Heavy Artillery*, 80; *CSN & CSMC Personnel*, M260, Roll 6; *1870 Norfolk., Virginia Census*, M593, Roll 1667.

257. *ORN*, 2, 1:310; *CSN & CSMC Personnel*, M260, Roll 6; *Vessel Papers*; *Richmond & James River Payrolls*; *15th Louisiana Infantry CSRs*, M320, Roll 270; Booth, *Louisiana Confederate Soldiers*, 3, bk. 1:1135.

258. *10th Louisiana Infantry CSRs*, M320, Roll 221; Booth, *Louisiana Confederate Soldiers*, 3, bk. 1:1160; *CSN & CSMC Personnel*, M260, Roll 6; *ORN*, 2, 1:309; *File NA-2*, M1091, Roll 18; *North Carolina Payrolls*; *Richmond & James River Payrolls*.

259. *ORN*, 2, 1:310; 1091, Roll 18; *Richmond & James River Payrolls*; *CSN Register*, 122; *ORN*, 2, 1:280; *CSN & CSMC Personnel*, M260, Roll 6.
260. *Shipping Articles*; *ORN*, 2, 1:310; *File NA-2,* M1091, Roll 18; *Richmond & James River Payrolls*; U.S. War Department, "CSRs of Confederate Soldiers Who Served in Organizations From the State of Virginia, 41st Btn. Cavalry," Microfilm Series, M324, Roll 203, RG 109, NA.
261. *3rd Alabama Infantry CSRs*, M311, Roll 109; *ORN*, 2, 1:309; *Vessel Papers*; *File NA-2,* M1091, Roll 18; *Richmond & James River Payrolls*; *Charleston Payrolls*; *Hospital & Prison Records*, M260, Roll 3.
262. Donnelly, *Confederate Enlisted Marines*, 64; *ORN*, 2, 1:310, 315; *Richmond & James River Payrolls*; *ORN*, 2, 1:280; *Wilmington Payrolls*; *Hospital & Prison Records*, M260, Roll 3.
263. *1st (Nelligan's) Louisiana Infantry CSRs*, M320, Roll 78; Booth, *Louisiana Confederate Soldiers*, 3, bk. 1:1200; *CSN & CSMC Personnel*, M260, Roll 6; *ORN*, 2, 1:309; *Vessel Papers*; *File NA-2,* M1091, Roll 18; *North Carolina Payrolls*; *Richmond & James River Payrolls*; U.S. War Department, "CSRs of Confederate Soldiers Who Served in Organizations from the State of Virginia, 10th Btn. Heavy Artillery," Microfilm Series, M324, Roll 230, RG 109, NA.
264. U.S. War Department, "CSRs of Confederate Soldiers Who Served in Organizations from the State of Virginia, 53rd Infantry," Microfilm Series, M324, Roll 949, RG 109, NA; Gregory, *53rd and 5th Btn. Virginia Infantry*, 177; *CSN & CSMC Personnel*, M260, Roll 6; *ORN*, 2, 309; *File NA-2,* M1091, Roll 18; *North Carolina Payrolls*; *Richmond & James River Payrolls.*
265. *CSN & CSMC Personnel*, M260, Roll 6; *ORN*, 2, 1:310; *Richmond & James River Payrolls*; *Cobb's Legion, Georgia Infantry CSRs*, M266, Roll 580.
266. U.S. War Department, "CSRs of Confederate Soldiers Who Served in Organizations from the State of Virginia, 14th Infantry," Microfilm Series, M324, Roll 550, RG 109, NA; Crews and Parrish, *14th Virginia Infantry*, 121; *CSN & CSMC Personnel*, M260, Roll 6; *ORN*, 2, 1:308; *Richmond & James River Payrolls*; *4th Btn. Virginia Infantry CSRs.*
267. Hartzler, *Marylanders in the Confederacy*, 214; U.S. War Department, "CSRs of Confederate Soldiers Who Served in Organizations from the State of Maryland, 1st Artillery," Microfilm Series, M321, Roll 9, RG 109, NA; *ORN*, 2, 1:308; *Vessel Papers*; *Richmond & James River Payrolls*; *CSN Register*, 124.
268. Donnelly, *Confederate Enlisted Marines*, 66; *ORN*, 2, 1:311; *Richmond & James River Payrolls*; *Hospital & Prison Records*, M260, Roll 3.
269. *Shipping Articles*; *CSN & CSMC Personnel*, M260, Roll 6; *ORN*, 2, 1:310; *Richmond & James River Payrolls*; Scharf, *Confederate States*, 715.

270. *ORN*, 2, 1:309; *Vessel Papers*; *Richmond & James River Payrolls*; *15th Louisiana Infantry CSRs*, M320, Roll 270; Booth, *Louisiana Confederate Soldiers*, 3, bk. 1:1238.
271. *ORN*, 2, 1:310; *Vessel Papers*; *Richmond & James River Payrolls*; *15th Louisiana Infantry CSRs*, M320, Roll 270; Booth, *Louisiana Confederate Soldiers*, 3, bk. 1:1209.
272. *Shipping Articles*; *CSN & CSMC Personnel*, M260, Roll 6; *ORN*, 2, 1:309; *File NA-2*, M1091, Roll 18; *North Carolina Payrolls*; *Richmond & James River Payrolls*; *4th Btn. Virginia Infantry CSRs*.
273. Donnelly, *CS Marine Corps*, 32, 49, 60, 146; Dudley, *Going South*, 55; *CSN Register*, 132; Donnelly, *Rebel Leathernecks*, 209–10.
274. *Wilmington Payrolls*; *ORN*, 2, 1:285; *CSN & CSMC Personnel*, M260, Roll 6; *ORN*, 2, 1:309; *File NA-2*, M1091, Roll 18; *North Carolina Payrolls*; *Richmond & James River Payrolls*.
275. *1860 Pasquotank Co., North Carolina Census*, M653, Roll 909; *ORN*, 2, 1:285; *Wilmington Payrolls*; *CSN & CSMC Personnel*, M260, Roll 6; *ORN*, 2, 1:310; *Richmond & James River Payrolls*; *Hospital & Prison Records*, M260, Roll 3.
276. *8th Alabama Infantry CSRs*, M311, Roll 177; *Shipping Articles*; *CSN & CSMC Personnel*, M260, Roll 6; *ORN*, 2, 1:309; *File NA-2*, M1091, Roll 18; *North Carolina Payrolls*; *Richmond & James River Payrolls*.
277. Willet, *Poquoson Watermen*, 194, 200; *32nd Virginia Infantry CSRs*, M324, Roll 785; Jensen, 194; *CSN & CSMC Personnel*, M260, Roll 6; *ORN*, 2, 1:309; *North Carolina Payrolls*; *Richmond & James River Payrolls*; *Virginia Pension Applications*; *R.E. Lee Soldiers' Home Application*; Markham, 18.
278. *9th Virginia Infantry CSRs*, M324, Roll 482; Hartzler, *Marylanders in the Confederacy*, 224; Trask, *9th Virginia Infantry*, 81; *CSN & CSMC Personnel*, M260, Roll 6; *ORN*, 2, 1:310; *Richmond & James River Payrolls*.
279. *1st North Carolina Infantry CSRs*, M270, Roll 96; Manarin, *North Carolina Troops*, 3:218; *ORN*, 2, 1:310; *Vessel Papers*; *North Carolina Payrolls*; *Richmond & James River Payrolls*.
280. Hamersly, *Complete General Navy Register*, 499; Dudley, *Going South*, 40; *CSN Register*, 135; Scharf, *Confederate States*, 113; *ORN*, 2, 1:308; *Richmond & James River Payrolls*; Coski, *Capital Navy*, 21, 27, 50, 74–75, 88, 100, 116, 121, 156, 164, 219, 246–47.
281. *13th North Carolina Infantry CSRs*, M270, Roll 217; Manarin, *North Carolina Troops*, 5:333; *CSN & CSMC Personnel*, M260, Roll 6; *ORN*, 2, 1:310; *Richmond & James River Payrolls*; U.S. War Department, "CSRs of Confederate Soldiers Who Served in Organizations from the State of

North Carolina, 57th Infantry," Microfilm Series, M270, Roll 530, RG 109, NA; Manarin, *North Carolina Troops*, 14:197.

282. *4th Georgia Infantry CSRs*, M266, Roll 186; *CSN & CSMC Personnel*, M260, Roll 6; *Richmond & James River Payrolls*; *Hospital & Prison Records*, M260, Roll 3.

283. *4th Georgia Infantry CSRs*, M266, Roll 186; *CSN & CSMC Personnel*, M260, Roll 6; *North Carolina Payrolls*; *Richmond & James River Payrolls.*

284. *Shipping Articles*; *ORN*, 2, 1:310; *Richmond & James River Payrolls*; U.S. Department of the Navy, "Mobile and Jackson Station, C.S. Navy," Microfilm Series, T829, Roll 167, RG 45, NA.

285. Donnelly, *Confederate Enlisted Marines*, 73; *ORN*, 2, 1:310; *Richmond & James River Payrolls.*

286. *1st North Carolina Infantry CSRs*, M270, Roll 96; Manarin, *North Carolina Troops*, 3:218; *CSN & CSMC Personnel*, M260, Roll 6; *ORN*, 2, 1:309; *Vessel Papers*; *File NA-2*, M1091, Roll 18; *North Carolina Payrolls*; *Richmond & James River Payrolls*; *ORN*, 1, 11:794.

287. *1st (Nelligan's) Louisiana Infantry CSRs*, M320, Roll 78; Booth, *Louisiana Confederate Soldiers*, 3, bk. 1:1068; *Hospital & Prison Records*, M260, Roll 3; *ORN*, 2, 1:309; *Vessel Papers*; *File NA-2*, M1091, Roll 18; *North Carolina Payrolls*; *Richmond & James River Payrolls*; *Charleston Payrolls*; *ORN*, 2, 1:274; *Wilmington Payrolls.*

288. *41st Virginia Infantry CSRs*, M324, Rolls 860, 866; Henderson, *41st Virginia Infantry*, 126; Fiveash, Virginia (Merrimac), 21–22; Weaver, *10th & 19th Btns.*, 146; *Kevill's Company* CSRs, M324, Roll 315; Weaver, *Assorted Heavy Artillery*, 80; Stewart, *History of Norfolk County*, 144; *R.E. Lee Soldiers' Home Application*; Markham, 18.

289. *2nd Louisiana Infantry CSRs*, M320, Roll 108; Booth, *Louisiana Confederate Soldiers*, 3, bk. 1:1081; *ORN*, 2, 1:310; *Richmond & James River Payrolls*; U.S. War Department, "CSRs of Confederate Soldiers Who Served in Organizations from the State of Maryland, 2nd Btn. Infantry," Microfilm Series, M321, Roll 20, RG 109, NA.

290. *Shipping Articles; CSN & CSMC Personnel*, M260, Roll 6; *ORN*, 2, 1:309; *File NA-2,* M1091, Roll 18; *North Carolina Payrolls*; *Richmond & James River Payrolls.*

291. *10th Louisiana Infantry CSRs*, M320, Roll 221; Booth, *Louisiana Confederate Soldiers*, 3, bk. 1:1100; *ORN*, 2, 1:309; *North Carolina Payrolls*; *Richmond & James River Payrolls*; *ORN*, 2, 1:278; *Hospital & Prison Records*, M260, Roll 3.

292. *Shipping Articles;* ORN, 2, 1:300, 292; *North Carolina Payrolls*; ORN, 2, 1:310; *Vessel Papers*; *Richmond & James River Payrolls*; *ORN*, 2, 1: 312; *Wilmington Payrolls.*

293. *1st North Carolina Infantry CSRs*, M270, Roll 96; *24th Btn. Virginia Partisan Rangers CSRs*; Manarin, *North Carolina Troops*, 3:218; *ORN*, 2, 1:310; *Vessel Papers*; *North Carolina Payrolls*; *Richmond & James River Payrolls*.
294. *1860 Warwick Co., Virginia Census*, M653, Roll 1382; *41st Virginia Infantry CSRs*, M324, Roll 866; Henderson, *41st Virginia Infantry*, 127; *CSN & CSMC Personnel*, M260, Roll 6; *Shipping Articles*; *Richmond & James River Payrolls*; *Hospital & Prison Records*, M260, Roll 3; *Virginia Pension Applications*.
Crews and Parrish, 127; ORN, 2, 1:310; *Vessel Papers*; *North Carolina Payrolls*; *Richmond & James River Payrolls*.
296. *Shipping Articles*; *ORN*, 2, 1:309; *Vessel Papers*; *File NA-2*, M1091, Roll 18; *North Carolina Payrolls*; *Richmond & James River Payrolls*; *Hospital & Prison Records*, M260, Roll 3.
297. *4th Georgia Infantry CSRs*, M266, Roll 186; Berent, "Georgians Crewed the *Virginia*," 124; *ORN*, 2, 1:310; *Richmond & James River Payrolls*; U.S. War Department, "CSRs of Confederate Soldiers Who Served in Organizations from the State of Maryland, 1st Cavalry," Microfilm Series, M321, Roll 4, RG 109, NA.
298. *10th Louisiana Infantry CSRs*, M320, Roll 222; Booth, *Louisiana Confederate Soldiers*, 3, bk. 1:1293; *ORN*, 2, 1:310; *CSN & CSMC Personnel*, M260, Roll 6; *Richmond & James River Payrolls*.
299. *ORN*, 2, 1:310; *Richmond & James River Payrolls*; Norris, *Confederate States' Ship* "Virginia," 204–33; *1860 City of Baltimore, Maryland Census*, M653, Roll 463; U.S. War Department, "CSRs of Confederate Soldiers Who Served in Organizations Raised Directly by the Confederate Government, Signal Corps, CSA," Microfilm Series, M258, Roll 119, RG 109, NA (hereafter cited as *Signal Corps CSRs*, followed by M258 and Roll Number); Beers, *The Confederacy*, 152–53, 235–236; Hartzler, *Marylanders in the Confederacy*, 234.
300. U.S. War Department, "CSRs of Confederate Soldiers Who Served in Organizations from the State of Georgia, 22nd Infantry," Microfilm Series, M266, Roll 345, RG 109, NA; Berent, "Georgians Crewed the *Virginia*," 124; *ORN*, 2, 1:310; *Richmond & James River Payrolls*.
301. *1860 Norfolk Co., Virginia Census*, M653, Roll 1366; Hamersly, *Complete General Navy Register*, 539; Dudley, *Going South*, 50; *CSN Register*, 145; *ORN*, 2, 1:308; White, *First Iron-Clad Naval Engagement*, 16; *Richmond & James River Payrolls*; *1870 Richmond Co., Virginia Census*, M593, Roll 1654; Stewart, *History of Norfolk County*, 144; Pollack, 193.
302. *ORN*, 2, 1:310; *Richmond & James River Payrolls*; U.S. War Department, "CSRs of Confederate Soldiers Who Served in Organizations from

the State of Virginia, Light Artillery, Nelson's and Page's Companies," Microfilm Series, M324, Roll 324, RG 109, NA (hereafter cited as *Nelson's and Page's Companies CSRs*); Moore, *Disbanded Light Artillery*, 121.

303. *1860 Washington Co., North Carolina Census*, M653, Roll 917; *1st North Carolina Infantry CSRs*, M270, Roll 96; Manarin, *North Carolina Troops*, 3:219; *ORN*, 2, 1:309; *Vessel Papers*; *Richmond & James River Payrolls*.

304. *ORN*, 2, 1:309; *File NA-2*, M1091, Roll 18; *Richmond & James River Payrolls*; *Charleston Payrolls*; *ORN*, 2, 1: 312.

305. *32nd Virginia Infantry CSRs*, M324, Roll 786; *ORN*, 2, 1:309; Jensen, 197.

306. U.S. War Department, "CSRs of Confederate Soldiers Who Served in Organizations from the State of North Carolina, 3rd Infantry," Microfilm Series, M270, Roll 130, RG 109, NA, (hereafter cited as *3rd North Carolina Infantry CSRs*, followed by M270 and Roll Number); Manarin, *North Carolina Troops*, 3:585; *CSN & CSMC Personnel*, M260, Roll 6; *ORN*, 2, 1:310; *Vessel Papers*; *North Carolina Payrolls*; *Richmond & James River Payrolls*; *39th Btn. Virginia Cavalry CSRs*; Driver, *1st, 39th & 24th Virginia Btns.*, 148.

307. *1860 Norfolk Co., Virginia Census*, M653, Roll 1366; *CSN Register*, 147; *ORN*, 2, 1:308; *Richmond & James River Payrolls*.

308. *ORN*, 2, 1:310; *Vessel Papers*; *North Carolina Payrolls*; *Richmond & James River Payrolls*.

309. *41st Virginia Infantry CSRs*, M324, Roll 867; *CSN & CSMC Personnel*, M260, Roll 6; *ORN*, 2, 1:310; *Richmond & James River Payrolls*; U.S. War Department, "CSRs of Confederate Soldiers Who Served in Organizations from the State of Virginia, 10th Cavalry," Microfilm Series, M324, Roll 107, RG 109, NA; Driver, *10th Virginia Cavalry*, 150; Henderson, *41st Virginia Infantry*, 130.

310. *12th Alabama Infantry CSRs*, M311, Roll 214; *CSN & CSMC Personnel*, M260, Roll 6; *ORN*, 2, 1:309; *File NA-2*, M1091, Roll 18; *Shipping Articles*; *Richmond & James River Payrolls*; *ORN*, 1, 17:864; *Wilmington Payrolls*; *Charleston Payrolls*; *CSN Register*, 152; *Mobile (Alabama) Daily Item*, April 26, 1910.

311. *Shipping Articles*; *New Orleans Payrolls*; *ORN*, 2, 1:309; *Vessel Papers*; *File NA-2*, M1091, Roll 18; *North Carolina Payrolls*; *Richmond & James River Payrolls*.

312. *12th Alabama Infantry CSRs*, M311, Roll 209; *File NA-2*, M1091, Roll 18; *Shipping Articles*; *ORN*, 2, 1:310; *Richmond & James River Payrolls*.

313. *S. Bassett French Sketches*; Tyler, *Encyclopedia of Virginia Biography*, 3:293; Hamersly, *Complete General Navy Register*, 506; *1860 Norfolk Co., Virginia Census*, M653, Roll 1366; Dudley, *Going South*, 41; *ORN*, 2, 1:308; *Richmond & James River Payrolls*; *CSN Register*, 152; Phillips, 196.

314. Donnelly, *Confederate Enlisted Marines*, 82; *ORN*, 2, 1:311; *Richmond & James River Payrolls*.

315. *1860 Norfolk Co., Virginia Census*, M653, Roll 1366; *9th Virginia Infantry CSRs*, M324, Roll 484; *CSN & CSMC Personnel*, M260, Roll 6; *ORN*, 2, 1:309; *File NA-2*, M1091, Roll 18; *Richmond & James River Payrolls*; *Shipping Articles*; *CSN Register*, 154; *1870 Norfolk Co., Virginia Census*, M593, Roll 1667; Trask, *9th Virginia Infantry*, 86, 112.
316. *1860 Norfolk Co., Virginia Census*, M653, Roll 1366; *6th Virginia Infantry CSRs*, M324, Roll 444; *CSN & CSMC Personnel*, M260, Roll 6; *ORN*, 2, 1:310; *Richmond & James River Payrolls*; U.S. War Department, "CSRs of Confederate Soldiers Who Served in Organizations from the State of Virginia, 1st Btn. Infantry," Microfilm Series, M324, Roll 369, RG 109, NA, (hereafter cited as *1st Btn. Virginia Infantry CSRs*).
317. *ORN*, 2, 1:310; *Richmond & James River Payrolls*; *Shipping Articles*; *Vessel Papers*; *Hospital & Prison Records*, M260, Roll 3; *15th Louisiana Infantry CSRs*, M320, Roll 271; Booth, *Louisiana Confederate Soldiers*, 3, bk. 2:189.
318. *14th North Carolina Infantry CSRs*, M270, Roll 228; *CSN & CSMC Personnel*, M260, Roll 6; Manarin, *North Carolina Troops*, 5:451; *ORN*, 2, 1:309; *File NA-2*, M1091, Roll 18; *Richmond & James River Payrolls*; U.S. War Department, "CSRs of Confederate Soldiers Who Served in Organizations from the State of North Carolina, 7th Btn. Cavalry," Microfilm Series, M270, Roll 42, RG 109, NA; Manarin, *North Carolina Troops*, 2:542; *6th North Carolina Cavalry CSRs*; Manarin, *North Carolina Troops*, 2:474.
319. *13th North Carolina Infantry CSRs*, M270, Roll 217; Manarin, *North Carolina Troops*, 5:353; *CSN & CSMC Personnel*, M260, Roll 6; *ORN*, 2, 1:309; *File NA-2*, M1091, Roll 18; *Richmond & James River Payrolls*.
320. *1st North Carolina Infantry CSRs*, M270, Roll 97; Manarin, *North Carolina Troops*, 3:219; *ORN*, 2, 1:310; *Vessel Papers*; *Richmond & James River Payrolls*.
321. *15th Louisiana Infantry CSRs*, M320, Roll 271; Booth, *Louisiana Confederate Soldiers*, 3, bk. 2:213; *ORN*, 2, 1:310; *Vessel Papers*; *Richmond & James River Payrolls*; Donnelly, *CS Marines Corps*, 162; Scharf, *Confederate States*, 794; *CSN Register*, 158.
322. Donnelly, *Confederate Enlisted Marines*, 85; *ORN*, 2, 1:310; *Richmond & James River Payrolls*.
323. *12th Virginia Infantry CSRs*, M324, Roll 529; Henderson, *12th Virginia Infantry*, 150; *ORN*, 2, 1:310; *Richmond & James River Payrolls*; *Graham's Company CSRs*; Moore, *Graham's, Jackson's, and Lurty's Artillery*, 121; *Virginia Pension Applications*.
324. Tyler, "Ashton Family," 117; Hartzler, *Marylanders in the Confederacy*, 251; Dudley, *Going South*, 52; Hamersly, *Complete General Navy Register*, 588; *ORN*,

2, 1:308; *Richmond & James River Payrolls*; *CSN Register*, 159; *Confederate Veteran*, June 1916, 274.

325. *1st (Colquitt's) Arkansas Infantry CSRs*, M317, Roll 51; *ORN*, 2, 1:310; *Vessel Papers*; *North Carolina Payrolls*; *Richmond & James River Payrolls*.

326. *ORN*, 2, 1:309; *Vessel Papers*; *File NA-2*, M1091, Roll 18; *Richmond & James River Payrolls*; *38th Btn. Virginia Light Artillery CSRs*, M324, Roll 257; Moore, *Fayette, Hampden, Thomas, and Blount's Artillery*, 167.

327. *Kevill's Company* CSRs, M324, Roll 315; Weaver, *Assorted Heavy Artillery*, 81; Fiveash, Virginia (Merrimac), 21–22.

328. *1st South Carolina Infantry CSRs*, M267, Roll 146; *1st (McCreary's) South Carolina Infantry CSRs*, M267, Roll 132; Salley, *South Carolina Troops*, 350–51, 366; *CSN & CSMC Personnel*, M260, Roll 6; *ORN*, 2, 1:309; *Shipping Articles*; *Richmond & James River Payrolls*; *Hospital & Prison Records*, M260, Roll 3.

329. *3rd North Carolina Infantry CSRs*, M270, Roll 131; Manarin, *North Carolina Troops*, 3:519; *CSN & CSMC Personnel*, M260, Roll 7; *ORN*, 2, 1:310; *Richmond & James River Payrolls*; *Hospital & Prison Records*, M260, Roll 3.

330. Hartzler, *Marylanders in the Confederacy*, 256; *CSN & CSMC Personnel*, M260, Roll 7; *ORN*, 2, 1:308; *File NA-2*, M1091, Roll 18; *North Carolina Payrolls*; *Richmond & James River Payrolls*; *Hospital & Prison Records*, M260, Roll 3.

331. *3rd Alabama Infantry CSRs*, M311, Roll 112; *Shipping Articles*; *ORN*, 2, 1:310; *Vessel Papers*; *Richmond & James River Payrolls*.

332. *CSN Register*, 169; *Richmond & James River Payrolls*; Rootes, "The Rootes Family," 211.

333. *1st North Carolina Artillery CSRs*, M270, Roll 58; Manarin, *North Carolina Troops*, 1:168; *Shipping Articles*; *ORN*, 2, 1:279, 309; *Vessel Papers*; *File NA-2*, M1091, Roll 18; *Richmond & James River Payrolls*; *ORN*, 1, 17:864; *Wilmington Payrolls*; *CSN Register*, 169.

334. *15th Louisiana Infantry CSRs*, M320, Roll 271; Booth, *Louisiana Confederate Soldiers*, 3, bk. 2:391; *Wilmington Payrolls*; *ORN*, 2, 1:310; *Vessel Papers*; *File NA-2*, M1091, Roll 18; *Richmond & James River Payrolls*; *Charleston Payrolls*; *4th Btn. Virginia Infantry CSRs*.

335. *1st (Colquitt's) Arkansas Infantry CSRs*, M317, Roll 51; *ORN*, 2, 1:309; *Vessel Papers*; *File NA-2*, M1091, Roll 18; *North Carolina Payrolls*; *Richmond & James River Payrolls*; *R.E. Lee Soldiers' Home Application*; Markham, 20.

336. *1st North Carolina Infantry CSRs*, M270, Roll 97; Manarin, *North Carolina Troops*, 3:219; *ORN*, 2, 1:309; *Vessel Papers*; *File NA-2*, M1091, Roll 18; *Richmond & James River Payrolls*; *Charleston Payrolls*.

337. *1860 Pike Co., Georgia Census*, M653, Roll 134; *3rd Georgia Infantry CSRs*, M266, Roll 171; Berent, "Georgians Crewed the *Virginia*," 124; *CSN & CSMC Personnel*, M260, Roll 7; *ORN*, 2, 1:310; *Richmond & James River Payrolls*.
338. *10th Louisiana Infantry CSRs*, M320, Roll 222; Booth, *Louisiana Confederate Soldiers*, 3, bk. 2:422; *CSN & CSMC Personnel*, M260, Roll 7; *ORN*, 2, 1:309; *File NA-2*, M1091, Roll 18; *North Carolina Payrolls*; *Richmond & James River Payrolls*; *Hospital & Prison Records*, M260, Roll 3.
339. *ORN*, 2, 1:309; *Richmond & James River Payrolls*; *ORN*, 2, 1:276; *Shipping Articles*; *ORN*, 2, 1:294–96; *Hospital & Prison Records*, M260, Roll 3.
340. *6th Virginia Infantry CSRs*, M324, Roll 445; Cavanaugh, *6th Virginia Infantry*, 122; *ORN*, 2, 1:309; *Vessel Papers*; *File NA-2*, M1091, Roll 18; *Richmond & James River Payrolls*; *1st Btn. Virginia Infantry CSRs*; Driver, *1st, 39th & 24th Virginia Btns.*, 122.
341. Donnelly, *Confederate Enlisted Marines*, 92; *ORN*, 2, 1:311; *Richmond & James River Payrolls*.
342. *17th North Carolina Infantry CSRs*, M270, Roll 250; Manarin, *North Carolina Troops*, 6:198; *ORN*, 2, 1:285, 310; *File NA-1*, M1091, Roll 17; *Richmond & James River Payrolls*; Manarin, *North Carolina Troops*, 6:758.
343. *12th Virginia Infantry CSRs*, M324, Roll 529; Henderson, *12th Virginia Infantry*, 153; *41st Virginia Infantry CSRs*, M324, Roll 868; Henderson, *41st Virginia Infantry*, 134; *ORN*, 2, 1:310; *Richmond & James River Payrolls*; U.S. War Department, "CSRs of Confederate Soldiers Who Served in Organizations from the State of Virginia, Light Artillery, Pegram's Company," Microfilm Series, M324, Roll 328, RG 109, NA; Weaver, *Branch, Harrington, and Staunton Artillery*, 47; *Virginia Pension Applications*.
344. *1st (McCreary's) South Carolina Infantry CSRs*, M267, Roll 133; Salley, *South Carolina Troops*, 367; *CSN & CSMC Personnel*, M260, Roll 7; *ORN*, 2, 1:309; *North Carolina Payrolls*; *Richmond & James River Payrolls*; *4th Btn. Virginia Infantry CSRs*.
345. *ORN*, 2, 1:309; *Vessel Papers*; *File NA-2*, M1091, Roll 18; *Hospital & Prison Records*, M260, Roll 4; *Richmond & James River Payrolls*; *ORN*, 1, 17:864, *Wilmington Payrolls*; *ORN*, 2, 1:304.
346. Donnelly, *Confederate Enlisted Marines*, 92-93; *ORN*, 2, 1:310; *Richmond & James River Payrolls*.
347. *S. Bassett French Sketches*; *CSN Register*, 174; Dudley, *Going South*, 52, *ORN*, 2, 1:308; *Richmond & James River Payrolls*; Stewart, *History of Norfolk County*, 658–61; Burial marker.
348. Donnelly, *Confederate Enlisted Marines*, 93; *ORN*, 2, 1:310; *Richmond & James River Payrolls*.

349. *ORN*, 2, 1:309; *File NA-2*, M1091, Roll 18; *North Carolina Payrolls*; *Richmond & James River Payrolls*.
350. *2nd Louisiana Infantry CSRs*, M320, Roll 110; Booth, *Louisiana Confederate Soldiers*, 3, bk. 2:493; *ORN*, 2, 1:309; *North Carolina Payrolls*; *Richmond & James River Payrolls*.
351. *3rd Alabama Infantry CSRs*, M311, Roll 112; *ORN*, 2, 1:309; *North Carolina Payrolls*; *Richmond & James River Payrolls*; *ORN*, 2, 1:304, 302, 278, 279.
352. *41st Virginia Infantry CSRs*, M324, Rolls 860, 868; Henderson, *41st Virginia Infantry*, 135; Fiveash, Virginia (Merrimac), 21–22; *Kevill's Company CSRs*, M324, Roll 315; Weaver, *Assorted Heavy Artillery*, 81; *1870 Norfolk Co., Virginia Census*, M593, Roll 1666; Stewart, *History of Norfolk County*, 145.
353. Hamersly, *Complete General Navy Register*, 639; Dudley, *Going South*, 43; *CSN Register*, 175; Scharf, *Confederate States*, 30; Coski, *Capital Navy*, 19; Jack, *Memoirs*, 12; *ORN*, 2, 1:308; *Richmond & James River Payrolls*; Shingleton, *John Taylor Wood*, 157; Burial marker.
354. *S. Bassett French Sketches*; *1860 Norfolk Co., Virginia Census*, M653, Roll 1366; *41st Virginia Infantry CSRs*, M324, Roll 868; Henderson, *41st Virginia Infantry*, 135.
355. *2nd Louisiana Infantry CSRs*, M320, Roll 110; Booth, *Louisiana Confederate Soldiers*, 3, bk. 2:526; *CSN & CSMC Personnel*, M260, Roll 7; *ORN*, 2, 1:309; *File NA-2*, M1091, Roll 18; *North Carolina Payrolls*; *Richmond & James River Payrolls*; Booth, *Louisiana Confederate Soldiers*, 3, bk. 2:528.
356. 1860 Montgomery Co. Alabama Census, M653, Roll 19; Donnelly, *Confederate Enlisted Marines*, 94; *ORN*, 2, 1:310; *Richmond & James River Payrolls*; *ORN*, 2, 1:277, 278, 302; *Wilmington Station*.
357. *13th North Carolina Infantry CSRs*, M270, Roll 218; Manarin, *North Carolina Troops*, 5:306; *CSN & CSMC Personnel*, M260, Roll 7; *ORN*, 2, 1:309; *North Carolina Payrolls*; *Richmond & James River Payrolls*; U.S. War Department, "CSRs of Confederate Soldiers Who Served in Organizations from the State of Virginia, Light Artillery, Hardwicke's Company," Microfilm Series, M324, Roll 309, RG 109, NA; Burial marker.
358. Hartzler, *Marylanders in the Confederacy*, 268; *ORN*, 2, 1:309; Curtis, *History of Famous Battle*, 12; White, *First Iron-Clad Naval Engagement*, 14; *Vessel Papers*; *Richmond & James River Payrolls*; *ORN*, 2, 1:298; *Charleston Payrolls*; *Richmond Times-Dispatch*, June 30, 1907.
359. *ORN*, 2, 1:309; *Vessel Papers*; *File NA-2*, M1091, Roll 18; *Richmond & James River Payrolls*.
360. Hamersly, *Complete General Navy Register*, 650; Dudley, *Going South*, 40; *CSN Register*, 179; Scharf, *Confederate States*, 113, 389; *ORN*, 2, 1:308; Jack,

Memoirs, 12; Curtis, *History of Famous Battle*, 7, 9; *Richmond & James River Payrolls*.

361. *R. Bolling Batte Papers*; *1860 Norfolk Co., Virginia Census*, M653, Roll 1366; Cavanaugh, *6th Virginia Infantry*, 124; *CSN Register*, 180; *ORN*, 2, 1:308; *Richmond & James River Payrolls*; Sinclair, *Two Years on the Alabama*, 300–02; *R.E. Lee Soldiers' Home Application*.
362. *1st (McCreary's) South Carolina Infantry CSRs*, M267, Roll 133; Salley, *South Carolina Troops*, 368; *CSN & CSMC Personnel*, M260, Roll 7; *ORN*, 2, 1:309; *North Carolina Payrolls*; *Richmond & James River Payrolls*.
363. *12th Virginia Infantry CSRs*, M324, Roll 530; Henderson, *12th Virginia Infantry*, 156; *CSN & CSMC Personnel*, M260, Roll 7; *ORN*, 2, 1:310; *Richmond & James River Payrolls*; U.S. War Department, "CSRs of Confederate Soldiers Who Served in Organizations from the State of Virginia, 16th Infantry," Microfilm Series, M324, Roll 579, RG 109, NA; Trask, *16th Virginia Infantry*, 116.
364. U.S. War Department, "CSRs of Confederate Soldiers Who Served in Organizations from the State of Virginia, Heavy Artillery, Jones' Company," Microfilm Series, M324, Roll 314, RG 109, NA; Weaver, *Assorted Heavy Artillery* Batteries, 139; *Nelson's & Page's Companies CSRs*; *Vessel Papers*; Moore, *Disbanded Light Artillery*, 127; *ORN*, 2, 1:310; *Richmond & James River Payrolls*; *Hospital & Prison Records*, M260, Roll 4.
365. Donnelly, *Confederate Enlisted Marines*, 97, 98; *ORN*, 2, 1:310; *Richmond & James River Payrolls*; *ORN*, 2, 1:280; *Wilmington Payrolls*.
366. *41st Virginia Infantry CSRs*, M324, Rolls 860, 868; Henderson, *41st Virginia Infantry*, 137; Fiveash, Virginia (Merrimac), 21–22; *Kevill's Company* C*SRs*, M324, Roll 315; Weaver, *Assorted Heavy Artillery*, 81.
367. Donnelly, *Confederate Enlisted Marines*, 99; *ORN*, 2, 1:310, 315; *Richmond & James River Payrolls*.
368. *4th Georgia Infantry CSRs*, M266, Roll 188; Berent, "Georgians Crewed the *Virginia*," 124; *CSN & CSMC Personnel*, M260, Roll 7; *ORN*, 2, 1:310; *North Carolina Payrolls*; *Richmond & James River Payrolls*; U.S. War Department, "CSRs of Confederate Soldiers Who Served in Organizations from the State of Georgia, 10th Infantry," Microfilm Series, M266, Roll 252, RG 109, NA; U.S. War Department, "CSRs of Confederate Soldiers Who Served in Organizations From the State of Georgia, 48th Infantry," Microfilm Series, M266, Roll 496, RG 109, NA; Henderson, *Georgia Confederate Soldiers*, 2:50, 5:166; *Florida Confederate Pension Application*.
369. *1860 Muhlenberg Co., Kentucky Census*, M653, Roll 388; Donnelly, *Confederate Enlisted Marines*, 99; *ORN*, 2, 1:311; *Richmond & James River Payrolls*.

370. *1860 Norfolk Co., Virginia Census*, M653, Roll 1366; *41st Virginia Infantry CSRs*, M324, Rolls 860, 868; Henderson, *41st Virginia Infantry*, 138; Fiveash, Virginia (Merrimac), 21–22; *Kevill's Company* CSRs, M324, Roll 315; Weaver, *Assorted Heavy Artillery*, 81.
371. *41st Virginia Infantry CSRs*, M324, Rolls 860, 868; Henderson, *41st Virginia Infantry*, 138; Fiveash, Virginia (Merrimac), 21–22; *Kevill's Company* CSRs, M324, Roll 315; Weaver, *Assorted Heavy Artillery*, 82; *Hospital & Prison Records*, M260, Roll 4.
372. *Shipping Articles*; *ORN*, 2, 1:309; *Vessel Papers*; *File NA-2,* M1091, Roll 18; *North Carolina Payrolls*; *Richmond & James River Payrolls*.
373. *1860 Newton Co., Georgia Census*, M653, Roll 133; *Cobb's Legion CSRs*, M266, Roll 589, RG 109, NA; *CSN & CSMC Personnel*, M260, Roll 7; *ORN*, 2, 1:309; *File NA-2,* M1091, Roll 18; *Richmond & James River Payrolls*; U.S. War Department, "CSRs of Confederate Soldiers Who Served in Organizations Raised Directly by the Confederate Government, Nitre and Mining Corps, CSA," Microfilm Series, M258, Roll 113, RG 109, NA.
374. Donnelly, *Confederate Enlisted Marines*, 100; *ORN*, 2, 1:311; *Richmond & James River Payrolls*.
375. *41st Virginia Infantry CSRs*, M324, Rolls 860, 868; Henderson, *41st Virginia Infantry*, 139; *Kevill's Company* CSRs, M324, Roll 315; Weaver, *Assorted Heavy Artillery*, 82; *Virginia Pension Applications*.
376. *ORN*, 2, 1:309; *CSN & CSMC Personnel*, M260, Roll 7; *Richmond & James River Payrolls*; *41st Virginia Infantry CSRs*, M342, Roll 868; *41st Virginia Infantry CSRs*, M324, Roll 868; Henderson, *41st Virginia Infantry*, 139; U.S. War Department, "CSRs of Confederate Soldiers Who Served in Organizations from the State of Virginia, 61st Infantry," Microfilm Series, M324, Roll 1030, RG 109, NA; *File NA-2,* M1091, Roll 18; Trask, *61st Virginia Infantry*, 91.
377. *ORN*, 2, 1:309; *Hospital & Prison Records*, M260, Roll 4; *Vessel Papers*; *File NA-2,* M1091, Roll 18; *North Carolina Payrolls*; *Richmond & James River Payrolls*; *CSN Register*, 188.
378. *41st Virginia Infantry CSRs*, M324, Rolls 860, 868; Henderson, *41st Virginia Infantry*, 140; Fiveash, Virginia (Merrimac), 21–22; Weaver, *Assorted Heavy Artillery*, 82; *1st Engineer Troops CSRs*, M258, Roll 96; 1870 City of Norfolk, Virginia Census, M593, Roll 1666.
379. *14th North Carolina Infantry CSRs*, M270, Roll 229; Manarin, *North Carolina Troops*, 5:433; *Vessel Papers*; *ORN*, 2, 1:310; *North Carolina Payrolls*; *Richmond & James River Payrolls*; U.S. War Department, "CSRs of Confederate

Soldiers Who Served in Organizations Raised Directly by the Confederate Government, 3rd Engineer Troops," Microfilm Series, M258, Roll 101, RG 109, NA.

380. *9th Virginia Infantry CSRs*, M324, Roll 485; Trask, *9th Virginia Infantry*, 93; *ORN*, 2, 1:309; *Vessel Papers*; *File NA-2*, M1091, Roll 18; *Hospital & Prison Records*, M260, Roll 4; *Richmond & James River Payrolls*; *Charleston Payrolls*.
381. *1860 Gloucester Co., Virginia Census*, M653, Roll 1347; U.S. War Department, "CSRs of Confederate Soldiers Who Served in Organizations from the State of Virginia, 34th Infantry," Microfilm Series, M324, Roll 817, RG 109, NA; Scott, *34th Virginia Infantry*, 140; White, *First Iron-Clad Naval Engagement*, 16; *Signal Corps CSRs*, M258, Roll 121; *5th Virginia Cavalry CSRs*; Driver, *5th Virginia Cavalry*, 258.
382. Paullin, "Tattnall, Josiah," in *Dictionary of American Biography*, 310–11; Spence, "Tattnall, Josiah," in *Dictionary of American Military Biography*, 3:1087–89; Thomas, "Tattnall, Josiah," in *Historical Times Illustrated Encyclopedia*, 742; Hamersly, *Complete General Navy Register*, 699; Dudley, *Going South*, 34; Berent, "Georgians Crewed the *Virginia*," 24; *CSN Register*, 191; Scharf, *Confederate States*, 628–29; U.S. Department of the Navy, *Dictionary of Naval Fighting Ships*, 7:53–57.
383. U.S. Department of the Navy, "Savannah Station and Fleet, C.S. Navy," Microfilm Series, T829, Roll 171, RG 45, NA; *ORN*, 2, 1:308; *Richmond & James River Payrolls*; *CSN Register*, 191.
384. *1st North Carolina Infantry CSRs*, M270, Roll 98; Manarin, *North Carolina Troops*, 3:232; *ORN*, 2, 1:309; *Vessel Papers*; *File NA-2*, M1091, Roll 18; *Richmond & James River Payrolls*.
385. *4th Georgia Infantry CSRs*, M266, Roll 188; Berent, "Georgians Crewed the *Virginia*," 124; *ORN*, 2, 1:309; *File NA-2*, M1091, Roll 18; *Richmond & James River Payrolls*; *Charleston Payrolls*; *ORN*, 1, 11:794; Dunaway, *Georgia Division Ancestor Roster*, 9:22.
386. *12th Virginia Infantry CSRs*, M324, Roll 531; Henderson, *12th Virginia Infantry*, 160; *CSN & CSMC Personnel*, M260, Roll 7; *ORN*, 2, 1:310; *Richmond & James River Payrolls*; U.S. War Department, "CSRs of Confederate Soldiers Who Served in Organizations from the State of Virginia, 1st Btn. Infantry, Local Defense," Microfilm Series, M324, Roll 371, RG 109, NA; U.S. War Department, "CSRs of Confederate Soldiers Who Served in Organizations from the State of Virginia, 15th Infantry," Microfilm Series, M324, Roll 568, RG 109, NA; Manarin, *15th Virginia Infantry*, 118.
387. *1850 Sumter Co., Alabama Census*, M432, Roll 15; Donnelly, *Rebel Leathernecks*, 202-3; *CSN Register*, 193; *ORN*, 2, 1:310; Jack, *Memoirs*, 12;

U.S. War Department, "CSRs of Confederate Soldiers Who Served in Organizations from the State of Alabama, 11th Infantry," Microfilm Series, M311, Roll 207, RG 109, NA; Donnelly, *CS Marines Corps*, 11, 13, 26, 48, 51, 141, 171.

388. *1860 Wake Co., North Carolina Census*, M653, Roll 916; U.S. War Department, "CSRs of Confederate Soldiers Who Served in Organizations from the State of North Carolina, 2nd Infantry," Microfilm Series, M270, Roll 114, RG 109, NA; Manarin, *North Carolina Troops*, 3:420; *ORN*, 2, 1:309; *Vessel Papers*; *File NA-2*, M1091, Roll 18; *North Carolina Payrolls*; *Richmond & James River Payrolls*; *Charleston Payrolls*; *Hospital & Prison Records*, M260, Roll 4.
389. *1850 Newton Co., Georgia Census*, M432, Roll 79; *1860 Newton Co., Georgia Census*, M653, Roll 133; *3rd Georgia Infantry CSRs*, M266, Roll 172; Berent, "Georgians Crewed the *Virginia*," 124–25; *CSN & CSMC Personnel*, M260, Roll 7; *ORN*, 2, 1:310; *North Carolina Payrolls*; *Richmond & James River Payrolls*; U.S. War Department, "CSRs of Confederate Soldiers Who Served in Organizations from the State of Georgia, 13th Cavalry," Microfilm Series, M266, Roll 55, RG 109, NA.
390. *Shipping Articles*; *ORN*, 2, 1:309; *Vessel Papers*; *File NA-2*, M1091, Roll 18; *Richmond & James River Payrolls*; *Charleston Payrolls*; *Hospital & Prison Records*, M260, Roll 4.
391. *12th Virginia Infantry CSRs*, M324, Roll 531; Henderson, *12th Virginia Infantry*, 161; *CSN & CSMC Personnel*, M260, Roll 7; *ORN*, 2, 1:310; *Richmond & James River Payrolls*; U.S. War Department, "CSRs of Confederate Soldiers Who Served in Organizations from the State of Virginia, 18th Btn. Artillery," Microfilm Series, M324, Roll 242, RG 109, NA; Chernault and Weaver, *18th and 20th Battalions*, 84.
392. *1st South Carolina Infantry CSRs*, M267, Roll 146; *1st (McCreary's) South Carolina Infantry CSRs*, M267, Roll 134; Salley, *South Carolina Troops*, 368; *CSN & CSMC Personnel*, M260, Roll 7; *ORN*, 2, 1:309; *Richmond & James River Payrolls*; *Charleston Payrolls*; *Hospital & Prison Records*, M260, Roll 4; *Florida Confederate Pension Application*.
393. *6th Virginia Infantry CSRs*, M324, Roll 447; Cavanaugh, *6th Virginia Infantry*, 130; *CSN & CSMC Personnel*, M260, Roll 7; *ORN*, 2, 1:309; *Vessel Papers*; *File NA-2*, M1091, Roll 18; *North Carolina Payrolls*; *Richmond & James River Payrolls*; *Charleston Payrolls*; *Hospital & Prison Records*, M260, Roll 4; *1870 Norfolk Co., Virginia Census*, M593, Roll 1666.
394. Donnelly, *Confederate Enlisted Marines*, 105; *ORN*, 2, 1:311; *Richmond & James River Payrolls*.

395. Hamersly, *Complete General Navy Register*, 723; Dudley, *Going South*, 52; *ORN*, 2, 1:308; Jack, *Memoirs*, 13; *Richmond & James River Payrolls*; *Wilmington Payrolls*; *CSN Register*, 199.
396. *2nd Louisiana Infantry CSRs*, M320, Roll 111; Booth, *Louisiana Confederate Soldiers*, 3, bk. 2:944; *CSN & CSMC Personnel*, M260, Roll 7; *ORN*, 2, 1:310; *Richmond & James River Payrolls*.
397. *ORN*, 2, 1:310; *Vessel Papers*; *Richmond & James River Payrolls* U.S. War Department, "CSRs of Confederate Soldiers Who Served in Organizations from the State of Kentucky, 2nd (Duke's) Cavalry," Microfilm Series, M319, Roll 15, RG 109, NA.
398. *1860 York Co., Virginia Census*, M653, Roll 1385; U.S. War Department, "CSRs of Confederate Soldiers Who Served in Organizations from the State of Virginia, 115th Militia," Microfilm Series, M324, Roll 1055, RG 109, NA; *CSN & CSMC Personnel*, M260, Roll 7; *ORN*, 2, 1:309; *North Carolina Payrolls*; *Richmond & James River Payrolls*; *ORN*, 2, 1:277; *1870 York Co., Virginia Census*, M593, Roll 1682.
399. *1st (McCreary's) South Carolina Infantry CSRs*, M267, Roll 134; Salley, *South Carolina Troops*, 423; *CSN & CSMC Personnel*, M260, Roll 7; *ORN*, 1, 17:43.
400. U.S. War Department, "CSRs of Confederate Soldiers Who Served in Organizations from the State of North Carolina, 8th Infantry," Microfilm Series, M270, Roll 187, RG 109, NA; Manarin, *North Carolina Troops*, 4:532; *CSN & CSMC Personnel*, M260, Roll 7; *ORN*, 2, 1:306, 309; *Vessel Papers*; *File NA-2*, M1091, Roll 18; *North Carolina Payrolls*; *Richmond & James River Payrolls*; *Charleston Payrolls*; *ORN*, 2, 1:312; *Hospital & Prison Records*, M260, Roll 4.
401. *6th Virginia Infantry CSRs*, M324, Roll 449; Cavanaugh, *6th Virginia Infantry*, 131; *ORN*, 2, 1:310; *Richmond & James River Payrolls*; *ORN*, 2, 1:305, 274; *Hospital & Prison Records*, M260, Roll 4.
402. *41st Virginia Infantry CSRs*; Henderson, *41st Virginia Infantry*, 144; Fiveash, Virginia (Merrimac), 21–22; *Kevill's Company* C*SRs*, M324, Roll 315; Weaver, *Assorted Heavy Artillery*, 82.
403. *1860 Norfolk Co., Virginia Census*, M653, Roll 1366; *9th Virginia Infantry CSRs*, M324, Roll 486; *ORN*, 2, 1:309; *Vessel Papers*; *File NA-2*, M1091, Roll 18; *North Carolina Payrolls*; *Richmond & James River Payrolls*; *1870 Norfolk Co., Virginia Census*, M593, Roll 1667; Trask, *9th Virginia Infantry*, 97, 114.
404. U.S. War Department, "CSRs of Confederate Soldiers Who Served in Organizations from the State of Virginia, 3rd Infantry," Microfilm Series, M324, Roll 395, RG 109, NA; Wallace, *3rd Virginia Infantry*, 108; *File NA-2*, M1091, Roll 18; *Shipping Articles*; *North Carolina Payrolls*; *Richmond & James River Payrolls*.

405. *13th North Carolina Infantry CSRs*, M270, Roll 220; Manarin, *North Carolina Troops*, 5:336; *ORN*, 2, 1:310; *Richmond & James River Payrolls*.
406. *ORN*, 2, 1:309; *Vessel Papers*; *File NA-2*, M1091, Roll 18; *Richmond & James River Payrolls*; *CSN Register*, 205.
407. *ORN*, 2, 1:309; *Vessel Papers*; *File NA-2*, M1091, Roll 18; *Richmond & James River Payrolls*; *Shipping Articles*; *Hospital & Prison Records*, M260, Roll 4.
408. Donnelly, *Confederate Enlisted Marines*, 110; *ORN*, 2, 1:310, 315; *Richmond & James River Payrolls*.
409. *ORN*, 2, 1:309; *Richmond & James River Payrolls*.
410. Donnelly, *Confederate Enlisted Marines*, 110; *ORN*, 2, 1:311; *Richmond & James River Payrolls*.
411. *1st (McCreary's) South Carolina Infantry CSRs*, M267, Roll 134; Salley, *South Carolina Troops*, 369; *CSN & CSMC Personnel*, M260, Roll 7; *ORN*, 2, 1:309; *File NA-2*, M1091, Roll 18; *North Carolina Payrolls*; *Richmond & James River Payrolls*; *Hospital & Prison Records*, M260, Roll 4.
412. *1st (McCreary's) South Carolina Infantry CSRs*, M267, Roll 134; *North Carolina Payrolls*; *Richmond & James River Payrolls*; U.S. War Department, "CSRs of Confederate Soldiers Who Served in Organizations from the State of South Carolina, 1st (Charleston) Btn. Infantry," Microfilm Series, M267, Roll 151, RG 109, NA.
413. *1st North Carolina Infantry CSRs*, M270, Roll 99; Manarin, *North Carolina Troops*, 3:154; *CSN & CSMC Personnel*, M260, Roll 7; *ORN*, 2, 1:309; *Vessel Papers*; *File NA-2*, M1091, Roll 18; *Richmond & James River Payrolls*.
414. White, *First Iron-Clad Naval Engagement*, 1–5; *CSN Register*, 209; *ORN*, 2, 1:308; *Richmond & James River Payrolls*; U.S. War Department, "CSRs of Confederate Soldiers Who Served in Organizations from the State of Georgia, 19th Btn. Infantry," Microfilm Series, M266, Roll 325, RG 109, NA; Stewart, *History of Norfolk County*, 979–81; Burial marker.
415. Donnelly, *Confederate Enlisted Marines*, 111–12; *ORN*, 2, 1:311; *Richmond & James River Payrolls*.
416. *ORN*, 2, 1:310; *North Carolina Payrolls*; *Richmond & James River Payrolls*; *1860 Norfolk Co., Virginia Census*, M653, Roll 1366; *1870 Norfolk Co., Virginia Census*, M593, Roll 1667.
417. *10th Louisiana Infantry CSRs*, M320, Roll 224; Booth, *Louisiana Confederate Soldiers*, 3, bk. 2:1094; *ORN*, 2, 1:309; *North Carolina Payrolls*; *Richmond & James River Payrolls*; *Hospital & Prison Records*, M260, Roll 4.
418. *CSN Register*, 212; White, *First Iron-Clad Naval Engagement*, 16; *1870 City of Norfolk, Virginia Census*, M593, Roll 1666.

419. *9th Virginia Infantry CSRs*, M324, Roll 487; *CSN & CSMC Personnel*, M260, Roll 7; *ORN*, 2, 1:309; *File NA-2*, M1091, Roll 18; *North Carolina Payrolls*; *Richmond & James River Payrolls*; *Hospital & Prison Records*, M260, Roll 4; Stewart, *History of Norfolk County*, 150; Trask, *9th Virginia Infantry*, 100, 114.
420. *15th Louisiana Infantry CSRs*, M320, Roll 272; Booth, *Louisiana Confederate Soldiers*, 3, bk. 2:1101; *ORN*, 2, 1:310; *Vessel Papers*; *Richmond & James River Payrolls*; *Shipping Articles*; *ORN*, 2, 1:290, 308, 291.
421. *8th Alabama Infantry CSRs*, M311, Roll 177; *ORN*, 2, 1:310; *Vessel Papers*; *Richmond & James River Payrolls*; *Shipping Articles*; *ORN*, 2, 1:292, 283; *Wilmington Payrolls*; *Hospital & Prison Records*, M260, Roll 4.
422. *9th Virginia Infantry CSRs*, M324, Roll 487; Trask, *9th Virginia Infantry*, 100; *CSN & CSMC Personnel*, M260, Roll 7; *ORN*, 2, 1:309; *North Carolina Payrolls*; *Richmond & James River Payrolls*; *Griffin's Company CSRs*; Nicholas and Servis, *Powhatan, Salem and Courtney*, 225; U.S. War Department, "CSRs of Confederate Soldiers Who Served in Organizations from the State of Virginia, 2nd Cavalry," Microfilm Series, M324, Roll 24, RG 109, NA; Driver, *2nd Virginia Cavalry*, 286.
423. *1st North Carolina Artillery CSRs*, M270, Roll 60; Manarin, *North Carolina Troops*, 1:170; *ORN*, 2, 1:283, 293, 310; *Vessel Papers*; *Richmond & James River Payrolls*; *ORN*, 2, 1:293, 283; *Charleston Payrolls*; *Shipping Articles*.
424. *3rd North Carolina Infantry CSRs*, M270, Roll 133; Manarin, *North Carolina Troops*, 1:170; *ORN*, 2, 1:293; 309; *Vessel Papers*; *File NA-1*, M1091, Roll 17; *North Carolina Payrolls*; *Richmond & James River Payrolls*; *ORN*, 2, 1:293, 283.
425. *10th Louisiana Infantry CSRs*, M320, Roll 224; Booth, *Louisiana Confederate Soldiers*, 3, bk. 2:1144; *ORN*, 2, 1:310; *Richmond & James River Payrolls*.
426. Faust, "Wood, John Taylor," in *Historical Times Illustrated Encyclopedia*, 840–41; Hamersly, *Complete General Navy Register*, 789; Dudley, *Going South*, 39; *ORN*, 2, 1:308; Jack, *Memoirs*, 12; *CSN Register*, 216; *Richmond & James River Payrolls*; Shingleton, *John Taylor Wood*, 2, 4, 5, 69, 161, 197, 203, 209, 224; Burial marker.
427. *13th North Carolina Infantry CSRs*, M270, Roll 220; Manarin, *North Carolina Troops*, 5:297; *CSN & CSMC Personnel*, M260, Roll 7; *ORN*, 2, 1:310; *North Carolina Payrolls*; *Richmond & James River Payrolls*; U.S. War Department, "CSRs of Confederate Soldiers Who Served in Organizations from the State of North Carolina, 8th Btn. Partisan Rangers," Microfilm Series, M270, Roll 45, RG 109, NA; Manarin, *North Carolina Troops*, 2:593; U.S. War Department, "CSRs of Confederate Soldiers Who Served in Organizations from the State of North Carolina, 66th Infantry," Microfilm Series, M270, Roll 563, RG 109, NA.

428. *1860 Henrico Co., Virginia Census*, M653, Roll 1353; White, *First Iron-Clad Naval Engagement*, 16; *CSN Register*, 217.

429. *13th North Carolina Infantry CSRs*, M270, Roll 220; Manarin, *North Carolina Troops*, 5:326; *ORN*, 2, 1:310; *North Carolina Payrolls*; *Richmond & James River Payrolls*; U.S. War Department, "CSRs of Confederate Soldiers Who Served in Organizations from the State of North Carolina, 4th Cavalry," Microfilm Series, M270, Roll 31, RG 109, NA.

430. *6th Virginia Infantry CSRs*, M324, Roll 450; Cavanaugh, *6th Virginia Infantry*, 139; *File NA-2*, M1091, Roll 18; *ORN*, 2, 1:309; *North Carolina Payrolls*; *Richmond & James River Payrolls*; *ORN*, 1, 11:794; *Charleston Payrolls*; *Hospital & Prison Records*, M260, Roll 4; *Confederate Veteran*, February 1917, 87.

431. U.S. War Department, "CSRs of Confederate Soldiers Who Served in Organizations from the State of North Carolina, 12th Infantry," Microfilm Series, M270, Roll 210, RG 109, NA; Manarin, *North Carolina Troops*, 5:173; *Wilmington Station*; *ORN*, 2, 1:309; *File NA-2*, M1091, Roll 18; *North Carolina Payrolls*; *Richmond & James River Payrolls ORN*, 1, 17:864; *Wilmington Payrolls*.

432. *Confederate Veteran*, July 1908, 348.

433. Cline, "Ironclad Ram *Virginia*," 243–49.

434. *Confederate Veteran*, November 1916, 512; U.S. War Department, "CSRs of Confederate Soldiers Who Served in Organizations from the State of Louisiana, Green's Artillery Company," Microfilm Series, M320, Roll 51, RG 109, NA; U.S. War Department, "CSRs of Confederate Soldiers Who Served in Organizations from the State of Louisiana, 2nd Cavalry," Microfilm Series, M320, Roll 9, RG 109, NA.

435. *Shipping Articles*.

436. *41st Virginia Infantry CSRs*, M324, Roll 864; Henderson, *41st Virginia Infantry*, 115; Weaver, *Assorted Heavy Artillery*, 78.

437. *Confederate Veteran*, October 1916, 463.

438. Ibid., September 1913, 431.

439. *4th Georgia Infantry CSRs*, M266, Roll 186; *Shipping Articles*.

440. *Kevill's Company CSRs*; Weaver, *Assorted Heavy Artillery*, 81.

441. *1st South Carolina Infantry CSRs*, M267, Roll 146; *1st (McCreary's) South Carolina Infantry CSRs*, M267, Roll 133; Salley, *South Carolina Troops*, 366; Holland, *24th Virginia Cavalry*, 132; *CSN & CSMC Personnel*, M260, Roll 7.

442. *CSN & CSMC Personnel*, M260, Roll 7.

443. *1st South Carolina Infantry CSRs* , M267, Roll 146; Salley, *South Carolina Troops*, 367; *CSN & CSMC Personnel*, M260, Roll 7.

Bibliography

Archives and Manuscripts

Georgia Department of Archives and History. Atlanta, Georgia.
James Keenan Letter, March 10, 1862.

Library of Congress, Washington, DC.
Stephen R. Mallory Diary.

Library of Virginia. Richmond, Virginia.
Application for Aid to Citizens of Virginia Wounded and Maimed during the Late War, While Serving as Soldiers or Marines, Commonwealth of Virginia Act of 1884. State Government Records Collection.
Pension Applications, Confederate Veterans and Widows, Commonwealth of Virginia Acts of 1888, 1900 and 1902. State Government Records Collection.
R. Bolling Batte Papers, 1918–1992. Personal Papers Collection.
Robert E. Lee Camp Confederate Soldiers' Home (Richmond, Virginia), Applications for Admission, 1884–1941. State Government Records Collection.
S. Bassett French Biographical Sketches, 1820–1898. Personal Papers Collection.

Massachusetts Historical Society. Boston, Massachusetts.
Everett Collection.

National Archives. Washington, D.C.
Bureau of the Census. Bureau of the Census Records. Record Group 29.

Papers Pertaining to Vessels of or Involved with the Confederate States of America, File V-5, CSS *Virginia*. Microfilm Series, M909, Roll 30.
Subject File of the Confederate States Navy, 1861-1865. Microfilm Series, M1091.
File HA, Miscellaneous, Box 160, "Narrative of H. Ashton Ramsay, Chief Engineer, Confederate States Steamer Merrimack, during her engagements in Hampton Roads, 1862." (M1091, Roll 13).
File NA, CSS *Alabama*–CSS *Neuse* (M1091, Roll 17).
File NA, CSS *New Orleans*–CSS *Yorktown* (M1091, Roll 18).

Population Schedules of the Seventh Census of the United States of America, 1850. Microfilm Series, M432.
Baton Rouge Parish, Louisiana (M432, Roll 229).
Carteret County, North Carolina (M432, Roll 623).
Duval County, Florida (M432, Roll 58).
Muscogee and Newton Counties, Georgia (M432, Roll 79).
Norfolk County, Virginia (M432, Roll 964).
Rutherford County, Tennessee (M432, Roll 894).
Sumter County, Alabama (M432, Roll 15).

Population Schedules of the Eight Census of the United States of America, 1860. Microfilm Series, M653.
Accomac County, Virginia (M653, Roll 1330).
Beaufort County, North Carolina (M653, Roll 887).
Bibb County, Georgia (M653, Roll 111).
Bladen County, North Carolina (M653, Roll 888).
Carteret County, North Carolina (M653, Roll 890).
Chatham County, Georgia (M653, Roll 115).
Chesterfield County, Virginia (M653, Roll 1340).
City of Baltimore, Maryland (M653, Rolls 459, 463, 466).
City of New Orleans, Louisiana (M653, Rolls 415-417, 419).
Currituck County, North Carolina (M653, Roll 895).
Dinwiddie County, Virginia (M653, Roll 1342).
Elizabeth City County, Virginia (M653, Roll 1343).
Gloucester County, Virginia (M653, Roll 1347).
Henrico County, Virginia (M653, Roll 1353).

Macon County, Alabama (M653, Roll 14).
Mobile County, Alabama (M653, Roll 17).
Montgomery County, Alabama (M653, Roll 19).
Muhlenberg County, Kentucky (M653, Roll 388).
Nansemond County, Virginia (M653, Roll 1365).
Newton County, Georgia (M653, Roll 133).
Norfolk County, Virginia (M653, Roll 1366).
Pasquotank County, North Carolina (M653, Roll 909).
Perquimans County, North Carolina (M653, Roll 909).
Pike County, Georgia (M653, Roll 134).
Prince William County, Virginia (M653, Roll 1373).
Roanoke County, Virginia (M653, Roll 1375).
Wake County, North Carolina (M653, Roll 916).
Warwick County, Virginia (M653, Roll 1382).
Washington County, North Carolina (M653, Roll 917).
York County, Virginia (M653, Roll 1385).

Population Schedules of the Ninth Census of the United States of America, 1870. Microfilm Series, M593.
Anson and Caswell Counties, North Carolina (M593, Roll 1122).
Carteret County, North Carolina (M593, Roll 1127).
Dinwiddie County, Virginia (M593, Roll 1643).
Franklin County, Kentucky (M593, Roll 462).
Norfolk County, Virginia (M593, Rolls 1666, 1667).
Richmond County, Virginia (M593, Roll 1654).
York County, Virginia (M593, Roll 1682).

Southern Historical Collection. Louis Round Wilson Library. University of North Carolina. Chapel Hill, North Carolina.
Franklin Buchanan Letterbook, 1861–1863.
French Forrest Letterbook.
John Taylor Wood Papers.

U.S. Department of the Navy. Office of Naval Records and Library. Naval Records Collection of the Office of Naval Records and Library. Record Group 45.
Charleston Station and Fleet, C.S. Navy (T829, Roll 166).
1861–1862, Officers & Men, North Carolina Station, C.S. Navy (T829, Roll 170).

1861–1865, Shipping Articles of C.S. Navy (T829, Roll 173).
Miscellaneous Records of the Office of Naval Records and Library. Microfilm Series, T829.
Mobile and Jackson Station, C.S. Navy (T829, Roll 167).
New Orleans Station and Fleet, C.S. Navy (T829, Roll 169).
Richmond Station & James River Squadron, C.S. Navy (T829, Roll 172).
Savannah Station and Fleet, C.S. Navy (T829, Roll 171).
Wilmington Station, Marine Corps, and Miscellaneous (T829, Roll 165).

U.S. War Department. War Department Collection of Confederate Records. Record Group 109.
Compiled Service Records of Confederate Soldiers Who Served in Organizations from the State of Alabama. Microfilm Series, M311.
3rd Alabama Infantry (M311, Rolls 103-104, 106, 109, 112).
8th Alabama Infantry (M311, Rolls 171, 174-177, 181).
11th Alabama Infantry (M311, Roll 207).
12th Alabama Infantry (M311, Rolls 209-211, 213, 214).
35th Alabama Infantry (M311, Roll 362).
Compiled Service Records of Confederate Soldiers Who Served in Organizations from the State of Arkansas. Microfilm Series, M317.
1st (Colquitt's) Arkansas Infantry (M317, Rolls 46-47, 51).
Compiled Service Records of Confederate Soldiers Who Served in Organizations Raised Directly by the Confederate Government. Microfilm Series, M258.
1st Confederate Engineer Troops (M258, Rolls 93, 96).
3rd Confederate Engineer Troops (M258, Roll 101).
Nitre and Mining Corps, CSA (M258, Roll 113).
Signal Corps, CSA (M258, Rolls 119, 121).
Compiled Service Records of Confederate Soldiers Who Served in Organizations from the State of Georgia. Microfilm Series, M266.
3rd Georgia Infantry (M266, Rolls 167, 169, 171, 172).
4th Georgia Infantry (M266, Rolls 180, 184, 186, 188).
5th Georgia Infantry (M266, Roll 197).
13th Georgia Cavalry (M266, Roll 55).
19th Battalion Georgia Infantry (State Guards) (M266, Roll 325).
22nd Georgia Infantry (M266, Roll 345).
Cobb's Legion, Georgia Cavalry (M266, Roll 589).
Cobb's Legion, Georgia Infantry (M266, Rolls 580, 581).

Compiled Service Records of Confederate Soldiers Who Served in Organizations from the State of Kentucky. Microfilm Series, M319.

2nd (Duke's) Kentucky Cavalry (M319, Roll 15).

Compiled Service Records of Confederate Soldiers Who Served in Organizations from the State of Louisiana. Microfilm Series, M320.

1st Louisiana Heavy Artillery (M320, Roll 35).

1st (Nelligan's) Louisiana Infantry (M320, Rolls 72, 73, 75-78).

2nd Louisiana Cavalry (M320, Roll 9).

2nd Louisiana Infantry (M320, Rolls 102, 106, 108, 110-111).

5th Louisiana Infantry (M320, Roll 152).

8th Louisiana Infantry (M320, Roll 187).

10th Louisiana Infantry (M320, Rolls 219, 221-222, 224).

14th Louisiana Infantry (M320, Roll 257).

15th Louisiana Infantry (M320, Rolls 267-272).

C.S. Zouave Battalion Louisiana Infantry (M320, Roll 405).

Louisiana Artillery, Green's Company (M320, Roll 51).

Compiled Service Records of Confederate Soldiers Who Served in Organizations from the State of Maryland. Microfilm Series, M321.

1st Maryland Artillery (M321, Roll 9).

1st Maryland Cavalry (M321, Roll 4).

2nd Battalion Maryland Infantry (M321, Roll 20).

Compiled Service Records of Confederate Soldiers Who Served in Organizations from the State of Mississippi. Microfilm Series, M269.

11th Mississippi Infantry (M269, Roll 195).

12th Mississippi Infantry (M269, Roll 204).

Compiled Service Records of Confederate Soldiers Who Served in Organizations from the State of North Carolina. Microfilm Series, M270.

1st North Carolina Artillery (M270, Rolls 52, 53, 58, 60).

1st North Carolina Infantry (M270, Rolls 90-92, 94-99).

2nd North Carolina Infantry (M270, Roll 114).

3rd North Carolina Cavalry (M270, Roll 20).

3rd North Carolina Infantry (M270, Rolls 130-131, 133).

4th North Carolina Cavalry (M270, Roll 31).

6th North Carolina Cavalry (M270, Roll 40).

7th Battalion North Carolina Cavalry (M270, Roll 42).

8th Battalion North Carolina Partisan Rangers (M270, Roll 45).

8th North Carolina Infantry (M270, Roll 187).

10th Battalion North Carolina Heavy Artillery (M270, Roll 83).

12th North Carolina Infantry (M270, Roll 210).
13th North Carolina Infantry (M270, Rolls 212, 214, 216-218, 220).
14th North Carolina Infantry (M270, Rolls 223-226, 228-229).
17th North Carolina Infantry (Microfilm Series, M270, Roll 250).
26th North Carolina Infantry (M270, Roll 326).
57th North Carolina Infantry (M270, Roll 530).
66th North Carolina Infantry (M270, Roll 563).

Compiled Service Records of Confederate Soldiers Who Served in Organizations from the State of South Carolina. Microfilm Series, M267.
1st (Butler's) South Carolina Infantry (M267, Roll 111).
1st (Charleston) Battalion South Carolina Infantry (M267, Roll 151).
1st (McCreary's) South Carolina Infantry (M267, Rolls 126-129, 132-134).
1st South Carolina Artillery (M267, Roll 58).
1st South Carolina Infantry (M267, Rolls 145-146).
27th South Carolina Infantry (M267, Roll 357).

Compiled Service Records of Confederate Soldiers Who Served in Organizations from the State of Texas. Microfilm Series, M323.
4th Texas Infantry (M323, Roll 285).

Compiled Service Records of Confederate Soldiers Who Served in Organizations from the State of Virginia. Microfilm Series, M324.
1st Battalion Virginia Infantry (M324, Roll 369).
1st Battalion Virginia Infantry, Local Defense (M324, Roll 371).
2nd Virginia Cavalry (M324, Roll 24).
2nd Virginia Infantry (M324, Roll 373).
3rd Virginia Infantry (M324, Roll 395).
4th Battalion Virginia Infantry, Local Defense (M324, Roll 416).
5th Battalion Virginia Infantry (M324, Rolls 431-432).
5th Virginia Cavalry (M324, Roll 52).
6th Virginia Infantry (M324, Rolls 438, 440, 444-445, 447, 449-450).
9th Virginia Infantry (M324, Rolls 475-476, 478-482, 484-487).
10th Battalion Virginia Heavy Artillery (M324, Roll 230).
10th Virginia Cavalry (M324, Roll 107).
12th Virginia Infantry (M324, Rolls 518, 520-523, 529-531).
14th Virginia Infantry (M324, Rolls 550).
15th Virginia Infantry (M324, Rolls 568).
16th Virginia Infantry (M324, Roll 579).
17th Virginia Infantry (M324, Roll 586).
18th Virginia Infantry (M324, Roll 598).
18th Battalion Virginia Heavy Artillery (M324, Roll 242).

19th Battalion Virginia Heavy Artillery (M324, Rolls 243, 246).
20th Virginia Cavalry (M324, Roll 163).
24th Virginia Partisan Rangers (M324, Roll 180).
32nd Virginia Infantry (M324, Rolls 780-782, 785-786).
34th Virginia Infantry (M324, Roll 817).
38th Battalion Virginia Light Artillery (M324, Roll 257).
39th Battalion Virginia Cavalry (M324, Roll 199).
41st Battalion Virginia Cavalry (M324, Roll 203).
41st Virginia Infantry (M324, Rolls 860-868).
53rd Virginia Infantry (M324, Roll 949).
57th Virginia Infantry (M324, Roll 980).
61st Virginia Infantry (M324, Roll 1030).
115th Virginia Militia (M324, Roll 1055).
Virginia Heavy Artillery, Epes' Company (M324, Roll 293).
Virginia Heavy Artillery, Jones' Company (M324, Roll 314).
Virginia Heavy Artillery, Kevill's Company (M324, Roll 315).
Virginia Horse Artillery, Graham's Company (M324, Roll 304).
Virginia Light Artillery, Griffin's Company (M324, Roll 307).
Virginia Light Artillery, Hardwicke's Company (M324, Roll 309).
Virginia Light Artillery, Nelson's & Page's Companies (M324, Roll 324).
Virginia Light Artillery, Pegram's Company (M324, Roll 328).
Records Relating to Confederate Naval and Marine Personnel. Microfilm Series, M260.
Hospital and Prison Records of Naval Personnel (M260, Rolls 1-4).
Reference Cards and Papers Relating to Naval and Marine Personnel (M260, Rolls 5-7).

Virginia Historical Society. Richmond, Virginia.
Catesby ap Roger Jones, "Report of Ordnance Experiments at Jamestown, 12 October 1861."
Catesby ap Roger Jones Papers.
Robert Dabney Papers, Minor Family Papers Collection.

Virginia War Museum. Newport News, Virginia.
Minnish, J.W. "Reminiscences Relating to the Siege of Yorktown."

REFERENCE WORKS

Beers, Henry Putney. *The Confederacy: A Guide to the Archives of the Confederate States of America.* Washington, D.C.: National Archives and Records Administration, 1998. First published as *Guide to the Archives of the Government of the Confederate States of America.* Washington, D.C.: National Archives and Records Administration, 1968.

Hamersly, Thomas H.S., ed. *Complete General Navy Register of the United States of America from 1776 to 1887.* New York: T.H.S. Hamersly, 1888.

Hewett, Janet B., ed. *The Roster of Confederate Soldiers, 1861–1865.* 16 vols. Wilmington, NC: Broadfoot Publishing Company, 1996.

"Proceedings of the Confederate Congress." *Southern Historical Society Papers 44–52* (1923–1959).

Roberts, Robert B. *Encyclopedia of Historic Forts: The Military, Pioneer, and Trading Posts of the United States.* New York: Macmillan Publishing Co., 1988.

Sifakis, Stewart. *Compendium of the Confederate Armies: Florida and Arkansas.* New York: Facts On File, 1992.

———. *Compendium of the Confederate Armies: Kentucky, Maryland, Missouri, the Confederate Units and the Indian Units.* New York: Facts On File, 1995.

———. *Compendium of the Confederate Armies: Louisiana.* New York: Facts On File, 1995.

———. *Compendium of the Confederate Armies: Texas.* New York: Facts On File, 1995.

U.S. Department of the Navy. *Dictionary of American Naval Fighting Ships.* 8 vols. Washington, D.C.: U.S. Government Printing Office, 1959.

———. *Official Records of the Union and Confederate Navies in the War of Rebellion.* 30 vols. Washington, D.C.: Government Printing Office, 1894–1922.

———. *Register of Officers of the Confederate States Navy, 1861–1865, as Complied and Revised by the Office of Naval Records and Library, United States Navy Department 1931, from All Available Data.* With a new introduction by John M. Carroll. Washington, D.C.: Government Printing Office, 1931. Reprint, Mattituck, NY: J.M. Carroll & Co., 1983.

U.S. War Department. *The War of Rebellion: A Compilation of the Official Records of the Union and Confederate Armies.* 128 vols. Washington, D.C.: Government Printing Office, 1880–1902.

Wallace, Lee A., Jr. *A Guide to Virginia Military Organizations, 1861–1865.* Rev. 2nd ed. Lynchburg, VA: H.E. Howard, Inc., 1986.

PRIMARY SOURCES

Brooke, John Mercer. "The Plan and Construction of the *Merrimac.*" In *Battles and Leaders of the Civil War.* Vol. 1. Edited by Robert Underwood Johnson and Clarence Clough Buel. New York: Century Co., 1887.

———. "The *Virginia* or *Merrimac*: Her Real Projector." *Southern Historical Society Papers 14* (January 1891): 3–34.

Cannon, LeGrand Bouton. *Personal Reminiscences of the Rebellion, 1801–1865.* New York, 1895.

Cline, William R. "The Ironclad Ram *Virginia.*" *Southern Historical Society Papers 32* (December 1904): 243–49.

Colston, Raleigh E. "Watching the *Merrimac.*" In *Battles and Leaders of the Civil War*. Vol. 1. Edited by Robert Underwood Johnson and Clarence Clough Buel. New York: Century Co., 1887.

Curtis, Richard. *History of the Famous Battle Between the Iron-clad* Merrimac, *C.S.N and the Iron-clad* Monitor *and the* Cumberland *and the* Congress *of the U.S. Navy, March the 8th and 9th, 1862, as Seen by a Man at the Gun.* N.p., 18--? Reprint, Hampton, VA: Houston Print and Publishing Co., 1957.

Drake, William F. "CSS *Virginia.*" In *Voices of the Civil War: The Peninsula.* Edited by Paul Mathless. Alexandria, VA: Time-Life Books, 1997.

Early, Jubal Anderson. *Narrative of the War Between the States.* New York: Da Capo Press, 1991.

Eggleston, John R. "Captain Eggleston's Narrative of the Battle of the *Merrimac.*" *Southern Historical Society Papers 40* (1916): 166–78.

Ericsson, John. "The Building of the *Monitor.*" In *Battles and Leaders of the Civil War.* Vol. 1. Edited by Robert Underwood Johnson and Clarence Clough Buel. New York: Century Co., 1887.

———. "The Monitors." *Century Illustrated Monthly Magazine* (December 1885): 280–99.

Forrest, Douglas French. *Odyssey in Gray: A Diary of Confederate Service, 1863–1865.* Edited by William N. Still Jr. Richmond: Virginia State Library, 1979.

———. "An Odyssey in Gray: Selections from a Diary of Confederate Naval Life with the CSS *Rappahannock.*" Edited by William N. Still Jr. *Virginia Cavalcade* 29, no. 3 (Winter 1980): 124–29.

Foute, Robert Chester. "Echoes from Hampton Roads." *Southern Historical Society Papers* 19 (January 1891): 246–51.

Fox, Gustavus Vasa. *Confidential Correspondence of Gustavus Vasa Fox, Assistant Secretary of the Navy, 1861–1865.* Edited by R.M. Thompson and R. Wainwright. 2 vols. New York: Naval History Society, 1918–1919.

Gautier, Ange Simon. "Combat Naval de Hampton-Roads (Etats-Unis), 8 et 9 Mars 1862." *Revue Maritime et Coloniale* (April 1862): 807.

Goldsborough, Louis M. "Narrative of Rear Admiral Goldsborough, U.S. Navy." *U.S. Naval Institute Proceedings* 54 (July 1933): 1022–34.

Greene, Samuel Dana. "I Fired the First Gun and Thus Commenced the Great Battle." *American Heritage* (June 1957): 103.

———. "In the *Monitor*'s Turret." In *Battles and Leaders of the Civil War*. Vol. 1. Edited by Robert Underwood Johnson and Clarence Clough Buel. New York: Century Co., 1887.

———. "Manuscript." *U.S. Naval Academy Trident* (Spring 1942): 42–44.

Higgins, John F. "Brilliant Career of the *Merrimac*." *Confederate Veteran* (August 1900): 356–57.

Jack, Eugenius A. *Memoirs of E.A. Jack; Steam Engineer, CSS* Virginia. Edited by Alan B. Flanders and Neale O. Westfall. White Stone, VA: Brandyland Publishers, 1998.

Johnston, Joseph E. *Narrative of Military Operations During the Civil War*. New York: D. Appleton and Co., 1990.

Jones, Catesby ap Roger. "Services of the *Virginia*." *Southern Historical Society Papers* 11 (January 1883): 65–75.

Kell, John M. *Recollections of a Naval Life*. Washington, D.C.: Neale Publishing Company, 1900.

Lewis, Samuel. "Life on the *Monitor*: A Seaman's Story of the Fight with the *Merrimac*; Lively Experiences Inside the Famous 'Cheesebox on a Raft.'" In *Campfire Sketches and Battlefield Echoes of '61–'65*. Edited by William C. King and William P. Derby. Springfield, MA, 1883.

Littlepage, Hardin Beverly. "Career of the *Merrimack*." *Confederate Veteran* (March 1894): 86.

———. "A Midshipman aboard the *Virginia*." With an introduction by Jon Nielson. *Civil War Times Illustrated* (April 1974): 4–6, 8, 10–11, 42–47.

———. "A Midshipman Abroad." *Civil War Times Illustrated* (June 1974): 19–26.

———. "*Merrimac* vs. *Monitor*: A Midshipman's Account of the Battle with the 'Cheese Box." In *Campfire Sketches and Battlefield Echoes of '61–'65*. Edited by William C. King and William P. Derby. Springfield, MA, 1883.

———. "The Career of the *Merrimac-Virginia*: With Some Personal History." In *Voices of the Civil War: The Peninsula*. Edited by Paul Mathless. Alexandria, VA: Time-Life Books, 1997.

———. "With the Crew of the *Virginia*." *Civil War Times Illustrated* (May 1974): 36–43.

McClellan, George Brinton. *McClellan's Own Story*. New York: Charles L. Webster Company, 1887.

McDonald, Joseph. "How I Saw the *Monitor-Merrimac* Fight." *New England Magazine* (July 1907): 548.

Norris, William. *The Story of the Confederate States' Ship* "Virginia" *(Once* Merrimac*): Her Victory Over the* Monitor*; Born March 7th, Died May 10th, 1862*. Baltimore, MD: John B. Piet, 1879. Reprint, *Southern Historical Society Papers* 41 (September 1916): 204–33.

Parker, William Harwar. *Recollections of a Naval Officer, 1841–1865*. New York: Charles Scribner's Sons, 1883.

———. "Letter from Captain Parker." *Southern Historical Society Papers* 11 (1883): 34–39.

Peters, William H. *Recollections of Facts and Circumstances Connected with the Destruction of the Navy Yard at Portsmouth, Virginia, April 1861*. Portsmouth, VA: self-published, 1873.

Phillips, Dinwiddie Brazier. "The Career of the Iron-Clad *Virginia*, (formerly the *Merrimac*), Confederate States Navy, March–May, 1862." Collection of the Virginia Historical Society. Richmond: Virginia Historical Society, 1887.

———. "The Career of the *Merrimac*." *Southern Bivouac* (March 1887): 598–608.

Porter, David Dixon. *Incidents and Anecdotes of the Civil War*. New York: D. Appleton & Company, 1885.

Ramsay, Henry Ashton. "The Most Famous of Sea Duels, The Story of the *Merrimac*'s Engagement with the *Monitor*, and the Events that Preceded and Followed the Fight, Told by a Survivor." *Harper's Weekly*, February 10, 1912, 11–12.

———. "Wonderful Career of the *Merrimac*." *Confederate Veteran* (July 1907): 310–13.

Reblen, William "U.S.S. *Cumberland*." In *Voices of the Civil War: The Peninsula*. Edited by Paul Mathless. Alexandria, VA: Time-Life Books, 1997.

Selfridge, Thomas O., Jr. *Memoirs of Thomas O. Selfridge, Jr.: Rear Admiral, U.S.N.* New York: G.P. Putnam's Sons, 1924.

———. "The *Merrimac* and the *Cumberland*." *Cosmopolitan* (June 1893): 180.

Semmes, Raphael. *Memoirs of Service Afloat during the War Between the States*. Baltimore, MD: Kelly, Piet & Company, 1869. Reprint, New York: J.P. Kennedy & Sons, 1924.

Sinclair, Arthur. "How the *Merrimac* Fought the *Monitor*." *Heart's Magazine* (December 1913): 884–94.

———. *Two Years on the Alabama*. Boston: Lee and Shepard Publishers, 1895.

Stimers, Alban C. "An Engineer aboard the *Monitor*." *Civil War Times Illustrated* (April 1970): 35.

Stuyvesant, Moses S. "How the *Cumberland* Went Down." *War Papers and Reminiscences, 1861–1865, Read Before the Missouri Commandery, Military Order of the Loyal Legion of the United States* (MOLLUS). St. Louis: Missouri Commandery MOLLUS, 1892.

Welles, Gideon. *Diary of Gideon Welles, Secretary of the Navy under Lincoln and Johnson*. Edited by Howard K. Beale. 3 vols. New York: W.W. Norton Company, 1960.

———. "The First Iron-Clad *Monitor*." *Annals of War*. Philadelphia: J.B. Lippincott, 1879.

White, Elsberry Valentine. *The First Iron-Clad Naval Engagement in the World*. New York: J.S. Ogilvie Publishing Co., 1906.

Wood, John Taylor. "The First Fight of the Ironclads; March 9, 1862." In *Battles and Leaders of the Civil War*. Vol. 1. Edited by Robert Underwood Johnson and Clarence Clough Buel. New York: Century Co., 1887.

SECONDARY SOURCES

Alexander, William T., Captain, USNR (Ret.), and Colonel Joseph H. Alexander, USMC (Ret.). "From Ironclads to Infantry" *Naval History* 5, no. 2 (Summer 1991): 9–15.

Alger, F.S. "*Congress* and the *Merrimac*." *New England Magazine* (February 1899): 687–93.

Anderson, Bern. *By Sea and by River: The Naval History of the Civil War*. New York: Alfred A. Knopf Co., 1962.

Anderson, Mrs. John H. "Survivor of the First Battle Between Ironclads." *Confederate Veteran* (May 1928): 189.

Armstrong, Richard L. *19th and 20th Virginia Cavalry*. Lynchburg, VA: H.E. Howard, Inc., 1994.

Ballard, G.A. "British Battleships of the 1870s: The *Warrior* and *Black Prince*." *Mariner's Mirror* (April 1930): 168–86.

Barnard, J.G. *Notes on Sea Coast Defense*. New York: D. Van Nostrand, 1861.

Barthell, Edward E., Jr. *The Mystery of the Merrimack*. Muskegon, MI: Dana Printers Company, 1959.

Baxter, James Phinney, III. *The Introduction of the Ironclad Warship*. Cambridge, MA: Harvard University Press, 1933.

Bearss, Edwin C. *River of Lost Opportunities: The Civil War on the James River, 1861–1862*. Lynchburg, VA: H.E. Howard Inc., 1995.

Beese, Sumner B. *CA. S. Ironclad* Virginia *and U.S. Ironclad* Monitor. Newport News, VA: Mariner's Museum, 1996.

Berent, Irwin Mark. "Georgians Crewed the Virginia." *Georgia Genealogical Society Quarterly* (Fall 1979): 123–26.

Bergeron, Arthur W., Jr. "Buchanan, Franklin." In *Dictionary of American Military Biography*. Vol. 1.Westport, CT: Greenwood Press, 1984.

Booth, Andrew B., comp. *Records of Louisiana Confederate Soldiers and Louisiana Confederate Commands*. 4 vols. New Orleans: Commissioner, Louisiana Military Records, 1920.

Broadwater, John D. "Ironclads at Hampton Roads: CSS *Virginia*, the Confederacy's Formidable Warship." *Virginia Cavalcade* (Winter 1984): 100–13.

Brooke, George M., Jr. "*John M. Brooke.*" 2 vols. PhD. diss., University of North Carolina, 1955.

———. *John M. Brooke, Naval Scientist and Educator*. Charlottesville: University Press of Virginia, 1980.

Campbell, R. Thomas. *Fire and Thunder: Exploits of the Confederate States Navy*. Shippensburg, PA: Burd Street Press, 1997.

Catton, Bruce. "When the *Monitor* Met the *Merrimac*." *New York Times Magazine*, March 4, 1962.

Cavanaugh, Michael A. *6th Virginia Infantry*. Lynchburg, VA: H.E. Howard, Inc., 1988.

Chernault, Tracy, and Jeffrey C. Weaver. *18th and 20th Battalions of Virginia Heavy Artillery*. Lynchburg, VA: H.E. Howard, Inc., 1995.

Connor, John W. "The North's Fiasco at Norfolk Navy Yard." *Military Images* (November–December 1998): 30–31.

Coski, John M. *Capital Navy: The Men, Ships, and Operations of the James River Squadron*. Campbell, CA: Savas Woodbury Publishers, 1996.

Crew, R. Thomas, Jr., and Benjamin H. Trask. *Grimes' Battery, Grandy's Battery and Huger's Battery Virginia Artillery*. Lynchburg, VA: H.E. Howard, Inc., 1995.

Crews, Edward R., and Timothy A. Parrish. *14th Virginia Infantry*. Lynchburg, VA: H.E. Howard, Inc., 1995.

Cunningham, Sumner A., ed. "Deaths in Stonewall Camp, Portsmouth, Virginia." *Confederate Veteran* (May 1912): 238.

———. "Hunter Davidson." *Confederate Veteran* (June 1913): 307.

Dahlgren, Madeline V. *Memoir of John A. Dahlgren*. Boston: J.R. Osgood & Company, 1882.

Daly, Robert W., ed. *Aboard the U.S.S.* Monitor*: 1862: The Letters of Acting Paymaster William Frederick Keeler, U.S. Navy to his Wife, Anna*. Annapolis, MD: United States Naval Institute Press, 1964.

———. *How the* Merrimac *Won: The Strategic Story of the CSS* Virginia. New York: Crowell Inc, 1957.

Davis, William C. *Duel between the First Ironclads.* Garden City, NJ: Doubleday & Company, 1975.

Delafield, Richard, Alfred Mordecai and George McClellan. *Report Published by Secretary of War of Military Commission to Europe, 1854–1856.* 3 vols. Washington, D.C.: Government Printing Office, 1857–1860.

Delaney, Norman C. "Brooke, John Mercer"; "Buchanan, Franklin"; "Butt, Walter R."; "Davidson, Hunter"; "Jones, Catesby ap Roger"; and "Porter, John Luke." In *Historical Times Illustrated Encyclopedia of the Civil War*. New York: Harper & Row, Publishers, 1986.

Demaree, Albert L. "Our Navy's Worst Headache, The *Merrimack*." *United States Naval Institute Proceedings* 78 (March 1962): 66–83.

Dew, Charles B. *Ironmaker to the Confederacy: Joseph R. Anderson and the Tredegar Iron Works*. New Haven, CT: Yale University Press, 1966.

Donnelly, Ralph W. *The Confederate States Marine Corps: The Rebel Leathernecks*. Shippensburg, PA: White Mane Publishing Co. 1989.

———. *The History of the Confederate States Marine Corps*. Washington, NC: self-published, 1976.

———. *Service Record of Confederate Enlisted Marines*. Washington, NC: self-published, 1979.

Driver, Robert J., Jr. *1st Battalion Virginia Infantry, 39th Battalion Virginia Cavalry, and 24th Battalion Virginia Partisan Rangers*. Lynchburg, VA: H.E. Howard, Inc., 1996.

———. *2nd Virginia Cavalry*. Lynchburg, VA: H.E. Howard, Inc., 1995.

———. *5th Virginia Cavalry*. Lynchburg, VA: H.E. Howard, Inc., 1997.

———. *10th Virginia Cavalry*. Lynchburg, VA: H.E. Howard, Inc., 1992.

Dudley, William S. *Going South: U.S. Navy Officer Resignations & Dismissals on the Eve of the Civil War*. Washington, D.C.: Naval Historical Foundation, 1981.

Dunaway, Sarah Owen, ed. *Georgia Division United Daughters of the Confederacy Ancestor Roster*. 10 vols. Atlanta: Georgia Division United Daughters of the Confederacy, 1994.

Durkin, Joseph T. *Stephen R. Mallory: Confederate Navy Chief.* Chapel Hill: University of North Carolina Press, 1954.

Ellis, David Roberts. *The* Monitor *of the Civil War*. Annville, PA, 190-?.

Faust, Patricia L. "Jones, Catesby ap Roger"; and "Wood, John Taylor." In *Historical Times Illustrated Encyclopedia of the Civil War*. New York: Harper & Row, Publishers, 1986.

Fiveash, Joseph G. "The *Virginia's* Great Fight on the Water." *Southern Historical Society Papers* 34 (1906): 316–26.

———. Virginia *(*Merrimac*)*-Monitor *Engagement and a Complete Story of Operations in Hampton Roads and Adjacent Waters*. Norfolk, VA: Fiveash Publishing Co., 1907.

Flanders, Alan B. *The* Merrimac*: The Story of the Conversion of the U.S.S.* Merrimac *into the Confederate Ironclad Warship C.S.S.* Virginia. Portsmouth, VA: Navy Shipyard Museum, 1982.

———. *John L. Porter, Naval Constructor of Destiny*. White Stone, VA: Brandyland Publishers, 2000.

Florida Confederate Pension Application Files. Record Group 137, Series 587. Bureau of Archives and Records Management. Tallahassee, FL: Department of State, 1999. Database on-line. Available at http://www.dos.state.fl.us/dlis/barm/Florida_CSA_Pension_Files.htm.

Fort, W.B. "First Submarine in the Confederate Navy." *Confederate Veteran* (October 1918): 459–460.

Franklin, S.R. *Memoirs of a Rear Admiral*. New York: Harper & Brothers, 1892.

Frye, Dennis E. *2nd Virginia Infantry*. 3rd ed. Lynchburg, VA: H.E. Howard, Inc., 1984.

Great Rebellion Scrapbooks. Vol. 9, *Military Order of the Loyal Legion (MOLLUS)*. Boston: Massachusetts Commandery, MOLLUS, 1893.

Gregory, G. Howard. *38th Virginia Infantry*. Lynchburg, VA: H.E. Howard, Inc., 1988.

———. *53rd Virginia Infantry and 5th Battalion Virginia Infantry*. Lynchburg, VA: H.E. Howard, Inc., 1999.

Griffiths, Oliver W. "The New War Steamers." *United States Nautical Magazine* (April 1855): 298–310.

Hagerman, George M. "Lord of the Turtle Boats." *U.S. Naval Institute Proceedings* 93 (December 1967): 66–75.

Halsy, Ashley, Jr. "The Plan to Capture the *Monitor*: Seal the Turtle in its Shell." *Civil War Times Illustrated* (June 19660: 28–31.

Hartzler, Daniel D. *Marylanders in the Confederacy*. Westminster, MD: Family Line Publications, 1986.

Henderson, Lillian, ed. *Roster of the Confederate Soldiers of Georgia, 1861–1865*. 6 vols. Hapeville, GA: Longino & Porter, 1959–64.

Henderson, William D. *12th Virginia Infantry*. Lynchburg, VA: H.E. Howard, Inc., 1984.

———. *41st Virginia Infantry*. Lynchburg, VA: H.E. Howard, Inc., 1986.

Hoehling, A.A. *Thunder at Hampton Roads*. New York: Da Capo Press, 1993.

Hogg, Ian V. *A History of Artillery*. London: Hamlyn Publishing Group, 1974.

Holland, Darryl. *24th Virginia Cavalry*. Lynchburg, VA: H.E. Howard, Inc., 1997.

Hollyday, Lamar. "The *Virginia* and the *Monitor*." *Confederate Veteran* (October 1922): 380–82.

Isherwood, Benjamin F. *Experimental Researches in Steam Engineering*. 2 vols. Philadelphia: J.B. Lippincott, 1860.

James, Edward Wilson, ed. *The Lower Norfolk County Virginia Antiquary*. 6 vols. Baltimore, MD: Friedenwald Co. Printers, 1905.

Jensen, Les. *32nd Virginia Infantry*. Lynchburg, VA: H.E. Howard, Inc., 1990.

Johnson, Robert Erwin. *Rear Admiral John Rodgers, 1812–1882*. Annapolis, MD: United States Naval Institute Press, 1967.

Jones, Charles C., Jr. *The Life and Services of Commodore Josiah Tattnall*. Savannah, GA: Morning News Steam Printing House, 1878.

Jones, Lewis Hampton. *Captain Roger Jones of London and Virginia*. Albany, NY: J. Munsell's Sons, 1966.

Jones, Robert A. "Aftermath of an Ironclad." *Civil War Times Illustrated* (October 1972): 21–23.

Jones, T. Catesby. "The Iron-Clad *Virginia*." *Virginia Magazine of History and Biography* 49 (October 1941): 297–303.

Jones, Virgil Carrington. *The Civil War at Sea*. 3 vols. New York: Holt, Rinehart and Winston, 1960–1962.

Journal of the Congress of the Confederate States of America, 1861–1865. 7 vols. Washington, D.C.: Government Printing Office, 1904.

Langdon, Robert M. "Josiah Tattnall—Blood Is Thicker than Water." *United States Naval Institute Proceedings* 85 (June 1959): 156–58.

Lewis, Charles Lee. *Admiral Franklin Buchanan*. Baltimore, MD: Norman Remington Company, 1929.

———. "Buchanan, Franklin." *Dictionary of American Biography*. Vol. 3. New York: Charles Scribner's Sons, 1929.

Long, John S. "The Gosport Affair." *Journal of Southern History* 23 (May 1957): 155–72.

Lull, Edward P. *History of the United States Navy Yard at Gosport*. Washington, D.C.: Government Printing Office, 1874.

Luraghi, Raimondo. *A History of the Confederate Navy*. Translated by Paolo E. Coletta. Annapolis, MD: United States Naval Institute Press, 1996.

Mahan, Alfred Thayer. *From Sail to Steam: Recollections of a Naval Life*. New York: Harper and Brothers, 1907.

Manarin, Louis H. *15th Virginia Infantry*. Lynchburg, VA: H.E. Howard, Inc., 1990.

Manarin, Louis H., and Weymouth T. Jordan, comps. *North Carolina Troops, 1861–1865; A Roster*. 18 vols. Raleigh: North Carolina State Department of Archives and History, 1966.

Mann, Samuel A. "New Light on the Great Drewry's Bluff Fight." *Southern Historical Society Papers* 34 (1906): 85–93.

Markham, Jerald H., comp. *List of Confederate Veterans Buried in Hollywood Cemetery from Camp Lee Soldiers' Home, 1894–1946*. Lynchburg, VA: self-published, 1988.

Martin, Charles. "The Sinking of the *Congress* and *Cumberland* by the *Merrimac*." *Personal Recollections of the War of Rebellion: Papers Read Before the New York Commandery, Military Order of the Loyal Legion of the United States (MOLLUS)*. New York: New York Commandery MOLLUS, 1891–1912.

McIntire, William T. "U.S.S. *Congress*." In *Voices of the Civil War: The War Peninsula*. Edited by Paul Mathless. Alexandria, VA: Time-Life Books, 1997.

Moore, Frank, ed. *Rebellion Retold: A Diary of American Events*. 12 vols. New York: G.P. Putnam & Company, 1861–1868.

Moore, Robert H., II. *Graham's Petersburg, Jackson's Kanawha, and Lurty's Roanoke Horse Artillery*. Lynchburg, VA: H.E. Howard, Inc., 1996.

———. *Miscellaneous Disbanded Virginia Light Artillery*. Lynchburg, VA: H.E. Howard, Inc., 1997.

———. *Richmond Fayette, Hampden, Thomas, and Blount's Lynchburg Artillery*. Lynchburg, VA: H.E. Howard, Inc., 1991.

Murphy, Terence V. *10th Virginia Infantry*. Lynchburg, VA: H.E. Howard, Inc., 1989.

Newton, Virginius. Merrimac *or* Virginia. Richmond, VA: William Ellis Jones Company, 1907.

———. "The *Merrimac* or *Virginia*." *Southern Historical Society Papers* 20 (January 1892): 1–26.

Nicholas, Richard L., and Joseph Servis. *Powhatan, Salem and Courtney Henrico Artillery*. Lynchburg, VA: H.E. Howard, Inc., 1997.

Niven, John. *Gideon Welles, Lincoln's Secretary of the Navy*. New York: Oxford University Press, 1973.

O'Neil, Charles. "Engagement between the *Cumberland* and *Merrimack*." *U.S. Naval Institute Proceedings* 48 (June 1922): 893.

Osborne, William H. *History of the Twenty-ninth Regiment of Massachusetts.* Boston: J. Putnam, 1877.

Paullin, Charles O. "Tattnall, Josiah." *Dictionary of American Biography*. Vol. 18. New York: Charles Scribner's Sons, 1930.

Pollack, Edward. *Sketch Book of Portsmouth, Virginia: Its People and Its Trade.* Portsmouth, VA: self-published, 1886.

Porter, John W.H. *A Record of Events in Norfolk Co., Virginia., from April 19th, 1861, to May 10th, 1861, with a History of the Soldiers and Sailors of Norfolk County, Norfolk City, and Portsmouth Who Served in the Confederate States Army or Navy*. Portsmouth, VA: W.A. Fiske, 1892.

Preston, Robert L. "Did the *Monitor* or *Merrimac* Revolutionize Naval Warfare?" *William and Mary Quarterly* (July 19150: 58–66.

Quarstein, John V. *The Battle of the Ironclads*. Charleston, SC: Tempus Publishing Inc., 1999.

———. *Hampton and Newport News in the Civil War: War Comes to the Peninsula.* Lynchburg, VA: H.E. Howard, Inc., 1998.

Quarstein, John V., and Dennis Mroczowski. *Fort Monroe: The Key to the South.* Charleston, SC: Tempus Publishing Inc., 2000.

Rae, Thomas W. "The Little *Monitor* Saved Our Lives." *American History Illustrated* (July 1966): 34.

Ransom, Thomas. "The *Monitor* and the *Merrimac* in Hampton Roads." *Hobbies* (September 1959): 111.

Reaney, Henry. "How the Gun-Boat *Zouave* sided the *Congress*." In *Battles and Leaders of the Civil War*. Vol. 1. Edited by Robert Underwood Johnson and Clarence Clough Buel. New York: Century Co., 1887.

———. "The *Monitor* and *Merrimac*." *War Papers Read Before the Commandery of the State of Michigan, Military Order of the Loyal Legion of the United States (MOLLUS)*. Detroit: Michigan Commandery, MOLLUS, 1893–1898.

Riggs, David F. *Embattled Shrine: Jamestown in the Civil War*. Shippensburg, PA: White Mane Publishing Company, Inc., 1997.

Ripley, Warren. *Artillery and Ammunition of the Civil War*. New York: Promontory Press, 1970.

Robertson, James I. *18th Virginia Infantry*. Lynchburg, VA: H.E. Howard, Inc., 1984.

Rodman, J.T. *Reports of Experiments on the Properties of Metals for Cannons, and the Qualities of Cannon Powder.* Washington, D.C.: Government Printing Office, 1861.

Rootes, T.R. "The Rootes Family." *Virginia Magazine of History and Biography* (October 1896): 204–11.

Salley, A.S., Jr., comp. *South Carolina Troops in Confederate Service*. Vol. 1. Columbia, SC: R.L. Bryan Company, 1913.

Scharf, J. Thomas. *History of the Confederate States Navy from Its Organization to the Surrender of Its Last Vessel*. New York: Rogers & Sherwood, 1887. Reprint, New York: Gramercy Books, Random House Publishing, Inc., 1996.

Scott, Johnny L. *34th Virginia Infantry*. Appomattox, VA: H.E. Howard, Inc., 1999.

Sears, Stephen W. *To the Gates of Richmond: The Peninsula Campaign*. New York: Ticknor & Fields, 1992.

———, ed. *The Civil War Papers of George B. McClellan*. New York: Da Capo Press, 1992.

Shingleton, Royce Gordon. *John Taylor Wood, Sea Ghost of the Confederacy*. Athens: University of Georgia Press, 1979.

Shipp, J.F. "The Famous Battle of the Hampton Roads." *Confederate Veteran* (July 1916): 305–07.

Shippen, Edward. "Notes on the *Congress-Merrimac* Fight." *Century Illustrated Monthly Magazine* (August 1885): 642.

The Sinking of the Cumberland *by the Ironclad* Merrimac *off Newport News, Virginia. March 8, 1862*. New York: Currier and Ives, 1862.

Soley, James Russell. "The Navy in the Peninsular Campaign." In *Battles and Leaders of the Civil War*. Vol. 2. Edited by Robert Underwood Johnson and Clarence Clough Buel. New York: Century Co., 1887.

Spence, Vernon Gladden. "Tattnall, Josiah." In *Dictionary of American Military Biography*. Vol. 3. Westport, CT: Greenwood Press, 1984.

Stewart, William H., ed. *History of Norfolk County, Virginia and Representative Citizens*. Chicago: Biographical Publishing Co., 1902.

Stiles, Kenneth L. *4th Virginia Cavalry*. Lynchburg, VA: H.E. Howard, Inc., 1985.

Still, William N., Jr. *Iron Afloat: The Story of the Confederate Armourclad*. Nashville, TN: Vanderbilt University Press, 1971. Reprint, Columbia: University of South Carolina Press, 1985.

———, ed., *The Confederate States Navy: The Ships, Men, and Organization, 1861–1865*. Annapolis, MD: United States Naval Institute Press, 1997.

Sublett, Charles W. *57th Virginia Infantry*. Lynchburg, VA: H.E. Howard, Inc., 1985.

Swartz, Oretha. "Franklin Buchanan: A Study in Divided Loyalties." *United States Naval Institute Proceedings* 88 (December 1962): 62–66.

Templeton, J.A. "The Last Roll—W.G. Huddleston." *Confederate Veteran* (December 1929): 468.

Texler, Harrison A. *The Confederate Ironclad* Virginia *(Merrimac)*. Chicago: University of Chicago Press, 1938.

Thomas, Emory M. "Tattnall, Josiah." In *Historical Times Illustrated Encyclopedia of the Civil War*. New York: Harper & Row, Publishers, 1986.

Tindall, William. "The True Story of the *Virginia* and the *Monitor*." *Virginia Magazine of History and Biography* (January 1923): 1–38; (April 1923): 89–145.

Trask, Benjamin H. *9th Virginia Infantry*. 2nd ed. Lynchburg, VA: H.E. Howard, Inc., 1984.

———. *16th Virginia Infantry*. Lynchburg, VA: H.E. Howard, Inc., 1986.

———. *61st Virginia Infantry*. Lynchburg, VA: H.E. Howard, Inc., 1988.

Tyler, Lyon Gardiner, ed. "Ashton Family." *William and Mary Quarterly Historical Magazine* (October 1898): 117.

———. *Encyclopedia of Virginia Biography*. Vol. 3, New York: Lewis Historical Publishing Co., 1915.

Viele, Egbert L. "A Trip with Lincoln, Chase, and Stanton." *Scribner's Monthly* (October 1878): 819.

Wallace, Lee A., Jr. *3rd Virginia Infantry*. Lynchburg, VA: H.E. Howard, Inc., 1980.

———. *17th Virginia Infantry*. Lynchburg, VA: H.E. Howard, Inc., 1990.

Weaver, Jeffrey C. *10th and 19th Battalions of Heavy Artillery*. Lynchburg, VA: H.E. Howard, Inc., 1996.

———. *Branch, Harrington, and Staunton Hill Artillery*. Lynchburg, VA: H.E. Howard, Inc., 1997.

———. *Brunswick Rebel, Johnson, Southside, United, James City, Lunenburg, Rebel, Pamunkey Heavy Artillery and Young's Harborguard*. Lynchburg, VA: H.E. Howard, Inc., 1996.

Weinart, Richard P., Jr., and Robert Arthur. *Defender of the Chesapeake: The Story of Fort Monroe*. Shippensburg, PA: White Mane Publishing Company, 1989.

Willet, Albert James, Jr. *Poquoson Watermen*. Easley, SC: Southern Historical Press, Inc., 1988.

Wingfield, J.H.D. "Thanksgiving Service on the *Virginia*, March 10, 1862." *Southern Historical Society Papers* 14 (March 1891): 248–51.

Wise, John S. *The End of an Era*. Boston: Putnam and Sons, 1902.

Woodward, C. Vann, ed. *Mary Chesnut's Civil War*. New Haven: Yale University Press, 1981.

Woodward, Joseph H., II. *Alabama Blast Furnaces*. N.p.: self-published, 1940.

Periodicals

Baltimore American, May 11, 1862.
Boston Daily Evening Transcript, June 15, 1855.
Boston Journal, March 13, 1862.
Charleston [South Carolina] *Mercury*, October 30, 1861; March 19, 1862.
Columbus [Georgia] *Enquirer-Sun*, September 10–11, 1901.
Confederate Veteran, July 1908; July 1910; November 1910; May 1911; November 1916; February 1917; January 1925; April 1928; October 1930; November 1931.
Daily Press [Newport News, Virginia], March 9, 1862.
Daily Southern Argus [Norfolk, Virginia], April 17, 1852.
Lynchburg Virginian, September 12, 1861.
Mobile Daily Item, April 26, 1910.
Mobile Register, August 11, 1861; February 14, 1862.
New York Herald, April 15, 1862.
New York Times, April 22, 1856; February 14, 1862; March 14, 1862.
Norfolk [Virginia] *Beacon* [Virginia], July 20, 1824.
Norfolk [Virginia] *Day Book*, March 10, 1862.
Norfolk [Virginia] *Herald*, October 23, 1820; February 9, 1825; August 11, 1837.
Richmond Daily Enquirer, April 22, 1861.
Richmond Examiner, May 21, 1862; July 1, 1862.
Richmond Times-Dispatch, June 30, 1907.
Richmond Whig, May 22, 1862.
Scientific American, June 6, 1862.
Sumter Republican [Americus, Georgia], April 11, 1862.
Virginian Pilot and Norfolk Landmark, March 3, 1926.

Index

C

N

O

P

Q

R

S

Y

Z

About the Author

John V. Quarstein is an award-winning historian, preservationist, lecturer and author. He served as director of the Virginia War Museum for over thirty years and, after retirement, continues to work as a historian for the city of Newport News. He is in demand as a speaker throughout the nation.

Quarstein is the author of fourteen books, including the companion volume to *The CSS* Virginia, *The* Monitor *Boys*. He has produced, narrated and written six PBS documentaries, including the *Civil War in Hampton Roads* series, which was awarded a 2007 Silver Telly.

John Quarstein is the recipient of over twenty national and state awards, such as the United Daughters of the Confederacy's Jefferson Davis Gold Medal in 1999. Besides his lifelong interest in Tidewater Virginia history, Quarstein is an avid duck hunter and decoy collector. He lives on Old Point Comfort in Hampton, Virginia, and on his family's Eastern Shore farm near Chestertown, Maryland.

In 1987, The Mariners' Museum was designated by NOAA, on behalf of the federal government, as the repository for artifacts and archives from the USS *Monitor*. Working jointly with NOAA and the U.S. Navy, the museum has received more than 1,200 artifacts from the *Monitor*, including the steam engine, propeller and revolving gun turret, all now permanently housed in the state-of-the-art USS *Monitor* Center.

See www.marinersmuseum.org